VOLUME EIGHTY SIX

ADVANCES IN VIRUS RESEARCH

Mycoviruses

VOLUME EIGHTY SIX

ADVANCES IN VIRUS RESEARCH

Mycoviruses

Edited by

SAID A. GHABRIAL
Department of Plant Pathology, University of Kentucky
Lexington, Kentucky, USA

AMSTERDAM • BOSTON • HEIDELBERG • LONDON
NEW YORK • OXFORD • PARIS • SAN DIEGO
SAN FRANCISCO • SINGAPORE • SYDNEY • TOKYO
Academic Press is an imprint of Elsevier

Academic Press is an imprint of Elsevier
525 B Street, Suite 1800, San Diego, CA 92101-4495, USA
225 Wyman Street, Waltham, MA 02451, USA
The Boulevard, Langford Lane, Kidlington, Oxford, OX51GB, UK
32, Jamestown Road, London NW1 7BY, UK
Radarweg 29, PO Box 211, 1000 AE Amsterdam, The Netherlands

First edition 2013

Library of Congress Cataloging-in-Publication Data
A catalog record for this book is available from the Library of Congress

British Library Cataloguing-in-Publication Data
A catalogue record for this book is available from the British Library

ISBN: 978-0-12-394315-6
ISSN: 0065-3527

For information on all Academic Press publications visit our website at store.elsevier.com

Printed and bound by CPI Group (UK) Ltd, Croydon, CR0 4YY

Transferred to digital print 2012

ELSEVIER
Book Aid International
Working together to grow libraries in developing countries
www.elsevier.com • www.bookaid.org

CONTENTS

Contributors ix
Preface xiii

1. Viruses and Prions of *Saccharomyces cerevisiae* **1**

Reed B. Wickner, Tsutomu Fujimura, and Rosa Esteban

1. Introduction 2
2. L-A Virus and the Killer Satellite dsRNAs 3
3. Biology of the *S. cerevisiae* dsRNA Viruses 12
4. Yeast Narnaviruses: 20S and 23S RNAs 13
5. Yeast Prions 20
6. Biology of Yeast Prions 26
7. Prospects 26

Acknowledgments 28
References 28

2. Multiplexed Interactions: Viruses of Endophytic Fungi **37**

Xiaodong Bao and Marilyn J. Roossinck

1. Introduction 37
2. Classification of Fungal Endophytes 38
3. Viruses of Fungal Endophytes 40
4. Roles of Viruses in Endophytic Fungi 48
5. Practical Roles for Endophyte Viruses 53
6. Summary 54

Acknowledgments 54
References 54

3. 3D Structures of Fungal Partitiviruses **59**

Max L. Nibert, Jinghua Tang, Jiatao Xie, Aaron M. Collier, Said A. Ghabrial, Timothy S. Baker, and Yizhi J. Tao

1. Introduction to Partitiviruses 60
2. Partitivirus Capsid Structures 63
3. Partitivirus RdRp and dsRNA Structures 70
4. Comparisons with Other Bisegmented dsRNA Viruses 74
5. Comparisons with Other Fungal Viruses with Encapsidated dsRNA Genomes 77
6. Proposed Revisions to Partitivirus Taxonomy 79

Acknowledgments 81
References 81

4. Chrysovirus Structure: Repeated Helical Core as Evidence of Gene Duplication **87**

José R. Castón, Daniel Luque, Josué Gómez-Blanco, and Said A. Ghabrial

1. Introduction 88
2. dsRNA Virus Core Proteins 90
3. Penicillium chrysogenum Virus Capsid Structure 92
4. Cryphonectria nitschkei Chrysovirus Capsid Structure 95
5. Comparison of Chrysovirus and Totivirus CPs: Evolutionary Relationships 99
6. dsRNA Organization Within the Capsid 101
7. Conclusions and Prospects 104

Acknowledgments 104
References 105

5. Hypovirus Molecular Biology: From Koch's Postulates to Host Self-Recognition Genes that Restrict Virus Transmission **109**

Angus L. Dawe and Donald L. Nuss

1. Introduction 110
2. Molecular Characterization of Hypovirulence-Associated dsRNAs 110
3. The *Hypoviridae* 113
4. Technical Challenges of Mycovirus Research: Development of the Hypovirus Experimental System and Completion of Koch's Postulates 115
5. Hypovirus Translation and Gene Expression 117
6. Impact of Hypovirus Infection on the Fungal Host 121
7. Fungal Antiviral Defense Mechanisms 128
8. Concluding Remarks 138

References 140

6. The Family *Narnaviridae*: Simplest of RNA Viruses **149**

Bradley I. Hillman and Guohong Cai

1. Introduction 150
2. Discovery and Early Work on the *Narnaviridae* with Emphasis on Selected Systems 151
3. Early Work Leading to Description of Mitoviruses 152
4. Relationships Among Members of the *Narnaviridae* 158
5. Population Biology of Members of the *Narnaviridae* 164

6. Other Genome Features and Molecular Manipulation of Members of the *Narnaviridae* 166
7. Conclusions and Future Directions 169
References 170

7. Viruses of the White Root Rot Fungus, *Rosellinia necatrix* **177**

Hideki Kondo, Satoko Kanematsu, and Nobuhiro Suzuki

1. Introduction 178
2. Viruses of *R. necatrix* 186
3. Expansion of Host Ranges of *R. necatrix*-Infecting Viruses 199
4. Conclusions and Prospects 203
Acknowledgments 206
References 206

8. Viruses of the Plant Pathogenic Fungus *Sclerotinia sclerotiorum* **215**

Daohong Jiang, Yanping Fu, Li Guoqing, and Said A. Ghabrial

1. Introduction 216
2. Mycoviruses of *S. sclerotiorum* 218
3. The Interaction Between Mycovirus and *S. sclerotiorum* 238
4. Coinfection of *S. sclerotiorum* with Multiple Mycoviruses 240
5. The Potential Use of Mycoviruses to Control Sclerotinia Diseases 241
6. Conclusions and Prospects 242
Acknowledgments 243
References 244

9. Viruses of Botrytis **249**

Michael N. Pearson and Andrew M. Bailey

1. Introduction 250
2. Molecular Characterization and Phylogenetic Relationships of Botrytis Viruses 254
3. Effects of Viruses on Botrytis 259
4. Virus Transmission 264
5. Conclusions and Future Research 267
References 269

10. Insight into Mycoviruses Infecting *Fusarium* Species **273**

Won Kyong Cho, Kyung-Mi Lee, Jisuk Yu, Moonil Son, and Kook-Hyung Kim

1. Introduction 274
2. Isolation and Characterization of dsRNA Mycoviruses Infecting *Fusarium* Species 275

3. Diversity of Fusarium Mycoviruses Based on Genome Organization 277
4. Identification of Fungal Host Factors by Proteomics and Transcriptomics 281
5. Transmission of Fusarium Mycovirus by Protoplast 283
6. Future Directions for Research on Mycovirus–Host Interactions 285
Acknowledgments 285
References 285

11. Viruses of *Helminthosporium* (*Cochlioblus*) *victoriae* **289**

Said A. Ghabrial, Sarah E. Dunn, Hua Li, Jiatao Xie, and Timothy S. Baker

1. Introduction 290
2. Historical Perspectives 291
3. Viruses of *H. victoriae* 293
4. Host Genes Upregulated by Virus Infection 306
5. HvV190S Capsid Structure 314
6. Concluding Remarks 320
Acknowledgments 321
References 321

12. Phytophthora Viruses **327**

Guohong Cai and Bradley I. Hillman

1. Introduction to Oomycetes—Phylogeny, Habitats, and Properties 328
2. History of Oomycete Virus Research 330
3. Properties of Four Viruses of *P. infestans* 333
4. Transmission of Phytophthora Viruses 344
5. Molecular Manipulations of Phytophthora Viruses 345
6. Conclusions, Future Directions, and Closing Remarks 345
References 346

Index *351*
Color plate section at the end of the book

CONTRIBUTORS

Andrew M. Bailey
School of Biological Sciences, University of Bristol, Woodland Road, Bristol, BS8 1UG, United Kingdom

Timothy S. Baker
Department of Chemistry and Biochemistry, and Division of Biological Sciences, University of California, San Diego, La Jolla, California, USA

Xiaodong Bao
Department of Plant Pathology and Environmental Microbiology, Center for Infectious Disease Dynamics, Pennsylvania State University, University Park, Pennsylvania, USA

Guohong Cai
Department of Plant Biology and Pathology, School of Environmental and Biological Sciences, Rutgers, The State University of New Jersey, New Brunswick, New Jersey, USA

José R. Castón
Department of Structure of Macromolecules, Centro Nacional Biotecnología/CSIC, Campus de Cantoblanco, Madrid, Spain

Won Kyong Cho
Department of Agricultural Biotechnology, Center for Fungal Pathogenesis and Research Institute for Agriculture and Life Sciences, Seoul National University, Seoul, Republic of Korea

Aaron M. Collier
Department of Biochemistry and Cell Biology, Rice University, Houston, Texas, USA

Angus L. Dawe
Department of Biology, New Mexico State University, Las Cruces, New Mexico, USA

Sarah E. Dunn
Department of Chemistry and Biochemistry, University of California, San Diego, La Jolla, California, USA

Rosa Esteban
Instituto de Biología Funcional y Genómica, CSIC/Universidad de Salamanca, Salamanca, Spain

Yanping Fu
Provincial Key Lab of Plant Pathology of Hubei Province, College of Plant Science and Technology, Huazhong Agricultural University, Wuhan, Hubei Province, P. R. China

Tsutomu Fujimura
Instituto de Biología Funcional y Genómica, CSIC/Universidad de Salamanca, Salamanca, Spain

Said A. Ghabrial
Department of Plant Pathology, University of Kentucky, Lexington, Kentucky, USA

Josué Gómez-Blanco
Department of Structure of Macromolecules, Centro Nacional Biotecnología/CSIC, Campus de Cantoblanco, Madrid, Spain

Li Guoqing
The State Key Lab of Agricultural Microbiology, and Provincial Key Lab of Plant Pathology of Hubei Province, College of Plant Science and Technology, Huazhong Agricultural University, Wuhan, Hubei Province, P. R. China

Bradley I. Hillman
Department of Plant Biology and Pathology, School of Environmental and Biological Sciences, Rutgers, The State University of New Jersey, New Brunswick, New Jersey, USA

Daohong Jiang
The State Key Lab of Agricultural Microbiology, and Provincial Key Lab of Plant Pathology of Hubei Province, College of Plant Science and Technology, Huazhong Agricultural University, Wuhan, Hubei Province, P. R. China

Satoko Kanematsu
Apple Research Station, Institute of Fruit Tree Science, NARO, Morioka, Iwate, Japan

Kook-Hyung Kim
Department of Agricultural Biotechnology, Center for Fungal Pathogenesis and Research Institute for Agriculture and Life Sciences, Seoul National University, Seoul, Republic of Korea

Hideki Kondo
Institute of Plant Science and Resources (IPSR), Okayama University, Chuou, Kurashiki, Okayama, Japan

Kyung-Mi Lee
Department of Agricultural Biotechnology, Center for Fungal Pathogenesis and Research Institute for Agriculture and Life Sciences, Seoul National University, Seoul, Republic of Korea

Hua Li
Plant Pathology Department, University of Kentucky, Lexington, Kentucky, USA

Daniel Luque
Department of Structure of Macromolecules, Centro Nacional Biotecnología/CSIC, Campus de Cantoblanco, Madrid, Spain

Max L. Nibert
Department of Microbiology & Immunobiology, Harvard Medical School, Boston, Massachusetts, USA

Donald L. Nuss
Institute for Bioscience and Biotechnology Research, University of Maryland, Rockville, Maryland, USA

Michael N. Pearson
School of Biological Sciences, The University of Auckland, Auckland, New Zealand

Marilyn J. Roossinck
Department of Plant Pathology and Environmental Microbiology, Center for Infectious Disease Dynamics, Pennsylvania State University, University Park, Pennsylvania, USA

Moonil Son
Department of Agricultural Biotechnology, Center for Fungal Pathogenesis and Research Institute for Agriculture and Life Sciences, Seoul National University, Seoul, Republic of Korea

Nobuhiro Suzuki
Institute of Plant Science and Resources (IPSR), Okayama University, Chuou, Kurashiki, Okayama, Japan

Jinghua Tang
Department of Chemistry & Biochemistry and Division of Biological Sciences, University of California–San Diego, La Jolla, California, USA

Yizhi J. Tao
Department of Biochemistry and Cell Biology, Rice University, Houston, Texas, USA

Reed B. Wickner
Laboratory of Biochemistry and Genetics, National Institute of Diabetes Digestive and Kidney Diseases, National Institutes of Health, Bethesda, Maryland, USA

Jiatao Xie
Department of Plant Pathology, University of Kentucky, Lexington, Kentucky, USA

Jisuk Yu
Department of Agricultural Biotechnology, Center for Fungal Pathogenesis and Research Institute for Agriculture and Life Sciences, Seoul National University, Seoul, Republic of Korea

PREFACE

The year 1962 marks the first discovery of a fungal virus, in cultivated mushroom (Hollings, 1962)[1], and is considered the birth date of fungal virology. The belated discovery of mycoviruses may reflect the fact that latent infections are common. There were, however, some clues to suggest the existence of fungal viruses prior to 1962, for example, the discovery in 1959 of the transmissible disease of *Helminthosporium victoriae* thought to have a viral etiology. This year, 2012, we celebrate the 50th anniversary of the discovery of mycoviruses, and it is indeed timely to prepare a thematic issue on mycoviruses for the *Advances in Virus Research* (AVR).

Considerable progress in our knowledge of mycoviruses has emerged during the past 50 years. Fungal viruses occur widely in all major taxa of fungi and they have diverse genomes including positive-sense single-stranded RNA (classified into five families: *Alphaflexiviridae, Barnaviridae, Gammaflexivirida, Hypoviridae*, and *Narnaviridae*), double-stranded RNA (also classified into five families: *Chrysoviridae, Endornaviridae, Partitiviridae, Reoviridae*, and *Totiviridae*), and recently geminivirus-like circular ssDNA genomes (unclassified).

Although progress in molecular characterization of mycoviruses was fast and plentiful information was generated on the diversity of genome organization and relationships to viruses of other organisms, only limited progress was made in understanding their biology including host range, infection and replication processes, and their interactions with their hosts. A major reason for this was the fact that fungal viruses are not readily amenable to conventional infectivity assays. However, recent technological advances allowed for transfection of fungal protoplasts with purified virions or with full-length viral transcripts and transformation with full-length cDNA clones of viral RNA. These developments allowed for the completion of Koch's postulates and for expanding the host ranges of desired mycoviruses as well as for engineering mycoviruses for improved biological control potential. The development of a hypovirus reverse genetics system, the first for a mycovirus, is credited for much of the progress in this aspect.

Studies of viruses that infect phytopathogenic fungi are gaining in strength and intensity because of the current convincing evidence that mycoviruses are

[1] Hollings, M. 1962. Viruses associated with a die-back disease of cultivated mushroom. Nature (London) 196, 962-965.

responsible for debilitation and/or hypovirulence phenotypes in many economically important plant pathogens. From agricultural perspectives, mycoviruses may contribute to sustainable agriculture as biological control agents. Presently, control of plant pathogenic fungi is a formidable task because of the lack of appropriate disease control strategies. In addition to the health hazards and the risks to the environment, the use of fungicides is often cost prohibitive.

Study of virus/host interactions is a main area of modern virology. With the recent advances in fungal genomics and availability of mycovirus reverse genetics systems, it is now possible to use mycoviruses as tools to explore the physiology of their fungal hosts. Some fungal virus systems have contributed not only to the study of mycoviruses but also to the study of many areas in biology including virus–host cell interactions and general virology. The best example here is the yeast Saccharomyces cerevisiae virus-L-A (ScV-L-A), the prototype of the family *Totiviridae* and the best-studied dsRNA mycovirus biochemically and at the molecular and structural levels. The discovery of the yeast killer phenotype and establishing that secreted killer proteins are encoded by satellite dsRNAs depended on ScV-L-A for replication, and encapsidation have substantially strengthened our knowledge in many areas of biology and provided deeper understanding of essential cellular mechanisms such as posttranslational preproprotein processing in the secretory pathway.

The ability to purify well-characterized mycoviruses, particularly those in the families *Chrysoviridae*, *Partitiviridae*, and *Totiviridae* in good quantity and high quality, encouraged the collaboration between structure biology and mycovirology laboratories, and as a result, novel mycovirus capsid structural features were revealed.

In summary, the more we know about fungal viruses, the more we discredit the idea that all viruses are evil. Mycoviruses are indeed the good guys and hold great promise for exploitation as biological control agents. Mycovirus-encoded killer toxins are becoming more interesting with respect to possible applications in biomedicine and production of transgenic plants with broad resistance to plant pathogens. Furthermore, mycoviruses are of common occurrence in endophytic fungi with potential exciting mutualistic roles in the complex interactions with endophytes and plants. These topics and others are discussed in this volume of AVR by the world leaders in the field of fungal virology.

I would like to thank Professor Karl Maramorosch for inviting me to edit this thematic volume; all colleagues who have enthusiastically prepared comprehensive, up to date and stimulating reviews; and the technical staff who were supportive in the production of this book.

SAID A. GHABRIAL
Editor

CHAPTER ONE

Viruses and Prions of *Saccharomyces cerevisiae*

Reed B. Wickner[*,1], **Tsutomu Fujimura**[†] **and Rosa Esteban**[†]
[*]Laboratory of Biochemistry and Genetics, National Institute of Diabetes Digestive and Kidney Diseases, National Institutes of Health, Bethesda, Maryland, USA
[†]Instituto de Biología Funcional y Genómica, CSIC/Universidad de Salamanca, Salamanca, Spain
[1]Corresponding author: e-mail address: wickner@helix.nih.gov

Contents

1. Introduction 2
2. L-A Virus and the Killer Satellite dsRNAs 3
   2.1 Virion structure 3
   2.2 L-A- and M-encoded proteins 4
   2.3 The replication cycle 4
   2.4 Transcription 7
   2.5 Translation 7
   2.6 L-A cap-snatching 9
   2.7 Encapsidation 11
   2.8 Replication 12
3. Biology of the *S. cerevisiae* dsRNA VIRUSES 12
4. Yeast Narnaviruses: 20S and 23S RNAs 13
   4.1 Launching systems of narnaviruses 15
   4.2 *cis*-Acting signals 17
   4.3 *SKI1* antiviral activity 18
   4.4 Replication intermediates 19
5. Yeast Prions 20
   5.1 The range of yeast prions 22
   5.2 Most yeast prions are amyloid filaments 23
   5.3 Prion variants 23
   5.4 In-register parallel architecture of prions can explain heritable conformation 24
   5.5 Chaperones and yeast prions 25
   5.6 Transmission barriers: Interspecies and intraspecies 25
6. Biology of Yeast Prions 26
7. Prospects 26
Acknowledgments 28
References 28

Advances in Virus Research, Volume 86 2013, Published by Elsevier Inc.
ISSN 0065-3527
http://dx.doi.org/10.1016/B978-0-12-394315-6.00001-5

Abstract

Saccharomyces cerevisiae has been a key experimental organism for the study of infectious diseases, including dsRNA viruses, ssRNA viruses, and prions. Studies of the mechanisms of virus and prion replication, virus structure, and structure of the amyloid filaments that are the basis of yeast prions have been at the forefront of such studies in these classes of infectious entities. Yeast has been particularly useful in defining the interactions of the infectious elements with cellular components: chromosomally encoded proteins necessary for blocking the propagation of the viruses and prions, and proteins involved in the expression of viral components. Here, we emphasize the L-A dsRNA virus and its killer-toxin-encoding satellites, the 20S and 23S ssRNA naked viruses, and the several infectious proteins (prions) of yeast.

1. INTRODUCTION

Here we review the viruses and prions of *Saccharomyces cerevisiae*, including the dsRNA viruses that resemble the cores of mammalian dsRNA viruses, the "naked RNA" single-stranded RNA viruses that are closest in sequence to some RNA bacteriophage, and the yeast prions that are self-propagating amyloids of various chromosomally encoded proteins. These infectious elements of yeast share a mode of transmission by cell–cell mating; none are known to include an extracellular phase. The dsRNA and ssRNA viruses lack the outer virion layer typical of the corresponding viruses of other organisms that use an extracellular route as the primary mode of transmission. The retrotransposons of yeast, Ty1, Ty2, . . ., Ty5 (not reviewed here, but see: Beauregard, Curcio, & Belfort, 2008; Maxwell & Curcio, 2007), likewise resemble the cores of retroviruses, with no *env* gene and no extracellular phase, but with a similar intracellular life cycle. *S. cerevisiae* has also been used as a host for plant and animal viruses (Gancarz, Hao, He, Newton, & Ahlquist, 2011; Kovalev, Pogany, & Nagy, 2012; Zhao & Frazer, 2002).

Biologically, the yeast viruses resemble other virus groups that do not have a known extracellular phase. The Endornaviruses (Fukuhara et al., 2006; Roossinck, Sabanadzovic, Okada, & Valverde, 2011) are large (~14 kb) dsRNA unencapsidated replicons in plants and fungi. Partitiviruses are encapsidated dsRNA viruses of plants and fungi, also without an extracellular phase (Ghabrial et al., 2011). An array of Hypoviruses moderating the pathogenicity of *Cryphonectria parasitica* for chestnut trees are strictly intracellular RNA viruses (Nuss, 2005). Moreover, endogenous retroviruses and some Herpesviruses are unable to spread via a cell-free intermediate.

2. L-A VIRUS AND THE KILLER SATELLITE dsRNAs

L-A, as the type species of *Totiviridae*, is a 4.6-kb single-segment dsRNA virus with an icosahedral coat comprising a single major coat protein (called Gag, for reasons that will be evident), and a Gag–Pol fusion protein formed by ribosomal frameshifting and including the RNA-dependent RNA polymerase. L-A is stably maintained in cells, without any apparent slowing of growth, but its presence in only 15 of 70 wild *S. cerevisiae* (Nakayashiki, Kurtzman, Edskes, & Wickner, 2005) indicates that it must have a net detrimental effect.

Several satellite dsRNAs are found in some strains, called M1, M2, M28, or Mlus, each encoding a secreted protein toxin and immunity to the toxin (Schmitt & Breinig, 2006). M dsRNAs encode a preprotoxin, which is processed by cleavage to produce the mature toxin. Preprotoxin ORFs are not related with host-encoded genes, with the exception of klus prepotoxin, which shows a high degree of conservation with *S. cerevisiae* YFR020W ORF, of unknown function, suggesting an evolutionary relationship (Rodriguez-Cousino et al., 2011).

These toxins kill other cells by different mechanisms. Among them, K1 and K28 have been studied in detail. They interact with sensitive cells through receptors in the cell wall or the plasma membrane. Ionophoric K1 disrupts cytoplasmic membrane function by forming cation-selective ion channels (Martinac et al., 1990). K28 toxin, in contrast, blocks DNA synthesis and arrests cells in the early S phase of the cell cycle (Schmitt, Klavehn, Wang, Schonig, & Tipper, 1996).

2.1. Virion structure

The X-ray structure of the L-A virus (Naitow, Canady, Wickner, & Johnson, 2002; Tang et al., 2005) shows an icosahedral structure with "$T=2$" symmetry, a supposedly forbidden mode (Fig. 1.1A). In fact, the structure is a $T=1$ structure (60 units) with an assymmetric dimer of Gag as the unit (Caston et al., 1997; Esteban & Wickner, 1986; Naitow et al., 2002). The dsRNA is packed in layers inside the virion, and pores at the fivefold axes allow access of nucleotides to the interior for RNA synthesis and the escape of the new viral (+) strands to the cytoplasm. A groove in the virion surface includes residue His154, the site of attachment of 7mGMP removed from cellular mRNAs (Fig. 1.1B; see section 2.6).

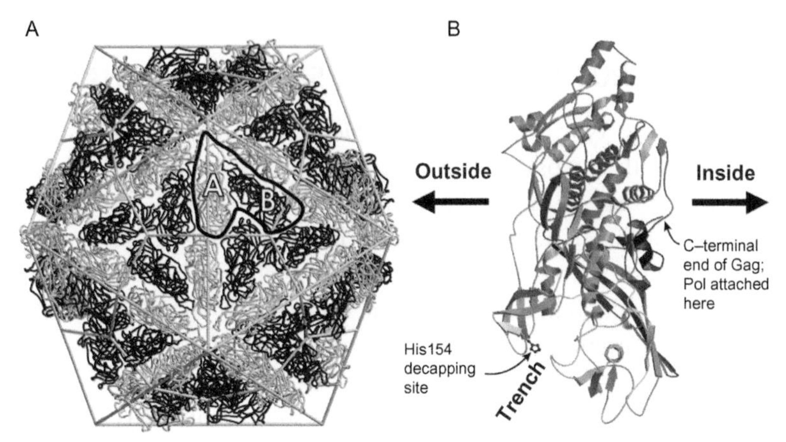

Figure 1.1 L-A virion structure. (A) "$T=2$" architecture with two nonidentical forms of the Gag major coat protein (a and b). (B) Structure of a single monomer. Note that the trench with His154, the site of cap attachment, is outside the particles, while the C-terminus of Gag, which is extended as Gag–Pol in one or two subunits per particle, is internal. *Modified from Naitow et al. (2002).* (See Page 1 in Color Section at the back of the book.)

2.2. L-A- and M-encoded proteins

L-A has two long ORFs, called Gag and Pol for their analogy with retroviruses (Fig. 1.2). The 5′ Gag ORF encodes the major coat protein and Pol, overlapping with the 3′ end of the Gag ORF, encodes the RNA-dependent RNA polymerase, expressed only as a fusion protein with Gag. Virus particles contain only 1–2 copies of Gag–Pol, made by a −1 ribosomal frameshift similar in mechanism to that of many retroviruses (Dinman, Icho, & Wickner, 1991).

M dsRNA encodes a preprotoxin, which is processed by cleavage after dibasic residues by Kex2p followed by removal of the dibasic residues by Kex1p (Bostian, Jayachandran, & Tipper, 1983; Cooper & Bussey, 1989; Julius, Brake, Blair, Kunisawa, & Thorner, 1984; Leibowitz & Wickner, 1976) (Fig. 1.2). This processing pattern was immediately recognized to resemble that of precursors of insulin and several other human hormones, and led to the discovery of the corresponding (homologous) processing enzymes (Steiner, Smeekens, Ohagi, & Chan, 1992).

2.3. The replication cycle

The L-A virus replicates entirely inside the coat. New (+) strands are made inside the coat from the dsRNA template by a conservative mechanism and are extruded from the particles where they are translated and/or

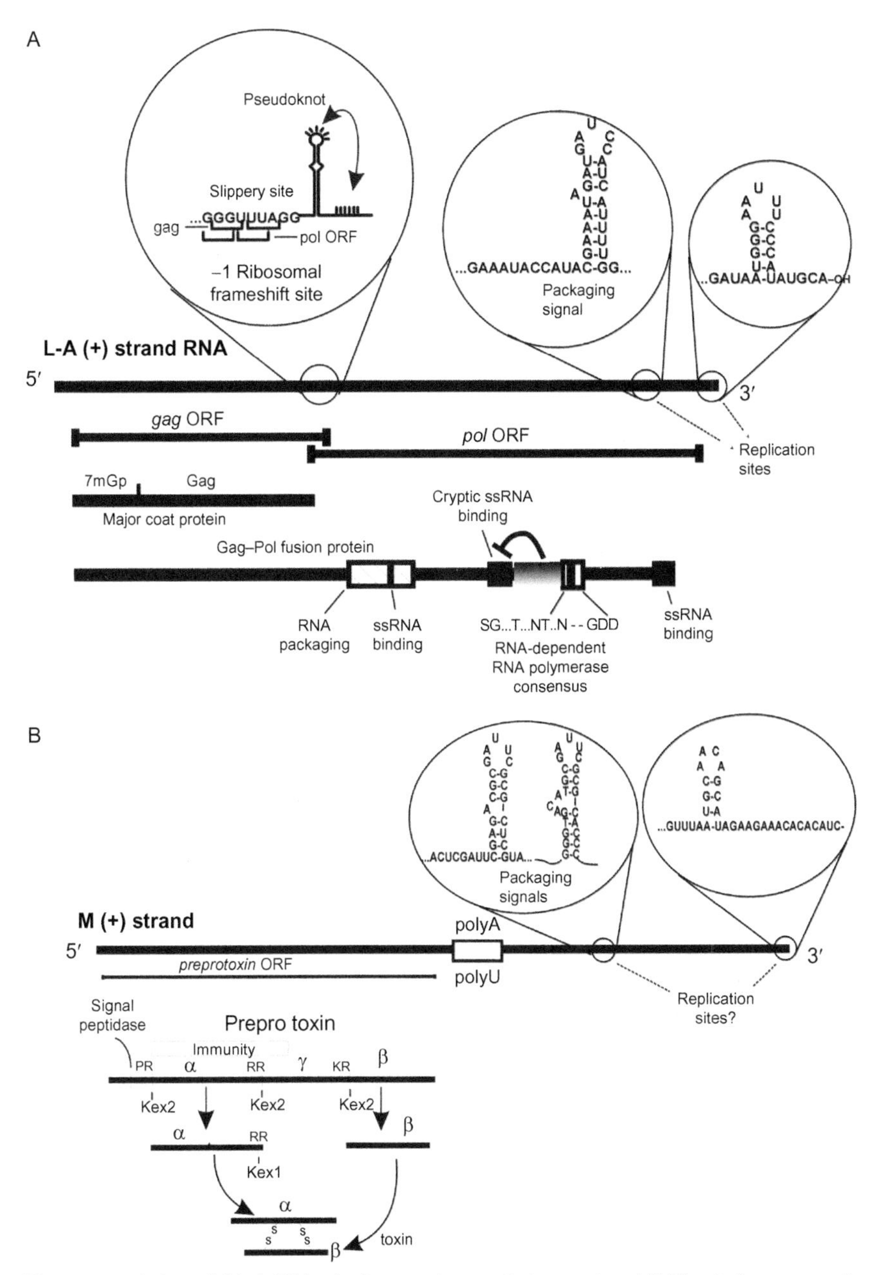

Figure 1.2 L-A and M dsRNA *cis* sites and encoded proteins. (A) The L-A genome is shown with the sites of ribosomal frameshifting, the packaging site (and the overlapping internal replication enhancer), and the 3′ replication site. The His154 cap attachment site is shown as 7mGp. (B) The M1 satellite genome is shown with packaging sites, replication sites, the encoded preprotoxin and its posttranslational processing pathway.

encapsidated to form new particles (Fig. 1.3). Newly encapsidated (+) strands are converted to dsRNA form within the particles to complete the cycle. Note that the (+) and (−) strands are made at different points of the replication cycle.

Examination of virus particles carrying a deletion mutant of L-A (called X) or carrying the M1 satellite dsRNA suggested a "head-full replication" model, in which a single (+) strand is encapsidated, and it replicates within the virion until the particle is full of dsRNA (Esteban & Wickner, 1986, 1988) (Fig. 1.3). Then new (+) strands are extruded from the particles.

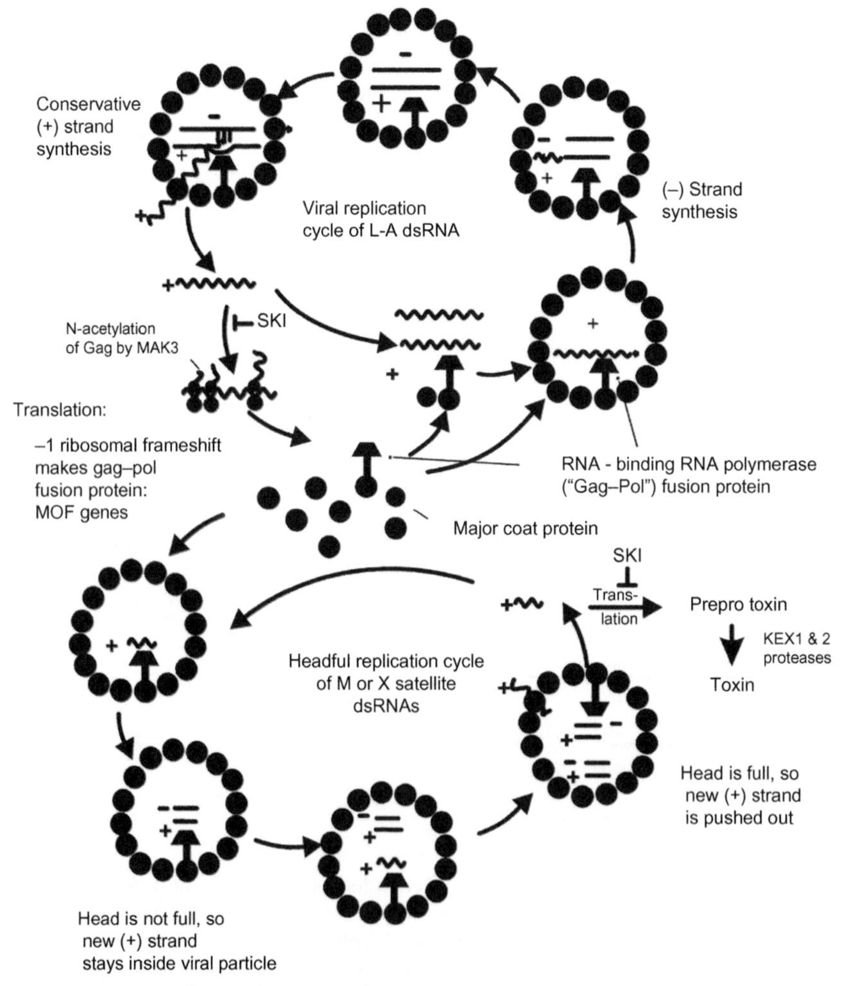

Figure 1.3 L-A and M replication cycles.

Indeed, M1 particles with a single dsRNA molecule often retain newly made (+) strands, while those with two M1 dsRNA molecules per particle export all of their new (+) strand transcripts (Esteban & Wickner, 1986). Moreover, it has been directly shown that a single (+) strand is initially encapsidated (Fujimura, Esteban, Esteban, & Wickner, 1990).

2.4. Transcription

The (+) strand synthesis process goes on inside the particles and is a conservative reaction, meaning parental strands stay together (Fujimura, Esteban, & Wickner, 1986; Herring & Bevan, 1974). Detailed studies of the transcription, replication, and packaging of the L-A virus employed an *in vitro* system in which the enzymes were supplied by empty particles, produced by exposure of virions to low-salt conditions (Fujimura & Wickner, 1988a, 1988b). These opened empty particles specifically bind viral (+) ssRNA in a reaction whose specificity was shown to be the same as the packaging reaction (see section 2.7), can convert these (+) strands to dsRNA form (the replication reaction, see section 2.8) and can transcribe these dsRNA molecules or dsRNA added initially to the particles.

2.5. Translation

2.5.1 Ribosomal frameshifting

The 5′ Gag and 3′ Pol open reading frames overlap by 130 nt. Most ribosomes translating Gag terminate at the Gag termination codon, but about 1% of ribosomes translating Gag perform a −1 ribosomal frameshift at a special sequence into the Pol open reading frame and synthesize a Gag–Pol fusion protein (Dinman et al., 1991; Fujimura & Wickner, 1988a; Icho & Wickner, 1989). The frameshift signal consists of a "slippery site" whose sequence is . . .G GGU UUA GG. . . with the Gag reading frame indicated, and a pseudoknot just 3′ to this slippery site. The pseudoknot makes the ribosomes pause with the underlined codons in the ribosomal A and P sites. The slippery site is such that pairing of the nonwobble bases is still correct if the ribosome shifts back one base. This happens to about 1% of the ribosomes, and when they resume translation in the new frame, they extend the polypeptide to make the Gag–Pol fusion protein (Dinman et al., 1991). This mechanism serves to (a) encode two proteins from one RNA segment, conserving space, and (b) fix the ratio of Gag to Gag–Pol proteins by adjusting the structure of the frameshift site, a ratio that is critical for the efficient propagation of the virus (Dinman & Wickner, 1992), and

(c) strongly suggests that the Pol domain is properly incorporated on the inside of the viral particle when the Gag domain is incorporated into the particles [note that the C-terminus of the Gag domain faces the inside of the particles (Naitow et al., 2002)].

2.5.2 60S ribosomal subunits and SKI2, SKI3, etc

Mutants in any of a wide array of genes that are partially deficient in 60S subunits result in loss of M dsRNA and decreased copy number of L-A dsRNA (e.g., Edskes, Ohtake, & Wickner, 1998; Ohtake & Wickner, 1995). Because the viral mRNAs lack a 3′ polyA, and polyA may have a special role in the 60S subunit joining step (Kahvejian, Svitkin, Sukarieh, M'Boutchou, & Sonenberg, 2005; Munroe & Jacobson, 1990; Searfoss, Dever, & Wickner, 2001), mutants deficient in 60S subunits may tip the balance of translation away from the viral transcripts and toward the cellular mRNAs (Edskes et al., 1998).

The *SKI* (superkiller) genes (Benard, Carroll, Valle, & Wickner, 1998; Ridley, Sommer, & Wickner, 1984; Toh-e, Guerry, & Wickner, 1978) include *SKI1/XRN1*, a 5′ exoribonuclease specific for uncapped RNAs with a prominent role in mRNA degradation (Larimer, Hsu, Maupin, & Stevens, 1992), while *SKI4* and *SKI6* are subunits of the "exosome" complex, originally named so because it was believed to be composed of exoribonucleases (Mitchell, Petfalski, Shevchenko, Mann, & Tollervey, 1997), but now known to be devoid of such activities (Chlebowski, Tomecki, Lopez, Seraphin, & Dziembowski, 2011). *SKI2*, *SKI3*, and *SKI8* encode a cytoplasmic complex (Brown, Bai, & Johnson, 2000). *SKI2*, *SKI3*, *SKI6*, *SKI7*, and *SKI8* together block the expression of non-polyA mRNAs, as judged by electroporation experiments (Benard et al., 1998; Benard, Carroll, Valle, & Wickner, 1999; Masison et al., 1995). While all are involved in the 3′ exonuclease degradation of mRNA (Jacobs-Anderson & Parker, 1998), kinetic experiments suggest that translation effects may also be involved. *SLH1* is a paralog of *SKI2*, and the *ski2 slh1* double mutant expresses Cap+ poly(A)− mRNAs with the same kinetics as it does Cap+ polyA+ mRNAs (Searfoss & Wickner, 2000). This result indicates that the translation apparatus does not inherently need polyA.

2.5.3 N-acetylation of Gag is needed for viral assembly

MAK3, *MAK10*, and *MAK31* are all necessary for L-A and M propagation, and together they form an *N*-acetyltransferase complex with Mak3p as the catalytic subunit (Rigaut et al., 1999; Sommer & Wickner, 1982;

Tercero & Wickner, 1992; Toh-e & Sahashi, 1985). Mak3p recognizes the four N-terminal residues of L-A Gag, and acetylates the initiator methionine, a modification necessary for viral assembly (Tercero, Dinman, & Wickner, 1993; Tercero, Riles, & Wickner, 1992; Tercero & Wickner, 1992). Homologs of the Mak3–Mak10–Mak31 *N*-acetyltransferase complex are now known in many organisms including a human complex (e.g., Starheim et al., 2009).

2.6. L-A cap-snatching

The prominent feature of eukaryotic mRNA is the presence of the cap structure (m^7GpppX—) at the 5′ end. The structure is crucial for the efficient translation and stability of mRNA. In the cell and for most viruses, the cap is installed on mRNA cotranscriptionally by three enzymatic reactions (Shuman, 1995; Venkatesan, Gershowitz, & Moss, 1980). RNA triphosphatase eliminates the 5′ γ phosphate from a nascent PolII transcript. Guanylyltransferase forms a Gp-enzyme intermediate with GTP and transfers Gp to the diphosphorylated 5′ end of the transcript to form a 5′-5′ triphosphate linkage. Finally, methyltransferase methylates the guanine base at N7. The influenza virus installs its mRNA with a cap structure in a nonconventional mechanism (cap snatching) (Boivin, Cusack, Ruigrok, & Hart, 2010; Plotch, Bouloy, Ulmanen, & Krug, 1981). The trimeric viral polymerase binds to the cap structure of the host mRNA, cleaves the RNA endonucleolytically 10–13-nt downstream, and utilizes the capped fragment as a primer to transcribe the viral genome. Recently, L-A virus has been found to furnish its transcript with a cap structure by a novel cap-snatching mechanism (Fujimura & Esteban, 2011). L-A only transfers the m^7Gp moiety from the host mRNA to the diphosphorylated 5′ end of the viral transcript. Further, unlike the influenza virus, L-A utilizes Gag to catalyze the reaction.

In 1992, Sonenberg and coworkers found that Gag of L-A virus covalently binds to the cap structure of mRNA (Blanc, Goyer, & Sonenberg, 1992). Subsequent studies showed that m^7Gp derived from mRNA decapping is attached covalently to His-154 of Gag (Blanc, Ribas, Wickner, & Sonenberg, 1994). Gag with a mutation at His-154, when expressed from a vector, could support replication of M1; interestingly, however, its expression (killer toxin production) was severely affected by the mutation (Blanc et al., 1994). L-A and M1 transcripts have diphosphate at their 5′ ends (Fujimura & Esteban, 2010). Isolated L-A virions can transfer the m^7Gp moiety from mRNA to the 5′ end of the L-A transcript, thus forming the authentic cap structure found in the yeast (Fig. 1.4) (Fujimura & Esteban, 2011). The α and β phosphates at the 5′ end of the transcripts are conserved in the triphosphate linkage of the product. A mutation at His-154 abolished the cap transfer reaction. Because the toxin production is severely affected by the mutation, it indicates that capping is essential for the efficient expression of the viral transcript. The N7 methylation of the 5′ terminal G is essential for cap donor activity, and the smallest molecule with donor activity is the dinucleotide cap analog m^7GpppG (Fujimura & Esteban, 2012). The cap acceptor needs to be 5′ diphosphorylated. A 5′ tri- or monophosphorylated viral transcript does not function as cap acceptor. Although L-A virions can utilize exogenously added viral transcripts as cap acceptors, the capping reaction requires the viral polymerase actively engaging in transcription (Fujimura & Esteban, 2012). Because the polymerase is confined inside the virion, whereas the cap-snatching site is located on the cytoplasmic surface of the virion (see next paragraph), it indicates coordination between the transcription and cap-snatching sites. This coordination may minimize the risk of accidental capping of nonviral RNA when the polymerase is dormant. The physical separation of the cap-snatching site from the transcription site may ensure that not all of the transcripts are capped. This will allow the virus to synthesize two types of transcripts: one, noncapped transcripts presumably destined for encapsidation (genomic ssRNA) and the other, capped transcripts for translation (mRNA).

Structural studies of L-A virions have revealed a trench on the outer surface of Gag that includes His-154 (Naitow et al., 2002; Tang et al., 2005). The trenches are located in the Gag asymmetric dimer close to the icosahedral two-

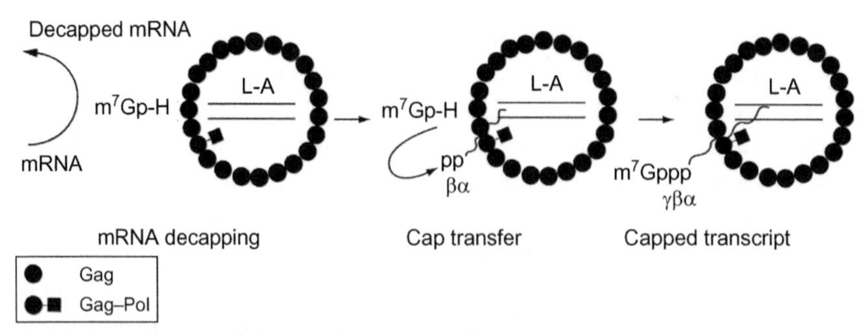

Figure 1.4 Schematic diagram of L-A cap-snatching mechanism. Gag of an L-A virion decaps mRNA and forms an intermediate with m^7Gp through His-154. Then m^7Gp is transferred to the diphosphorylated 5′ end of the viral transcript to form a 5′-5′ triphosphate linkage.

and threefold symmetric axes. L-A transcripts are made inside the virion and presumably released to the cytoplasm through one of the pores located at the fivefold axes (Naitow et al., 2002). His-154 is located at the tip of a loop (residues 144–163) that is part of the upper rim of the trench. Upon m^7GDP binding, the rim moves inwardly and forms a closed conformation (Tang et al., 2005). Guanylyltransferase also contains a trench that can adopt either open or closed conformation during the mRNA capping reaction (Chu et al., 2011; Hakansson, Doherty, Shuman, & Wigley, 1997). Further, secondary structure elements around the trench of Gag resemble those of guanylyltransferase (Naitow et al., 2002). The latter enzyme forms a Gp-enzyme intermediate with GTP and transfers Gp to the diphosphorylated 5′ termini of PolII transcripts. Thus the capping reaction of L-A resembles that of guanylyltransferase. These similarities suggest a convergent evolution.

In the trench, Tyr-150, Asp-152, Tyr-452, and Tyr-538 are located close to the bound m^7GDP, which suggests their involvement in cap recognition. Mutagenesis analyses indicate that these residues are crucial for decapping activity (Tang et al., 2005). The eminent feature of cap-binding proteins is the presence of two aromatic amino acids in their cap-binding pockets (Chu et al., 2011; Hakansson et al., 1997). The aromatic rings of these amino acids sandwiched the m^7G aromatic ring of the cap structure by stacking. The delocalization of the positive charge arising from the N7 methylation of the guanine contributes to strong interactions between the π-electrons of the stacked aromatic rings (Hu, Gershon, Hodel, & Quiocho, 1999). Tyr-452 and Tyr-538 of L-A Gag sandwich m^7GDP by a potential stacking (Tang et al., 2005). Further, the cap-snatching reaction requires N7 methylation of the cap for donor activity (Fujimura & Esteban, 2012). It is likely that the L-A virus utilizes a mechanism similar to that of cap-binding proteins to recognize mRNA for cap snatching. Fungal totiviruses (members of the genus *Totivirus*), including L-BC, share the same (or similar) four amino acids aforesaid as well as His-154 at comparable positions in their capsid proteins (Fujimura & Esteban, 2011). Gag of L-BC has been observed to possess decapping activity (Blanc et al., 1992). Therefore, it is likely that the cap-snatching mechanism of L-A is widespread among totiviruses of fungi.

2.7. Encapsidation

Viral (+) strands, made in the transcription reaction, are extruded from the particles where they may serve as mRNAs and/or be encapsidated by Gag and Gag–Pol to make new viral particles. The RNA encapsidation signal is located 400 nt from the 3′ end of the L-A (+) strand and consists of a stem-loop with a critical A residue protruding from the 5′ side of the stem (Esteban, Fujimura, & Wickner, 1988; Fujimura et al., 1990). The encapsidation signal is recognized by the N-proximal part of the Pol part of Gag–Pol, resulting in packaging of a single (+) strand in each virus particle (Fujimura, Ribas, Makhov, & Wickner, 1992; Ribas, Fujimura, & Wickner, 1994).

2.8. Replication

The viral (+) strands are converted to dsRNA form by the RNA-dependent RNA polymerase in a reaction involving recognition of an internal site, largely overlapping with the packaging site, and synthesis beginning at the 3′ end of the template (Esteban, Fujimura, & Wickner, 1989). The polymerase apparently binds first to the internal site, and then begins synthesis at the 3′ end, without sliding along the template (Fujimura & Wickner, 1992).

3. BIOLOGY OF THE *S. CEREVISIAE* dsRNA VIRUSES

L-BC is a 4.6-kb dsRNA virus group with sequence similarity to L-A, but clearly constituting a distinct replicon (Park, Lopinski, Masuda, Tzeng, & Bruenn, 1996; Sommer & Wickner, 1982). L-BC and L-A replicate stably in the same cells, but while L-A supports the M dsRNA satellites, L-BC does not. L-A requires *MAK3*, *MAK10*, and *MAK31*, the *N*-acetyltransferase encoding genes, but L-BC does not, and the *clo1* mutation that results in loss of L-BC does not affect L-A (Wesolowski & Wickner, 1984; Wickner, 1980; Wickner & Toh-e, 1982).

Wild isolates of L-A show a variety of genetic activities, defined by their ability to support the propagation of M dsRNAs, and their interaction with each other (Wickner, 1980; Wickner & Toh-e, 1982). Although cDNA clones have been used extensively in defining the activities of L-A-encoded functions, these activities appear to resist study because it has not yet been possible to launch the L-A virus from transcripts of the cDNA clone.

Mutants in (at least) *ski2*, *ski3*, and *ski8* are cold-sensitive for growth, and actually die at low temperature if they carry M dsRNA (Ridley et al., 1984). Although these *ski* mutants have a great deal more dsRNA when carrying only L-A than when they are also carrying M, they are not cold-sensitive (Ball, Tirtiaux, & Wickner, 1984; Ridley et al., 1984). The molecular basis of this phenotype has not been found, but it suggests that the presence of the toxin-immunity-encoding M dsRNAs is not an undiluted advantage. It has recently been shown that *S. cerevisiae* can support an RNAi system if only

Dicer and Argonaut are imported from *S. castetllii* (Drinnenberg et al., 2009), and shown that the M and L-A dsRNAs are lost from the constructed strains (but not L-BC) (Drinnenberg, Fink, & Bartel, 2011). However, M dsRNAs are rather rare in wild strains (Nakayashiki et al., 2005; Young & Yagiu, 1978), indicating that there must be a significant disadvantage resulting from carrying it that more than makes up for the obvious advantage. This remains an unsolved problem.

4. YEAST NARNAVIRUSES: 20S AND 23S RNAs

Most laboratory strains of *S. cerevisiae* harbor 20S RNA and fewer strains 23S RNA. These viruses are small positive-stranded RNA viruses belonging to the family *Narnaviridae* and encode only their RNA-dependent RNA polymerases. The RNA genomes are not packed into a massive protein capsid, but form ribonucleoprotein complexes with their cognate RNA polymerases in the host cytoplasm. 20S RNA was first found in 1971 by Kadowaki and Halvorson as an RNA species accumulated under sporulation conditions (Kadowaki & Halvorson, 1971). The name was based on its mobility relative to that of 18S and 25S rRNAs in a gel. Garvik and Haber (1978) established that 20S RNA is a cytoplasmically inherited genetic element and that the accumulation of 20S RNA is independent of the sporulation process of the host. In 1984, Wesolowski and Wickner reported two heat-inducible, cytoplasmically inherited dsRNAs, W and T (Wesolowski & Wickner, 1984). When 20S RNA (Matsumoto & Wickner, 1991) and W (Rodriguez, Esteban, & Esteban, 1991) were independently cloned and sequenced, it was evident that they were identical but in two different molecular forms, ssRNA and dsRNA. Subsequently, T dsRNA was also characterized by sequencing and its ssRNA form (23S RNA) was identified (Esteban, Rodriguez, & Esteban, 1992).

The 20S RNA and 23S RNA genomes are small (2514 and 2891 nt long, respectively) (Rodriguez-Cousino, Solorzano, Fujimura, & Esteban, 1998) and rich in GC. Both RNAs possess 5-nt inverted repeats (5′ GGGGC. . .GCCCC—OH) at the 5′ and 3′ termini. Each RNA has a single open reading frame that spans almost the entire viral genome (Fig. 1.5A). The 5′ nontranslating regions are only 12 nt in 20S RNA and 6 nt in 23S RNA. 20S RNA and 23S RNA encode 91-kDa (p91) and 104-kDa proteins (p104), respectively (Esteban et al., 1992; Matsumoto & Wickner, 1991; Rodriguez et al., 1991), both with consensus sequences for RNA-dependent RNA polymerases and most closely related to those of the positive-stranded RNA coliphages such as Qβ.

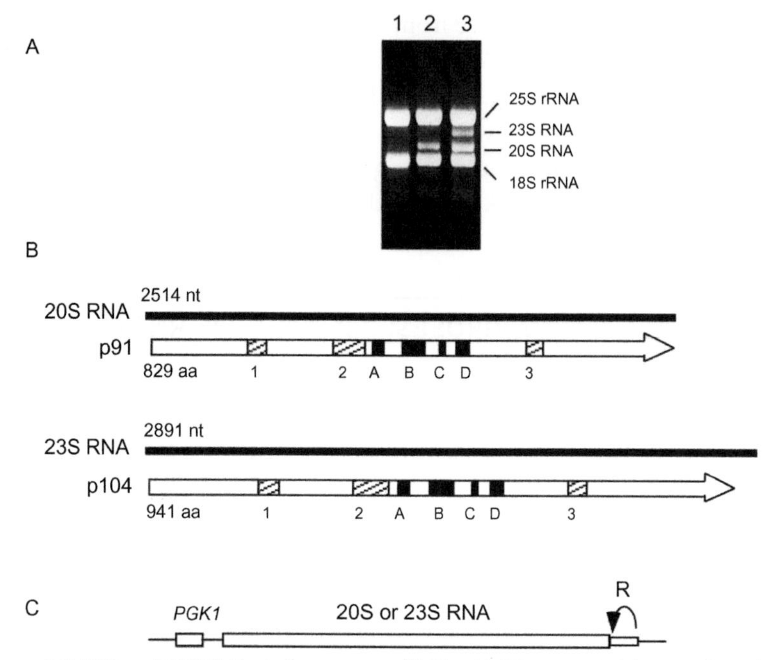

Figure 1.5 20S and 23S RNA viral genomes. (A) Total RNA extracted from induced cells containing no viruses (lane 1), 20S RNA alone (lane 2), or 20S and 23S RNAs (lane 3). Ethidium bromide staining of an agarose gel is shown. 18S and 25S rRNAs are indicated. (B) 20S and 23S RNA genomes and the encoded proteins, p91 and p104. p91 and p104 contain the consensus sequences (A–D) for RNA-dependent RNA polymerases, which are most closely related to those of RNA coliphages. 1–3 indicate amino acid stretches conserved between p91 and p104. (C) Diagram of a launching plasmid. The complete sequence of the 20S or 23S RNA genome is inserted downstream of the constitutive *PGK1* promoter (*PGK1*). The hepatitis delta virus ribozyme (R) is fused directly to the 3′ end of the viral genome. Thin lines indicate sequences derived from the vector.

20S and 23S RNAs migrate in sucrose gradients almost as naked RNAs, and phenol treatment hardly changes their mobility (Esteban, Fujimura, Garcia-Cuellar, & Esteban, 1994; Garcia-Cuellar, Esteban, Fujimura, Rodriguez-Cousino, & Esteban, 1995; Widner, Matsumoto, & Wickner, 1991). The mass of most viruses consists of capsids and lipid layers, indicating that 20S and 23S RNAs are not encapsidated. However, p91 and p104 are associated with 20S RNA and 23S RNA, respectively (Esteban et al., 1994; Garcia-Cuellar et al., 1995) in a 1:1 protein:RNA stoichiometry in the host cytoplasm (Solorzano, Rodriguez-Cousino, Esteban, & Fujimura, 2000). Preliminary data indicate that the resting complexes of 20S RNA contain no host proteins. Pull-down experiments detected no proteins other than p91 in metabolically labeled resting complexes. Further, p91 can form a resting complex with 20S RNA in *E. coli* in the absence of yeast proteins (Vega, 2010).

4.1. Launching systems of narnaviruses

20S and 23S RNA viruses can be generated from vectors (Esteban & Fujimura, 2003; Esteban, Vega, & Fujimura, 2005). In both cases, the complete sequences of the viral genomes are placed downstream of the constitutive *PGK1* promoter (Fig. 1.5B). The efficiency of virus launching is high. 20S RNA can be generated in more than 70% of colonies of the yeast transformed with the vector. It is critical to directly fuse the hepatitis delta virus ribozyme to the 3′ end of the viral genome. In the absence of the ribozyme, the vectors failed to generate the viruses, and insertion of a few Gs between the viral 3′ end and the ribozyme greatly reduced viral generation. In the latter case, the virus, once generated, was indistinguishable from the authentic one because the extra Gs at the 3′ end were eliminated during the launching process. The launching vectors also contained nonviral sequences (about 40 nt) between the major transcription initiation site of the promoter and the viral 5′ end. These sequences were also eliminated during the launching process and, as will be discussed later, the Ski1 5′ exonuclease plays the major role in their elimination.

The narnaviruses have a terminal repair system (Esteban et al., 2005; Fujimura & Esteban, 2004a, 2004b). The elimination of the last and penultimate Cs at the viral 3′ end in the vector, or changing them to the other nucleotides did not affect virus generation. However, the viruses generated were found to have recovered these terminal Cs. It indicates that these terminal Cs are essential for replication, but that the presence of a repair system makes them dispensable for virus launching (Fig. 1.6A). By contrast, modifications at the third or fourth C from the 3′ end abolished virus generation. Although the fifth G is part of the 5-nt inverted repeat, this G is not essential for replication, and a modified nucleotide at this position can be retained in the generated viruses. The terminal repair could be done by host enzymes, or by the viral RNA polymerase. A modified nucleotide at the 3′ terminus may be removed by an exonuclease and then the correct nucleotides might be installed by the CCA-adding enzyme. The positive and negative strands of 20S and 23S RNAs contain stem-loop structures adjacent to the 3′ ends, which resemble the so-called top-half domain of tRNA (Fig. 1.6A). The CCA-adding enzyme recognizes the top-half domain of tRNA precursors and adds CCA to their 3′ ends in a nontemplated fashion. That 20–30% of

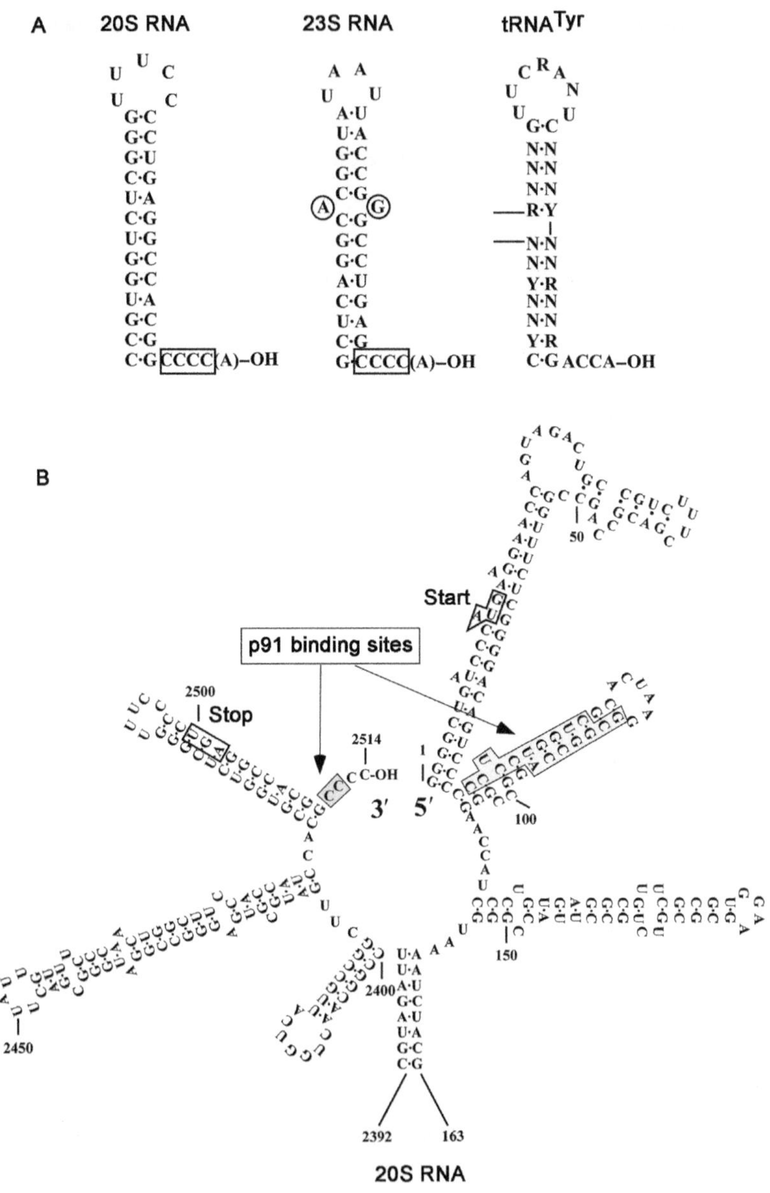

Figure 1.6 *cis*-Acting signals. (A) The 3′ terminal sequences of 20S and 23S RNAs with *cis*-acting signals. For comparison, the top-half domain of $tRNA^{Tyr}$ is shown. (B) The 5′ and 3′ terminal regions of the 20S RNA genome. The 5′ and 3′ *cis* sites are indicated by lightly shaded boxes. The initiation (Start) and termination (Stop) codons of p91 are also indicated.

the positive and negative strands of W and T dsRNAs have a nontemplated A at the 3′ termini (Rodriguez-Cousino & Esteban, 1992) is consistent with the possible involvement of this enzyme in the repair. Alternatively, the terminal repair could be done by the viral polymerase during replication. RdRps are also known to add a nontemplated A at the 3′ end of the products. Although 20S RNA and 23S RNA genomes have 5-nt inverted repeats at both termini, it is unlikely that the 5′ complementary sequence is used as a template to correct modifications at the 3′ end because both the modification at the 3′ end and the compensatory mutation at the 5′ end were corrected to the wild-type sequences without negative effects on virus generation (Fujimura & Esteban, 2004a, 2004b).

20S RNA can also be generated from the negative strands expressed from a vector (Esteban et al., 2005). Because p91 cannot be decoded from the negative strand, virus generation required p91 to be expressed from a second plasmid. As modifications or deletions at the 3′ terminal or penultimate positions did not affect virus generation, it is likely that the same repair mechanism observed on the positive strand also operates on the negative strand.

4.2. *cis*-Acting signals

In a resting complex, p91 interacts with 20S RNA at least at three different sites: the 5′ end, the 3′ end, and, to a lesser extent, the internal sites (Fujimura & Esteban, 2007). The 5′ end site is located in the second stem-loop structure from the 5′ end (Fig. 1.6B). The stem structure is important for complex formation but the loop sequence is not. It has been observed that there is a tight correlation between complex formation and virus generation among those mutants, suggesting that a stable complex formation is a prerequisite for virus generation. The 3′ end site is located at the very end of the 3′ terminus, partially overlapping with the 5-nt inverted repeat. The third and fourth Cs from the 3′ end are important for complex formation but the 3′ terminal and penultimate Cs as well as the fifth G were dispensable. As the third and fourth Cs, but not the fifth G, were essential for replication, there is again a good correlation between replication and complex formation at the 3′ end site. It is not known whether the 3′ end site extends further inside, as in the 23S RNA virus. The internal site is located in the middle of the 20S RNA molecule between nt 1253 and 1513. Mutations at the 5′ or 3′ end site reduced complex formation to a basal level (10–20%) compared with the unmodified RNA. However, the effects of these mutations are not additive, that is, the combination of mutations at these two sites does not reduce complex formation further down. It suggests that the internal site is responsible for the basal level of complex formation. Secondary structure prediction reveals intramolecular long distance interactions among the 5′ end, 3′ end, and internal regions, which brings the three sites close together (Fujimura & Esteban, 2007). This may allow a p91 molecule to simultaneously interact with these three sites in the resting complex.

In the case of the 23S RNA/p104 complex, only the 3′ end site has been analyzed in detail (Fujimura & Esteban, 2004a, 2004b). The 3′ end site is bipartite, consisting of the third and fourth Cs from the 3′ end and a stem-loop structure adjacent to the 3′ end (Fig. 1.6A). Like 20S RNA, the fifth G from the 3′ end is dispensable for complex formation and also for virus generation. The stem-loop structure contains a mismatched pair of purines in the middle of the stem. This mismatch is important for both complex formation and virus generation. Any combination of purines in the mismatch supported both complex formation and virus generation, however, eliminating one of the purines or changing them with pyrimidines abolished both activities. The fact that 20S RNA has no mismatched pair at the stem-loop structure adjacent to the 3′ end may contribute to the specificity of complex formation in these viruses.

4.3. *SKI1* antiviral activity

During virus launching, the 20S RNA and 23S RNA genomes in the vector are transcribed from the *PGK1* promoter by PolII. The transcripts possess not only nonviral sequences (about 40 nt) from the vector but also the cap structure at the 5′ ends (Fig. 1.5B). The generated viruses, however, have no extra nucleotides at the 5′ ends. These extra nucleotides are eliminated during virus generation. mRNA degradation in eukaryotes usually begins with the shortening of the poly(A) tail at the 3′ end, followed by decapping at the 5′ end by the Dcp1/Dcp2 decapping enzyme (Parker & Song, 2004; Wilusz, Wormington, & Peltz, 2001). Because decapping is a crucial step in mRNA degradation, numerous proteins (Lsm1p-7p, Pat1p, Dhh1p, etc.) are involved in this reaction. Then the decapped RNA is degraded by the *SKI1/XRN1* 5′ exonuclease. Alternatively, deadenylated RNA is digested by the 3′ exonuclease exosome (Jacobs-Anderson & Parker, 1998; Mitchell et al., 1997). In yeast, the *SKI1/XRN1* 5′ exonuclease plays a major role in mRNA degradation. It was found that the launching plasmid failed to generate the 20S RNA virus in *ski1Δ* or *dhh1Δ* strains (Esteban, Vega, & Fujimura, 2008), signifying the importance of the mRNA degradation pathway in virus launching. When the 5′ nonviral sequence was reduced from 47 to 9 nt, the plasmid generated the 20S RNA virus in *ski1Δ* strains indicating that the *SKI1* 5′ exonuclease is largely responsible for eliminating the long 5′ extra sequence. Interestingly, these genes did not affect replication of 20S RNA virus when the endogenous virus was introduced by a cytoplasmic mixing from a donor strain. This confirms that these genes are involved in the 5′ processing of the transcripts but not in viral replication *per se*.

Because of the 5′ exonuclease activity, *SKI1* is involved in the host's defense against the L-A virus (Masison et al., 1995). The copy number of L-A increases several-fold in *ski1Δ* strains and the virus can be easily cured from the cell by overexpression of the *SKI1* gene. By contrast, endogenous 20S RNA virus is fairly insensitive to overexpression of *SKI1* (Esteban et al., 2008). The positive strand of L-A has no prominent structures at its 5′ end, whereas 20S RNA (and also 23S RNA) has a strong secondary structure at the 5′ end (Fig. 1.6B). Further, the first 4 nt at the 5′ ends of both 20S and 23S RNAs are Gs, and these nucleotides are buried at the bottom of the stem. It is known that the progression of the 5′ *SKI1* exonuclease is blocked by a cluster of Gs and also by a strong secondary structure. It is likely that these features confer on 20S RNA insensitivity to the antiviral activity of *SKI1*. Indeed, when mutations were introduced to the stem to destabilize the structure, 20S RNA became vulnerable to *SKI1* activity (Esteban et al., 2008). These observations indicate that *SKI1* is a potent weapon for the host to fight against RNA virus infection and suggest that 20S and 23S RNAs have evolved to develop elaborate secondary structures at the 5′ ends, partly as a countermeasure against *SKI1* because the viruses have no protective capsid. The third and fourth Cs from the 3′ ends are not only essential for replication but also important for resting complex formation in 20S and 23S RNA viruses. It strongly suggests that p91 and p104 directly interact with these terminal nucleotides of the cognate RNAs. These interactions may be important to protect the viral genomes from degradation by the exosome. The copy numbers of 20S and 23S RNAs increase several-fold in *ski2*, *ski3*, *ski4*, *ski7*, or *ski8* mutants (Matsumoto, Fishel, & Wickner, 1990; Ramirez-Garrastacho & Esteban, 2011), which suggests that the viruses have some stages in their replication cycles that are vulnerable to the 3′ exosome. This group of Ski proteins also blocks expression specifically of non-polyA mRNAs (Benard et al., 1999; Brown & Johnson, 2001; Masison et al., 1995), a feature common to L-A, 20S and 23S RNA.

4.4. Replication intermediates

In vegetative growing cells, the copy number of 20S RNA is very low (5–20 copies per cell) (Matsumoto et al., 1990). Under sporulation conditions (1% K acetate), it reaches up to 20,000 copies per cell, predominantly in the form of resting complexes with the positive strands. Lysates prepared from these cells contain a minor amount of replication intermediates that possess actinomycin D- and α-ammanitin-insensitive RNA polymerase activity (Garcia-Cuellar, Esteban, & Fujimura, 1997). The intermediate consists of a full-length negative strand template, a positive strand of less than unit length, and p91 (Fujimura, Solorzano, & Esteban, 2005). It is not known whether the intermediate also contains host proteins. The synthesis of RNA results from chain-elongation of the positive strand. The *de novo* synthesis of 20S RNA has not been observed in the lysates. Upon completion of RNA synthesis, the positive strand product and the negative strand template are released from the intermediate. The released strands can be retained on a negatively charged membrane, which suggests that they are still associated with proteins. The RNAs in the replication intermediate have largely a single-stranded RNA backbone. Deproteination with phenol converts them to a dsRNA form. Therefore, W is not a replication intermediate of 20S RNA but a byproduct. Figure 1.7 shows a model of the replication intermediates of 20S RNA (Esteban & Fujimura, 2006). The initiation process of intermediate formation is not known. The intermediate may contain two p91 molecules: one interacting with the negative strand at the 3′ and 5′ end regions and the other polymerizing the positive strand and also concurrently interacting with its 5′ *cis* site. These interactions may protect both RNAs from degradation by exonucleases. Further, these interactions may prevent both RNAs from annealing in a long stretch beyond the polymerization site, thus keeping the RNA backbone largely single stranded. Because each p91 has distinct interactions with the respective strands in the intermediate, this may condition their partition when the positive and negative strands are released. In the cell, the negative strand, as soon as released, may be recruited again to form a new replication intermediate, because the majority of negative strands are found in the form of positive strand-synthesizing intermediates.

Although a resting complex of the negative strand as well as a negative strand-synthesizing intermediate apparently are also constituents of the 20S RNA replication cycle, these complexes have not been analyzed yet because of their low abundance in the cell.

5. YEAST PRIONS

In addition to the nucleic acid-containing viruses, yeast can harbor any of several infectious proteins or prions, conceptually similar to the mammalian prions (Table 1.1). Most yeast prions are based on self-propagating amyloids,

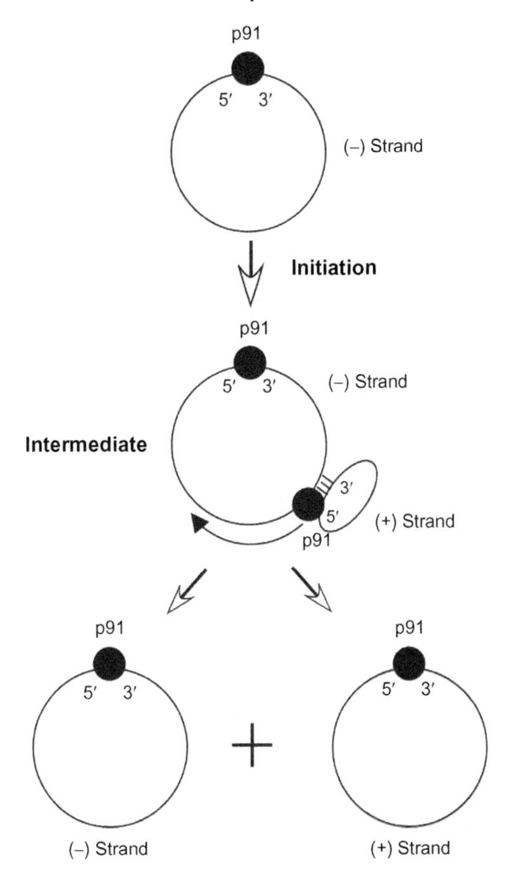

Figure 1.7 Model for 20S RNA positive strand synthesis. The intermediate may contain two p91 molecules (filled circles). For simplicity, the interaction of p91 with the internal *cis* site of the positive strand is omitted.

although one is a protease that can activate its own inactive precursor protein (Roberts & Wickner, 2003). Amyloid is a linear polymer of a single protein species, with a largely beta-sheet structure in which the beta strands are perpendicular to the long axis of the filament. Surprisingly, a single prion protein can be the basis of any of many different prion variants that differ genetically and biologically. Based on the in-register parallel beta-sheet architecture of several yeast prions, a mechanism has been proposed for the templating of amyloid conformation that must underlie this stable inheritance of prion variant.

Table 1.1 Amyloid-based yeast prions

Prion	Protein	Protein function	Phenotype of prion	Ref.
[URE3]	Ure2p	Negative regulator of catabolism of poor nitrogen sources	Inappropriate derepression of enzymes and transporters for poor N sources	Wickner (1994)
[PSI+]	Sup35p	Translation termination subunit	Inappropriate readthrough of translation termination codons	Wickner (1994)
[PIN+]	Rnq1p	Unknown	Rare priming of other prion formation	Derkatch, Bradley, Hong, and Liebman (2001)
[SWI+]	Swi1p	Chromatin-remodeling complex subunit	Poor growth on raffinose, glycerol, or galactose	Du et al. (2008)
[OCT+]	Cyc8p	Transcription repression factor subunit	Defective transcription repression	Patel et al. (2009)
[MOT+]	Mot3p	Transcription repressor of aerobic-repressed genes	Inappropriate derepression of "anaerobic" genes	Alberti et al. (2009)
[ISP+]	Sfp1p	Positive transcription factor of translation-related genes	Antisuppression	Rogoza et al. (2010)
[MOD+]	Mod5p	tRNA isopentenyltransferase	Partial inactivation of enzyme	Suzuki et al. (2012)

5.1. The range of yeast prions

[URE3] and [PSI+] were the first yeast prions found, each long known as a mysterious nonchromosomal genetic element assumed to be nucleic acid based, but not correlated with known RNA or DNA molecules. Based on their genetic properties, [URE3] and [PSI+] were shown to be prions of Ure2p, a regulator of nitrogen catabolism, and Sup35, a subunit of the translation termination factor (Wickner, 1994). These prions produce a phenotype similar to that of partial deficiency of the corresponding protein.

[PIN+] is a prion of Rnq1p, a protein of no known function. The [PIN+] phenotype is the ability to rarely cross-seed other prions, originally [PSI+]. Several other new prions were discovered based on their ability to produce a Pin-like effect, including [SWI+] (Du, Park, Yu, Fan, & Li, 2008), a prion of the chromatin-remodeling factor Swi1p, [OCT+], a prion of the transcription repressor Cyc8p (Patel, Gavin-Smyth, & Liebman, 2009), and [MOD+], a prion of the tRNA isopentenyltransferase Mod5p (Suzuki, Shimazu, & Tanaka, 2012). Another, [MOT+], a prion of Mot3p, was found in a general screen of Q/N-rich proteins (Alberti, Halfmann, King, Kapila, & Lindquist, 2009).

5.2. Most yeast prions are amyloid filaments

Each of the prions listed in Table 1.1 is an amyloid of the corresponding protein. Each prion protein has a domain (the prion domain) which is responsible for the prion properties of the whole protein, and constitutes the part of the protein that forms amyloid. The Ure2p prion domain changes from unstructured (Pierce, Baxa, Steven, Bax, & Wickner, 2005) to beta sheet (Baxa et al., 2005; Taylor, Cheng, Williams, Steven, & Wickner, 1999), the remainder of the molecule remaining largely unchanged in structure (Bai, Zhou, & Perrett, 2004). Solid-state NMR studies combined with electron microscopic data showed that amyloids of the prion domains of Ure2p, Sup35p, and Rnq1p are parallel in-register beta sheets, with the sheets folded lengthwise along the long axis of the filaments (Fig. 1.8) (Baxa et al., 2007; Shewmaker, Wickner, & Tycko, 2006; Wickner, Dyda, & Tycko, 2008).

5.3. Prion variants

Remarkably, a single protein, with a single sequence, can be the basis of several (perhaps many) different prion variants, with different biological properties and different amyloid conformations. Most variants are quite stably propagated (though some variants are unstable and are lost or change to another variant). Different prion variants are due to the same protein sequence adopting (and propagating) different amyloid conformations (Bessen & Marsh, 1994; Tanaka, Chien, Naber, Cooke, & Weissman, 2004). How does a prion protein adopt any of several/many different conformations and reliably transmit that conformation to new molecules joining the ends of the filaments? This has long been the central puzzle of prions, and because proteins were believed to be unable to *template* their conformation, it was argued that prions must have a nucleic acid component, just like other elements of heredity.

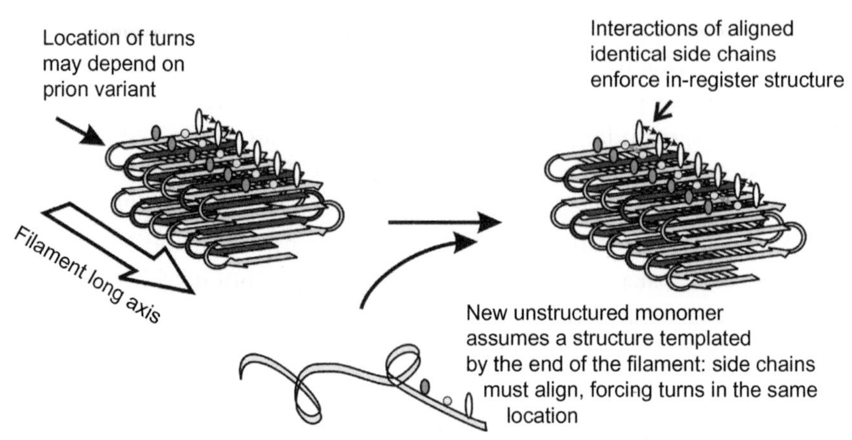
Location of turns
may depend on
prion variant
Interactions of aligned
identical side chains
enforce in-register structure
Filament long axis
New unstructured monomer
assumes a structure templated
by the end of the filament: side chains
must align, forcing turns in the same
location

Figure 1.8 Yeast prion parallel in-register beta-sheet architecture and hypothesized mechanism of variant information propagation. Parallel in-register beta sheets are characterized by lines of identical amino acid side chains extending the length of the filaments. These yeast prion beta sheets are known to be folded along the filament long axis as shown. It is proposed that different prion variants have the folds in different locations. The parallel structure is maintained by the favorable interactions of identical side chains. The energetic advantage in these interactions force a new (unstructured) monomer joining the end of the filament to become in-register, thereby ensuring that the folds/turns of this monomer are in the same locations as those of previous molecules in the filament. This ensures inheritance of prion variant information (conformation) by templating, analogous to DNA templating of sequence information. (See Page 1 in Color Section at the back of the book.)

5.4. In-register parallel architecture of prions can explain heritable conformation

The parallel in-register architecture greatly restricts the possible structures. The beta sheet of the prion domain is known in each case to be folded lengthwise, but the locations of the folds are not known. It is hypothesized that the folds are at different locations in different prion variants, and that the resulting different conformations have different biochemical and biological properties as a result (Wickner, Edskes, Shewmaker, & Nakayashiki, 2007; Wickner et al., 2010) (Fig. 1.8). Most importantly, this hypothesis provides a means to explain how a given architecture can be stably propagated and transmitted to new molecules joining the end of the filament. The same energetically favorable interactions between aligned identical amino acid side chains that keep the structure in-register will direct a newly joining molecule to adopt the same conformation as those already in the filament, and thus to have its turns (folds) at the same locations (Wickner et al., 2007, 2010) (Fig. 1.8). There is currently no other model to explain prion variant information propagation. The only prion for which a detailed atomic structure has been determined, that of HET-s of the filamentous fungus *Podospora anserina* (Saupe, 2011), has a beta-helix structure (Wasmer et al., 2008), and this prion does not have prion variants.

5.5. Chaperones and yeast prions

Because protein conformation is such a central aspect of prions, it has been found that chaperones play several crucial roles in their generation and propagation. Hsp104, acting with Hsp70s and Hsp40s, is important in prion replication (Chernoff, Lindquist, Ono, Inge-Vechtomov, & Liebman, 1995), breaking long filaments into shorter ones, perhaps by extracting protein monomers from the middle of the filaments (Kryndushkin, Alexandrov, Ter-Avanesyan, & Kushnirov, 2003; Ness, Ferreira, Cox, & Tuite, 2002; Paushkin, Kushnirov, Smirnov, & Ter-Avanesyan, 1996). All of the known amyloid-based yeast prions require Hsp104 for their propagation. Overproduction of Hsp104 cures the [PSI+] prion, apparently by a different, but as yet unknown, mechanism (reviewed by Reidy & Masison, 2011). The cytoplasmic Hsp70s, Ssa1p, and Ssa2p, as well as the Hsp40s, Ydj1p, and Sis1p, are also critical for prion propagation (Jung, Jones, Wegrzyn, & Masison, 2000; Moriyama, Edskes, & Wickner, 2000; Newnam, Wegrzyn, Lindquist, & Chernoff, 1999; Sondheimer, Lopez, Craig, & Lindquist, 2001). Hsp70s are ATPases, whose ATP/ADP states determine the binding and release of substrates. These states are regulated by the Hsp40s, by nucleotide exchange factors and other components, with a clear correlation between ATP/ADP state and prion-propagating ability (Sharma & Masison, 2009).

5.6. Transmission barriers: Interspecies and intraspecies

Early recognition that sheep scrapie could be transmitted to goats, but only with difficulty, gave rise to the species barrier concept, shown to be largely based on differences between donor and recipient in the prion protein sequence (Prusiner et al., 1990). A similar phenomenon is known in yeast for [PSI+] (Chen, Newnam, & Chernoff, 2007) and for [URE3] (Edskes et al., 2011). Indeed, there are even intraspecies barriers to transmission of [PSI+] between different polymorphs of Sup35p within *S. cerevisiae*, which suggests that such barriers have been selected to protect cells from the ill effects of this prion (Bateman & Wickner, 2012). It is notable that the species barriers are prion variant-dependent, so that with the same donor–recipient pair of prion protein sequences, transmission may be efficient or near zero, depending on the particular prion variant/amyloid conformation whose transmission is being tested (Chen et al., 2010; Edskes, McCann, Hebert, & Wickner, 2009).

6. BIOLOGY OF YEAST PRIONS

Although yeast prions are proposed to benefit their host (Eaglestone, Cox, & Tuite, 1999; True & Lindquist, 2000), there is as yet no reproducible evidence for this conclusion (Namy et al., 2008). Prion variants differ dramatically in their properties, with the usual mild variants of [PSI+] and [URE3] used in laboratory studies showing little discernible effect on growth, but other, probably more common variants being lethal or nearly so (McGlinchey, Kryndushkin, & Wickner, 2011). To estimate the overall benefit or detriment of yeast prions, a survey of 70 wild strains was carried out. Although the 2-μm DNA plasmid, estimated by two groups to confer a ~1–3% slowing of growth on various strains, was found in 38 of 70 wild strains, none of these strains had [URE3] or [PSI+] and only 11 had [PIN+] (Nakayashiki et al., 2005). This suggests that even the mild variants of each of these prions confer a net overall detriment of >1% on their hosts.

Other lines of evidence indicate that the [URE3] and [PSI+] prions are detrimental. Cells infected with either prion show a stress response with induction of Hsp104 and Hsp70s, indicating that the cells consider prion infection a stress (Jung et al., 2000; Schwimmer & Masison, 2002). The prion domains have nonprion functions, a central role in mRNA turnover for the Sup35p prion domain and an important role in stabilization of Ure2p for its prion domain (Hoshino, Imai, Kobayashi, Uchida, & Katada, 1999; Hosoda et al., 2003; Shewmaker, Mull, Nakayashiki, Masison, & Wickner, 2007). Thus, these domains are not conserved to allow prion formation, but for their normal function. In fact, prion formation is not conserved. Although a range of species conserve some sequence in the Ure2p prion domain, several of these are unable to form [URE3] as tested in *S. cerevisiae*, but some species that do not show sequence conservation can form the prion (Edskes et al., 2011; Safadi, Talarek, Jacques, & Aigle, 2011). Thus, prion formation appears to be a sporadic condition. Indeed, prion formation occurs stochastically, and often under conditions which make matters worse for the cell, suggesting that it is not adaptive.

7. PROSPECTS

Although a large body of work has accumulated on the yeast viruses and prions, many interesting problems remain to be solved. Each of these elements is found in a limited array of wild strains, which indicates that they must be a net detriment to the host, probably beyond simply the use of some energy and material resources. Elucidation of the nature of these interactions would be of broad interest. The M dsRNAs use the coat supplied by L-A, and several lines of evidence suggest that L-A has first call on its own encoded coat proteins, but clear proof of this is lacking. The newly elucidated cap-stealing by L-A helps explain the ability of viral mRNAs to be translated, but they still lack the 3′ polyA, a stringent requirement for translation in eukaryotes. Moreover, the *ski1/xrn1* gene was first detected by its "superkiller" phenotype (Toh-e et al., 1978), which suggests that cap-stealing by L-A is not 100% efficient. Do Ski2, Ski3, Ski8 affect translation or only $3' \rightarrow 5'$ decay?

How and why are 20S and 23S RNAs amplified under sporulation conditions (10,000×!), the phenomenon that led to their first detection (Kadowaki & Halvorson, 1971)? Can the virus-launching vectors be put to use for biotechnologic purposes? Narnaviruses exist not in the form of a conventional virion structure with a protective capsid, but as ribonucleoprotein complexes in the host cytoplasm. A fine structural analysis of these complexes is essential to understand the unique mode of narnavirus existence. Resting complexes seem to have no host proteins. But, are there any host proteins directly involved in narnavirus replication? 20S and 23S RNAs have strong secondary structures at the 5′ ends. This leads to the question of how p91 and p104 are translated, cap-dependent, or -independent? In the latter case, these structures may serve as ribosome-entry sites.

While yeast prions seem to have a parallel in-register architecture, the heterogeneity of amyloids made *in vitro* has so far precluded detailed elucidation of structures, so that the hypothesis that prion variants differ in the locations of the folds in the sheet (see above) has yet to be verified. Moreover, the basis of the different properties of yeast prions is not yet understood. Yeast prions can be lethal (McGlinchey et al., 2011), but the variety of lethal prion variants and the mechanisms of the pathology they produce are just beginning to be explored. The fungal prion [Het-s] is so far unique in having demonstrated benefit to the cell (Coustou, Deleu, Saupe, & Begueret, 1997), but it also produces a (pathologic) meiotic drive phenomenon (Dalstra, Swart, Debets, Saupe, & Hoekstra, 2003), clouding the issue. Suggestions that yeast prions are beneficial (Eaglestone et al., 1999; True & Lindquist, 2000) have met with an enthusiastic reception not yet justified by the data. Yeast prions are important models for the increasingly common amyloidoses and promise to continue to reveal important clues to understanding and dealing with these diseases.

ACKNOWLEDGMENTS

R. W. was supported by the Intramural Program of the National Institute of Diabetes and Digestive and Kidney Diseases. T. F. and R. E. were supported by Grant BFU2010-15768 from the Spanish Ministry of Education and Science.

REFERENCES

Alberti, S., Halfmann, R., King, O., Kapila, A., & Lindquist, S. (2009). A systematic survey identifies prions and illuminates sequence features of prionogenic proteins. *Cell*, *137*, 146–158.

Bai, M., Zhou, J. M., & Perrett, S. (2004). The yeast prion protein Ure2 shows glutathione peroxidase activity in both native and fibrillar forms. *The Journal of Biological Chemistry*, *279*, 50025–50030.

Ball, S. G., Tirtiaux, C., & Wickner, R. B. (1984). Genetic control of L-A and L-BC dsRNA copy number in killer systems of *Saccharomyces cerevisiae*. *Genetics*, *107*, 199–217.

Bateman, D. A., & Wickner, R. B. (2012). [PSI+] prion transmission barriers protect *Saccharomyces cerevisiae* from infection: Intraspecies 'species barriers'. *Genetics*, *190*, 569–579.

Baxa, U., Cheng, N., Winkler, D. C., Chiu, T. K., Davies, D. R., Sharma, D., et al. (2005). Filaments of the Ure2p prion protein have a cross-beta core structure. *Journal of Structural Biology*, *150*, 170–179.

Baxa, U., Wickner, R. B., Steven, A. C., Anderson, D., Marekov, L., Yau, W.-M., et al. (2007). Characterization of β-sheet structure in Ure2p1-89 yeast prion fibrils by solid state nuclear magnetic resonance. *Biochemistry*, *46*, 13149–13162.

Beauregard, A., Curcio, M. J., & Belfort, M. (2008). The take and give between retrotransposable elements and their hosts. *Annual Review of Genetics*, *42*, 587–617.

Benard, L., Carroll, K., Valle, R. C. P., & Wickner, R. B. (1998). Ski6p is a homolog of RNA-processing enzymes that affects translation of non-poly(A) mRNAs and 60S ribosomal subunit biogenesis. *Molecular and Cellular Biology*, *18*, 2688–2696.

Benard, L., Carroll, K., Valle, R. C. P., & Wickner, R. B. (1999). The Ski7 antiviral protein is an EF1-α homolog that blocks expression of non-poly(A) mRNA in *Saccharomyces cerevisiae*. *Journal of Virology*, *73*, 2893–2900.

Bessen, R. A., & Marsh, R. F. (1994). Distinct PrP properties suggest the molecular basis of strain variation in transmissible mink encephalopathy. *Journal of Virology*, *68*, 7859–7868.

Blanc, A., Goyer, C., & Sonenberg, N. (1992). The coat protein of the yeast double-stranded RNA virus L-A attaches covalently to the cap structure of eukaryotic mRNA. *Molecular and Cellular Biology*, *12*, 3390–3398.

Blanc, A., Ribas, J. C., Wickner, R. B., & Sonenberg, N. (1994). His154 is involved in the linkage of the *Saccharomyces cerevisiae* L-A double-stranded RNA virus *gag* protein to the cap structure of mRNAs and is essential for M_1 satellite virus expression. *Molecular and Cellular Biology*, *14*, 2664–2674.

Boivin, S., Cusack, S., Ruigrok, R. W., & Hart, D. J. (2010). Influenza A virus polymerase: Structural insights into replication and host adaptation mechanisms. *The Journal of Biological Chemistry*, *285*, 28411–28417.

Bostian, K. A., Jayachandran, S., & Tipper, D. J. (1983). A glycosylated protoxin in killer yeast: Models for its structure and maturation. *Cell*, *32*, 169–180.

Brown, J. T., Bai, X., & Johnson, A. W. (2000). The yeast antiviral proteins Ski2p, Ski3p and Ski8p exist as a complex in vivo. *RNA*, *6*, 449–457.

Brown, J. T., & Johnson, A. W. (2001). A cis-acting element known to block 3′ mRNA degradation enhances expression of polyA-minus mRNA in wild-type yeast cells and phenocopies a *ski* mutant. *RNA*, *7*, 1566–1577.

Caston, J. R., Trus, B. L., Booy, F. P., Wickner, R. B., Wall, J. S., & Steven, A. C. (1997). Structure of L-A virus: A specialized compartment for the transcription and replication of double-stranded RNA. *The Journal of Cell Biology, 138*, 975–985.

Chen, B., Bruce, K. L., Newnam, G. P., Gyoneva, S., Romanyuk, A. V., & Chernoff, Y. O. (2010). Genetic and epigenetic control of the efficiency and fidelity of cross-species prion transmission. *Molecular Microbiology, 76*, 1483–1499.

Chen, B., Newnam, G. P., & Chernoff, Y. O. (2007). Prion species barrier between the closely related yeast proteins is detected despite coaggregation. *Proceedings of the National Academy of Sciences of the United States of America, 104*, 2791–2796.

Chernoff, Y. O., Lindquist, S. L., Ono, B.-I., Inge-Vechtomov, S. G., & Liebman, S. W. (1995). Role of the chaperone protein Hsp104 in propagation of the yeast prion-like factor [psi^+]. *Science, 268*, 880–884.

Chlebowski, A., Tomecki, R., Lopez, M. E., Seraphin, B., & Dziembowski, A. (2011). Catalytic properties of the eukaryotic exosome. *Advances in Experimental Medicine and Biology, 702*, 63–78.

Chu, C., Das, K., Tyminski, J. R., Bauman, J. D., Guan, R., Qiu, W., et al. (2011). Structure of the guanylyltransferase domain of human mRNA capping enzyme. *Proceedings of the National Academy of Sciences of the United States of America, 108*, 10104–10108.

Cooper, A., & Bussey, H. (1989). Characterization of the yeast KEX1 gene product: A carboxypeptidase involved in processing secreted precursor proteins. *Molecular and Cellular Biology, 9*, 2706–2714.

Coustou, V., Deleu, C., Saupe, S., & Begueret, J. (1997). The protein product of the *het-s* heterokaryon incompatibility gene of the fungus *Podospora anserina* behaves as a prion analog. *Proceedings of the National Academy of Sciences of the United States of America, 94*, 9773–9778.

Dalstra, H. J. P., Swart, K., Debets, A. J. M., Saupe, S. J., & Hoekstra, R. F. (2003). Sexual transmission of the [Het-s] prion leads to meiotic drive in *Podospora anserina*. *Proceedings of the National Academy of Sciences of the United States of America, 100*, 6616–6621.

Derkatch, I. L., Bradley, M. E., Hong, J. Y., & Liebman, S. W. (2001). Prions affect the appearance of other prions: The story of *[PIN]*. *Cell, 106*, 171–182.

Dinman, J. D., Icho, T., & Wickner, R. B. (1991). A −1 ribosomal frameshift in a double-stranded RNA virus of yeast forms a gag-pol fusion protein. *Proceedings of the National Academy of Sciences of the United States of America, 88*, 174–178.

Dinman, J. D., & Wickner, R. B. (1992). Ribosomal frameshifting efficiency and *gag/gag-pol* ratio are critical for yeast M_1 double-stranded RNA virus propagation. *Journal of Virology, 66*, 3669–3676.

Drinnenberg, I. A., Fink, G. R., & Bartel, D. P. (2011). Compatibility with killer explains the rise of RNAi-deficient fungi. *Science, 333*, 1592.

Drinnenberg, I. A., Weinberg, D. E., Xie, K. T., Mower, J. P., Wolfe, K. H., Fink, G. R., et al. (2009). RNAi in budding yeast. *Science, 326*, 544–550.

Du, Z., Park, K.-W., Yu, H., Fan, Q., & Li, L. (2008). Newly identified prion linked to the chromatin-remodeling factor Swi1 in *Saccharomyces cerevisiae*. *Nature Genetics, 40*, 460–465.

Eaglestone, S. S., Cox, B. S., & Tuite, M. F. (1999). Translation termination efficiency can be regulated in *Saccharomyces cerevisiae* by environmental stress through a prion-mediated mechanism. *The EMBO Journal, 18*, 1974–1981.

Edskes, H. K., Engel, A., McCann, L. M., Brachmann, A., Tsai, H.-F., & Wickner, R. B. (2011). Prion-forming ability of Ure2 of yeasts is not evolutionarily conserved. *Genetics, 188*, 81–90.

Edskes, H. K., McCann, L. M., Hebert, A. M., & Wickner, R. B. (2009). Prion variants and species barriers among *Saccharomyces* Ure2 proteins. *Genetics, 181*, 1159–1167.

Edskes, H. K., Ohtake, Y., & Wickner, R. B. (1998). Mak21p of *Saccharomyces cerevisiae*, a homolog of human CAATT-binding protein, is essential for 60S ribosomal subunit biogenesis. *The Journal of Biological Chemistry, 273*, 28912–28920.
Esteban, R., & Fujimura, T. (2003). Launching the yeast 23S RNA Narnavirus shows 5′ and 3′ cis-acting signals for replication. *Proceedings of the National Academy of Sciences of the United States of America, 100*, 2568–2573.
Esteban, R., & Fujimura, T. (2006). K. L. Heffereon (Ed.), *Recent advances in RNA virus replication* (pp. 171–194). Kerala, India: Transworld Research Network.
Esteban, L. M., Fujimura, T., Garcia-Cuellar, M., & Esteban, R. (1994). Association of yeast viral 23S RNA with its putative RNA-dependent RNA polymerase. *The Journal of Biological Chemistry, 269*, 29771–29777.
Esteban, R., Fujimura, T., & Wickner, R. B. (1988). Site-specific binding of viral plus single-stranded RNA to replicase-containing open virus-like particles of yeast. *Proceedings of the National Academy of Sciences of the United States of America, 85*, 4411–4415.
Esteban, R., Fujimura, T., & Wickner, R. B. (1989). Internal and terminal cis-acting sites are necessary for in vitro replication of the L-A double-stranded RNA virus of yeast. *The EMBO Journal, 8*, 947–954.
Esteban, L. M., Rodriguez, C. N., & Esteban, R. (1992). T double-stranded RNA (dsRNA) sequence reveals that T and W dsRNAs form a new RNA family in Saccharomyces cerevisiae. Identification of 23 S RNA as the single-stranded form of T dsRNA. *The Journal of Biological Chemistry, 267*, 10874–10881.
Esteban, R., Vega, L., & Fujimura, T. (2005). Launching of the yeast 20S RNA narnavirus by expressing the genomic or antigenomic viral RNA *in vivo*. *The Journal of Biological Chemistry, 280*, 33725–33734.
Esteban, R., Vega, L., & Fujimura, T. (2008). 20S RNA narnavirus defies the antiviral activity of *SKI1/XRN1* in *Saccharomyces cerevisiae*. *The Journal of Biological Chemistry, 283*, 25812–25820.
Esteban, R., & Wickner, R. B. (1986). Three different M1 RNA-containing viruslike particle types in Saccharomyces cerevisiae: in vitro M1 double-stranded RNA synthesis. *Molecular and Cellular Biology, 6*, 1552–1561.
Esteban, R., & Wickner, R. B. (1988). A deletion mutant of L-A double-stranded RNA replicates like M_1 double-stranded RNA. *Journal of Virology, 62*, 1278–1285.
Fujimura, T., & Esteban, R. (2004a). Bipartite 3′-cis-acting signal for replication in yeast 23S RNA virus and its repair. *The Journal of Biological Chemistry, 279*, 13215–13223.
Fujimura, T., & Esteban, R. (2004b). The bipartite 3′-cis-acting signal for replication is required for formation of a ribonucleoprotein complex in vivo between the viral genome and its RNA polymerase in yeast 23 S RNA virus. *The Journal of Biological Chemistry, 279*, 44219–44228.
Fujimura, T., & Esteban, R. (2007). Interactions of the RNA polymerase with the viral genome at the 5′ and 3′ ends contribute to 20S RNA narnavirus persistence in yeast. *The Journal of Biological Chemistry, 282*, 19011–19019.
Fujimura, T., & Esteban, R. (2010). Yeast double-stranded RNA virus L-A deliberately synthesizes RNA transcripts with 5′-diphosphate. *The Journal of Biological Chemistry, 285*, 22911–22918.
Fujimura, T., & Esteban, R. (2011). Cap-snatching mechanism in yeast L-A double-stranded RNA virus. *Proceedings of the National Academy of Sciences of the United States of America, 108*, 17667–17671.
Fujimura, T., & Esteban, R. (2012). Cap snatching of yeast L-A double-stranded RNA virus can operate in *trans* and requires viral polymerase actively engaging in transcription. *The Journal of Biological Chemistry, 287*, 12797–12804.
Fujimura, T., Esteban, R., Esteban, L. M., & Wickner, R. B. (1990). Portable encapsidation signal of the L-A double-stranded RNA virus of *S. cerevisiae*. *Cell, 62*, 819–828.

Fujimura, T., Esteban, R., & Wickner, R. B. (1986). In vitro L-A double-stranded RNA synthesis in virus-like particles from Saccharomyces cerevisiae. *Proceedings of the National Academy of Sciences of the United States of America, 83*, 4433–4437.

Fujimura, T., Ribas, J. C., Makhov, A. M., & Wickner, R. B. (1992). Pol of gag-pol fusion protein required for encapsidation of viral RNA of yeast L-A virus. *Nature, 359*, 746–749.

Fujimura, T., Solorzano, A., & Esteban, R. (2005). Native replication intermediates of the yeast 20S RNA virus have a single-stranded RNA backbone. *The Journal of Biological Chemistry, 280*, 7398–7406.

Fujimura, T., & Wickner, R. B. (1988a). Gene overlap results in a viral protein having an RNA binding domain and a major coat protein domain. *Cell, 55*, 663–671.

Fujimura, T., & Wickner, R. B. (1988b). Replicase of L-A virus-like particles of Saccharomyces cerevisiae. in vitro conversion of exogenous L-A and M1 single-stranded RNAs to double-stranded form. *The Journal of Biological Chemistry, 263*, 454–460.

Fujimura, T., & Wickner, R. B. (1992). Interaction of two cis sites with the RNA replicase of the yeast L-A virus. *The Journal of Biological Chemistry, 267*, 2708–2713.

Fukuhara, T., Koga, R., Aoki, N., Yuki, C., Yamamoto, N., Oyama, N., et al. (2006). The wide distribution of endornaviruses, large double-stranded RNA replicons with plasmid-like properties. *Archives of Virology, 151*, 995–1002.

Gancarz, B., Hao, L., He, Q., Newton, M. A., & Ahlquist, P. (2011). Systematic identification of novel, essential host genes affecting bromovirus RNA replication. *PLoS One, 6*, e23988.

Garcia-Cuellar, M. P., Esteban, R., & Fujimura, T. (1997). RNA-dependent RNA polymerase activity associated with the yeast viral p91/20S RNA ribonucleoprotein complex. *RNA, 3*, 27–36.

Garcia-Cuellar, M. P., Esteban, L. M., Fujimura, T., Rodriguez-Cousino, N., & Esteban, R. (1995). Yeast viral 20S RNA is associated with its cognate RNA-dependent RNA polymerase. *The Journal of Biological Chemistry, 270*, 20084–20089.

Garvik, B., & Haber, J. E. (1978). New cytoplasmic genetic element that controls 20S RNA synthesis during sporulation in yeast. *Journal of Bacteriology, 134*, 261–269.

Ghabrial, S. A., Nibert, M. L., Maiss, E., Lesker, T., Baker, T. S., & Tao, Y. Z. (2011). Partitiviridae. In A. M. Q. King, M. J. Adams, E. B. Carstens & E. J. Lefkowitz (Eds.), *Virus taxonomy: Ninth report of the international committee on taxonomy of viruses* (pp. 523–534). San Diego: Elsevier.

Hakansson, K., Doherty, A. J., Shuman, S., & Wigley, D. B. (1997). X-ray crystallography reveals a large conformation change during guanyl transfer by mRNA capping enzymes. *Cell, 89*, 545–553.

Herring, A. J., & Bevan, A. E. (1974). Virus-like particles associated with the double-stranded RNA species found in killer and sensitive strains of the yeast *Saccharomyces cerevisiae*. *Journal of General Virology, 22*, 387–394.

Hoshino, S., Imai, M., Kobayashi, T., Uchida, N., & Katada, T. (1999). The eukaryotic polypeptide chain releasing factor (eRF3/GSPT) carrying the translation termination signal to the 3′-poly(A) tail of mRNA. *The Journal of Biological Chemistry, 274*, 16677–16680.

Hosoda, N., Kobayashii, T., Uchida, N., Funakoshi, Y., Kikuchi, Y., Hoshino, S., et al. (2003). Translation termination factor eRF3 mediates mRNA decay through the regulation of deadenylation. *The Journal of Biological Chemistry, 278*, 38287–38291.

Hu, G., Gershon, P. D., Hodel, A. E., & Quiocho, F. A. (1999). mRNA cap recognition: Dominant role of enhanced stacking interactions between methylated bases and protein aromatic side chains. *Proceedings of the National Academy of Sciences of the United States of America, 96*, 7149–7154.

Icho, T., & Wickner, R. B. (1989). The double-stranded RNA genome of yeast virus L-A encodes its own putative RNA polymerase by fusing two open reading frames. *The Journal of Biological Chemistry, 264*, 6716–6723.

Jacobs-Anderson, J. S., & Parker, R. (1998). The 3′ to 5′ degradation of yeast mRNAs is a general mechanism for mRNA turnover that requires the SKI2 DEVH box protein and 3′ to 5′ exonucleases of the exosome complex. *The EMBO Journal, 17*, 1497–1506.

Julius, D., Brake, A., Blair, L., Kunisawa, R., & Thorner, J. (1984). Isolation of the putative structural gene for the lysine-arginine-cleaving endopeptidase required for the processing of yeast prepro-alpha factor. *Cell, 37*, 1075–1089.

Jung, G., Jones, G., Wegrzyn, R. D., & Masison, D. C. (2000). A role for cytosolic Hsp70 in yeast [PSI+] prion propagation and [PSI+] as a cellular stress. *Genetics, 156*, 559–570.

Kadowaki, K., & Halvorson, H. O. (1971). Appearance of a new species of ribonucleic acid during sporulation in *Saccharomyces cerevisiae*. *Journal of Bacteriology, 105*, 826–830.

Kahvejian, A., Svitkin, Y. V., Sukarieh, R., M'Boutchou, M. N., & Sonenberg, N. (2005). Mammalian poly(A)-binding protein is a eukaryotic translation initiation factor, which acts via multiple mechanisms. *Genes & Development, 19*, 104–113.

Kovalev, N., Pogany, J., & Nagy, P. D. (2012). A co-opted DEAD-box RNA helicase enhances tombusvirus plus-strand synthesis. *PLoS Pathogens, 8*, e1002537.

Kryndushkin, D. S., Alexandrov, I. M., Ter-Avanesyan, M. D., & Kushnirov, V. V. (2003). Yeast [PSI^{+}] prion aggregates are formed by small Sup35 polymers fragmented by Hsp104. *The Journal of Biological Chemistry, 278*, 49636–49643.

Larimer, F. W., Hsu, C. L., Maupin, M. K., & Stevens, A. (1992). Characterization of the XRN1 gene encoding a 5′ -> 3′ exoribonuclease: Sequence data and analysis of disparate protein and mRNA levels of gene-disrupted cells. *Gene, 120*, 51–57.

Leibowitz, M. J., & Wickner, R. B. (1976). A chromosomal gene required for killer plasmid expression, mating, and spore maturation in Saccharomyces cerevisiae. *Proceedings of the National Academy of Sciences of the United States of America, 73*, 2061–2065.

Martinac, B., Zhu, H., Kubalski, A., Zhou, X. L., Culbertson, M., Bussey, H., et al. (1990). Yeast K1 killer toxin forms ion channels in sensitive yeast spheroplasts and in artificial liposomes. *Proceedings of the National Academy of Sciences of the United States of America, 87*, 6228–6232.

Masison, D. C., Blanc, A., Ribas, J. C., Carroll, K., Sonenberg, N., & Wickner, R. B. (1995). Decoying the cap- mRNA degradation system by a dsRNA virus and poly(A)- mRNA surveillance by a yeast antiviral system. *Molecular and Cellular Biology, 15*, 2763–2771.

Matsumoto, Y., Fishel, R., & Wickner, R. B. (1990). Circular single-stranded RNA replicon in *Saccharomyces cerevisiae*. *Proceedings of the National Academy of Sciences of the United States of America, 87*, 7628–7632.

Matsumoto, Y., & Wickner, R. B. (1991). Yeast 20 S RNA replicon. Replication intermediates and encoded putative RNA polymerase. *The Journal of Biological Chemistry, 266*, 12779–12783.

Maxwell, P. H., & Curcio, M. J. (2007). Host factors that control long terminal repeat retrotransposons in *Saccharomyces cerevisiae*: Implications for the regulation of mammalian retroviruses. *Eukaryotic Cell, 6*, 1069–1080.

McGlinchey, R., Kryndushkin, D., & Wickner, R. B. (2011). Suicidal [PSI+] is a lethal yeast prion. *Proceedings of the National Academy of Sciences of the United States of America, 108*, 5337–5341.

Mitchell, P., Petfalski, E., Shevchenko, A., Mann, M., & Tollervey, D. (1997). The exosome, a conserved eukaryotic RNA processing complex containing multiple 3′→5′ exoribonucleases. *Cell, 91*, 457–466.

Moriyama, H., Edskes, H. K., & Wickner, R. B. (2000). [URE3] prion propagation in Saccharomyces cerevisiae: Requirement for chaperone Hsp104 and curing by overexpressed chaperone Ydj1p. *Molecular and Cellular Biology, 20*, 8916–8922.

Munroe, D., & Jacobson, A. (1990). mRNA poly(A) tail, a 3′ enhancer of translation initiation. *Molecular and Cellular Biology, 10*, 3441–3455.

Naitow, H., Canady, M. A., Wickner, R. B., & Johnson, J. E. (2002). L-A dsRNA virus at 3.4 Angstroms resolution reveals particle architecture and mRNA decapping mechanism. *Nature Structural Biology, 9*, 725–728.

Nakayashiki, T., Kurtzman, C. P., Edskes, H. K., & Wickner, R. B. (2005). Yeast prions [URE3] and [PSI^+] are diseases. *Proceedings of the National Academy of Sciences of the United States of America, 102*, 10575–10580.

Namy, O., Galopier, A., Martini, C., Matsufuji, S., Fabret, C., & Rousset, C. (2008). Epigenetic control of polyamines by the prion [PSI^+]. *Nature Cell Biology, 10*, 1069–1075.

Ness, F., Ferreira, P., Cox, B. S., & Tuite, M. F. (2002). Guanidine hydrochloride inhibits the generation of prion "seeds" but not prion protein aggregation in yeast. *Molecular and Cellular Biology, 22*, 5593–5605.

Newnam, G. P., Wegrzyn, R. D., Lindquist, S. L., & Chernoff, Y. O. (1999). Antagonistic interactions between yeast chaperones Hsp104 and Hsp70 in prion curing. *Molecular and Cellular Biology, 19*, 1325–1333.

Nuss, D. L. (2005). Hypovirulence: Mycoviruses at the fungal-plant interface. *Nature Reviews. Microbiology, 3*, 632–642.

Ohtake, Y., & Wickner, R. B. (1995). Yeast virus propagation depends critically on free 60S ribosomal subunit concentration. *Molecular and Cellular Biology, 15*, 2772–2781.

Park, C., Lopinski, J. D., Masuda, J., Tzeng, T.-H., & Bruenn, J. A. (1996). A second double-stranded RNA virus from yeast. *Virology, 216*, 451–454.

Parker, R., & Song, H. (2004). The enzymes and control of eukaryotic mRNA turnover. *Nature Structural & Molecular Biology, 11*, 121–127.

Patel, B. K., Gavin-Smyth, J., & Liebman, S. W. (2009). The yeast global transcriptional co-repressor protein Cyc8 can propagate as a prion. *Nature Cell Biology, 11*, 344–349.

Paushkin, S. V., Kushnirov, V. V., Smirnov, V. N., & Ter-Avanesyan, M. D. (1996). Propagation of the yeast prion-like [psi^+] determinant is mediated by oligomerization of the *SUP35*-encoded polypeptide chain release factor. *The EMBO Journal, 15*, 3127–3134.

Pierce, M. M., Baxa, U., Steven, A. C., Bax, A., & Wickner, R. B. (2005). Is the prion domain of soluble Ure2p unstructured? *Biochemistry, 44*, 321–328.

Plotch, S. J., Bouloy, M., Ulmanen, I., & Krug, R. M. (1981). A unique cap(m7GpppXm)-dependent influenza virion endonuclease cleaves capped RNAs to generate the primers that initiate viral RNA transcription. *Cell, 23*, 847–858.

Prusiner, S. B., Scott, M., Foster, D., Pan, K.-M., Groth, D., Mirenda, C., et al. (1990). Transgenic studies implicate interactions between homologous PrP isoforms in scrapie prion replication. *Cell, 63*, 673–686.

Ramirez-Garrastacho, M., & Esteban, R. (2011). Yeast RNA viruses as indicators of exosome activity: Human exosome hCsl4p participates in RNA degradation in *Saccharomyces cerevisiae*. *Yeast, 28*, 821–832.

Reidy, M., & Masison, D. C. (2011). Modulation and elimination of yeast prions by protein chaperones and co-chaperones. *Prion, 5*, 245–249.

Ribas, J. C., Fujimura, T., & Wickner, R. B. (1994). Essential RNA binding and packaging domains of the Gag-Pol fusion protein of the L-A double-stranded RNA virus of *Saccharomyces cerevisiae*. *The Journal of Biological Chemistry, 269*, 28420–28428.

Ridley, S. P., Sommer, S. S., & Wickner, R. B. (1984). Superkiller mutations in Saccharomyces cerevisiae suppress exclusion of M_2 double-stranded RNA by L-A-HN and confer cold sensitivity in the presence of M and L-A-HN. *Molecular and Cellular Biology, 4*, 761–770.

Rigaut, G., Shevchenko, A., Rutz, B., Wilm, M., Mann, M., & Seraphin, B. (1999). A generic protein purification method for protein complex characterization and proteome exploration. *Nature Biotechnology, 17*, 1030–1032.

Roberts, B. T., & Wickner, R. B. (2003). A class of prions that propagate via covalent autoactivation. *Genes & Development, 17*, 2083–2087.

Rodriguez, C. N., Esteban, L. M., & Esteban, R. (1991). Molecular cloning and characterization of W double-stranded RNA, a linear molecule present in *Saccharomyces cerevisiae*. Identification of its single-stranded RNA form as 20 S RNA. *The Journal of Biological Chemistry, 266*, 12772–12778.

Rodriguez-Cousino, N., & Esteban, R. (1992). Both yeast W double-stranded RNA and its single-stranded form 20S RNA are linear. *Nucleic Acids Research, 20*, 2761–2766.

Rodriguez-Cousino, N., Maqueda, M., Ambrona, J., Zamora, E., Esteban, R., & Ramirez, M. (2011). A new wine *Saccharomyces cerevisiae* killer toxin (Klus), encoded by a double-stranded RNA virus, with broad antifungal activity is evolutionarily related to a chromosomal host gene. *Applied and Environmental Microbiology, 77*, 1822–1832.

Rodriguez-Cousino, N., Solorzano, A., Fujimura, T., & Esteban, R. (1998). Yeast positive-strand virus-like RNA replicons. 20S and 23S RNA terminal nucleotide sequences and 3′ end secondary structures resemble those of RNA coliphages. *The Journal of Biological Chemistry, 273*, 20363–20371.

Rogoza, T., Goginashvili, A., Rodionova, S., Ivanov, M., Viktorovskaya, O., Rubel, A., et al. (2010). Non-mendelian determinant [ISP+] in yeast is a nuclear-residing prion form of the global transcriptional regulator Sfp1. *Proceedings of the National Academy of Sciences of the United States of America, 107*, 10573–10577.

Roossinck, M. J., Sabanadzovic, S., Okada, R., & Valverde, R. A. (2011). The remarkable evolutionary history of endornaviruses. *The Journal of General Virology, 92*, 2674–2678.

Safadi, R. A., Talarek, N., Jacques, N., & Aigle, M. (2011). Yeast prions: Could they be exaptations? The *URE2*/[URE3] system in *Kluyveromyces lactis*. *FEMS Yeast Research, 11*, 151–153.

Saupe, S. J. (2011). The [Het-s] prion of *Podospora anserina* and its role in heterokaryon incompatibility. *Seminars in Cell & Developmental Biology, 22*, 460–468.

Schmitt, M. J., & Breinig, F. (2006). Yeast viral killer toxins: Lethality and self-protection. *Nature Reviews. Microbiology, 4*, 212–221.

Schmitt, M. J., Klavehn, P., Wang, J., Schonig, I., & Tipper, D. J. (1996). Cell cycle studies on the mode of action of yeast K28 killer toxin. *Microbiology, 142*, 2655–2662.

Schwimmer, C., & Masison, D. C. (2002). Antagonistic interactions between yeast [PSI+] and [URE3] prions and curing of [URE3] by Hsp70 protein chaperone Ssa1p but not by Ssa2p. *Molecular and Cellular Biology, 22*, 3590–3598.

Searfoss, A., Dever, T. E., & Wickner, R. B. (2001). Linking the 3′ poly(A) tail to the subunit joining step of translation initiation: Relations of Pab1p, eIF5B (Fun12p) and Ski2p-Slh1p. *Molecular Biology of the Cell, 21*, 4900–4908.

Searfoss, A. M., & Wickner, R. B. (2000). 3′ poly(A) is dispensable for translation. *Proceedings of the National Academy of Sciences of the United States of America, 97*, 9133–9137.

Sharma, D., & Masison, D. C. (2009). Hsp70 structure, function, regulation and influence on yeast prions. *Protein & Peptide Letters, 16*, 571–581.

Shewmaker, F., Mull, L., Nakayashiki, T., Masison, D. C., & Wickner, R. B. (2007). Ure2p function is enhanced by its prion domain in *Saccharomyces cerevisiae*. *Genetics, 176*, 1557–1565.

Shewmaker, F., Wickner, R. B., & Tycko, R. (2006). Amyloid of the prion domain of Sup35p has an in-register parallel β-sheet structure. *Proceedings of the National Academy of Sciences of the United States of America, 103*, 19754–19759.

Shuman, S. (1995). Capping enzyme in eukaryotic mRNA synthesis. *Progress in Nucleic Acid Research and Molecular Biology, 50*, 101–129.

Solorzano, A., Rodriguez-Cousino, N., Esteban, R., & Fujimura, T. (2000). Persistent yeast single-stranded RNA viruses exist in vivo as genomic RNA:RNA polymerase complexes in 1:1 stoichiometry. *The Journal of Biological Chemistry, 275*, 26428–26435.

Sommer, S. S., & Wickner, R. B. (1982). Yeast L dsRNA consists of at least three distinct RNAs; evidence that the non-Mendelian genes [HOK], [NEX] and [EXL] are on one of these dsRNAs. *Cell*, *31*, 429–441.

Sondheimer, N., Lopez, N., Craig, E. A., & Lindquist, S. (2001). The role of Sis1 in the maintenance of the [RNQ+] prion. *The EMBO Journal*, *20*, 2435–2442.

Starheim, K. K., Gromyko, D., Evjenth, R., Ryningen, A., Varhaug, J. E., Lillehaug, J. R., et al. (2009). Knockdown of human N^{α}-terminal acetyltransferase complex C leads to p53-dependent apoptosis and aberrant human Arl8b localization. *Molecular and Cellular Biology*, *29*, 3569–3581.

Steiner, D. F., Smeekens, S. P., Ohagi, S., & Chan, S. J. (1992). The new enzymology of precursor processing endoproteases. *The Journal of Biological Chemistry*, *267*, 23435–23438.

Suzuki, G., Shimazu, N., & Tanaka, M. (2012). A yeast prion, Mod5, promotes acquired drug resistance and cell survival under environmental stress. *Science*, *336*, 355–359.

Tanaka, M., Chien, P., Naber, N., Cooke, R., & Weissman, J. S. (2004). Conformational variations in an infectious protein determine prion strain differences. *Nature*, *428*, 323–328.

Tang, J., Naitow, H., Gardner, N. A., Kolesar, A., Tang, L., Wickner, R. B., et al. (2005). The structural basis of recognition and removal of cellular mRNA 7-methyl G 'caps' by a viral capsid protein: A unique viral response to hose defense. *Journal of Molecular Recognition*, *18*, 158–168.

Taylor, K. L., Cheng, N., Williams, R. W., Steven, A. C., & Wickner, R. B. (1999). Prion domain initiation of amyloid formation *in vitro* from native Ure2p. *Science*, *283*, 1339–1343.

Tercero, J. C., Dinman, J. D., & Wickner, R. B. (1993). Yeast *MAK3* N-acetyltransferase recognizes the N-terminal four amino acids of the major coat protein (*gag*) of the L-A double-stranded RNA virus. *Journal of Bacteriology*, *175*, 3192–3194.

Tercero, J. C., Riles, L. E., & Wickner, R. B. (1992). Localized mutagenesis and evidence for post-transcriptional regulation of MAK3, a putative N-acetyltransferase required for dsRNA virus propagation in *Saccharomyces cerevisiae*. *The Journal of Biological Chemistry*, *267*, 20270–20276.

Tercero, J. C., & Wickner, R. B. (1992). MAK3 encodes an N-acetyltransferase whose modification of the L-A *gag* N-terminus is necessary for virus particle assembly. *The Journal of Biological Chemistry*, *267*, 20277–20281.

Toh-e, A., Guerry, P., & Wickner, R. B. (1978). Chromosomal superkiller mutants of Saccharomyces cerevisiae. *Journal of Bacteriology*, *136*, 1002–1007.

Toh-e, A., & Sahashi, Y. (1985). The *PET18* locus of *Saccharomyces cerevisiae*: A complex locus containing multiple genes. *Yeast*, *1*, 159–172.

True, H. L., & Lindquist, S. L. (2000). A yeast prion provides a mechanism for genetic variation and phenotypic diversity. *Nature*, *407*, 477–483.

Vega, L. (2010). Salamanca, Spain, University of Salamanca. Ph.D.

Venkatesan, S., Gershowitz, A., & Moss, B. (1980). Modification of the 5′ end of mRNA. Association of RNA triphosphatase with the RNA guanylyltransferase-RNA (guanine-7-)methyltransferase complex from vaccinia virus. *The Journal of Biological Chemistry*, *255*, 903–908.

Wasmer, C., Lange, A., Van Melckebeke, H., Siemer, A. B., Riek, R., & Meier, B. H. (2008). Amyloid fibrils of the HET-s(218–279) prion form a beta solenoid with a triangular hydrophobic core. *Science*, *319*, 1523–1526.

Wesolowski, M., & Wickner, R. B. (1984). Two new double-stranded RNA molecules showing non-mendelian inheritance and heat inducibility in *Saccharomyces cerevisiae*. *Molecular and Cellular Biology*, *4*, 181–187.

Wickner, R. B. (1980). Plasmids controlling exclusion of the K_2 killer double-stranded RNA plasmid of yeast. *Cell, 21*, 217–226.
Wickner, R. B. (1994). [URE3] as an altered *URE2* protein: Evidence for a prion analog in *S. cerevisiae*. *Science, 264*, 566–569.
Wickner, R. B., Dyda, F., & Tycko, R. (2008). Amyloid of Rnq1p, the basis of the [PIN^+] prion, has a parallel in-register β-sheet structure. *Proceedings of the National Academy of Sciences of the United States of America, 105*, 2403–2408.
Wickner, R. B., Edskes, H. K., Shewmaker, F., & Nakayashiki, T. (2007). Prions of fungi: Inherited structures and biological roles. *Nature Reviews. Microbiology, 5*, 611–618.
Wickner, R. B., Shewmaker, F., Edskes, H., Kryndushkin, D., Nemecek, J., McGlinchey, R., et al. (2010). Prion amyloid structure explains templating: How proteins can be genes. *FEMS Yeast Research, 10*, 980–991.
Wickner, R. B., & Toh-e, A. (1982). [HOK], a new yeast non-Mendelian trait, enables a replication-defective killer plasmid to be maintained. *Genetics, 100*, 159–174.
Widner, W. R., Matsumoto, Y., & Wickner, R. B. (1991). Is 20S RNA naked? *Molecular and Cellular Biology, 11*, 2905–2908.
Wilusz, C. J., Wormington, M., & Peltz, S. W. (2001). The cap-to-tail guide to mRNA turnover. *Nature Reviews. Molecular Cell Biology, 2*, 237–246.
Young, T. W., & Yagiu, M. (1978). A comparison of the killer character in different yeasts and its classification. *Antonie Van Leeuwenhoek, 44*, 59–77.
Zhao, K. N., & Frazer, I. H. (2002). *Saccharomyces cerevisiae* is permissive for replication of bovine papilloma virus type 1. *Journal of Virology, 76*, 12265–12273.

CHAPTER TWO

Multiplexed Interactions: Viruses of Endophytic Fungi

Xiaodong Bao and Marilyn J. Roossinck[1]
Department of Plant Pathology and Environmental Microbiology, Center for Infectious Disease Dynamics, Pennsylvania State University, University Park, Pennsylvania, USA
[1]Corresponding author: e-mail address: mjr25@psu.edu

Contents

1. Introduction 37
2. Classification of Fungal Endophytes 38
3. Viruses of Fungal Endophytes 40
   3.1 Viruses of Class 1 endophytes 40
   3.2 Viruses of Class 2 and Class 3 endophytes 45
   3.3 Viruses of Class 4 endophytes 47
4. Roles of Viruses in Endophytic Fungi 48
5. Practical Roles for Endophyte Viruses 53
6. Summary 54
Acknowledgments 54
References 54

Abstract

Mycoviruses have been detected from all four classes of fungal endophytes. The virus species richness is probably extremely high in endophytes. The incidence and diversity of mycoviruses may be affected by transmission modes, virus–fungus–plant interactions, and endophyte population structures. Endophyte viruses are unlikely to be strong antagonists to their fungal hosts and can clearly play mutualistic roles in the multiplex symbioses with endophytes and plants under some environmental conditions. A better understanding of fungal endophyte viruses will help prospects of future applications for sustainable agriculture.

1. INTRODUCTION

Symbiosis is defined as two or more dissimilar entities living together intimately (deBary, 1879). This can take different forms ranging from antagonistic to mutualistic, and is almost certainly ubiquitous in all life forms, and

Advances in Virus Research, Volume 86
ISSN 0065-3527
http://dx.doi.org/10.1016/B978-0-12-394315-6.00002-7

constitutes a critical component of evolution. The partners in symbiosis have been identified from eukaryotes, prokaryotes, archea, and viruses (Márquez, Redman, Rodriguez, & Roossinck, 2007; Moran, Degnan, Santos, Dunbar, & Ochman, 2005; Rodriguez, White, Arnold, & Redman, 2009; Roossinck, 2010; Wegley et al., 2004). Recent studies suggest that, in some cases, parasitic symbioses are the results of mutualism breakdown (Sachs & Simms, 2006). While many forms of symbiosis have been studied and described with mathematical models, few have considered multiplex interactions in cooperative evolution (Sachs, Mueller, Wilcox, & Bull, 2004). In this chapter, we will review some aspects of the mutualistic symbiotic mycoviruses in the fungi that are known as endophytes.

Although strictly speaking fungal endophytes include all fungi that reside within plants as symbionts, this term is generally used to describe mutualistic plant–microbe interactions. However, when plants and associated fungi interact by simultaneous cooperation and competition, based on environmental conditions, the outcome may shift between mutualism and antagonism in a continuum (Neuhauser & Fargione, 2004; Nuismer, Gomulkiewicz, & Morgan, 2003; Saikkonen, Faeth, Helander, & Sullivan, 1998). A pathogen can survive asymptomatically inside the host even for years before causing any significant disease (Clay & Schardl, 2002; Wilson, 1995). Mutualistic and commensalistic endophytes have been extensively studied since the 1970s, and this review describes the viruses of these endophytes.

As with most viruses, our understanding of mycoviruses began from pathogenic viruses. The first described mycovirus caused dieback disease in cultivated mushrooms (Hollings, 1962). However, most known mycoviruses are believed to be benign to their host, at least under experimental conditions, although many studies have only tested a few traits, such as mycelium growth and sporulation rates, or fungal heat tolerance *in vitro* (Aoki et al., 2009; Herrero, Pérez-Sánchez, Oleaga, & Zabalgogeazcoa, 2011; Zabalgogeazcoa, Benito, Ciudad, Criado, & Eslava, 1998). The potential mutualistic roles of mycoviruses in nature are poorly studied.

2. CLASSIFICATION OF FUNGAL ENDOPHYTES

Fungal endophytes have been classified into four groups based on their life histories (Rodriguez et al., 2009). The Class 1 endophytes are the clavicipitaceous endophytes associated with warm- and cool-season grasses. They are host specific, mainly in the grass family Poaceae and rarely in

Cyperaceae, and are often vertically transmitted through seeds. Metabolites produced by Class 1 endophytes can enhance plant resistance to biotic and abiotic stresses, such as the production of ergot as a defensive agent to deter herbivores, or confer drought tolerance in cool-season grasses (Kuldau & Bacon, 2008). Because of their interesting protective roles on host grasses, Class 1 endophytes have been well studied and frequently used to represent the entire endophyte group. In fact, Class 1 endophytes are from only a small group in the family Clavicipitaceae, with phylogenetic relationships to animal-associated species in the same family. It is possible that animal pathogenic fungal ancestors are the origin of Class 1 endophytes prior to a series of host-jumping events to grasses (Spatafora, Sung, Sung, Hywel-Jones, & White, 2007).

Compared with Class 1 endophytes, Class 2 endophytes are highly diverse and comprised of species from Pezizomycotina (Ascomycota) to Agaricomycotina and Pucciniomycotina (Basidiomycota) (Rodriguez et al., 2009). They colonize roots, stems, leaves, or the whole plant. They can be vertically or horizontally transmitted. Some Class 2 endophytes, typically found as predominant species in plants under high stresses, confer habitat-adapted benefits to the host plant, including tolerance to biotic stress, heat, or salinity (Redman, Sheehan, Stout, Rodriguez, & Henson, 2002; Rodriguez et al., 2008; Rodriguez, Redman, & Henson, 2004).

Class 3 endophytes are distinct from the other endophytes in that they restrict their colonization to above-ground plant tissues, forming localized infection mainly on leaves and twigs of host plants, but occasionally also the reproductive organs (Arnold, 2007). This group is also extremely diverse, is horizontally transmitted, and contains many Basidiomycetous species (Arnold, Henk, Eells, Lutzoni, & Vilgalys, 2007). It is not always easy to distinguish foliage endophytes from foliage epiphytes, which grow upon the leaf surface. Some fungal species can be found as Class 2 or Class 3 endophytes under different ecological conditions.

Class 4 endophytes are found in the rhizosphere, another common habitat of microorganisms with high biodiversity. The dark septate endophytes represent most of the currently known root fungal endophytes, characterized by their melanized septa and restricted association with plant roots (Sieber & Grünig, 2006). These endophytes are prone to be misidentified as mycorrhizal fungi or other species with dark septa that coexist in the rhizosphere (Jumpponen, 2001; Rodriguez et al., 2009). The Class 4 endophytes have a broad host range, with over 600 plant species known as hosts (Jumpponen & Trappe, 1998), although the number of formally described anamorphic taxa is relatively small and not well understood, partially because they lack the fungal reproduction structures used for classification (Sieber & Grünig, 2006). Some consider the arbuscular mycorrhizal fungi as endophytes as well; a recent paper describes the first putative virus from these mycorrhizal fungi found by dsRNA analysis. The dsRNA encodes a virus-like RdRp but is not related to known families of mycoviruses. Fungal strains without the virus seem to have greater plant growth enhancement (Ikeda, Shimura, Kitahara, Masuta, & Ezawa, 2012).

3. VIRUSES OF FUNGAL ENDOPHYTES

As with most fungi, viruses are common in endophytic fungi (Table 2.1). Because endophytes are not as well studied as pathogenic fungi, their viruses are also less well known. In addition, endophytes can be very difficult to culture, and in some cases, their viruses are rapidly lost in culture. Hence, our current understanding of viruses of endophytic fungi is very limited.

3.1. Viruses of Class 1 endophytes

Epichloë festucae is a typical Class 1 endophyte with balanced vertical and horizontal transmission and mixed asymptomatic and pathogenic life cycles (Craven, Hsiau, Leuchtmann, Hollin, & Schardl, 2001). Two dsRNA elements of 5.2 and 3.2 kb were detected in one out of four isolates of *E. festucae* from asymptomatic red fescue plants (*Festuca rubra*). The larger dsRNA is associated with isometric virus particles of an estimated diameter of 50 nm, while no particles have been detected for the smaller element (Zabalgogeazcoa et al., 1998). The larger element was later named as Epichloë festucae virus 1 (EfV1) (Romo, Leuchtmann, García, & Zabalgogeazcoa, 2007). EfV1 is in the genus *Victorivirus* with a genome of 5109 bp, although the associated particles are larger than those of other known viruses in this genus. Two putative ORFs are found in the genome that overlap by four nucleotides. The virus seems to be common since it was detected in two different host populations at an incidence of 36.4%, and also found in another five populations. Vertical transmission of EfV1 was tested and no virus was detected in the progeny of sexual spores, suggesting that transmission may rely solely on asexual reproduction of the fungus (Romo et al., 2007). A virus related to EfV1 was isolated from an unrelated entomopathogenic fungus, *Tolypocladium cylindrosporum* (Hypocreales, Sordariomycetes) (Herrero & Zabalgogeazcoa, 2011), with amino acid sequence identity of 35% (expect $= 8e^{-125}$), suggesting an historical host-jumping event.

Table 2.1 The mycoviruses and putative viruses of fungal endophytes[a]

Endophyte host	Endophyte family (order)	Virus or putative virus[b]	Virus family or proposed family	Virus sigla
Atkinsonella hypoxylon	Clavicipitaceae	Atkinsonella hypoxylon virus	*Partitiviridae*	AhV
Alternaria alternata	Pleosporaceae	Alternaria alternata virus 1	Unassigned	AaV-1
A. alternata	Pleosporaceae	A. alternata endornavirus 1[c]	*Endornaviridae*	
A. alternata	Pleosporaceae	A. alternata totivirus 1[c]	*Totiviridae*	
A. alternata	Pleosporaceae	A. alternata totivirus 2[c]	*Totiviridae*	
A. alternata	Pleosporaceae	A. alternata chrysovirus 1[c]	*Chrysoviridae*	
A. alternata	Pleosporaceae	A. alternata chrysovirus 2[c]	*Chrysoviridae*	
A. alternata	Pleosporaceae	A. alternata partitivirus 3[c]	*Partitiviridae*	
A. alternata	Pleosporaceae	A. alternata hypovirus 1[c]	*Hypoviridae*	
Alternaria triticina	Pleosporaceae	A. triticinia chrysovirus 3[c]	*Chrysoviridae*	
A. triticina	Pleosporaceae	A. triticinia mitovirus 1[c]	*Narnaviridae*	
Beauveria bassiana	Clavicipitaceae	Beauveria bassiana totivirus (6.0 kbp dsRNA)[c]	*Totiviridae*	
B. bassiana	Clavicipitaceae	B. bassiana partitivirus (1.9–2.0 kbp dsRNA)[c]	*Partitiviridae*	
Cercospora	Mycosphaerellaceae	Cercospora endornavirus 1[c]	*Endornaviridae*	
Cercospora	Mycosphaerellaceae	Cercospora partitivirus 3[c]	*Partitiviridae*	
Cercospora	Mycosphaerellaceae	Cercospora CThTV-like virus 2[c]	Unassigned	

Continued

Table 2.1 The mycoviruses and putative viruses of fungal endophytes—cont'd

Endophyte host	Endophyte family (order)	Virus or putative virus	Virus family or proposed family	Virus sigla
Cladosporium	Davidiellaceae	Cladosporium partitivirus 1[c]	*Partitiviridae*	
Cladosporium	Davidiellaceae	Cladosporium partitivirus 3[c]	*Partitiviridae*	
Curvularia	Pleosporaceae	Curvularia endornavirus 1[c]	*Endornaviridae*	
Curvularia	Pleosporaceae	Curvularia chrysovirus 2[c]	*Chrysoviridae*	
Curvularia	Pleosporaceae	Curvularia chrysovirus 3[c]	*Chrysoviridae*	
Curvularia	Pleosporaceae	Curvularia CThTV-like virus 1[c]	Unassigned	
Curvularia inaequalis	Pleosporaceae	C. inaequalis chrysovirus (3.4–4.5 kbp dsRNA)[c]	*Chrysoviridae*	
Curvularia protuberata	Pleosporaceae	Curvularia thermal tolerance virus	Unassigned	CThTV
Drechslera biseptata	Pleosporaceae	1.0–1.5 kbp dsRNA	Unassigned	
Epichloë festucae	Clavicipitaceae	Epichloë festucae virus 1	*Totiviridae*	EfV 1
E. festucae	Clavicipitaceae	3.2 kbp dsRNA	Unassigned	
Fusarium culmorum	Nectriaceae	F. culmorum chrysovirus[c] (3–4.4 kbp dsRNA)	*Chrysoviridae*	
Gaeumannomyces cylindrosporus	Magnaporthaceae	30 nm VLP	Unassigned	
G. graminis	Magnaporthaceae	2.6 kbp dsRNA	Unassigned	

Mastigobasidium intermedium	Leucosporidiaceae	Mastigobasidium intermedium totivirus[c] (1.4–5.7 kbp dsRNA)	*Totiviridae*
Penicillium canescens	Trichocomaceae	1.6–1.8 kbp dsRNA	Unassigned
Penicillium sp.	Trichocomaceae	Penicillium sp. chrysovirus[c] (3.5–3.8–4.5 kbp dsRNA)	*Chrysoviridae*
Phoma	(Pleosporales)	Phoma endornavirus 1[c]	*Endornaviridae*
Phoma	(Pleosporales)	Phoma totivirus 4[c]	*Totiviridae*
Phoma	(Pleosporales)	Phoma totivirus 5[c]	*Totiviridae*
Phoma	(Pleosporales)	Phoma chrysovirus 2[c]	*Chrysoviridae*
Phoma	(Pleosporales)	Phoma chrysovirus 4[c]	*Chrysoviridae*
Phoma	(Pleosporales)	Phoma partitivirus 1[c]	*Partitiviridae*
Preussia	Sporormiaceae	Phoma partitivirus 3[c]	*Partitiviridae*
Rhizoctonia bataticola	Ceratobasidiaceae	R. bataticola totivirus[c] (6.9 kbp dsRNA)	*Totiviridae*
Stemphylium solani	Pleosporaceae	Stemphylum solani totivirus 3[c]	*Totiviridae*
S. solani	Pleosporaceae	S. solani chrysovirus 2[c]	*Chrysoviridae*
S. solani	Pleosporaceae	S. solani chrysovirus 3[c]	*Chrysoviridae*
S. solani	Pleosporaceae	S. solani partitivirus 2[c]	*Partitiviridae*
S. solani	Pleosporaceae	S. solani mitovirus 2[c]	*Narnaviridae*

Continued

Table 2.1 The mycoviruses and putative viruses of fungal endophytes—cont'd

Endophyte host	Endophyte family (order)	Virus or putative virus	Virus family or proposed family	Virus sigla
Tolypocladium cylindrosporum	Ophiocordycipitaceae	*T. cylindrosporum* totivirus 1[c]	*Totiviridae*	TcV1
T. cylindrosporum	Ophiocordycipitaceae	*T. cylindrosporum* chrysovirus 2[c]	*Chrysoviridae*	TcV2
T. cylindrosporum	Ophiocordycipitaceae	*T. cylindrosporum* virus 3[c]	Unassigned	TcV3
Torrubiella confragosa	Clavicipitaceae	Torrubiella confragosa partitivirus[c] (2–2.3 kbp dsRNA)	*Partitiviridae*	
T. confragosa	Clavicipitaceae	T. confragosa totivirus[c] (6.0 kbp dsRNA)	*Totiviridae*	
Valsa sp.	Valsaceae	Valsa totivirus[c] (4.5 kbp dsRNA)	*Totiviridae*	

[a]See text for pertinent references to the viruses/putative viruses listed in this table.
[b]Putative viruses with the same number (e.g., Curvularia endornavirus 1 and A. alternata endornavirus 1, and so forth) share at least 90% nt identity (over $\geq$150 bases) even though they were derived from different fungal host species.
[c]These are putative viruses, based on sequence similarity to known viruses.

EfV1 was not the first virus collected from Clavicipitaceous fungi. Atkinsonella hypoxylon virus (AhV), the type species of the *Partitivirus* genus, was reported from the pathogenic fungus *Atkinsonella hypoxylon*. The virus was detected from the fungal isolates of choke disease on *Danthonia spicata* (poverty oatgrass) (Oh & Hillman, 1995). Among six examined isolates, five had dsRNA elements, and at least two sets of dsRNAs were similar. One virus strain AhV-2H has three dsRNA genomic fragments of 2180, 2135, and 1790 bp. The dsRNA1 encodes a 78-kDa putative polymerase, and dsRNA2 encodes a 74-kDa putative coat protein. The dsRNA3 does not encode any obvious ORFs and may be a satellite segment (Oh & Hillman, 1995). The viruses of *A. hypoxylon* could be more diverse than those of *E. festucae*, but the sample size of *A. hypoxylon* is too small to draw any conclusion. It would be interesting to ask if the differences between the transmission strategies used by the antagonistic *A. hypoxylon* and the mutualistic *E. festucae* have any effects on the virus population structures.

Although the fungal host *A. hypoxylon* is referred to as a pathogen, the initial experimental results suggested it could be mainly vertically transmitted through seeds (Clay, 1994), violating the prediction that mutualism and commensalism are favored by vertically transmitted symbionts (Ewald, 1987; Fine, 1975). In a later study, however, higher levels of horizontal transmission were shown in natural populations by the use of molecular markers (Kover, Dolan, & Clay, 1997).

3.2. Viruses of Class 2 and Class 3 endophytes

Some of the Class 2 endophytes confer habitat-adapted benefits. For example, *Curvularia protuberata* was isolated from panic grass growing in geothermal soils with temperatures up to 65 °C (Redman et al., 2002), while *Fusarium culmorum* strains collected from coastal dunegrass confer salt tolerance to plants (Rodriguez et al., 2008). Márquez et al. (2007) reported a virus in *C. protuberata* that is necessary for conferring the heat tolerance in this virus–fungus–plant three-way mutualistic symbiosis. The Curvularia thermal tolerance virus (CThTV) has two dsRNA genome segments of about 2.2 and 1.8 kbp. The 2.2-kbp segment encodes two putative ORFs that are similar to different elements of known RdRps. The two ORFs overlap such that a single protein could be expressed through frame shifting. The 1.8-kb fragment has two or more ORFs that encode a coat protein and other protein(s) of unknown functions. Under experimental conditions, CThTV is vertically transmitted through fungal conidial spores at nearly 100%, and can be horizontally transmitted by anastomosis (Márquez et al., 2007).

Although more than one putative mycovirus has been found in the salt habitat-adapted *F. culmorum* their functions in the system have not been determined. In other isolates of *F. culmorum*, dsRNA elements of 3 and 4.4 kbp were detected (Herrero, Márquez, & Zabalgogeazcoa, 2009).

CThTV is the only well-characterized virus from the genus *Curvularia*, although other dsRNA elements have been detected in surveys, suggesting viruses are common in this endophyte. An isolate of *Curvularia inaequalis* collected from *Ammophila arenaria* (Marram grass) contained two dsRNA elements of 3.4 and 4.5 kbp (Herrero et al., 2009). In another study, Feldman, Morsy, and Roossinck (2012) surveyed endophytic fungi collected from the Tallgrass Prairie Preserve in northwestern Oklahoma, using cDNA cloning and 454 sequencing methods. Asymptomatic western ragweed (*Ambrosia psilostachya*) and a common parasitic partner, dodder (*Cuscuta cuspidata*), were collected as pairs and assayed first for fungi, and subsequently the fungi were analyzed for dsRNAs. Twenty-five sets of viral sequences were assessed from 20 fungal isolates out of a total number of 225 endophytes isolated from the ragweed and dodder pairs. A large number (165) of the endophytes isolated in this study were from the order Pleosporales, including species in the genera *Alternaria* (63), *Curvularia* (13), *Phoma* (26), and *Stemphylium* (25). Similar virus sequences were detected from endophytic fungi from different genera in the same order, and sometime from different orders. Four sets of dsRNA sequences from *Curvularia* spp. were tentatively identified as belonging to endornaviruses (one set), chrysoviruses (two sets), and CThTV-like viruses (one set). Of these the endornavirus and the second set of chrysovirus-like dsRNAs were also identified from both *Phoma* spp. and *Alternaria alternata* by sequence similarities (at least 90% nt identity over $\geq$150 bases). Species members from the genus *Phoma* are ubiquitous and often found as Class 2 endophytes that occupy similar ecological niches as *Curvularia* spp. Additional putative viruses were also detected from *Phoma* in the same study, including a putative partitivirus, two putative chrysoviruses, and two putative totiviruses. One of the putative chrysovirus dsRNAs was also found in *Curvularia*, *A. alternata*, and *Stemphylium solani*, while the putative partitivirus was also found in *Cladosporium* (Feldman et al., 2012).

The species in the genus *Cladosporium* may have different lifestyles, including endophytes that colonize different plant tissues from leaves to roots (Santamaría & Diez, 2005). Two putative partitiviruses were detected in genus *Cladosporium*. Both putative partitiviruses are distinct from the previously described Cladosporium fulvum T-1 virus, isolated from a plant pathogenic *Cladosporium* (McHale et al., 1992).

A. alternata is widely distributed on more than 300 plant hosts. It is an antagonist on many plant hosts with host/nonhost selective toxins related to the pathogenicity. It also may be a commensal or mutualistic Class 3 endophyte. The fungus mainly associates with plant leaves and twigs (Márquez, Bills, Acuña, & Zabalgogeazcoa, 2010; Santamaría & Diez, 2005), but it has also been detected as a root endophyte (Class 2) on orchids (Bayman & Otero, 2006). Seven putative mycoviruses were found in the ca. 60 *A. alternata* isolates from the Oklahoma tallgrass prairie samples (Feldman et al., 2012) including an endornavirus, two totiviruses, two chrysoviruses, a hypovirus, and a partitivirus. Three of these seven viruses (designated Alternaria alternata putative endornavirus 1, A. alternata putative chrysovirus 3, and A. alternata putative partitivirus 3) were found in other related genera. These seven viruses are distinct from either of the *A. alternata* viruses previously described on plant pathogenic isolates, which are A. alternata REAL virus, a 6.0-kb retrotransposon (Kaneko, Tanaka, & Tsuge, 2000), and A. alternata virus 1, an unclassified virus with four genomic dsRNA (Aoki et al., 2009). On *S. solani*, a leaf associated fungus, five viruses were detected and two of them were shared with other endophyte hosts.

Despite being grouped in different endophyte classes, *Curvularia* and *Alternaria* are sister anamorphic genera in the family Pleosporaceae. In a five-loci phylogenetic analysis, these two genera were placed in the same robustly supported clade (Zhang et al., 2009). There are some other common plant-associated fungi in this group, including *Drechslera* and *Stemphylium* that contain dsRNA elements (Feldman et al., 2012; Herrero et al., 2009). It is interesting that a large number of viruses or viral dsRNAs have been detected from the endophytes in Pleosporaceae (Table 2.1). The detected viruses and putative viruses are from diverse species, and it is common to find similar viruses in related or distinct host genera. It suggests that at least some viruses are generalists that are horizontally transmissible with a broad host range, even though transmission has not been widely documented. This could be a reflection of the lifestyles of viruses in Class 2 and Class 3 endophytes. Using metagenomics approaches allows us to see some scenarios that have not been detected before.

3.3. Viruses of Class 4 endophytes

The dark septate endophytes are poorly characterized (Sieber & Grünig, 2006), and we have very limited knowledge about their associated mycoviruses. Our current understanding of endophyte viruses is heavily dependent on the activities of very few virologists working in this area. *Phialophora graminicola* (teleomorph *Gaeumannomyces cylindrosporus*), which was previously known as *Phialophora radicicola* var. *graminicola*, is a dark septate endophyte that functions as an antagonist of the take-all disease pathogen *Gaeumannomyces graminis* (anamorph *Phialophora radicicola* var. *radicicola*), and hence a mutualist of plants. Thirty nanometer virus-like particles (VLPs) were reported in one out of 30 *P. graminicola* isolates. The VLPs had sedimentation coefficients of 115S and 79S, and four components. These VLPs in *P. graminicola* are serologically different from the commonly observed VLPs in *P. radicicola* and *Gaeumannomyces graminis* var. *graminis* (Rawlinson & Muthyalu, 1975). A few viruses were reported from other *Phialophora* spp., two of which have similar dsRNA segments to that from *P. graminicola* but were not detected by serological similarity (Buck, 1986). *P. radicicola* may also be considered as an endophyte (Sieber & Grünig, 2006), but some authors argue that it should be excluded because its dark septate endophyte appearance has been only observed in cropping systems (Jumpponen & Trappe, 1998). Additional viruses from *G. graminis* as a pathogen were reported by Buck and colleagues, including a totivirus, two partitiviruses, and an unassigned dsRNA virus (Buck, 1986). Recently, a 2.6-kbp dsRNA was detected in *G. graminis* endophyte isolates that were associated with the perennial grass *Holcus lanatus* (Herrero et al., 2009). The incidence of viruses in the *P. radicicola*/*G. graminis* group was high in several studies (Herrero et al., 2009; Rawlinson & Muthyalu, 1974a, 1974b).

4. ROLES OF VIRUSES IN ENDOPHYTIC FUNGI

The incidence of dsRNAs in surveys of fungi ranges from a few to 100% and varies even among different fungal populations of the same species. From the data reviewed in this chapter, viruses in Class 1 endophytes are relatively common (although with low species diversities), but this may simply be a reflection of the much more extensive studies of Class 1 endophytes. It is unknown if that is also true for Class 4 endophytes. The viruses and dsRNA elements found in other endophytes are more diverse but incidence is generally lower, based on two surveys. Many factors may affect the fungal virus incidence in a particular host, including vertical transmission rate (through asexual and sexual spores), horizontal transmission rate, fungus/virus communication, and environmental stresses. Also, some viruses seem to be readily lost in culture, suggesting that most survey results are artificially low. Although it is rare to detect any beneficial or detrimental traits conferred by mycoviruses, there are likely some subtle effects that are difficult to demonstrate statistically with small sample sizes (van Diepeningen, Debets, & Hoekstra, 2006). So far, only a very few traits have been tested in most studies, such as growth rate, reproduction rate, pathogenicity, or heat tolerance. Given the persistent nature of many of these viruses, it seems likely that they play different roles to benefit their partners outside of the laboratory (Fig. 2.1).

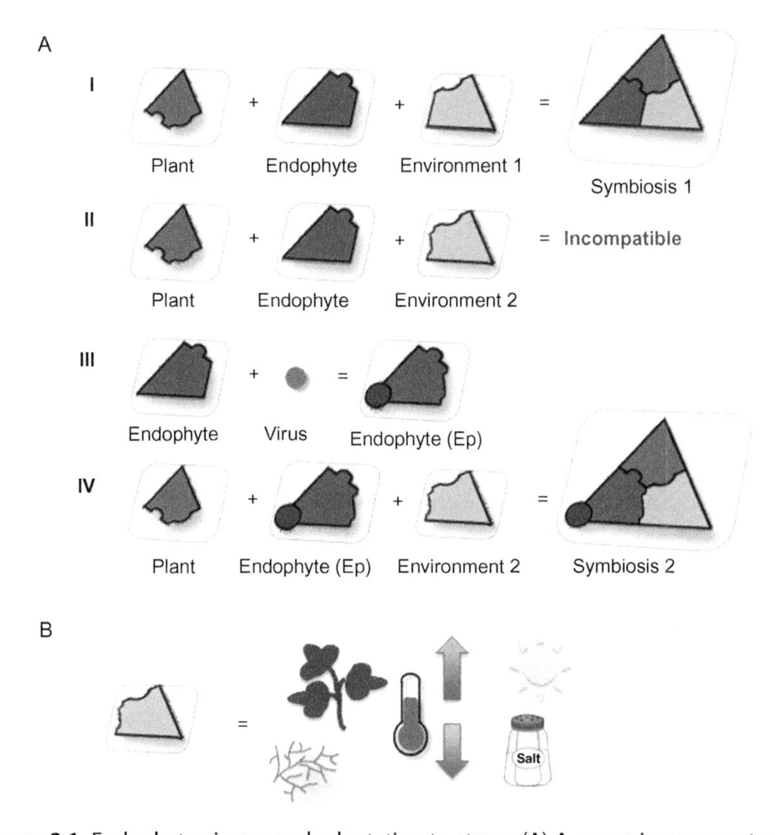

Figure 2.1 Endophyte viruses and adaptation to stress. (A) A mycovirus may act as an epigenetic element in beneficial interactions. (I) A successful "symbiosis triangle" with the three elements of plant, endophyte, and environment. (II) A change of environment may cause the beneficial interaction to collapse. (III) A symbiotic virus modifies the expression profiles of the endophyte. (IV) A new "symbiosis triangle" with changed environment, plant, endophyte, and its virus with epigenetic effects. (B) Changes in the environment may include biotic stress, such as invasive plant competitors or pathogen infections; or abiotic stress, such as heat, drought, high UV radiation, or salt or heavy metal soil contamination. (See Page 2 in Color Section at the back of the book.)

Beneficial viruses are probably more common than previously realized. Mutualistic viruses have been described from a number of different prokaryotic and eukaryotic hosts (Roossinck, 2011b). So far, CThTV is the only example established to be mutualistic in a symbiotic system with plants and fungal endophytes. This is probably because the endophyte viruses have been overlooked in previous studies. Only a few fungal endophyte viruses have been recognized and investigated for their potential functions in symbiotic systems. In nature, it's very likely that viruses are very common in endophytes as well as in other fungi. Even with an average virus incidence of only 10% in endophyte populations, and given the endophyte viruses may not be host specific, the virus species richness was still higher than that of endophytes, suggesting a hyperdiversity of mycoviruses in a natural ecosystem (Feldman et al., 2012). Considering the total estimated number of fungal species ranges from 1.5 million to as many as 5.1 million (Hawksworth, 1991, Blackwell, 2011), the number of virus species in endophytes is certainly extremely large. It is very likely that more mutualistic endophyte viruses will be described in the future. Potentially, mycoviruses are an indispensable genetic factor for endophytes and plant hosts, particularly in adaptation to harsh environments, where viruses can provide additional genetic information in an epigenetic manner. An endophyte population could lose its virus via imperfect vertical transmission, and the virus could be reintroduced by horizontal transmission. This trait would be favorable for the endophyte populations experiencing highly variable environmental conditions along temporal and/or spatial dimensions.

By using CThTV as an example to study how endophyte viruses may benefit the partners in a three-way symbiosis, a virus free strain of *C. protuberata* (cured by freezing/thawing) and a strain reinfected through anastomosis were compared with and without heat stress by using expressed sequence tags that were developed from subtracted cDNA libraries. The analyses suggested CThTV-infected *C. protuberata* stains differentially expressed many genes under heat stress, including those encoding osmoprotectants, melanin synthesis enzymes, and heat shock proteins (Morsy, Oswald, He, Tang, & Roossinck, 2010). In another study, biochemical contents were compared by using of FTIR and Raman spectromicroscopy with the wild *C. protuberata* strain (with CThTV) with a virus free type culture strain (CpATCC) that has no virus (Isenor, Kaminskyj, Rodriguez, Redman, & Gough, 2010). In free-living fungal isolates, the distribution of mannitol, another potential osmoprotectant, was different between the two stains. The mechanism that CThTV virus has evolved to protect its fungal host from the fatal damages of heat stress still requires further investigation to show how the viral element(s) may directly or indirectly manipulate the expression profiles of the fungal host that confers thermal tolerance with plants.

In a natural system, the subtle balance among plants, endophytes, and endophyte viruses are complicated stories. One of the problems is that we cannot assume that any beneficial traits observed on Petri dishes constitute a mutualistic relationship in nature, and vice versa. For example, if a virus promotes the growth rate of it fungal host, this aggressive growing may require more energy from the plant and result in disease. Based on host population structures, sometimes this could drive the disease-causing fungus to extinction, and maybe also the host plant (deCastro & Bolker, 2005; Fisher et al., 2012).

Mycoviruses are thought to be transmitted vertically. An infection exclusively maintained by vertical transmission cannot be persistent if the parasite reduces the fitness of hosts (Fine, 1975). A pathogenic parasite can only maintain a high prevalence in host populations when both vertical and horizontal transmission rates are sufficiently high (Lipsitch, Nowak, Ebert, & May, 1995), or the virulent strain is being protected by other less virulent vertically transmitted strains (Lipsitch, Siller, & Nowak, 1996). The studies on the pathogenic fungus *A. hypoxylon* provided an empirical example on this theory (Clay, 1994; Kover et al., 1997). Unlike monogenetic populations such as cultivated mushrooms or the fungal pathogens of crops, most endophytes survive in a highly diverse ecosystem, which leaves the viruses less chance for horizontal transmission by anastomosis. Cryptic viruses in plants also lack known methods of horizontal transmission. They are benign symbionts and maintained at low titer with vertical transmission (Roossinck, 2010). This type of commensalism is also possible for endophytes and their viruses, although the cost of harboring virus for endophytes may be greater than for plants since the ratios of host/parasite biomasses are quite different.

Vertical transmission rates of mycoviruses do not always reach 100%, especially during host sexual reproduction (Chu et al., 2004; Ihrmark, Stenstrom, & Stenlid, 2004; Pearson, Beever, Boine, & Arthur, 2009). If there is no horizontal transmission, an imperfect vertical transmission will eventually drive the viruses to extinction. But if some beneficial traits from the mycovirus are favored by natural selection, the virus incidence can be stable at equilibrium. Alternatively, it's also possible but less likely that the current virus populations are far from equilibrium because their endophyte hosts were affected by historical climate changes (Rodriguez & Redman, 2005; Rodriguez et al., 2004). Another explanation is that virus-infected endophytes could be maintained as metapopulations in different patches with variable selection pressures. A stage-based model describes a possible metapopulation structure when assuming a positive selection in at least one of the patches (Saikkonen, Ion, & Gyllenberg, 2002). According to this model, viruses can go extinct in some populations and be reintroduced to other empty patches, resulting in a global dynamic balance. This hypothesis may be better applied to some natural areas with patchy environmental conditions. Although it makes the most sense to assume that endophyte viruses are mutualistic at least under certain conditions, it not necessary to force the viruses to confer strong benefits to endophyte hosts to compensate the average costs to host populations. In another words, if some of patches work as reservoirs of the viruses, that would be sufficient to maintain the viruses in the endophyte host populations. The generalist viruses that infected a broad range of hosts detected in Class 2 and Class 3 endophytes (Feldman et al., 2012) may imply rapid fluctuations in host patches (Elena, Agudelo-Romero, & Lalic, 2009).

Viruses may play an epigenetic role in their hosts (Roossinck, 2011a, 2011b). In rapidly changing environments, adaption through epigenetic effects can be an advantage compared with the gradual effects of Darwinian evolution upon accumulated genomic variations. There is already evidence showing the RNA nature of most mycoviruses facilitates their involvement in complicated host responses. The viral dsRNA in *Penicillium funiculosum* induces interferon for antiviral activities in animal cells (Lampson, Tytell, Field, Nemes, & Hilleman, 1967). Mycoviruses were shown to be correlated with the suppression of RNA silencing in their hosts (Hammond, Andrewski, Roossinck & Keller, 2008; Segers et al., 2007). The interactions between viral-encoded proteins and host metabolons are complicated and not yet well understood. Many members of the *Endornavirus* genus contain glycosyltransferase (GT) domains in their polyproteins, whose functions are unknown. The GT genes belong to three different superfamilies that were probably acquired independently and likely to be functional (Roossinck, Sabanadzovic, Okada, & Valverde, 2011). It was recently suggested that glycosylation could be another type of epigenetic regulation (Hanover, Krause, & Love, 2012; Lauc & Zoldoš, 2009). It is an open question if mycoviruses encode any proteinases that play a role in epigenetic modification in host cells, but endornaviruses all contain a single large ORF that would require a proteinase for generation of functional proteins.

5. PRACTICAL ROLES FOR ENDOPHYTE VIRUSES

Modern agriculture has dramatically enhanced plant productivity, at a huge ecological cost of global pollution, soil degradation, greenhouse gas emissions, and loss of biodiversity. It is well known that the plants need symbiotic microbes in nature to balance nutrition, and to survive under biotic and abiotic stresses (Andrews, Cripps, & Edwards, 2012; Tikhonovich & Provorov, 2011). Our current agricultural systems are expensively maintained by supplying artificial chemical products, which can harm natural ecosystems by releasing chemical pesticides, herbicides, fertilizers, and potential genetic contaminants. In addition, monocultures created by cropping are more susceptable to the emerging aggressive pathogens such as rust race Ug99 and can lead to the spread of diseases around the world (Fisher et al., 2012). Instead of the endless effort put into the battles against new pathogens, sustainable agriculture, including the use of beneficial microbes, provides a promising alternative, particularly for smallholder farmers who cannot afford the expense of agrichemicals and improved seeds.

Plants have been thriving on the earth for more than 400 million years with microbial symbionts (Rodriguez & Redman, 2008). Over the past few decades, we have started to understand some of the roles endophytes can play with their host. Unlike Class 1 endophytes, most fungal endophytes involved in habitat adaptation are generalists (i.e., they can colonize a very wide range of plants) (Higgins, Coley, Kursar, & Arnold, 2011). With better understanding of endophyte viruses, it will be important to think about a number of questions before applying an endophyte to sustainable agricultural system. (1) Is there any virus playing a role to enable the endophyte to confer habitat-adapted benefits to host? (2) Is a virus needed to prevent an endophyte from shifting from mutualism to antagonism? (3) If the application of an endophyte will dramatically change its original population structure and thus affect virus transmission strategy, can that becomes a reason to drive some benign virus into evolving toward parasitism or mutualism? (4) Are there any toxic by-products in the plant–endophyte interactions, which could be generated by either the fungus or the fungal virus? (5) Can the endophyte and/or its virus be readily delivered to a farming system?

6. SUMMARY

We are at the very beginning in our understanding of the viruses of fungal endophytes, even though a huge number probably exist in nature. From the few examples currently known, they are probably not very different from the mycoviruses in other fungi, but the lifestyles of host endophytes may affect the incidence and diversity of their symbiotic viruses. Mutualistic symbiosis is almost certainly not rare, and is likely a rule, in the multiplex interactions between viruses, endophytes and plants. Hypothetically, mycoviruses are a potential transportable factor that can provide their partners more flexibility for rapid adaptation, a favorable trait during temporal or spatial environmental changes. The knowledge of fungal endophyte viruses will be valuable for practicing sustainable agriculture particularly against the background of global climate changes.

ACKNOWLEDGMENTS

We thank Catherine Fearnhead in the Rothamsted Library for help in accessing research reports and the book editor for helpful suggestions. This work was supported by the National Science Foundation awards IOS-0950579 and IOS-1157148 and by Pennsylvania State University.

REFERENCES

Andrews, M., Cripps, M. G., & Edwards, G. R. (2012). The potential of beneficial microorganisms in agricultural systems. *Annals of Applied Biology*, *160*, 1–5.

Aoki, N., Moriyama, H., Kodama, M., Arie, T., Teraoka, T., & Fukuhara, T. (2009). A novel mycovirus associated with four double-stranded RNAs affects host fungal growth in *Alternaria alternata*. *Virus Research*, *140*, 179–187.

Arnold, A. E. (2007). Understanding the diversity of foliar endophytic fungi: Progress, challenges, and frontiers. *Fungal Biology Reviews*, *21*, 51–66.

Arnold, A. E., Henk, D. A., Eells, R. L., Lutzoni, F., & Vilgalys, R. (2007). Diversity and phylogenetic affinities of foliar fungal endophytes in loblolly pine inferred by culturing and environmental PCR. *Mycologia*, *99*, 185–206.

Bayman, P., & Otero, J. T. (2006). Microbial endophytes of orchid roots. In B. Schulz, C. Boyle & T. N. Sieber (Eds.), *Soil biology*, Vol. 9, (pp. 153–177). Berlin/Heidelberg Springer-Verlag.

Blackwell, M. (2011). The fungi: 1, 2, 3 . . . 5.1 million species? *American Journal of Botany*, *98*, 426–438.

Buck, K. W. (1986). Viruses of the wheat take-all fungus, *Gaeumannomyces gramanis* var. *Tritici*. In K. W. Buck (Ed.), *Fungal virology* (pp. 221–236). Boca Raton: CRC Press.

Chu, Y.-M., Lim, W.-S., Yea, S.-J., Cho, J.-D., Lee, Y.-W., & Kim, K.-H. (2004). Complexity of dsRNA mycovirus isolated from *Fusarium graminearum*. *Virus Genes*, *28*, 135–143.

Clay, K. (1994). Hereditary symbiosis in the grass genus *Danthonia*. *The New Phytologist*, *126*, 223–231.
Clay, K., & Schardl, C. (2002). Evolutionary origins and ecological consequences of endophyte symbiosis with grasses. *The American Naturalist*, *160*, S99–S127.
Craven, K. D., Hsiau, P. T. W., Leuchtmann, A., Hollin, W., & Schardl, C. L. (2001). Multigene phyologeny of *Epichloë* species, fungal symbionts of grasses. *Annals of the Missouri Botanical Garden*, *88*, 14–34.
deBary, H. A. (1879). *Die Erscheinung der Symbiose*. Strasburg: Verlag von Karl J. Trübner. Privately Published.
deCastro, F., & Bolker, B. (2005). Mechanisms of disease-induced extinction. *Ecology Letters*, *8*, 117–126.
Elena, S. F., Agudelo-Romero, P., & Lalic, J. (2009). The evolution of viruses in multi-host fitness landscapes. *The Open Virology Journal*, *3*, 1–6.
Ewald, P. W. (1987). Transmission modes and evolution of the parasitism-mutualism continuum. *Annals of the New York Academy of Sciences*, *503*, 295–307.
Feldman, T. S., Morsy, M. R., & Roossinck, M. J. (2012). Are communities of microbial symbionts more diverse than communities of macrobial hosts? *Fungal Biology*, *116*, 465–477.
Fine, P. E. (1975). Vectors and vertical transmission: An epidemiologic perspective. *Annals of the New York Academy of Sciences*, *266*, 173–194.
Fisher, M. C., Henk, D. A., Briggs, C. J., Brownstein, J. S., Madoff, L. C., McCraw, S. L., et al. (2012). Emerging fungal threats to animal, plant and ecosystem health. *Nature*, *484*, 186–194.
Hammond, T. M., Andrewski, M. D., Roossinck, M. J., & Keller, N. P. (2008). Aspergillus mycoviruses are targets and suppressors of RNA silencing. *Eukaryotic Cell*, *7*, 350–357.
Hanover, J. A., Krause, M. W., & Love, D. C. (2012). Bittersweet memories: Linking metabolism to epigenetics through O-GlcNAcylation. *Nature Reviews. Molecular Cell Biology*, *13*, 312–321.
Hawksworth, D. L. (1991). The fungal dimension of biodiversity: magnitude, significance, and conservation. *Mycological Research*, *95*, 641–655.
Herrero, N., Márquez, S. S., & Zabalgogeazcoa, I. (2009). Mycoviruses are common among different species of endophytic fungi of grasses. *Archives of Virology*, *154*, 327–330.
Herrero, N., Pérez-Sánchez, R., Oleaga, A., & Zabalgogeazcoa, I. (2011). Tick pathogenicity, thermal tolerance and virus infection in *Tolypocladium cylindrosporum*. *Annals of Applied Biology*, *159*, 192–201.
Herrero, N., & Zabalgogeazcoa, I. (2011). Mycoviruses infecting the endophytic and entomopathogenic fungus *Tolypocladium cylindrosporum*. *Virus Research*, *160*, 409–413.
Higgins, K. L., Coley, P. D., Kursar, T. A., & Arnold, A. E. (2011). Culturing and direct PCR suggest prevalent host generalism among diverse fungal endophytes of tropical forest grasses. *Mycologia*, *103*, 247–260.
Hollings, M. (1962). Viruses associated with a die-back disease of cultivated mushroom. *Nature*, *196*, 962–965.
Ihrmark, K., Stenstrom, E., & Stenlid, J. (2004). Double-stranded RNA transmission through basidiospores of Heterobasidion annosum. *Mycological Research*, *108*, 149–153.
Ikeda, Y., Shimura, H., Kitahara, R., Masuta, C., & Ezawa, T. (2012). A novel virus-like double-stranded RNA in an obligate biotroph arbuscular mycorrhizal fungus: A hidden player in mycorrhizal symbiosis. *Molecular Plant-Microbe Interactions*, *25*, 1005–1012.
Isenor, M., Kaminskyj, S. G. W., Rodriguez, R. J., Redman, R. S., & Gough, K. M. (2010). Characterization of mannitol in *Curvularia protuberata* hyphae by FTIR and Raman spectromicroscopy. *The Analyst*, *135*, 3249–3254.

Jumpponen, A. (2001). Dark septate endophytes—Are they mycorrhizal? *Mycorrhiza*, *11*, 207–211.

Jumpponen, A., & Trappe, J. M. (1998). Dark septate endophytes: A review of facultative biotrophic root-colonizing fungi. *The New Phytologist*, *140*, 295–310.

Kaneko, I., Tanaka, A., & Tsuge, T. (2000). *REAL*, an LTR retrotransposon from the plant pathogenic fungus *Alternaria alternata*. *Molecular and General Genetics*, *263*, 625–634.

Kover, P. X., Dolan, T. E., & Clay, K. (1997). Potential versus actual contribution of vertical transmission to pathogen fitness. *Proceedings of the Royal Society London Series B*, *264*, 903–909.

Kuldau, G., & Bacon, C. (2008). Clavicipitaceous endophytes: Their ability to enhance resistance of grasses to multiple stresses. *Biological Control*, *46*, 57–71.

Lampson, G. P., Tytell, A. A., Field, A. K., Nemes, M. M., & Hilleman, M. R. (1967). Inducers of interferon and host resistance I. Double-stranded RNA from extracts of Penicillium funiculosum. *Proceedings of the National Academy of Sciences of the United States of America*, *58*, 782–789.

Lauc, G., & Zoldoš, V. (2009). Epigenetic regulation of glycosylation could be a mechanism used by complex organisms to compete with microbes on an evolutionary scale. *Medical Hypotheses*, *73*, 510–512.

Lipsitch, M., Nowak, M. A., Ebert, D., & May, R. M. (1995). The population dynamics of vertically and horizontally transmitted parasites. *Proceedings of the Royal Society London Series B*, *260*, 321–327.

Lipsitch, M., Siller, S., & Nowak, M. A. (1996). The evolution of virulence in pathogens with vertical and horizontal transmission. *Evolution*, *50*, 1729–1741.

Márquez, S. S., Bills, G. F., Acuña, L. D., & Zabalgogeazcoa, I. (2010). Endophytic mycobiota of leaves and roots of the grass Holcus lanatus. *Fungal Diversity*, *41*, 115–123.

Márquez, L. M., Redman, R. S., Rodriguez, R. J., & Roossinck, M. J. (2007). A virus in a fungus in a plant—Three way symbiosis required for thermal tolerance. *Science*, *315*, 513–515.

McHale, M. T., Roberts, I. N., Noble, S. M., Beaumont, C., Whitehead, M. P., Seth, D., et al. (1992). CfT-I: An LTR-retrotransposon in *Cladosporium fulvum*, a fungal pathogen of tomato. *Molecular and General Genetics*, *233*, 337–347.

Moran, N. A., Degnan, P. H., Santos, S. R., Dunbar, H. E., & Ochman, H. (2005). Inaugural article: The players in a mutualistic symbiosis: Insects, bacteria, viruses, and virulence genes. *Proceedings of the National Academy of Sciences of the United States of America*, *102*, 16919–16926.

Morsy, M. R., Oswald, J., He, J., Tang, Y., & Roossinck, M. J. (2010). Teasing apart a three-way symbiosis: Transcriptome analyses of *Curvularia protuberata* in response to viral infection and heat stress. *Biochemical and Biophysical Research Communications*, *401*, 225–230.

Neuhauser, C., & Fargione, J. E. (2004). A mutualism—Parasitism continuum model and its application to plant-mycorrhizae interactions. *Ecological Modelling*, *177*, 337–352.

Nuismer, S. L., Gomulkiewicz, R., & Morgan, M. T. (2003). Coevolution in temporally variable environments. *The American Naturalist*, *162*, 195–204.

Oh, C.-S., & Hillman, B. I. (1995). Genome organization of a partitivirus from the filamentous ascomycete *Atkinsonella hypoxylon*. *The Journal of General Virology*, *76*, 1461–1470.

Pearson, M. N., Beever, R. E., Boine, B., & Arthur, K. (2009). Mycoviruses of filamentous fungi and their relevance to plant pathology. *Molecular Plant Pathology*, *10*, 115–128.

Rawlinson, C. J., & Muthyalu, G. (1974). Similar viruses in *Gaeumannomyces* spp. and *Phialophora* spp. Report of the Rothamsted Experimental Station, pp. 228–229.

Rawlinson, C. J., & Muthyalu, G. (1974). Take-all fungi and associated viruses from other countries. Report of the Rothamsted Experimental Station, p. 228.

Rawlinson, C. J., & Muthyalu, G. (1975). Relationship of viruses in *Gaeumannomyces* spp. and *Phialophora* spp. Report of the Rothamsted Experimental Station, p. 256.

Redman, R. S., Sheehan, K. B., Stout, R. G., Rodriguez, R. J., & Henson, J. M. (2002). Thermotholerance generated by plant/fungal symbiosis. *Science*, *298*, 1581.

Rodriguez, R. J., Henson, J., VanVolkenburgh, E., Hoy, M., Wright, L., Beckwith, F., et al. (2008). Stress tolerance in plants via habitat-adapted symbiosis. *The ISME Journal*, *2*, 404–416.

Rodriguez, R., & Redman, R. (2005). Balancing the generation and elimination of reactive oxygen species. *Proceedings of the National Academy of Sciences of the United States of America*, *102*, 3175–3176.

Rodriguez, R., & Redman, R. (2008). More than 400 million years of evolution and some plants still can't make it on their own: Plant stress tolerance via fungal symbiosis. *Journal of Experimental Botany*, *59*, 1109–1114.

Rodriguez, R. J., Redman, R. S., & Henson, J. M. (2004). The role of fungal symbioses in the adaptation of plants to high stress environments. In *Mitigation and adaptation strategies for global change*, Vol. 9, (pp. 261–272). Amsterdam: Kluwer Academic Publishers.

Rodriguez, R. J., White, J. F., Jr., Arnold, A. E., & Redman, R. S. (2009). Fungal entophytes: Diversity and functional roles. *The New Phytologist*, *182*, 314–330.

Romo, M., Leuchtmann, A., García, B., & Zabalgogeazcoa, I. (2007). A totivirus infection the mutualistic fungal endophyte *Epichloë festucae*. *Virus Research*, *124*, 38–43.

Roossinck, M. J. (2010). Lifestyles of plant viruses. *Philosophical Transactions of the Royal Society B*, *365*, 1899–1905.

Roossinck, M. J. (2011a). Changes in population dynamics in mutualistic *versus* pathogenic viruses. *Viruses*, *3*, 12–19.

Roossinck, M. J. (2011b). The good viruses: Viral mutualistic symbioses. *Nature Reviews. Microbiology*, *9*, 99–108.

Roossinck, M. J., Sabanadzovic, S., Okada, R., & Valverde, R. A. (2011). The remarkable evolutionary history of endornaviruses. *The Journal of General Virology*, *92*, 2674–2678.

Sachs, J. L., Mueller, U. G., Wilcox, T. P., & Bull, J. J. (2004). The evolution of cooperation. *Quarterly Reviews of Biology*, *79*, 135–160.

Sachs, J. L., & Simms, E. L. (2006). Pathways to mutualism breakdown. *Trends in Ecology & Evolution*, *21*, 585–592.

Saikkonen, K., Faeth, S. H., Helander, M., & Sullivan, T. J. (1998). Fungal endophytes: A continuum of interactions with host plants. *Annual Review of Ecology and Systematics*, *29*, 319–343.

Saikkonen, K., Ion, D., & Gyllenberg, M. (2002). The persistence of vertically transmitted fungi in grass metapopulations. *Proceedings of the Royal Society of London Series B*, *269*, 1397–1403.

Santamaría, O., & Diez, J. J. (2005). Fungi in leaves, twigs and stem bark of *Populus tremula* from northern Spain. *Forest Pathology*, *35*, 95–104.

Segers, G. C., Zhang, X., Deng, F., Sun, Q., & Nuss, D. L. (2007). Evidence that RNA silencing functions as an antiviral defence mechanism in fungi. *Proceedings of the National Academy of Science, USA*, *104*, 12902–12908.

Sieber, T. N., & Grünig, C. R. (2006). Biodiversity of fungal root-endophyte communities and populations, in particular of the dark septate endophyte *Phialocephala fortinii* s.l. In B. Schulz, C. Boyle & T. N. Sieber (Eds.), *Soil biology*, Vol. 9, (pp. 107–132).

Spatafora, J. W., Sung, G.-H., Sung, J.-M., Hywel-Jones, N. L., & White, J. F., Jr. (2007). Phylogenetic evidence for an animal pathogen origin of ergot and the grass endophytes. *Molecular Ecology*, *16*, 1701–1711.

Tikhonovich, I. A., & Provorov, N. A. (2011). Microbiology is the basis of sustainable agriculture. *Annals of Applied Biology*, *159*, 155–168.

van Diepeningen, A. D., Debets, A. J. M., & Hoekstra, R. F. (2006). Dynamics of dsRNA mycoviruses in black Aspergillus populations. *Fungal Genetics and Biology*, *43*, 446–452.

Wegley, L., Yu, Y., Breitbart, M., Casas, V., Kline, D. I., & Rohwer, F. (2004). Coral-associated Archaea. *Marine Ecology Progress Series*, *273*, 89–96.

Wilson, D. (1995). Endophyte: The evolution of a term, and clarification of its use and definition. *Oikos*, *73*, 274–276.

Zabalgogeazcoa, I., Benito, E. P., Ciudad, A. G., Criado, B. G., & Eslava, A. P. (1998). Double-stranded RNA and virus-like particles in the grass endophyte *Epichloë festucae*. *Mycological Research*, *102*, 914–918.

Zhang, Y., Schoch, C. L., Fournier, J., Crous, P. W., deGruyter, J., Woudenberg, J. H. C., et al. (2009). Multi-locus phylogeny of *Pleosporales*: A taxonomic, ecological and evolutionary re-evaluation. *Studies in Mycology*, *64*, 85–102.

CHAPTER THREE

3D Structures of Fungal Partitiviruses

Max L. Nibert[*,1,2], Jinghua Tang[†], Jiatao Xie[‡], Aaron M. Collier[§], Said A. Ghabrial[‡,2], Timothy S. Baker[†,2] and Yizhi J. Tao[§,2]

[*]Department of Microbiology and Immunobiology, Harvard Medical School, Boston, Massachusetts, USA
[†]Department of Chemistry and Biochemistry, and Division of Biological Sciences, University of California–San Diego, La Jolla, California, USA
[‡]Department of Plant Pathology, University of Kentucky, Lexington, Kentucky, USA
[§]Department of Biochemistry and Cell Biology, Rice University, Houston, Texas, USA
[1]Corresponding author: e-mail address: mnibert@hms.harvard.edu
[2]Co-senior authors

Contents

1. Introduction to Partitiviruses 60
2. Partitivirus Capsid Structures 63
3. Partitivirus RdRp and dsRNA Structures 70
4. Comparisons with Other Bisegmented dsRNA Viruses 74
5. Comparisons with Other Fungal Viruses with Encapsidated dsRNA Genomes 77
6. Proposed Revisions to Partitivirus Taxonomy 79
Acknowledgments 81
References 81

Abstract

Partitiviruses constitute one of the nine currently recognized families of viruses with encapsidated, double-stranded (ds)RNA genomes. The partitivirus genome is bisegmented, and each genome segment is packaged inside a separate viral capsid. Different partitiviruses infect plants, fungi, or protozoa. Recent studies have shed light on the three-dimensional structures of the virions of three representative fungal partitiviruses. These structures include a number of distinctive features, allowing informative comparisons with the structures of dsRNA viruses from other families. The results and comparisons suggest several new conclusions about the functions, assembly, and evolution of these viruses.

Since May 2008, we have reported the three-dimensional (3D) structures of three fungal viruses from the dsRNA virus family *Partitiviridae*: Penicillium stoloniferum virus S (PsV-S), Penicillium stoloniferum virus F (PsV-F), and

Advances in Virus Research, Volume 86
ISSN 0065-3527
http://dx.doi.org/10.1016/B978-0-12-394315-6.00003-9

Fusarium poae virus 1 (FpV1) (Ochoa et al., 2008; Pan et al., 2009; Tang, Ochoa, et al., 2010; Tang, Pan, et al., 2010). This work has represented the collaborative efforts of four laboratories, whose principal investigators are noted as co-senior authors of this review. The work was first begun in late 2006 upon recognition of the fact that no 3D structures of partitiviruses had been reported at that time, and indeed ours remain the only such structures reported through the present. As discussed below, the three fungal partitiviruses we have analyzed most likely represent two different taxonomic genera, leaving partitiviruses from several other genera yet to be analyzed for structural comparisons across the whole family. Among those partitiviruses from other genera are ones that infect distinct hosts, namely, plants and the apicomplexan protozoan *Cryptosporidium*. Thus, a good deal of structural diversity within this family may remain to be discovered. In this review, after briefly introducing the partitiviruses, we summarize the major structural features of the three analyzed strains, compare these 3D structures to those of other encapsidated dsRNA viruses, and discuss the implications of these structures for viral functions, assembly, and evolution.

1. INTRODUCTION TO PARTITIVIRUSES

Fungal, plant, and protozoan partitiviruses have bisegmented genomes, comprising two distinct, linear dsRNA molecules, each 1.4–2.3 kbp in length for a total genome length of 3.1–4.4 kbp (reviewed by Ghabrial, Ochoa, Baker, & Nibert, 2008, Ghabrial et al., 2011). The two genome segments encode two proteins: one, the viral coat protein (CP) and, the other, the viral RNA-dependent RNA polymerase (RdRp). Notably, each of these segments is packaged inside a separate, though presumably identical, viral capsid, meaning that not only is the genome bisegmented but also the infectious unit is at minimum biparticulate. Partitiviruses are thought to undergo efficient, natural transmission between host cells only through intracellular means, such as during cell division (mitosis, meiosis) or cell–cell fusion (hyphal anastomosis), which allow transfer of multiple virus particles to each new cell. Natural transmission by cell-penetrating insect or other vectors that feed on virus-infected host cells, as known for many plant viruses, is also conceivable but has yet to be demonstrated for partitiviruses. Encapsidated satellite dsRNAs, which are dependent on helper virus for replication, are associated with some partitiviruses, including PsV-F and FpV1 (Compel, Papp, Bibo, Fekete, & Hornok, 1999; Kim, Choi, & Lee, 2005).

Taxonomically, the family *Partitiviridae* currently comprises four genera: *Partitivirus*, *Alphacryptovirus*, *Betacryptovirus*, and *Cryspovirus* (Ghabrial et al., 2008, 2011; Nibert, Woods, Upton, & Ghabrial, 2009; Fig. 3.1). Recognized members of the genus *Partitivirus* infect fungi, those of the genera *Alphacryptovirus* and *Betacryptovirus* infect plants, and those of the genus *Cryspovirus* infect protozoa (*Cryptosporidium* species). As discussed below, however, this current taxonomic and classification scheme has become subject to several criticisms as the number of partitivirus isolates and genome sequences have grown in recent years. Thus, some reworking of the genera and their characteristic features within this family appears warranted.

Penicillium stoloniferum viruses PsV-S and PsV-F are both members of the genus *Partitivirus* and can coinfect the saprophytic ascomycete *Penicillium stoloniferum*. Each encodes an ~47-kDa CP and an ~62-kDa RdRp, but the amino acid (aa) sequences of these proteins are readily distinguishable, exhibiting only 19% and 27% identity, respectively (Kim et al., 2005, Kim, Kim, & Kim, 2003; Tang, Ochoa, et al., 2010). They are also distinguishable by serological reactivities (Bozarth, Wood, & Mandelbrot, 1971). The respective names PsV-S and PsV-F reflect the relative electrophoretic mobilities of their particles on agarose gels: S, slow; F, fast (Bozarth et al., 1971). The genome of PsV-S comprises a 1754-bp dsRNA1, encoding the 539-aa RdRp, and a 1582-bp dsRNA2, encoding the 434-aa CP (Kim et al., 2003), whereas the genome of PsV-F comprises a 1677-bp dsRNA1, encoding the 538-aa RdRp, and a 1500-bp dsRNA2, encoding the 420-aa CP (Kim et al., 2005). PsV-F, but not PsV-S, contains at least one satellite segment, the 677-bp dsRNA3, which is unrelated in sequence to the other two segments (Kim et al., 2005; Tang, Pan, et al., 2010). Although both viruses can coinfect *P. stoloniferum*, the CP of each associates only with itself in forming the respective capsids (Buck & Kempson-Jones, 1974) and packages only its own RNAs (Bozarth et al., 1971). Purified virions of both viruses exhibit semiconservative transcription activity (Buck, 1978; Pan et al., 2009), reflecting that the viral RdRp is packaged into virions along with the CP and genome segments, as is characteristic of other encapsidated dsRNA viruses.

Fusarium poae virus 1 is also a member of the genus *Partitivirus* and can infect the phytopathogenic ascomycete *Fusarium poae*, one cause of Fusarium head blight in cereal grains worldwide. This virus was originally named FUPO-1 (Compel et al., 1999) but was later renamed FpV1 for consistency with existing names for other partitiviruses (Ghabrial et al., 2008, 2011). The genome of FpV1 comprises a 2203-bp dsRNA1, encoding the

Encapsidated dsRNA viruses

Genome segments	Family / Subfamily / Genus	Capsids
1	***Totiviridae***	T=1(120)
	Giardiavirus	
	Leishmaniavirus	
	Totivirus	
	Trichomonasvirus	
	Victorivirus	
2	*Birnaviridae*	T=13
	Avibirnavirus	
	Aquabirnavirus	
	Blosnabirnavirus	
	Entomobirnavirus	
	Megabirnaviridae	Not reported
	Megabirnavirus	
	Partitiviridae	T=1(120) (boxed) **PsV-S**, **PsV-F**, **FpV1**
	Alphacryptovirus	
	Betacryptovirus	
	Cryspovirus	
	Partitivirus	
	Picobirnaviridae	T=1(120)
	Picobirnavirus	
3	*Cystoviridae*	T=1(120) + T=13
	Cystovirus	
4	***Chrysoviridae***	T=1(60)
	Chrysovirus	
	Quadriviridae	Not reported
	Quadrivirus	
9–12	***Reoviridae***	T=1(120) + T=13
	Sedoreovirinae	
	Cardoreovirus	
	Mimoreovirus	
	Orbivirus	
	Phytoreovirus	
	Rotavirus	
	Seadornavirus	
	Spinareovirinae	
	Aquareovirus	
	Coltivirus	
	Cypovirus	
	Dinovernavirus	
	Fijivirus	
	Idnoreovirus	
	Mycoreovirus	
	Orthoreovirus	
	Oryzavirus	

Figure 3.1 Taxonomy and properties of encapsidated dsRNA viruses. Taxa that include fungal viruses are bolded. Icosahedral symmetries of the viral capsids are indicated; for $T=1$ capsids, the number in parentheses indicates whether there are 60 or 120 subunits in each capsid. Cystoviruses and reoviruses have two concentric icosahedral capsids with different symmetries as indicated; the $T=1$ capsid of each of these viruses is the one that encloses the genome (inner capsid). Viruses in certain reovirus genera may lack parts of the $T=13$ outer capsid. The $T=1(120)$ capsid structure of partitiviruses is boxed to emphasize the focus of this review, and the particular partitiviruses for which structures are described (PsV-S, etc.) are indicated beside the box.

673-aa (78-kDa) RdRp, and a 2185-bp dsRNA2, encoding the 637-aa (70-kDa) CP (Compel et al., 1998). Thus, the genome of FpV1 is ~30% longer than that of PsV-S and PsV-F, and the CP of FpV1 is ~50% larger (by molecular mass) than that of PsV-S and PsV-F. In these regards, FpV1 is similar to the prototype species of genus *Partitivirus*, *Atkinsonella hypoxylon virus* (Oh & Hillman, 1995; strain abbreviation AhV). Indeed, sequence-based phylogenetic comparisons have revealed that FpV1 and AhV belong to the same subclade of partitivirus isolates, whereas PsV-S and PsV-F are distantly related within a distinct subclade (Boccardo & Candresse, 2005; Crawford et al., 2006; Ghabrial et al., 2008, 2011; Tang, Ochoa, et al., 2010; Willenborg, Menzel, Vetten, & Maiss, 2009). These findings represent some of the compelling evidence that a taxonomic reworking of the family *Partitiviridae* is warranted, including to divide the current genus *Partitivirus* into at least two new genera. These findings, along with the fact that the level of sequence identity among the CPs of PsV-S, PsV-F, and FpV1 is quite low (identity scores between FpV1 and PsV-S or PsV-F are, respectively, 14% or 13%, versus 19% between PsV-S and PsV-F), additionally led us to predict that a 3D structure determination for FpV1 virions may reveal unique variations on the capsid architectures first revealed from our studies of PsV-S and PsV-F. FpV1 also appears to contain at least one, ~550-bp satellite segment (Compel et al., 1998; Tang, Ochoa, et al., 2010), for which a sequence has not yet been reported.

Although much remains unknown about various other aspects of partitivirus infection, a few of these are addressed in the structure-oriented discussions below.

2. PARTITIVIRUS CAPSID STRUCTURES

Prior to our recent 3D structure determinations, partitivirus virions were known from negative-stain transmission electron microscopy to be isometric and small, on the order of 30–40 nm in diameter (Bozarth et al., 1971; Buck & Kempson-Jones, 1973; Compel et al., 1998; Crawford et al., 2006). In addition, they appeared to have a single-layer capsid (not the double- or triple-layer capsids of more complex dsRNA viruses from the families *Cystoviridae* and *Reoviridae*; Fig. 3.1) and some short surface protuberances rising above the contiguous shell region. The genomic dsRNA appeared to be centrally enclosed by this shell, as inferred from the presence of less dense "empty" particles, from which the central dsRNA had seemingly been lost (or had never been present) and which were therefore more completely penetrated by negative stain. Stoichiometric estimates based on sedimentation- and gel-based molecular weights of whole particles and CP subunits suggested the presence of ~120 CP molecules and ~1 RdRp molecule per particle in the case of fungal partitivirus PsV-S (Buck & Kempson-Jones, 1973, 1974). Based on these earlier observations and in light of more recent structure determinations for other dsRNA viruses with genome-enclosing capsids comprising 120 icosahedrally arranged subunits of their respective CPs or inner-capsid proteins (ICPs; Cheng et al., 1994; Dunn et al., 2013; Grimes et al., 1998; Huiskonen et al., 2006; McClain, Settembre, Temple, Bellamy, & Harrison, 2010; Naitow, Tang, Canady, Wickner, & Johnson, 2002; Reinisch, Nibert, & Harrison, 2000; Tang et al., 2008; Yu, Jin, & Zhou, 2008; Zhou et al., 2001), the expectation had quite reasonably been that the partitivirus capsid is another related example of one of these 120-subunit structures with $T=1$ symmetry and an icosahedral asymmetric unit (IAU) comprising a CP dimer. This is in fact what our recent 3D structure determinations have shown, though with some new aspects of interest for describing partitivirus capsids, in particular, and 120-subunit capsids of other dsRNA viruses, in general.

For our structural analyses of partitivirus virions, we began with a coinfected culture of *P. stoloniferum*, from which we differentially isolated the virions of PsV-S and PsV-F. These purified virions were then subjected to transmission electron cryomicroscopy (cryo-TEM) and 3D image reconstruction. The analysis of PsV-S progressed most quickly, and so the cryo-TEM structure of PsV-S was the first to be reported, at a nominal resolution of 7.3 Å (Ochoa et al., 2008; Fig. 3.2). In the meantime, crystals of PsV-F virions had been grown and were found to be amenable to high-resolution data collection. As a result, the second report contained not only a cryo-TEM structure of PsV-F at a nominal resolution of 8.0 Å but also a crystal structure at 3.3 Å, the latter of which provided a nearly complete atomic model for PsV-F CP (Pan et al., 2009; Fig. 3.3). For the third report, the cryo-TEM structures of both PsV-S and PsV-F were carefully refined to nominal resolutions of 4.5–4.7 Å, making use of the fitted PsV-F crystal structure for assessing progress during the refinement process, and the final PsV-S cryo-TEM map and PsV-F crystal structure were used to generate a nearly complete atomic model for PsV-S CP (Tang, Pan, et al., 2010; Fig. 3.3). To complete our studies of partitivirus structures to date, we lastly isolated FpV1 virions from a culture of *F. poae* and determined a cryo-TEM structure at a nominal resolution of 5.0 Å (Tang, Ochoa, et al., 2010;

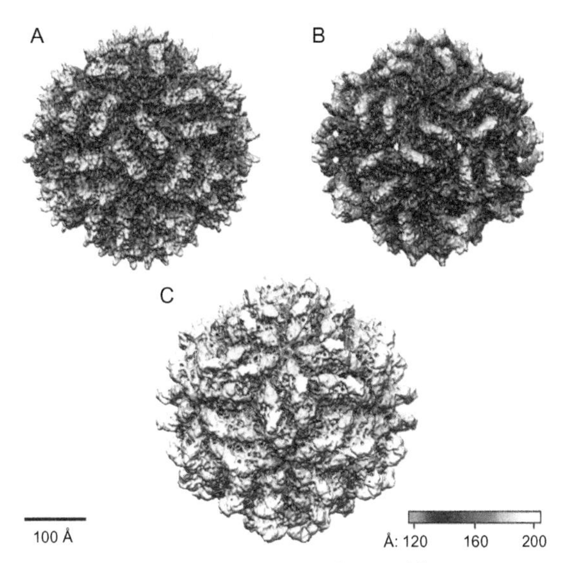

Figure 3.2 Fungal partitivirus virion structures obtained by cryo-TEM and icosahedral 3D image reconstruction. Radially color-coded surface views are shown for (A) PsV-S, (B) PsV-F, and (C) FpV1 (Pan et al., 2009; Tang, Ochoa, et al., 2010; Tang, Pan, et al., 2010). The structures are shown at the same scale (see bar at lower left), with the same radial color map (lower right; radii in Å) applied to each. (See Page 2 in Color Section at the back of the book.)

Fig. 3.2). With this set of new 3D structures in hand, we could then draw a variety of new conclusions regarding both conserved and variable elements of partitivirus structure.

The outermost diameter of each partitivirus virion ranges from 350 and 370 Å for PsV-S and PsV-F, respectively, to 420 Å for FpV1 (Fig. 3.2). The innermost diameter of the contiguous capsid shell of each virus ranges from 250 Å for both PsV-S and PsV-F to 270 Å for FpV1. Combining these two sets of data, the radial span of each capsid ranges from 100 and 120 Å for PsV-S and PsV-F, respectively, to 150 Å for FpV1. The larger diameter and span of the FpV1 capsid correlates with the ~50% larger molecular mass of its CP. In addition, the larger innermost diameter of the FpV1 capsid, and thus the larger volume of central cavity enclosed by it, correlates with the >30% greater length of each FpV1 genome segment.

The capsid of each partitivirus is formed from 120 icosahedrally arranged subunits of the respective CP. In each virus, these 120 subunits fall into two categories, reflecting two non-quasi-equivalent positions within the capsid shell (Fig. 3.3A and B). Sixty of the subunits, the so-called *A* subunits,

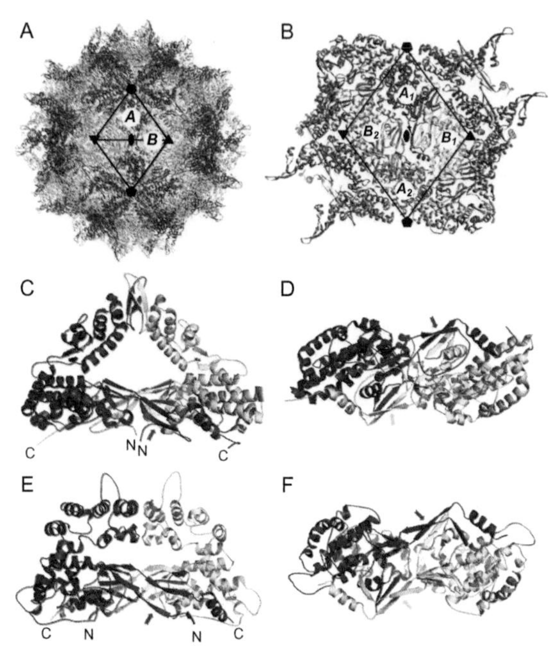

Figure 3.3 Fungal partitivirus capsid and CP structures obtained by X-ray crystallography or by cryo-TEM and homology modeling. (A) Cα trace of the PsV-F capsid structure viewed along an I2 axis. *A* and *B* subunits are colored red and yellow, respectively. I2, I3, and I5 axes are marked with symbols (oval, triangles, and pentagons) and connected by lines. (B) A CP dodecamer with two quasisymmetric *A–B* dimers (A_1–B_1 and A_2–B_2) differentially colored (red:yellow and magenta:orange, respectively). A_1–B_1 and A_2–B_2 extend in antiparallel fashion along either side of an I2 axis, forming a dimer of dimers (tetramer) that is largely confined within the diamond shape formed by lines connecting the I3 and I5 axes. (C–F) Views of the PsV-F quasisymmetric *A–B* dimer Cα trace (C and D) and the PsV-S quasisymmetric *A–B* dimer Cα trace (E and F) as viewed from the side (C and E) and from beneath (i.e., from inside the particle) (D and F). *A* and *B* subunits are colored red and yellow, respectively. In (C) and (E), visible N-and C-termini are labeled. Red and yellow arrows point toward β-strands involved in domain swapping, as discussed in the text. Cα traces for PsV-F and PsV-S are, respectively, from Pan et al. (2009) and Tang, Pan, et al. (2010). (See Page 3 in Color Section at the back of the book.)

approach and surround each icosahedral fivefold (I5) axis. The other 60 subunits, the so-called *B* subunits, approach and surround each icosahedral threefold (I3) axis. Both the *A* and *B* subunits approach each icosahedral twofold (I2) axis, though *A*:*A* contacts across this axis appear to predominate, as discussed further below.

One of the routine questions when describing a 120-subunit $T=1$ structure is which of the *A* and *B* subunits might be best identified as forming the

IAU, of which there are 60 such A–B dimers in the whole capsid. From previously determined structures for other dsRNA viruses (Castón et al., 1997; Grimes et al., 1998; Huiskonen et al., 2006; McClain et al., 2010; Naitow et al., 2002; Reinisch et al., 2000; Tang et al., 2008; Yu et al., 2008; Zhou et al., 2001), three good options are evident for any given A subunit (Fig. 3.4A). In two of these options (B_1 and B_2 in Fig. 3.4A), the B subunit is approximately parallel to and side-by-side with the chosen A subunit and is thus asymmetrically positioned relative to that A. In each of these two options, the resulting A–B dimer is relatively compact, and the surface area buried between the two subunits is similarly large. In the third option, the B subunit is antiparallel to and end-to-end with the chosen A subunit and is thus quasisymmetrically positioned relative to that A (B_3 in Fig. 3.4, left).

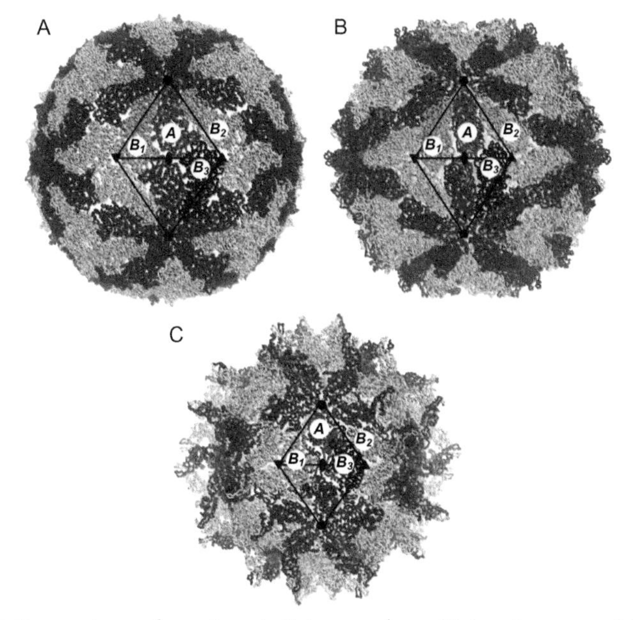

Figure 3.4 Comparison of reovirus, totivirus, and partitivirus inner capsid or capsid organizations. Shown are crystallography-derived Cα traces of (A) Bluetongue virus (family *Reoviridae*) inner capsid (Grimes et al., 1998), (B) Saccharomyces cerevisiae virus L-A (family *Totiviridae*) capsid (Naitow et al., 2002), and (C) PsV-F (family *Partitiviridae*) capsid (Pan et al., 2009). In each virus, A and B subunits are colored red and yellow, respectively. In addition, a chosen A subunit is labeled and colored magenta in each virus, and the three adjacent B subunits are colored green (B_1), cyan (B_2), and blue (B_3). The A–B_1 and A–B_2 dimers are asymmetric and the A–B_3 dimers are quasisymmetric in each case. *Note*: Different capsids are not shown at the same scale. (See Page 3 in Color Section at the back of the book.)

In this option, however, the resulting *A–B* dimer is a good deal more extended, that is, much less compact, with less buried surface area between the two subunits. In previous dsRNA virus structures, because of the compactness and greater buried surface area of each of the first two options, one of those types of asymmetric *A–B* dimer has been routinely chosen to represent the IAU. In partitivirus capsids, however, due in part to the smaller size of the CP shell domain, the quasisymmetric *A–B* dimer is not substantially less compact than the asymmetric *A–B* dimers, and indeed one of the asymmetric dimers is the least compact of the three options (Fig. 3.4, right). Thus, for the partitivirus capsids, choosing the quasisymmetric dimer to represent the IAU appears reasonable.

In fact, the atomic models of PsV-S and PsV-F help to address this question for partitiviruses in that there is a clear domain swapping within the shell regions of the *A* and *B* subunits in the quasisymmetric dimer of each virus (Pan et al., 2009; Tang, Pan, et al., 2010; Fig. 3.3B–F). This finding argues strongly that this particular type of *A–B* dimer is (i) likely to be the protomer for capsid assembly and (ii) thus well chosen to represent the IAU. Specifically, noteworthy features at this dimer interface within the capsid shell are four four-stranded β-sheets. In both PsV-S and PsV-F, two of these sheets consist of four β-strands from the same subunit, *A* or *B*, but the other two sheets are formed by two β-strands from each of the two different subunits, which thereby represents the domain swapping that suggests this dimer as the assembly promoter. Such domain swapping had not been seen in previous dsRNA virus capsid structures, but a very similar feature has also recently been found in a picobirnavirus (Duquerroy et al., 2009), as described more below.

There is in fact another striking feature in the partitivirus 3D structures that argues for the quasisymmetric *A–B* dimer to be the assembly promoter and thus well chosen to represent the IAU. This feature consists of a second, so-called "arch" domain in each CP subunit that represents an insertion in the middle of the shell domain of both PsV-S and PsV-S CP and protrudes on the particle surface (Pan et al., 2009; Tang, Pan, et al., 2010; Fig. 3.3C and E). An arch domain is also present in FpV1 CP, but the primary sequences that form it remain undefined in the absence of an atomic model to date (Tang, Ochoa, et al., 2010). In all three viruses, the arch domain of a given *A* subunit reaches up and over to make extensive additional contacts with, and only with, the arch domain of the *B* subunit within the quasisymmetric dimer (Fig. 3.3C and E). In both PsV-S and PsV-F, these contacts occur well above the shell surface, leaving an open solvent path underneath, and thus the contacting arch domains in each quasisymmetric *A–B* dimer indeed form an arch over the underlying shell domains, for a total of 60 such arches per particle. In FpV1, the arch domains are collapsed toward the shell domain surface such that little or no solvent path is left underneath, but extensive additional contacts with, and only with, the arch domains of the *A* and *B* subunits within the quasisymmetric dimer are nonetheless made. In each virus, the arch domain contacts thus increase the molecular interface of contacts between the two subunits within this dimer, presumably adding to its stability. Whether these roles in assembly and stability are the sole functions of these arches remains to be defined, but they are such striking structural features on the particle surfaces that some additional function, such as in anchoring to some intracellular host factor, seems likely.

An additional set of observations about the PsV-S and PsV-F capsids leads to another prediction about their assembly pathways. For any given quasisymmetric *A–B* dimer in the capsid, the adjacent such dimer against which it buries the largest surface area (~3500 Å^2) is the dimer immediately across the I2 axis from it (Figs. 3.3B and 3.4B). In comparison, contacts between any two such dimers around the I3 and I5 axes bury much smaller surface areas (~2000 Å^2). We have therefore proposed (Pan et al., 2009; Tang, Pan, et al., 2010) that the assembly of PsV-S and PsV-F capsids likely proceeds first from quasisymmetric *A–B* dimers (protomers) and next to antiparallel, symmetric dimers of these dimers (Fig. 3.3B). Thirty such diamond-shaped dimers of dimers (tetramers) would then interact further via threefold and fivefold symmetry contacts to complete assembly of the 120-subunit capsid. Notably, this proposed assembly pathway for partitiviruses is quite distinct from that involving pentamers of asymmetric *A–B* dimers (decamers) as a key intermediate as has been previously proposed for some other dsRNA viruses (Grimes et al., 1998; McClain et al., 2010), and indeed dimers of quasisymmetric *A–B* dimers and pentamers of asymmetric *A–B* dimers represent assembly pathways that are mutually exclusive. Assembly of the cystovirus inner capsid, on the other hand, proceeds through a stable, tetramer intermediate of *A–B* dimers (Kainov, Butcher, Bamford, & Tuma, 2003), suggesting that cystoviruses and partitiviruses may share similar assembly pathways for their genome-enclosing capsids. Returning to describe the contacts within each symmetric dimer of dimers in partitiviruses, substantial contacts occur between the *A* subunit from one protomer and the *B* subunit from the other protomer, but the predominant contacts are between the *A* subunits from the two protomers (Fig. 3.3B). Moreover, interestingly, it is the domain-swapping arm of each *A* subunit that mediates these predominate interactions, including those directly across each I2 axis in the assembled capsid.

It is worth noting that the partitivirus virions of each strain used for our 3D structure studies contained different dsRNA molecules: dsRNA1 or dsRNA2 and/or satellite(s). In preparing these particles for study, we did nothing extra to try to separate or enrich the particles according to specific dsRNA content. The fact that high-resolution capsid structures were nevertheless obtained, including the PsV-F crystal structure at 3.3-Å resolution, therefore indicates that the capsid of each virus does not vary to a substantial degree according to which dsRNA molecule is packaged. Instead, the overall structure of each capsid appears to be nearly identical, regardless of the respectively packaged dsRNA. Moreover, in the case of PsV-S and PsV-F, since these particles were differentially isolated from the same coinfected *P. stoloniferum* culture, the high-resolution capsid structures corroborate previous evidence that the CP of each virus associates only with itself in forming the respective capsids (Buck & Kempson-Jones, 1974), that is, that chimeric capsids are not formed to any substantial degree. Whether this specificity for self-interaction in complete capsids is maintained purely at the level of CP subunit–subunit interactions or is also aided by some sequestration of PsV-S and PsV-F CP subunits into separate compartments in coinfected cells remains an interesting question.

3. PARTITIVIRUS RdRp AND dsRNA STRUCTURES

The cryo-TEM and crystal structures described earlier tell us a great deal about the 120-subunit capsids of these fungal partitiviruses. But what about the other virion components: the RdRp and the dsRNA genome?

No structural elements attributable to the RdRp were identified in any of our partitivirus density maps, which is not surprising since there is thought to be only ~1 RdRp molecule per virion (Buck & Kempson-Jones, 1974). Furthermore, no structure of a partitivirus RdRp molecule purified in isolation has been reported to date. Thus, with regard to the RdRp, what we can say here is limited and speculative. First, conserved motifs common to the RdRps of other encapsidated dsRNA viruses (Bruenn, 1993) appear in the usual order in partitivirus RdRp sequences (Crawford et al., 2006; Ghabrial et al., 2011), suggesting that they adopt the right-hand fingers–palm–thumb configuration that is common to many RNA and DNA polymerases (Ortín & Parra, 2006). In the crystallized RdRps of other encapsidated dsRNA viruses, the catalytic hand domain spans between 440 and 590 aa (Butcher, Grimes, Makeyev, Bamford, & Stuart, 2001; Lu et al., 2008; Pan, Vakharia, & Tao, 2007; Tao, Farsetta, Nibert, & Harrison, 2002). Thus, since the overall sequence lengths of partitivirus RdRps approximate 510–680 aa, it seems likely that most of those sequences are devoted to forming the catalytic domain, with relatively limited regions of sequence available to form additional, large N- and C-terminal domains as seen in the larger RdRps of reoviruses and birnaviruses (Lu et al., 2008; Pan et al., 2007; Tao et al., 2002). The additional 64-aa C-terminal "priming" domain of cystovirus RdRp (Butcher et al., 2001), on the other hand, is probably small enough to have an equivalent in partitivirus RdRps.

Where might the RdRp be located within the partitivirus virion? Based on findings with other encapsidated dsRNA viruses (McClain et al., 2010; Sen et al., 2008; Zhang, Walker, Chipman, Nibert, & Baker, 2003), the partitivirus RdRp is likely to be located interior to the capsid, probably anchored by noncovalent interactions to the undersurface of the capsid shell. Importantly, of course, from that position it can access the RNA template for transcription. As to its specific location relative to the symmetry axes of the capsid, it is more difficult to say. In reoviruses, each copy of the RdRp is anchored nearest, and overlapping, an I5 axis of the inner capsid (McClain et al., 2010; Zhang et al., 2003). In cystovirus procapsids, however, each copy of the RdRp appears to be anchored near an I3 axis of the inner capsid, though it has been proposed that it might rotate nearer an I5 axis during particle maturation (Sen et al., 2008). In reoviruses, it furthermore appears that RNA transcripts exit through inner-capsid pores cotranscriptionally, and thus each copy of the RdRp is thought to be positioned near pentonal or peripentonal pores that allow this exit (Diprose et al., 2001; Lawton, Estes, & Prasad, 1997; Mendez, Weiner, She, Yeager, & Coombs, 2008; Yang et al., 2012; Zhang et al., 2003). From our partitivirus structures, we have seen that small ($\leq$5-Å diameter) pores through the capsid are found at both the I5 and the I3 axes, but these are too small to allow RNA transcript exit without some conformational rearrangement. Regarding such potential rearrangement, capsid elements surrounding each I5 axis of PsV-F appear to be flexible (based on crystallographic temperature factors) (Pan et al., 2009) and so this favors the idea that partitivirus transcripts exit through the I5 pores and that the RdRp is positioned near one of these sites.

One other important aspect of the partitivirus RdRp is that partitivirus transcription is semiconservative (Buck, 1978), meaning that the parental RNA plus strand is released while the progeny RNA plus strand is retained as part of the genomic duplex (i.e., until the next round of transcription when it is released, etc.). This is also the case for cystovirus transcription, but not for reovirus transcription, which is instead conservative (parental RNA plus strand retained as part of the genomic duplex, progeny RNA plus strand released). Models for reovirus transcription include complex steps for retaining and rewinding the parental plus strand during transcription (Lu et al., 2008; Tao et al., 2002), which are therefore not required as part of partitivirus transcription. Cystovirus transcription is thus likely the better model for partitivirus transcription (Butcher et al., 2001).

With regard to the packaged dsRNA genome structures of PsV-S, PsV-F, and FpV1, there is a bit more to say from the data, but speculation still dominates the interesting considerations. The cryo-TEM structures of all three partitiviruses are marked by concentric rings of RNA densities in the particle interior (Ochoa et al., 2008; Pan et al., 2009; Tang, Ochoa, et al., 2010; Tang, Pan, et al., 2010; Fig. 3.5). These rings have been seen in many other dsRNA viruses and are thought to reflect that the naked (i.e., not protein-coated) dsRNA is packed on average in locally parallel arrays and distributed evenly so as to minimize interhelix spatial and electrostatic conflicts, as first shown for dsDNA bacteriophages (Earnshaw & Harrison, 1977; Maniatis, Venable, & Lerman, 1974). One element of the partitivirus CPs not summarized earlier is that the N-terminal ~40 aa of PsV-S and PsV-F CP are not visualized in the high-resolution structures and thus do not appear in the atomic models (Pan et al., 2009; Tang, Pan, et al., 2010). Notably, though, the first aa visualized in each of these models lies on the undersurface of the capsid (Figs. 3.3C, E and 3.5B), suggesting that the N-terminal peptides plunge into the RNA density regions and thus likely contact the RNA. These RNA contacts may play roles in RNA packaging or synthesis and likely also define where the outermost layer (ring) of RNA is positioned relative to the undersurface of the capsid in the assembled virion. The more-internal RNA layers (rings) may then be subsequently positioned relative to this outer, protein-contacted layer. In the cryo-TEM structures of all three partitiviruses, but especially in PsV-F, these inwardly projecting N termini of the CP subunits (putatively identified as such in FpV1) are also represented by strands or bumps of density that span or enter the space between the capsid undersurface and the outer RNA ring, near the I2 axes (Fig. 3.5).

Given that partitivirus virions become transcriptionally active upon addition of NTPs (Buck, 1978; Pan et al., 2009), there are also interesting structural questions about the genomic and product RNA molecules as they are moved internally to and/or through the capsid shell during transcription.

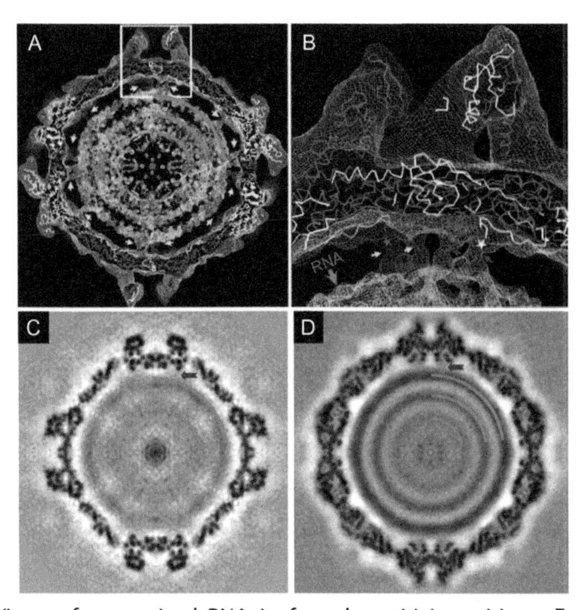

Figure 3.5 Views of genomic dsRNA in fungal partitivirus virions. Equatorial cross sections are shown for PsV-F, PsV-S, and FpV1 cryo-TEM maps. The Cα trace of PsV-F (*A* subunits, red; *B* subunits, yellow) is fitted into the cryo-TEM reconstruction of the PsV-F virion at 8.0-Å resolution. Magenta arcs highlight three rings of RNA density that are evident in the particle interior. Cryo-TEM densities corresponding to the disordered N termini of both subunits (not visible in the Cα traces) are indicated by cyan arrows. A close-up view of the boxed region in (A) is shown in (B). The ordered N-terminal ends of the *A* and *B* subunit Cα traces are indicated by red and yellow stars, respectively. The disordered N-termini from two adjacent *A*–*B* dimers, indicated by cyan and blue arrows, respectively, extend as tube-like densities from the end of the Cα traces into the underlying outer ring of RNA density. (C and D) Density projection images of thin, planar sections encompassing the equatorial regions of (C) PsV-S (Tang, Pan, et al., 2010) and (D) FpV1 (Tang, Ochoa, et al., 2010) are shown. Magenta arcs highlight two or three rings of RNA density evident in PsV-S or FpV1, respectively. Blue arrowheads indicate examples of close approach between the capsid undersurface and the outer RNA ring in both viruses. *(A) and (B) panels have been reproduced from Pan et al. (2009).* (See Page 4 in Color Section at the back of the book.)

Considering findings with other dsRNA virus RdRps (Butcher et al., 2001; Lu et al., 2008; Tao et al., 2002), our expectation is that the genomic segment of each partitivirus virion is anchored to the internally capsid-bound RdRp in a noncovalent fashion at the end of the dsRNA molecule that includes the 3′ end of the minus-strand RNA, that is, the site of transcription initiation. In this way, the "promoter" region of the minus-strand RNA remains near the RdRp throughout the transcription cycle and therefore in position to reinitiate transcription at the end of each cycle. In addition, again considering findings with other dsRNA viruses (Diprose et al., 2001; Lawton et al., 1997; Mendez et al., 2008; Yang et al., 2012; Zhang et al., 2003), we expect the partitivirus RdRp is properly positioned so that the parental plus strand, that is, the released product of semiconservative transcription, is directed toward and through a trans-capsid pore for extrusion to the particle exterior for subsequent use in either protein translation or RNA packaging into newly forming virions.

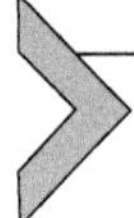

4. COMPARISONS WITH OTHER BISEGMENTED dsRNA VIRUSES

Until recently (see below), there were two other recognized families of dsRNA viruses with bisegmented genomes: *Birnaviridae* and *Picobirnaviridae* (Fig. 3.1). How do birnaviruses and picobirnaviruses compare to the partitiviruses, especially at a structural level? Birnaviruses are known to infect arthropods (insects) and vertebrates (fish, reptiles, and birds; Delmas, Mundt, Vakharia, & Wu, 2011), and picobirnaviruses are known to infect vertebrates (mammals; Delmas, 2011). Thus, their host ranges are quite distinct from those of partitiviruses. One fundamental correlate is that both birnaviruses and picobirnaviruses can undergo regular, extracellular transmission between cells in their complex animal hosts, as well as between host individuals, and thus their virions must contain both the molecular components and the dynamic capabilities for productive cell entry. This is unlike the case for partitiviruses as described earlier. In addition, as a consequence of undergoing extracellular transmission on a regular basis, both birnaviruses and picobirnaviruses are thought to be uniparticulate, that is, to package both of their essential genome segments within each infectious virion, which is again unlike partitiviruses.

Birnaviruses are oddities among encapsidated dsRNA viruses in that they do not contain a 120-subunit capsid. Instead, their single, genome-enclosing capsid is $T=13$, comprising 780 subunits of the main capsid protein VP2 (Coulibaly et al., 2005). This unusual symmetry is the same as that found in the outer capsids of reoviruses and cystoviruses, raising interesting questions with regard to the evolution of these viruses. In any case, the birnavirus capsid is larger and quite distinct from that of partitiviruses.

The story for picobirnaviruses, on the other hand, is quite different. In fact, the CP folds and assembled capsids of rabbit picobirnavirus (RaPBV), the only picobirnavirus for which a 3D structure has been reported to date

(Duquerroy et al., 2009), are quite similar to those of partitiviruses (Tang, Pan, et al., 2010; Fig. 3.6). Of particular note is that the RaPBV capsid exhibits domain swapping within the shell regions of the quasisymmetric *A–B* dimer (Fig. 3.6C), thereby arguing for this dimer to be the assembly protomer and well chosen to represent the IAU, as is also the case for partitiviruses. In addition, the RaPBV CP includes an internally inserted "protruding" domain, comparable to the arch domain of partitiviruses, which contributes additional contacts within, and only within, the quasisymmetric *A–B* dimer. In fact, additional domain swapping between the *A* and *B* subunits occurs within the protruding domain of RaPBV (Fig. 3.6C). Although the protruding domain contacts in RaPBV do not produce an arch as in PsV-S and PsV-F, they appear to be more similar to those in FpV1. Lastly, the assembly pathway of picobirnaviruses may be similar to that proposed for partitiviruses in that two quasisymmetric *A–B* dimers make extensive contacts across each I2 axis in RaPBV, forming

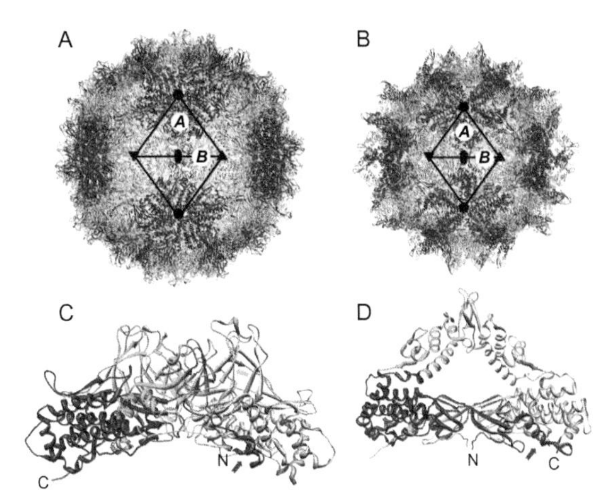

Figure 3.6 Comparison of picobirnavirus and partitivirus capsid organizations and CP folds. Shown are crystallography-derived Cα traces of (A) Rabbit picobirnavirus (RaPBV) virus-like particles (Duquerroy et al., 2009) and (B) partitivirus PsV-F virions (Pan et al., 2009). *A* and *B* subunits are colored red and yellow, respectively. I2, I3, and I5 axes are marked with symbols and connected with lines as in Fig. 3.3. The two capsids are shown at the same scale. Also shown are side views of the Cα traces of the quasisymmetric *A–B* dimer of (C) RaPBV and (D) PsV-F. The *B* subunit is colored gray in each. In the *A* subunit, the shell domain is colored red and the protruding/arch domain is colored green. Visible N- and C-termini are labeled for the *A* subunit of each virus. Red arrows point toward an α-helix involved in domain swapping in each virus, as discussed in the text. (See Page 4 in Color Section at the back of the book.)

compact, diamond-shaped symmetric dimers of dimers (tetramers) in which *A*:*B* contacts between the two dimers occur, but *A*:*A* contacts predominate, again largely through the domain-swapping arms in their shell regions (Fig. 3.6A and C). Thus, the picobirnavirus and partitivirus capsids, as perhaps the viruses themselves, appear to be closely related, probably sharing a unique, common ancestor.

Despite these similarities, it is important to remember that, given their capacity for extracellular transmission, picobirnavirus virions must include the molecular machinery for cell entry (Duquerroy et al., 2009), which partitivirus virions lack. Thus, it seems likely that either the picobirnavirus capsid evolved the capacities for receptor binding and membrane penetration from a common ancestor that lacked them or the partitivirus capsid lost the capacities for receptor binding and membrane penetration from a common ancestor that had them. Regions of the picobirnavirus capsid involved in receptor binding (presumably in the protruding domain) and membrane penetration (presumably hydrophobic and buried in the virion) remain undefined to date. Nonetheless, taking the perspective that partitiviruses likely evolved before picobirnaviruses because their hosts and life styles are simpler, the similar capsid structures of RaPBV and partitiviruses suggest that the path for a nonenveloped virus to acquire cell-entry functions including receptor binding and membrane penetration may be relatively uncomplicated, without a need for major architectural changes to the capsid. Note that partitivirus virions can in fact initiate *de novo* infection of fungal protoplasts when transfection methods that bypass the need for virion-inherent entry machinery are used (Kanematsu, Sasaki, Onoue, Oikawa, & Ito, 2010; Sasaki, Kanematsu, Onoue, Oyama, & Yoshida, 2006). Thus, the partitivirus virion appears to be a potentially infectious "payload" that simply lacks the necessary, built-in components for efficient payload delivery from outside cells.

Two other general properties of picobirnaviruses suggest several interesting questions upon comparison with partitiviruses. One is the presumed packaging of both of the two essential genome segments of picobirnaviruses into the same infectious particle, which is not the case for partitiviruses. What changes in the picobirnavirus or partitivirus CP might have been needed to evolve in order to accommodate this difference in packaging strategy? Are there two copies of packaged RdRp, one for each segment, in the picobirnavirus virion? Is there indeed enough room in the central cavity of the picobirnavirus virion to hold both segments in addition to probably two copies of RdRp? The other property in question is the style of transcription by picobirnavirus virions. Is it semiconservative as with partitiviruses, or is it instead conservative as with many other encapsidated dsRNA viruses? Because infectious, genome-containing picobirnavirus virions have been difficult to isolate in sufficient quantities to date (note that the RaPBV crystal structure is that of virus-like particles assembled after CP expression in insect cells), this relatively simple question remains unanswered.

A fourth family of bisegmented dsRNA viruses, *Megabirnaviridae*, has been recently recognized (Fig. 3.1). To date, it is represented by a single characterized strain, Rosellinia necatrix megabirnavirus 1, from the phytopathogenic ascomycete *Rosellinia necatrix*, which is the cause of white root rot in fruit trees worldwide (Chiba et al., 2009). The 3D structure of this virus has yet to be reported, but the virions appear to be somewhat larger than those of partitiviruses, ~50 nm in diameter. The genome is also larger, comprising 8931-bp dsRNA1 and 7180-bp dsRNA2, with both the CP and the RdRp encoded on dsRNA1. More detailed comparisons with partitiviruses await further characterizations.

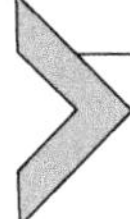

5. COMPARISONS WITH OTHER FUNGAL VIRUSES WITH ENCAPSIDATED dsRNA GENOMES

Other fungal viruses with encapsidated dsRNA genomes are currently found in four other families: *Totiviridae* (one genome segment, >4.5 kbp), *Chrysoviridae* (four segments, >11 kbp in total), *Quadriviridae* (four segments, >16 kbp in total), and *Reoviridae* (9–12 segments, >18 kbp in total; 11 or 12 segments in members of the genus *Mycoreovirus*; Fig. 3.1). Representative structures are available for three of these families (all but *Quadriviridae*; Lin et al., 2012), and indeed 3D structures of fungal virus strains are available for both totiviruses and chrysoviruses (Castón et al., 2006; Cheng et al., 1994; Gómez-Blanco et al., 2012; Luque et al., 2010; Naitow et al., 2002). All have genome-enclosing capsids or inner capsids with $T=1$ icosahedral symmetry, like partitiviruses, but despite this basic similarity, there are a number of differences.

Totivirus capsids and reovirus inner capsids are 120-subunit $T=1$ structures, like those of partitiviruses, but their subunit architectures are relatively thinner in the radial direction. Their respective CPs and ICPs are thus often described as "plate-like" and generally lack the sort of major protruding regions as possessed by partitivirus (and RaPBV) CPs. In addition, the totivirus CPs and reovirus ICPs are commonly described as having two or three different domains within their shell regions and are therefore more elongated

(Dunn et al., 2013; Grimes et al., 1998; McClain et al., 2010; Naitow et al., 2002; Reinisch et al., 2000; Tang et al., 2008; Yu et al., 2008; Zhou et al., 2001; see Fig. 3.4A), whereas partitiviruses are described as having a single shell domain, with a second domain protruding above the shell (Pan et al., 2009; Tang, Pan, et al., 2010; see Fig. 3.3C and E). Although totivirus CPs are generally somewhat larger in molecular mass (>70 kDa), reovirus ICPs are consistently even larger (>100 kDa). As a consequence of these different protein structures and sizes, the central cavity enclosed by totivirus capsids and reovirus inner capsids is a good deal larger than that enclosed by partitivirus capsids, consistent with the different amounts of dsRNA being packaged inside each. Other differences in describing the IAUs and potential assembly pathways of these viruses are discussed earlier.

The capsid of chrysoviruses is even more distinct (Gómez-Blanco et al., 2012; Luque et al., 2010). Though also exhibiting $T=1$ icosahedral symmetry, it is a "simple" $T=1$ formed from only 60 CP subunits per virion. The chrysovirus CP is relatively large, however, more similar in molecular mass (>100 kDa) to the ICPs of reoviruses, and has been shown to reflect a genetic duplication, such that each monomeric CP subunit contains two structurally similar shell domains. As a result, the chrysovirus capsid is in fact structurally similar to that of the 120-subunit dsRNA viruses, including partitiviruses. Since an atomic model of the chrysovirus CP is not available to date, it is not yet known with certainty whether the two shell domains in each monomer reside side by side or end to end within the capsid, adopting respectively either a more or a less compact form that would likely impact the pathway of chrysovirus capsid assembly.

An additional point of importance for comparison is that the genome-enclosing CPs or ICPs of totiviruses, partitiviruses, picobirnaviruses, cystoviruses, chrysoviruses, and reoviruses, all of which form $T=1$ structures, are also all relatively rich in α-helices within their shell domains (see Fig. 3.6, top). Furthermore, it appears from comparisons of these structures that certain key α-helices may be widely conserved in their approximate structural placements and interactions within many of these proteins (Dunn et al., 2013; Luque et al., 2010; Ochoa et al., 2008). This suggests that an ancient, ancestral helix-rich CP with the capacity for $T=1$ capsid assembly might be common to many or all of the encapsidated dsRNA viruses that have been studied to date, with the exception of birnaviruses. The main capsid protein of birnaviruses, in contrast, which forms a $T=13$ structure, is richer in β-sheets and related to the β-rich jelly-roll structures of many plus-strand RNA virus capsid subunits (Coulibaly et al., 2005).

6. PROPOSED REVISIONS TO PARTITIVIRUS TAXONOMY

Earlier in this review, we suggested that the family *Partitiviridae* is due for some taxonomic revisions, based on a recent accumulation of genomic sequence data and phylogenetic analyses that have called some of the previous taxonomic groupings into question (Boccardo & Candresse, 2005; Crawford et al., 2006; Ghabrial et al., 2008, 2011; Tang, Ochoa, et al., 2010; Willenborg et al., 2009). One such suggested revision, discussed earlier, is to divide the genus *Partitivirus* into two new genera (Fig. 3.7). Regarding the viruses whose structures we have determined, PsV-S and PsV-F would be placed in one of these new genera, whereas FpV1 would be placed in the other, along with the *Partitivirus* prototype strain AhV. It is helpful to note that the 3D structures of these viruses, emphasized in this review, concur with this division, in that the PsV-S and PsV-F capsids are more similar to one another than either is to the FpV1 capsid (Ochoa et al., 2008; Pan et al., 2009; Tang, Ochoa, et al., 2010, Tang, Pan, et al., 2010).

One interesting side note here is that the putative new genus containing FpV1 and AhV, based on phylogenetic results, would also contain at least two viruses that seem to have been isolated from plants, *Primula malacoides* and *Cannabis sativa* (Li, Tian, Du, Duns, & Chen, 2009; Ziegler, Matoušek, Steger, & Schubert, 2012; Fig. 3.7). Previously, host range has been considered an important determinant for drawing genus divisions, and thus fungal partitiviruses have been placed in the genus *Partitivirus* while plant partitiviruses have been placed in the genus *Alphacryptovirus* or the genus *Betacryptovirus* depending on certain phenotypic characteristics. Sequence-based phylogenetic results, however, including the ones described earlier, suggest that fungal versus plant host range may not be a proper criterion for drawing genus divisions for at least some partitiviruses. There remains the possibility that the partitiviruses of *P. malacoides* and *C. sativa* were in fact derived from contaminating fungi, perhaps pathogens or symbionts of these plants, but the other possible interpretation is that partitiviruses can transfer across the fungus–plant host boundary over evolutionary time periods, perhaps through cross-feeding insects or other vectors, or perhaps simply through the intimate association of certain fungi with their plant hosts. Indeed, recent evidence for integration into plant genomes by partitivirus sequences related to ones from fungi, suggesting horizontal gene transfer, also supports the notion of a common ancestor of fungal and plant partitiviruses, consistent with their possible inclusion in the same, modern genus (Chiba et al., 2011; Liu et al., 2011).

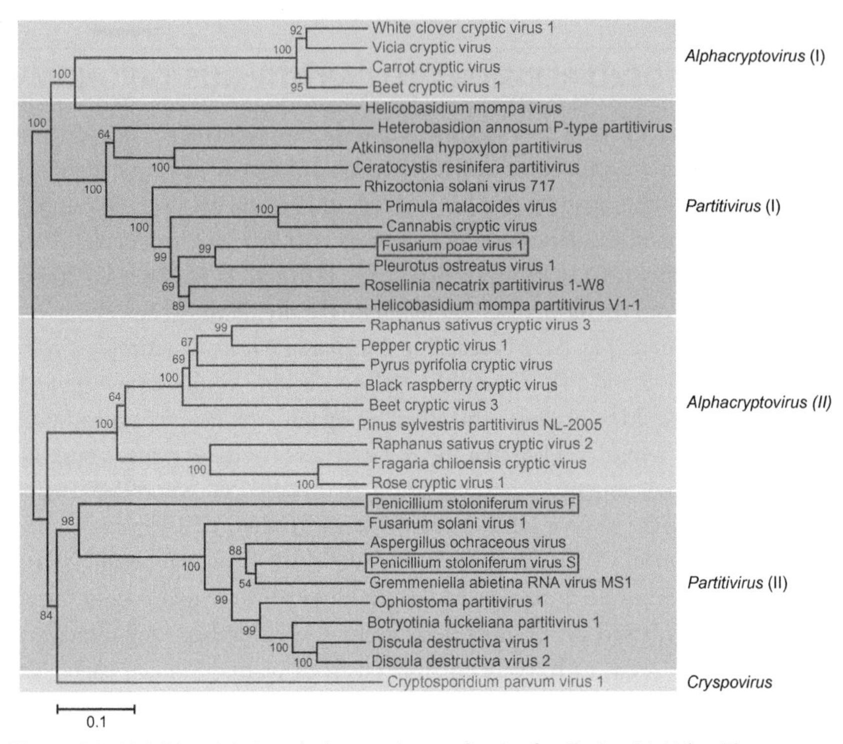

Figure 3.7 Neighbor-joining phylogenetic tree for the family *Partitiviridae*. The tree was constructed from complete aa sequences of RdRps of representative members and probable members of the family. The aa sequences were aligned using the program CLUSTAL X2, and the tree was generated for codon positions using the MEGA5 phylogenetic package. Bootstrap percentages out of 2000 replicates are indicated at the nodes. Green, blue, and yellow shading, respectively, highlight the assignments of individual viruses to the current genera *Alphacryptovirus*, *Partitivirus*, and *Cryspovirus*. The divisions of the *Alphacryptovirus* and *Partitivirus* regions into two sections each (I, II) reflect the four new genera, obtained by dividing the current genera, that are suggested by these and previous phylogenetic results as cited in the text. Red boxes indicate the three fungal partitiviruses for which 3D structures are described in recent reports and this review. GenBank accession numbers for the viruses in this figure are, top to bottom: YP_086754, ABN71237, ACL93278, YP_002308574, BAC23065, AF473549, NP_604475, YP_001936016, NP_620659, YP_00310476, AET80948, NP_624349, YP_227355, YP_392480, BAD32677, ACJ76981, ABC96789, BAA34783, ABU55400, AAB27624, AAY51483, ABB04855, AAZ06131, ABZ10945, YP_271922, BAA09520, ABV30675, AAN8683, NP_659027, CAJ31886, YP_001686789, AAG59816, NP_620301, and AAC47805. (For interpretation of the references to color in this figure legend, the reader is referred to the online version of this chapter.)

Another suggested revision based on phylogenetic results is to divide the genus *Alphacryptovirus* into two new genera as well (Boccardo & Candresse, 2005; Crawford et al., 2006; Ghabrial et al., 2008, 2011; Willenborg et al., 2009). In this case, it is again notable that one of the putative new genera, that including the *Alphacryptovirus* prototype white clover cryptic virus (Fig. 3.7), includes both plant isolates and at least one fungal isolate (Helicobasidium mompa virus in Fig. 3.7). Thus, the considerations noted above again pertain, suggesting that fungal versus plant host range may not be a proper criterion for drawing genus divisions for some partitiviruses. The status of the genus *Betacryptovirus* in the revised taxonomy remains unclear because no sequences from members of that genus have been reported to date. The genus *Cryspovirus*, containing isolates from the apicomplexan protozoan *Cryptosporidium*, on the other hand, appears to remain on solid footing as a divergent taxon within this family (Nibert et al., 2009; Fig. 3.7). In moving ahead to propose these taxonomic changes in a formal manner to the International Committee on Taxonomy of Viruses, additional 3D structures, from members of the other current and proposed genera, would be helpful.

The highly similar CP and capsid structures of RaPBV and partitiviruses contribute to raising another taxonomic question, namely, whether picobirnaviruses and partitiviruses should indeed have been separated into two families. Given several other fundamental properties that differ between these virus groups as described earlier, we consider their current separation into two families to remain quite appropriate. Nevertheless, addition of a superfamily designation to encompass both families *Picobirnaviridae* and *Partitiviridae*, for example, reflecting their putatively more recent common ancestor, might be something to consider.

ACKNOWLEDGMENTS

This work was supported in part by NIH grants R37-GM033050 and 1S10-RR020016 (T. S. B.), by Kentucky Science and Engineering Foundation grant KSEF-2178-RDE-013 (S. A. G.), and by Welch Foundation grant C-1565 (Y. J. T.). The San Diego Supercomputer Center provided access to TeraGrid computing, and support from the University of California–San Diego and the Agouron Foundation (T. S. B.) were used to establish and equip cryo-TEM facilities at the University of California–San Diego.

REFERENCES

Boccardo, G., & Candresse, T. (2005). Complete sequence of the RNA1 of an isolate of *White clover cryptic virus 1*, type species of the genus *Alphacryptovirus*. *Archives of Virology*, *150*, 399–402.

Bozarth, R. F., Wood, H. A., & Mandelbrot, A. (1971). The Penicillium stoloniferum virus complex: Two similar double-stranded RNA virus-like particles in a single cell. *Virology*, *45*, 516–523.

Bruenn, J. A. (1993). A closely related group of RNA-dependent RNA polymerases from double-stranded RNA viruses. *Nucleic Acids Research*, *21*, 5667–5669.

Buck, K. W. (1978). Semi-conservative replication of double-stranded RNA by a virion-associated RNA polymerase. *Biochemical and Biophysical Research Communications*, *84*, 639–645.

Buck, K. W., & Kempson-Jones, G. F. (1973). Biophysical properties of Penicillium stoloniferum virus S. *Journal of General Virology*, *18*, 223–235.

Buck, K. W., & Kempson-Jones, G. F. (1974). Capsid polypeptides of two viruses isolated from *Penicillium stoloniferum*. *Journal of General Virology*, *22*, 441–445.

Butcher, S. J., Grimes, J. M., Makeyev, E. V., Bamford, D. H., & Stuart, D. I. (2001). A mechanism for initiating RNA-dependent RNA polymerization. *Nature*, *410*, 235–240.

Castón, R. J., Luque, D., Trus, B. L., Rivas, G., Alfonso, C., González, J. M., et al. (2006). Three-dimensional structure and stoichiometry of Helminthosporium victoriae 190S totivirus. *Virology*, *347*, 323–332.

Castón, J. R., Trus, B. L., Booy, F. P., Wickner, R. B., Wall, J. S., & Steven, A. C. (1997). Structure of L-A virus: A specialized compartment for the transcription and replication of double-stranded RNA. *The Journal of Cell Biology*, *138*, 975–985.

Cheng, R. H., Caston, J. R., Wang, G.-J., Gu, F., Smith, T. J., Baker, T. S., et al. (1994). Fungal virus capsids, cytoplasmic compartments for the replication of double-stranded RNA, formed as icosahedral shells of asymmetric gag dimers. *Journal of Molecular Biology*, *244*, 255–258.

Chiba, S., Kondo, H., Tani, A., Saisho, D., Sakamoto, W., Kanematsu, S., et al. (2011). Widespread endogenization of genome sequences of non-retroviral RNA viruses into plant genomes. *PLoS Pathogens*, 7, e1002146.

Chiba, S., Salaipeth, L., Lin, Y. H., Sasaki, A., Kanematsu, S., & Suzuki, N. (2009). A novel bipartite double-stranded RNA mycovirus from the white root rot fungus *Rosellinia necatrix*: Molecular and biological characterization, taxonomic considerations, and potential for biological control. *Journal of Virology*, *83*, 12801–12812.

Compel, P., Papp, I., Bibo, M., Fekete, C., & Hornok, L. (1999). Genetic relationships and genome organization of double-stranded RNA elements of *Fusarium poae*. *Virus Genes*, *18*, 49–56.

Coulibaly, F., Chevalier, C., Gutsche, I., Pous, J., Navaza, J., Bressanelli, S., et al. (2005). The birnavirus crystal structure reveals structural relationships among icosahedral viruses. *Cell*, *120*, 761–772.

Crawford, L. J., Osman, T. A., Booy, F. P., Coutts, R. H., Brasier, C. M., & Buck, K. W. (2006). Molecular characterization of a partitivirus from *Ophiostoma himal-ulmi*. *Virus Genes*, *33*, 33–39.

Delmas, B. (2011). Picobirnaviridae. In A. M. Q. King, M. J. Adams, E. B. Carstens & E. J. Lefkowitz (Eds.), *Virus taxonomy: Ninth report of the international committee on taxonomy of viruses* (pp. 535–539). Oxford, UK: Elsevier.

Delmas, B., Mundt, E., Vakharia, V. N., & Wu, J. L. (2011). Birnaviridae. In A. M. Q. King, M. J. Adams, E. B. Carstens & E. J. Lefkowitz (Eds.), *Virus taxonomy: Ninth report of the international committee on taxonomy of viruses* (pp. 499–507). Oxford, UK: Elsevier.

Diprose, J. M., Burroughs, J. N., Sutton, G. C., Goldsmith, A., Gouet, P., Malby, R., et al. (2001). Translocation portals for the substrates and products of a viral transcription complex: The bluetongue virus core. *The EMBO Journal*, *20*, 7229–7239.

Dunn, S. E., Li, H., Cardone, G., Nibert, M. L., Ghabrial, S. A., and Baker, T. S. (2013). Three-dimensional structure of the genus *Victorivirus* prototype strain HvV190S suggests that the capsid proteins in all totiviruses may share a conserved core. PLOS Pathogens, in press.

Duquerroy, S., Da Costa, B., Henry, C., Vigouroux, A., Libersou, S., Lepault, J., et al. (2009). The picobirnavirus crystal structure provides functional insights into virion assembly and cell entry. *The EMBO Journal, 28*, 1655–1665.

Earnshaw, W. C., & Harrison, S. C. (1977). DNA arrangement in isometric phage heads. *Nature, 268*, 598–602.

Ghabrial, S. A., Nibert, M. L., Maiss, E., Lesker, T., Baker, T. S., & Tao, Y. J. (2011). Partitiviridae. In A. M. Q. King, M. J. Adams, E. B. Carstens & E. J. Lefkowitz (Eds.), *Virus taxonomy: Ninth report of the international committee on taxonomy of viruses* (pp. 523–534). Oxford, UK: Elsevier.

Ghabrial, S. A., Ochoa, W. F., Baker, T. S., & Nibert, M. L. (2008). Partitiviruses: General features. In B. W. J. Mahy & M. H. V. van Regenmortel (Eds.), (3rd ed.). *Encyclopedia of virology*, Vol. 4, (pp. 68–75). Oxford, UK: Elsevier.

Gómez-Blanco, J., Luque, D., González, J. M., Carrascosa, J. L., Alfonso, C., Trus, B., et al. (2012). Cryphonectria nitschkei virus 1 structure shows that the capsid protein of chrysoviruses is a duplicated helix-rich fold conserved in fungal double-stranded RNA viruses. *Journal of Virology, 86*, 8314–8318.

Grimes, J. M., Burroughs, J. N., Gouet, P., Diprose, J. M., Malby, R., Ziéntara, S., et al. (1998). The atomic structure of the bluetongue virus core. *Nature, 395*, 470–478.

Huiskonen, J. T., de Haas, F., Bubeck, D., Bamford, D. H., Fuller, S. D., & Butcher, S. J. (2006). Structure of the bacteriophage ϕ6 nucleocapsid suggests a mechanism for sequential RNA packaging. *Structure, 14*, 1039–1048.

Kainov, D. E., Butcher, S. J., Bamford, D. H., & Tuma, R. (2003). Conserved intermediates on the assembly pathway of double-stranded RNA bacteriophages. *Journal of Molecular Biology, 328*, 791–804.

Kanematsu, S., Sasaki, A., Onoue, M., Oikawa, Y., & Ito, T. (2010). Extending the fungal host range of a partitivirus and a mycoreovirus from *Rosellinia necatrix* by inoculation of protoplasts with virus particles. *Phytopathology, 100*, 922–930.

Kim, J. W., Choi, E. Y., & Lee, J. I. (2005). Genome organization and expression of the Penicillium stoloniferum virus F. *Virus Genes, 31*, 175–183.

Kim, J. W., Kim, S. Y., & Kim, K. M. (2003). Genome organization and expression of the Penicillium stoloniferum virus S. *Virus Genes, 27*, 249–256.

Lawton, J. A., Estes, M. K., & Prasad, B. V. V. (1997). Three-dimensional visualization of mRNA release from actively transcribing rotavirus particles. *Nature Structural Biology, 4*, 118–121.

Li, L., Tian, Q., Du, Z., Duns, G. J., & Chen, J. (2009). A novel double-stranded RNA virus detected in *Primula malacoides* is a plant-isolated partitivirus closely related to partitivirus infecting fungal species. *Archives of Virology, 154*, 565–572.

Lin, Y. H., Chiba, S., Tani, A., Kondo, H., Sasaki, A., Kanematsu, S., et al. (2012). A novel quadripartite dsRNA virus isolated from a phytopathogenic filamentous fungus, *Rosellinia necatrix*. *Virology, 426*, 42–50.

Liu, H., Fu, Y., Xie, J., Cheng, J., Ghabrial, S. A., Li, G., et al. (2011). Widespread horizontal gene transfer from double-stranded RNA viruses to eukaryotic nuclear genomes. *Journal of Virology, 84*, 11876–11887.

Lu, X., McDonald, S. M., Tortorici, M. A., Tao, Y. J., Vasquez-Del Carpio, R., Nibert, M. L., et al. (2008). Mechanism for coordinated RNA packaging and genome replication by rotavirus polymerase VP1. *Structure, 16*, 1678–1688.

Luque, D., González, J. M., Garriga, D., Ghabrial, S. A., Havens, W. M., Trus, B., et al. (2010). The T=1 capsid protein of Penicillium chrysogenum virus is formed by a repeated helix-rich core indicative of gene duplication. *Journal of Virology, 84*, 7256–7266.

Maniatis, T., Venable, J. H., Jr., & Lerman, L. S. (1974). The structure of psi DNA. *Journal of Molecular Biology, 84*, 37–64.

McClain, B., Settembre, E., Temple, B. R., Bellamy, A. R., & Harrison, S. C. (2010). X-ray crystal structure of the rotavirus inner capsid particle at 3.8 Å resolution. *Journal of Molecular Biology, 397*, 587–599.

Mendez, I. I., Weiner, S. G., She, Y. M., Yeager, M., & Coombs, K. M. (2008). Conformational changes accompany activation of reovirus RNA-dependent RNA transcription. *Journal of Structural Biology, 162*, 277–289.

Naitow, H., Tang, J., Canady, M., Wickner, R. B., & Johnson, J. E. (2002). L-A virus at 3.4 Å resolution reveals particle architecture and mRNA decapping mechanism. *Nature Structural Biology, 9*, 725–728.

Nibert, M. L., Woods, K. M., Upton, S. J., & Ghabrial, S. A. (2009). *Cryspovirus*: A new genus of protozoan viruses in the family *Partitiviridae*. *Archives of Virology, 154*, 1959–1965.

Ochoa, W. F., Havens, W. M., Sinkovits, R. S., Nibert, M. L., Ghabrial, S. A., & Baker, T. S. (2008). Partitivirus structure reveals a 120-subunit, helix-rich capsid with distinctive surface arches formed by quasisymmetric coat-protein dimers. *Structure, 16*, 776–786.

Oh, C. S., & Hillman, B. I. (1995). Genome organization of a partitivirus from the filamentous ascomycete *Atkinsonella hypoxylon*. *Journal of General Virology, 76*, 1461–1470.

Ortín, J., & Parra, F. (2006). Structure and function of RNA replication. *Annual Review of Microbiology, 60*, 305–326.

Pan, J., Dong, L., Lin, L., Ochoa, W. F., Sinkovits, R. S., Havens, W. M., et al. (2009). Atomic structure reveals the unique capsid organization of a dsRNA virus. *Proceedings of the National Academy of Sciences of the United States of America, 106*, 4225–4230.

Pan, J., Vakharia, V. N., & Tao, Y. J. (2007). The structure of a birnavirus polymerase reveals a distinct active site topology. *Proceedings of the National Academy of Sciences of the United States of America, 104*, 7385–7390.

Reinisch, K. M., Nibert, M. L., & Harrison, S. C. (2000). Structure of the reovirus core at 3.6-Å resolution. *Nature, 404*, 960–967.

Sasaki, A., Kanematsu, S., Onoue, M., Oyama, Y., & Yoshida, K. (2006). Infection of *Rosellinia necatrix* with purified viral particles of a member of *Partitiviridae* (RnPV1-W8). *Archives of Virology, 151*, 697–707.

Sen, A., Heymann, J. B., Cheng, N., Qiao, J., Mindich, L., & Steven, A. C. (2008). Initial location of the RNA-dependent RNA polymerase in the bacteriophage ϕ6 procapsid determined by cryo-electron microscopy. *The Journal of Biological Chemistry, 283*, 12227–12231.

Tang, J., Ochoa, W. F., Li, H., Havens, W. M., Nibert, M. L., Ghabrial, S. A., et al. (2010). Structure of Fusarium poae virus 1 shows conserved and variable elements of partitivirus capsids and evolutionary relationships to picobirnavirus. *Journal of Structural Biology, 172*, 363–371.

Tang, J., Ochoa, W. F., Sinkovits, R. S., Poulos, B. T., Ghabrial, S. A., Lightner, D. V., et al. (2008). Infectious myonecrosis virus has a totivirus-like, 120-subunit capsid, but with fiber complexes at the fivefold axes. *Proceedings of the National Academy of Sciences of the United States of America, 105*, 17526–17531.

Tang, J., Pan, J., Havens, W. M., Ochoa, W. F., Guu, T. S., Ghabrial, S. A., et al. (2010). Backbone trace of partitivirus capsid protein from electron cryomicroscopy and homology modeling. *Biophysical Journal, 99*, 685–694.

Tao, Y., Farsetta, D. L., Nibert, M. L., & Harrison, S. C. (2002). RNA synthesis in a cage—Structural studies of reovirus polymerase λ3. *Cell, 111*, 733–745.

Willenborg, J., Menzel, W., Vetten, H. J., & Maiss, E. (2009). Molecular characterization of two alphacryptovirus dsRNAs isolated from *Daucus carota*. *Archives of Virology*, *154*, 541–543.

Yang, C., Ji, G., Liu, H., Zhang, K., Liu, G., Sun, F., et al. (2012). Cryo-EM structure of a transcribing cypovirus. *Proceedings of the National Academy of Sciences of the United States of America*, *109*, 6118–6123.

Yu, X., Jin, L., & Zhou, Z. H. (2008). 3.88 Å structure of cytoplasmic polyhedrosis virus by cryo-electron microscopy. *Nature*, *453*, 415–419.

Zhang, X., Walker, S. B., Chipman, P. R., Nibert, M. L., & Baker, T. S. (2003). Reovirus polymerase λ3 localized by cryo-electron microscopy of virions at a resolution of 7.6 Å. *Nature Structural Biology*, *10*, 1011–1018.

Zhou, Z. H., Baker, M. L., Jiang, W., Dougherty, M., Jakana, J., Dong, G., et al. (2001). Electron cryomicroscopy and bioinformatics suggest protein fold models for rice dwarf virus. *Nature Structural Biology*, *8*, 868–873.

Ziegler, A., Matoušek, J., Steger, G., & Schubert, J. (2012). Complete sequence of a cryptic virus from hemp (*Cannabis sativa*). *Archives of Virology*, *157*, 383–385.

Willenborg, J., Maxwell, [illegible] Vervoort, H. J., [illegible] Maas, R. (20[illegible]). [illegible]

[illegible] (2014). [illegible] *Proceedings of the National Academy of Sciences of the United States of America*, [illegible]

[illegible] A structure of [illegible] polypeptide [illegible]

[illegible]

Zhou, [illegible] Baker, [illegible]

CHAPTER FOUR

Chrysovirus Structure: Repeated Helical Core as Evidence of Gene Duplication

José R. Castón[*,1]**, Daniel Luque**[*]**, Josué Gómez-Blanco**[*] **and Said A. Ghabrial**[†]

[*]Department of Structure of Macromolecules, Centro Nacional Biotecnología/CSIC, Campus de Cantoblanco, Madrid, Spain
[†]Department of Plant Pathology, University of Kentucky, Lexington, Kentucky, USA
[1]Corresponding author: e-mail address: jrcaston@cnb.csic.es

Contents

1. Introduction 88
2. dsRNA Virus Core Proteins 90
3. Penicillium chrysogenum Virus Capsid Structure 92
4. Cryphonectria nitschkei Chrysovirus Capsid Structure 95
5. Comparison of Chrysovirus and Totivirus CPs: Evolutionary Relationships 99
6. dsRNA Organization Within the Capsid 101
7. Conclusions and Prospects 104
Acknowledgments 104
References 105

Abstract

Chrysoviruses are double-stranded RNA viruses with a multipartite genome. Structure of two fungal chrysoviruses, Penicillium chrysogenum virus and Cryphonectria nitschkei chrysovirus 1, has been determined by three-dimensional cryo-electron microscopy analysis and in hydrodynamic studies. The capsids of both viruses are based on a $T=1$ lattice containing 60 subunits, remain structurally undisturbed throughout the viral cycle, and participate in genome metabolism. The capsid protein is formed by a repeated α-helical core, indicative of gene duplication. Whereas the chrysovirus capsid protein has two motifs with the same fold, most dsRNA virus capsid subunits consist of dimers of a single protein with similar folds. The arrangement of the chrysovirus α-helical core is conserved in the totivirus L-A capsid protein, suggesting a shared basic fold. The encapsidated genome is organized in concentric shells; whereas inner dsRNA shells are diffuse, the outermost layer is organized into a dodecahedral cage beneath the protein capsid. This genome ordering could constitute a framework for dsRNA transcription in the capsid interior and/or have a structural role for capsid stability.

Advances in Virus Research, Volume 86
ISSN 0065-3527
http://dx.doi.org/10.1016/B978-0-12-394315-6.00004-0

ABBREVIATIONS

3D three-dimensional
3DR three-dimensional reconstruction
CnCV1 Cryphonectria nitschkei chrysovirus 1
CnV1 Cryphonectria nitschkei virus 1 (abbreviation used in previous studies instead of CnCV1)
CP capsid protein
Cryo-EM cryo-electron microscopy
ds double-stranded
PcV Penicillium chrysogenum virus
RdRp RNA-dependent RNA polymerase
SSE secondary structure elements
T triangulation number

1. INTRODUCTION

Chrysoviruses are nonenveloped isometric virions characterized by a multipartite genome comprising four unrelated linear double-stranded (ds) RNA segments (Ghabrial & Castón, 2011). Penicillium chrysogenum virus (PcV) is the type species of the *Chrysoviridae*, a family of symptomless mycoviruses with a genome consisting of four monocistronic dsRNA segments (with genome sizes ranging from 2.4 to 3.6 kbp). Each segment is encapsidated separately in a similar particle (Jiang & Ghabrial, 2004), as deduced from calculations based on RNA:protein ratio and their molecular weights (Buck & Girvan, 1977; Wood & Bozarth, 1972). dsRNA-1 (3.6 kbp) encodes the RNA-dependent RNA polymerase (RdRp) (1117 amino acid residues with a molecular mass of 128,548 Da; there are 1 or 2 copies per virion), dsRNA-2 (3.2 kbp) encodes the capsid protein (CP) (982 amino acid residues, 108,806 Da), and dsRNA-3 and -4 (3 and 2.9 kbp) code for virion-associated proteins of unknown function (912 amino acid residues, 101,458 Da and 847 amino acid residues, 94,900 Da, respectively).

Chrysoviruses lack an extracellular route; virions accumulate in the cytoplasm and are transmitted intracellularly during cell division, formation of asexual spores, and cytoplasmic fusion following hyphal anastomosis. Based on multiple alignments of the conserved RdRp motifs, chrysoviruses appear to be more closely related to the monosegmented totiviruses than to the bisegmented partitiviruses (Ghabrial & Castón, 2011). The RdRp is incorporated into the capsid interior as a free protein, sharing certain analogies with the victorivirus Helminthosporium victoriae virus 190S virus (HvV190S) particle (the totivirus ScV-L-A incorporates the RdRp as a fusion protein with CP). Both HvV190S and ScV-L-A are members of the family *Totiviridae.*

This chapter centers on the structure of two chrysoviruses, Penicillium chrysogenum virus (PcV) and Cryphonectria nitschkei chrysovirus 1 (CnCV1). To understand the structural and functional implications of the chrysovirus capsid, as well as comparisons between chrysovirus and other dsRNA virus CP, we first provide brief overview of icosahedral capsid structure.

Viral capsids should not only be considered as mere static containers for nucleic acids in a vegetative state but also as dynamic, specialized macromolecular complexes that participate in numerous processes. Their multifunctional abilities are based on the intrinsic structural polymorphism of the CP, which allows them to acquire transient conformations (Dokland, 2000; Steven, Heymann, Cheng, Trus, & Conway, 2005) and to maintain slightly different intersubunit contacts that facilitate acquisition of distinct spatial conformations (Cardone, Purdy, Cheng, Craven, & Steven, 2009; Gertsman et al., 2009). Capsid structure is thus a standard for the analysis of nonequivalent interactions among identical subunits and the extensive use of symmetry (Harrison, 2007).

The protein lattice of icosahedral viruses is described by the traditional concept of quasi-equivalence introduced by Caspar and Klug (1962). Icosahedral capsids are defined by their triangulation number (T). The simplest capsids are formed of 60 identical subunits and assemble into 12 pentamers ($T=1$); those with more than 60 subunits assemble into pentamers and hexamers and cannot have identical, but only "quasi-equivalent" environments ($T>1$). The T number describes the number of different environments occupied by a subunit, and the rules of quasi-equivalence allow only certain values of T (those following the relationship $T=h^2+hk+k^2$, where h and k are nonnegative integers) (Baker, Olson, & Fuller, 1999). In theory, quasi-equivalence involves small differences in subunit interactions and conformations; in practice, viruses show broad variation in this principle, and the literature describes equivalent, quasi-equivalent, and nonequivalent capsids (Baker et al., 1999). A notable exception that bypasses the quasi-equivalence laws is the all-pentamer capsid of papilloma- and papovaviruses ($T=7$), formed by 72 pentamers rather than 12 pentamers and 60 hexamers (360 subunits would correspond to a disallowed "$T=6$") (Liddington et al., 1991; Rayment, Baker, Caspar, & Murakami, 1982).

Another striking exception is the $T=1$ inner core of dsRNA viruses, formed by 12 decamers in which dimers are the asymmetric unit, that is, these capsids would correspond to the forbidden "$T=2$" (Grimes et al., 1998; Naitow, Tang, Canady, Wickner, & Johnson, 2002; Reinisch, Nibert, & Harrison, 2000). A number of studies have firmly established that this architecture is a common feature of dsRNA viruses.

2. dsRNA VIRUS CORE PROTEINS

dsRNA viruses are a widely distributed group of viruses that infect practically all organisms, from the mammalian reoviruses to the bacteriophage ϕ6, and include plant and fungal viruses. The complexity of the capsid ranges from multilayered concentric (Grimes et al., 1998; Lu et al., 1998; Prasad et al., 1996; Shaw, Samal, Subramanian, & Prasad, 1996) to single shell capsid (Castón et al., 1997; Hill et al., 1999).

dsRNA viruses share numerous general architectural and functional principles that reveal common life cycle strategies. The genome of most dsRNA viruses is contained within a specialized icosahedrally symmetric $T=1$ capsid (referred to above as "$T=2$" capsids). These $T=1$ inner cores are functional complexes, as they transcribe the dsRNA and the ssRNA transcript is released for translation in the host cytoplasm. The inner core also takes part in dsRNA replication (minus strand synthesis), since the viral RdRp is frequently packaged as an integral component of the capsid. RdRp is incorporated as a replicative complex at the pentameric vertex (as in rotavirus capsids; Li, Baker, Jiang, Estes, & Prasad, 2009; Trask, McDonald, & Patton, 2012), as a fusion protein with the CP (as in the totivirus L-A; Castón et al., 1997), or as a separate, nonfused protein (as in the victorivirus Hv190SV; Castón et al., 2006; Ghabrial & Nibert, 2009). The structural integrity of the inner cores remains undisturbed after virus entry in the infected cell (Harrison, 2007), protecting the dsRNA from host defense mechanisms, and therefore constitutes a dsRNA-sequestering mechanism.

The first unambiguous $T=1$ layers were reported in association with the L-A and P4 totiviruses (Cheng et al., 1994), which infect the yeast *Saccharomyces cerevisiae* and the smut fungus *Ustilago maydis*, respectively. This capsid architecture, consisting of 60 asymmetric dimers of a single protein, has since been found as a ubiquitous feature in a broad spectrum of dsRNA viruses (Fig. 4.1). These capsids are described in members of (i) the family *Reoviridae* (Cheng et al., 2010; Lawton, Estes, & Prasad, 1997; McClain, Settembre, Temple, Bellamy, & Harrison, 2010; Yu, Jin, & Zhou, 2008;

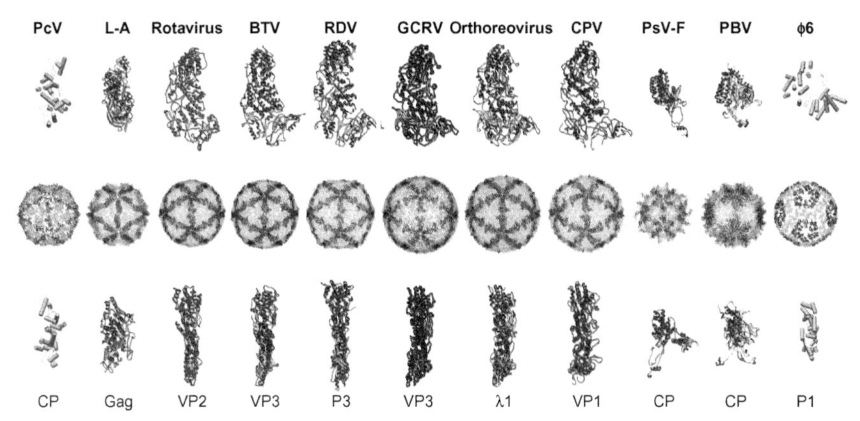

Figure 4.1 $T=1$ capsid protein X-ray and cryo-EM-based structures. $T=1$ capsids of *Penicillium chrysogenum* virus, L-A virus, rotavirus, bluetongue virus (BTV), rice dwarf virus (RDV), grass carp reovirus (GCRV), orthoreovirus, cytoplasmic polyhedrosis virus (CPV), Penicillium stoloniferum virus F (PsV-F), picobirnavirus (PBV), and ϕ6 phage viewed along a twofold axis of icosahedral symmetry (center row). PcV CP half-protein A, Gag (1m1c; 680 residues), VP2 (3kz4; 880 residues), VP3 (2btv; 901 residues), P3 (1uf2; 1019 residues), VP3 (3k1q; 1027 residues), λ1 (1ej6; 1275 residues), VP1 (3cnf; 1333 residues), PsV-F CP (3es5; 420 residues), PBV CP (2vf1; 590 residues), and P1 (emd-1206; 769 residues) shown from a top view (first row). Side views of the same structures are shown (bottom row; the $T=1$ shell exterior is at right). α-helices are represented by blue cylinders and β-sheets by yellow planks for cryo-EM-based structures. *Reproduced/adapted with permission from American Society for Microbiology [from* Journal of Virology, *84, 7256–7266, 2010, http://dx.doi.org/10.1128/JVI.00432-10].* (For interpretation of the references to color in this figure legend, the reader is referred to the online version of this chapter.)

Zhang, Walker, Chipman, Nibert, & Baker, 2003), which infect mostly higher eukaryotic systems; (ii) the family *Picobirnaviridae* (Duquerroy et al., 2009), a recently established taxonomic family whose members infect humans and other vertebrates; (iii) the family *Cystoviridae*, bacteriophages that infect the prokaryote *Pseudomonas syringae* (Huiskonen et al., 2006; Jaalinoja, Huiskonen, & Butcher, 2007); and (iv) the families *Totiviridae* (Castón et al., 2006, 1997; Tang et al., 2008) and *Partitiviridae* (Ochoa et al., 2008), whose members infect mainly unicellular and simple eukaryotes (fungi and protozoa). This stoichiometry and associated molecular architecture provide an optimal framework for the many activities related to RNA synthesis (Lawton, Estes, & Prasad, 2000; Patton & Spencer, 2000); the capsids organize the genome and replicative complex they contain and act as microcellular compartments or molecular sieves that isolate dsRNA molecules or replicative intermediates from host defense molecules.

At present, nine 120-subunit $T=1$ capsid layers have been resolved at atomic resolution (Fig. 4.1); these are the VP3 cores of the orbivirus bluetongue virus (BTV; Grimes et al., 1998) and the aquareovirus grass carp reovirus (GCRV; Cheng et al., 2010), λ1 cores of the reovirus in the genus *Orthoreovirus* (Reinisch et al., 2000), P3 of the rice dwarf virus (RDV; Nakagawa et al., 2003), VP1 of the cytoplasmic polyhedrosis virus (CPV; Yu et al., 2008), the VP2 inner capsid of rotavirus (McClain et al., 2010), the CP of *Penicillium stoloniferum* virus F (PsV-F, of the *Partitiviridae*; Pan et al., 2009), the CP of picobirnavirus (PBV; Duquerroy et al., 2009), and Gag of the yeast L-A totivirus (Naitow et al., 2002). Sequence similarities among these viruses are negligible and comparative analyses are inconclusive. All nine proteins are nevertheless predominantly α-helical. VP3, λ1, P3, VP1, and VP2 have a similar fold, PsV-F and PBV CP are also based on a similar fold, and Gag has a distinct tertiary structure bearing only a faint resemblance to the reovirus proteins (Reinisch, 2002).

Members of the *Birnaviridae* family are exceptions to the rule that all dsRNA viruses have a $T=1$ core formed by 60 CP dimers. Birnaviruses have a single $T=13$ shell (Coulibaly et al., 2005; Saugar et al., 2005; Vancini et al., 2012) surrounding a polyploid dsRNA genome (Luque, Rivas, et al., 2009) organized as a ribonucleoprotein complex (Luque, Saugar, et al., 2009).

So far, the three-dimensional (3D) structures of two chrysovirus capsids have been determined at subnanometer resolution by cryo-electron microscopy (cryo-EM) analysis, those of Penicillium chrysogenum virus and Cryphonectria nitschkei chrysovirus. PcV and CnCV1 capsids constitute partial exceptions to the most extended tendency among dsRNA viruses, the $T=1$ core with 60 equivalent dimers, as they have an authentic $T=1$ capsid formed by 60 copies of a single monomer.

3. PENICILLIUM CHRYSOGENUM VIRUS CAPSID STRUCTURE

Cryo-EM, combined with three-dimensional reconstruction (3DR) techniques and biophysical analysis, is used to establish the 3D structure and protein stoichiometry of the PcV capsid (Castón et al., 2003; Luque et al., 2010). These studies generally require concentrated homogeneous populations of viral particles, whose structural integrity must be preserved (e.g., salt is needed in buffers to purify full intact particles, which will otherwise be empty or broken).

3D Cryo-EM reconstruction of PcV has been calculated at 8-Å resolution (Luque et al., 2010; Fig. 4.2). The capsid diameter is 400 Å and the protein shell is 48 Å thick. The most prominent features of the $T=1$ capsid are 12 outwardly protruding pentons that correspond to the pentameric

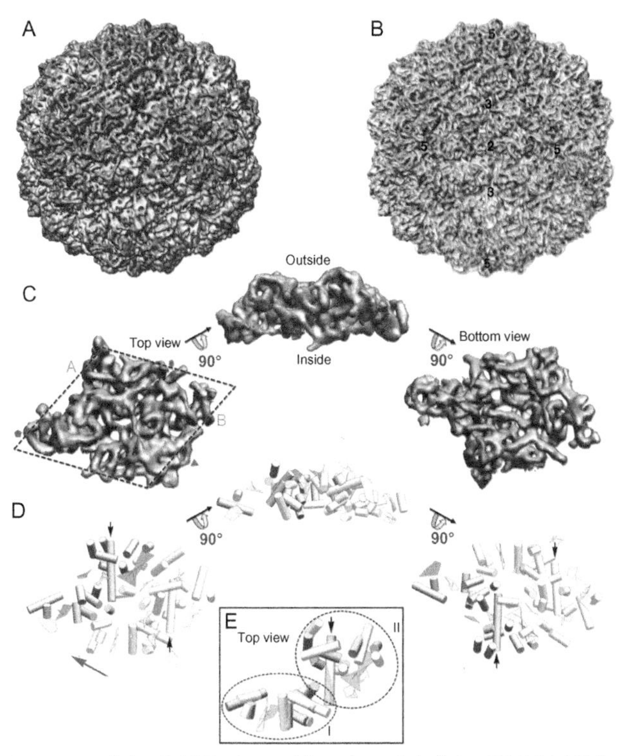

Figure 4.2 Structure of the PcV $T=1$ capsid and model of the CP fold. (A) Radially color-coded outer surface of the full capsid, viewed along an icosahedral twofold axis. (B) Surface-shaded virion capsid with CP elements in cyan and yellow; boundaries for two CP are outlined in red. Icosahedral symmetry axes are numbered. (C) Segmented asymmetric unit. The dashed line highlights the rhomboidal shape. Protein halves A (cyan) and B (yellow) are indicated. Red symbols indicate icosahedral symmetry axes. (D) PcV CP secondary structural elements (SSE), using color scheme and orientations as in (C): α-helices (cylinders) and β-sheets (planks). Red arrow indicates translation direction to superimpose half-protein B on A. Black arrows indicate the ~37-Å long α-helices of both PcV CP elements. (E) Superimposed views of conserved SSE in protein halves A and B; the relative spatial locations of 13 α-helices and two planar regions are very close. PcV protein elements are subdivided into domains I and II. *Reproduced/adapted with permission from American Society for Microbiology [from* Journal of Virology, 84, *7256–7266, 2010, http://dx.doi.org/10.1128/JVI.00432-10].* (See Page 5 in Color Section at the back of the book.)

positions of the icosahedral lattice (Fig. 4.2A). Structural asymmetric unit boundaries are established based on its compactness and its contacts with neighboring densities (Fig. 4.2B, red line). Numerous rod- and sheet-like densities (putative α-helices and β-sheets, respectively) are found on the relatively uneven outer surface, similar to the L-A virus. This contrasts with the smooth outer surface of reovirus particles, in which the CP has a plate-like structure.

CP monomers are formed by two ellipsoid-like structures with roughly similar morphology, although they differ slightly in size (Fig. 4.2C). The asymmetric unit is a rhomboidal prism, ~46 Å high and ~130 Å long. The smaller part makes up 43% of the total unit volume (blue; half-protein A) and the larger one represents 57% (yellow; half-protein B). These elements are arranged in two sets of five; five half-protein A structures surround the icosahedral fivefold axis, and five half-protein B structures are intercalated between them, forming a pseudo-decamer (Fig. 4.2B). This quaternary organization is similar to the 120-subunit $T=1$ shell in reo- and totiviruses (Fig. 4.1).

Analytical ultracentrifugation analysis of purified PcV virions allowed an independent estimation of copy number for the major CP (Castón et al., 2003). The predominant species (~80%) sediment with a standard sedimentation coefficient ($s_{20,w}$) of 135 ± 10S. This value is compatible with that calculated for a $T=1$ particle (60 monomers) with a frictional coefficient ratio of 1.30 ± 0.10.

PcV CP has a high α-helical content that includes 16 α-helices (and 5 β-sheets) in the half-protein A structure and 24 α-helices (and 7 β-sheets) in the half-protein B structure (Fig. 4.2D). Sequence-based secondary structure element (SSE) predictions of the CP correlate with the structural subunit, which also has a high α-helical content (Fig. 4.4). An unstructured central region that divides the PcV CP into two parts is located between amino acids ~550 and ~650, which might be the link between the two unequal halves of this protein.

Additional sequence analysis of the PcV CP suggested that it has two roughly similar halves, with conserved segments in both halves of the protein after the introduction of numerous gaps, although the importance of this observation was not initially evident (Castón et al., 2003). Further study showed that both PcV CP elements have a similar ~37-Å long α-helix that lies tangential to the capsid surface (Fig. 4.2D, black arrows). Careful inspection showed that both protein halves, although not identical, substantially resembled one another. Superimposition of one of the halves (treated as a rigid body) on the other by a ~45-Å translation (Fig. 4.2D; red arrow indicates translation direction) showed that the relative spatial locations of 13 α-helices and two planar regions were close and required only minor local adjustment (Fig. 4.2E). SSE preserved between both PcV CP halves consist of two domains: domain I has a characteristic bundle of four α-helices, located near the fivefold axis, and domain II has the longest α-helix, as well as other elements. Whereas the conserved core establishes lateral contacts with the neighboring units, the interface between halves A and B is less similar, with a larger number of variations (Luque et al., 2010).

Despite their lack of sequence similarity, the two PcV CP halves have a comparable structural signature, suggesting ancestral gene duplication. This peculiar building unit resembles the normal 120-subunit capsid monomer, and the PcV capsid might thus be considered a "pseudo-$T=2$" structure. Joined folds were first observed in the adenovirus trimeric capsomer (Roberts, White, Grutter, & Burnett, 1986); each monomer consists of two successive jelly-roll motifs, producing a pseudo-hexameric structure. Many other large dsDNA viruses have similar trimeric capsomers, including the Paramecium bursaria chlorella virus 1 (PBCV1; Nandhagopal et al., 2002) and bacteriophage PRD1 (Abrescia et al., 2004). The large subunit of comoviruses, ssRNA viruses that infect plants, is also made by fusion of 2 β-barrel domains (Lomonossoff & Johnson, 1991). In all these examples, there is a lack of sequence similarity between the joined motifs, similar to the PcV CP halves.

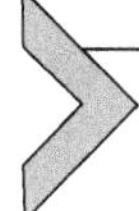

4. CRYPHONECTRIA NITSCHKEI CHRYSOVIRUS CAPSID STRUCTURE

The ~8-Å cryo-EM map of CnCV1 shows that the apparently unusual fold signature of PcV is a feature shared by chrysoviruses (Gómez-Blanco et al., 2012). As for the PcV capsid, the most prominent features of the 400-Å diameter $T=1$ capsid are 12 outwardly protruding pentons (Fig. 4.3A and C). Sedimentation velocity behavior of CnCV1 particles was also analyzed to determine their stoichiometry; the main species has a $s_{20,w}$ of 135S, corresponding to a true $T=1$ particle with a frictional coefficient ratio of 1.3 ± 0.05, similar to PcV (Gómez-Blanco et al., 2012). Similarities between the uneven outer surfaces of CnCV1 and PcV capsids are also clear (Fig. 4.3A and B). As with PcV, the asymmetric unit is formed by two similar ellipsoid-like structures (48% and 52% of total unit volume) (Fig. 4.3D; half-protein A, purple and half-protein B, green) arranged in two

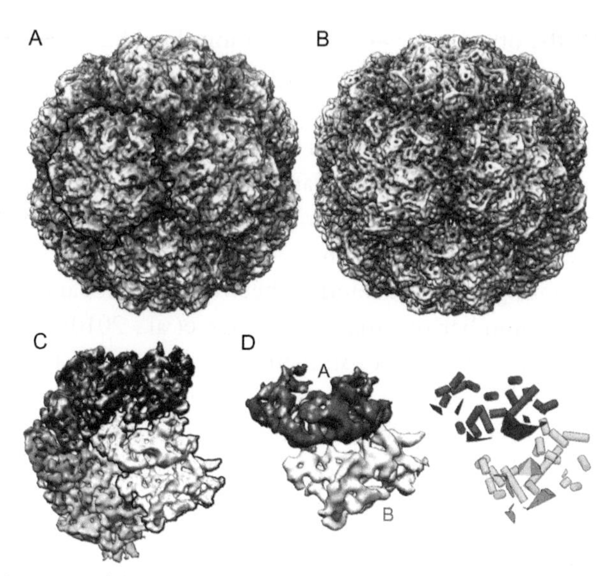

Figure 4.3 Structure of the CnCV1 $T=1$ capsid and model of the CP fold. (A and B) Radially color-coded outer surfaces of full capsids of (A) CnCV1 and (B) PcV, viewed along an icosahedral twofold axis. The boundary for a CnV1 pentamer is outlined in black. (C) Extracted pentamer from the CnCV1 capsid; the five CP structural subunits in a pentamer are indicated with colors. (D) Segmented CnCV1 CP monomer viewed from outside. (Left) Protein halves A (purple) and B (green) are indicated. (Right) CnCV1 asymmetric unit with putative α-helices (cylinders) and β-sheets (planks). (See Page 5 in Color Section at the back of the book.)

sets of five (Fig. 4.3C). Plane sequence similarity analysis of the CnCV1 CP showed no evidence of gene duplication, although the average identity between CP sequences of PcV (982 amino acids) versus CnCV1 (906 residues) was 25% (average similarity 40%) (Fig. 4.4). Multiple sequence alignment of CP amino acid sequences for several chrysoviruses (including CnCV1, PcV, Helminthosporium victoriae virus 145S, Amasya cherry disease-associated chrysovirus, Aspergillus fumigatus chrysovirus, Anthurium mosaic-associated virus, and Verticillium dahliae chrysovirus 1) indicated better alignment of the N-terminal than that of the C-terminal halves. PcV and CnCV1 SSE predictions of CP indicate a high α-helical content and align well, consistent with the cryo-EM maps. An unstructured central region also divides the CnCV CP into two parts, reflecting a structurally disordered region (amino acids ~550–610).

Structural comparison of CP secondary and tertiary structure is currently used as a criterion to establish relatedness, in the absence of sequence

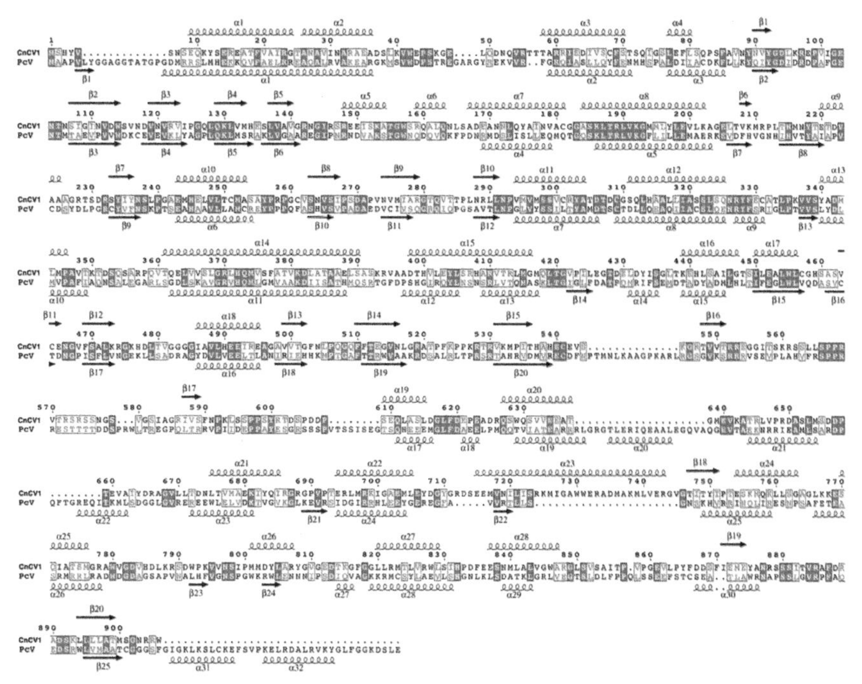

Figure 4.4 Sequence alignment and secondary structure consensus prediction of CnCV1 and PcV CP amino acid sequences. Several SSE prediction methods were used to test correlation with our structural subunit models. A consensus SSE prediction was obtained by simple majority at each sequence position. Identical residues, white on red background; partially conserved residues, red. Arrows represent β-chains and spirals, α-helices. *Reproduced with permission from American Society for Microbiology from* Journal of Virology, 86, *8314–8318, 2012, http://dx.doi.org/10.1128/JVI.00802-12].* (For interpretation of the references to color in this figure legend, the reader is referred to the online version of this chapter.)

similarity (Baker, Jiang, Rixon, & Chiu, 2005; Bamford, Grimes, & Stuart, 2005). The CnCV1 CP has 38 rod-like densities (putative α-helices), in addition to 12 planar regions (putative β-sheets); these SSE are also similar to PcV CP (Fig. 4.5A). Both CnCV1 CP halves also have a long α-helix tangential to the capsid surface, which is used as a reference for SSE comparisons (Fig. 4.5A, arrows). Superimposition of CnCV1 and PcV CP showed that the relative spatial locations of 24 α-helices and 7 β-sheets were close, indicating a similar structural signature of the asymmetric units. Comparable analysis of CnCV1 and PcV CP halves A and B (treated as rigid bodies) showed substantial resemblance, as 9 (CnCV1) and 13 α-helices (PcV) and 2 β-sheets were superimposed. These SSE constitute the CnCV1 and PcV conserved cores (Fig. 4.5A, bottom row). The result of these

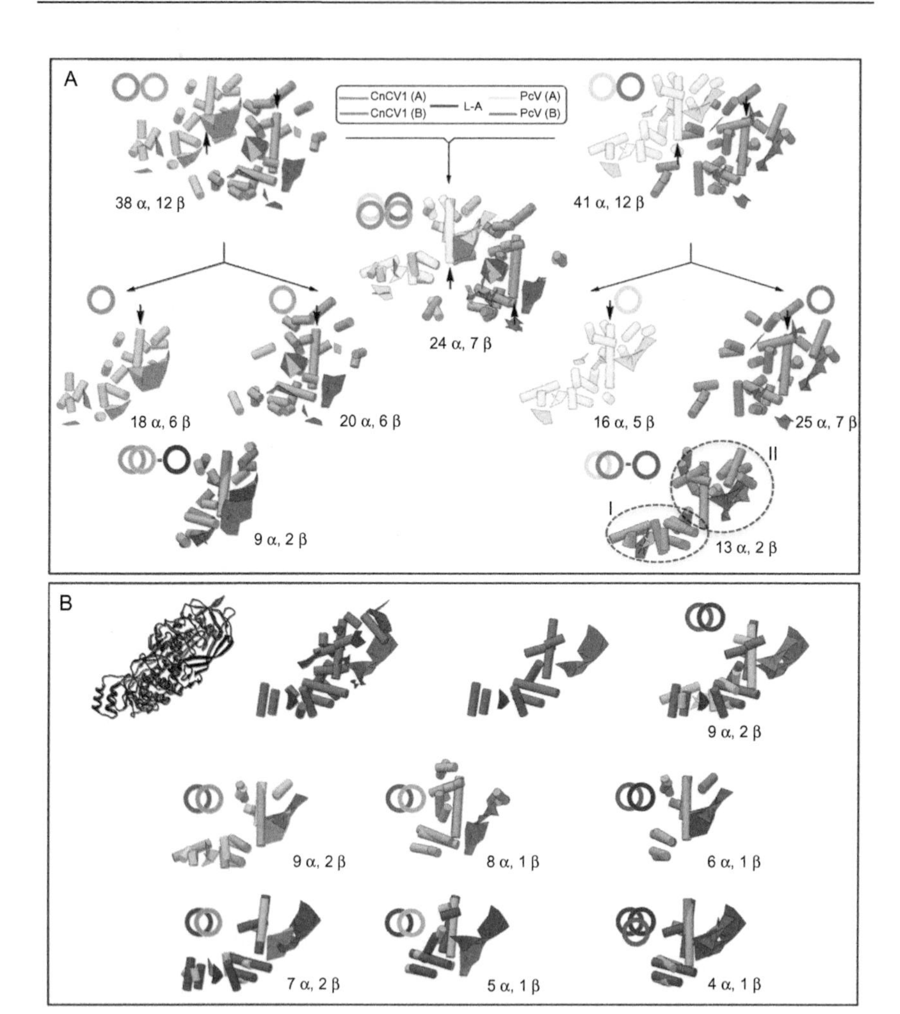

Figure 4.5 Structural comparison of CnCV, PcV, and L-A capsid proteins. (A) Superimposition (center) of CnCV1 CP SSE (half-protein A, purple and half-protein B, green) (left) on PcV CP SSE (half-protein A, blue and half-protein B, orange) (right). Total SSEs with close relative spatial locations are indicated (24 putative α-helices and 7 putative β-sheets). Black arrows indicate the ~25-Å (purple) and 35-Å (green) rod-like densities of both CnCV1 CP halves, and the two ~37-Å α-helices (blue and orange) of PcV CP elements. CnV1 and PcV CP are divided into their protein halves A and B (total SSE indicated, bottom). Circles and superimposed circles for each model (top) indicate structural superimpositions. Superimposed views of conserved SSE in CnCV1 protein halves A and B (conserved SSE, dark purple circle) (bottom row, left). Superimposition of conserved SSE in PcV protein halves A and B (conserved SSE, dark blue circle). The PcV conserved core is subdivided into domains I and II (bottom row, right). (B) L-A Gag structure. The polypeptide chain path (top view) is shown as a gradient from blue (N-terminus) to red (C-terminus) (top row, left); cylinder-and-plank representation of Gag (red), as for chrysovirus CP, with 17 α-helices and 10 β-sheets (center left); and

comparisons is consistent with the accepted idea that the 120-subunit $T=1$ layer architecture is conserved because common requirements apply for managing dsRNA transcription and replication.

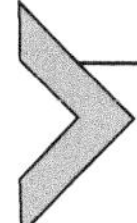

5. COMPARISON OF CHRYSOVIRUS AND TOTIVIRUS CPs: EVOLUTIONARY RELATIONSHIPS

With the exception of the ~430-amino acid CP of the partitivirus PsV-F, the PcV and CnCV1 CP structural halves are the smallest polypeptides of the dsRNA virus $T=1$ proteins (Gag is almost twice as large, and VP3, P3, CPS, and λ1 are almost four times as large as the PcV CP C-terminal half). These halves might represent a fold that evolved from the $T=1$ CP ancestral fold to the highly varied structures observed today. If this hypothesis is correct, structural "remnants" of the conserved chrysovirus motif might be found in the modern $T=1$ CP.

Although the fungal totivirus L-A has a genome of a single segment (Wickner, 1996) and the chrysovirus genome consists of four monocistronic segments, the PcV and L-A virus CP can be compared for several structural and functional reasons. These include (i) PcV and L-A viruses both infect simple eukaryotes and lack an extracellular phase in their life cycles, (ii) they share not only an uneven outer surface but also a roughly similar side view (Fig. 4.1), and (iii) PcV and L-A CPs have no outer shell, whereas *Reoviridae* $T=1$ cores act as a template to prime the assembly of $T=13$ surrounding capsid. Overlaying L-A Gag (Fig. 4.5B, top row) on either of the PcV halves maintains the same spatial arrangement in the shell and shows many SSE with similar spatial distributions, including 9 α-helices (of a total of 13) and 2 β-sheets. These structural parallels are found in the Gag N-terminus

selected Gag SSE (9 α-helices and 2 β-sheets) (center right); superimposed views of helical and planar regions of selected L-A Gag SSE and conserved PcV SSE (dark blue) (right). Superimposition of the PcV conserved core (dark blue) with CnCV half-protein A (left), CnCV1 half-protein B (center), or CnCV1 conserved core (dark purple) (middle row). Superimposition of the L-A Gag conserved core (red) and CnCV1 half-protein A (left), and CnCV1 half-protein B (center). Motif conserved in CnCV1 and PcV halves with L-A Gag (right) (bottom row). These structural matchings preserve the spatial orientation of CnCV1, PcV, and L-A capsid protein structural units within their capsids. *Reproduced/adapted with permission from American Society for Microbiology [from* Journal of Virology, 84, *7256–7266, 2010, http://dx.doi.org/10.1128/JVI.00432-10 and from* Journal of Virology, 86, *8314–8318, 2012, http://dx.doi.org/10.1128/JVI.00802-12].* (See Page 6 in Color Section at the back of the book.)

(residues 1–414), whereas the entire C-terminal region (residues 415–680) would be a terminal insertion to this conserved fold. This structural matching, in which CP orientations are maintained (i.e., the outer and inner surfaces coincide), indicates a shared motif for these two fungal virus $T=1$ CP. Comparison of the PcV conserved core with each CnCV1 half showed that the longest α-helix and other SSE (located mainly in domain II) are shared; furthermore, CnCV1 and PcV conserved cores are matched for six rods and one large curved surface (Fig. 4.5B, center row). CnCV1 and PcV CP halves have a similar structural signature and might thus derive from a gene duplication event. If that is the case, CnCV1 CP half-protein B is less structurally conserved and has diverged more rapidly than half-protein A.

Overlaying the L-A Gag structure on either CnCV1 halves showed many SSE with similar spatial distribution, including 7 or 5 α-helices (of 9) and 2 or 1 β-sheets (Fig. 4.5B, bottom row). Intersection among L-A Gag, CnCV1, and PcV conserved cores indicates a shared motif for these three fungal viruses. Systematic structural comparison of toti- and chrysovirus CP implies the gradual loss of nonshared SSE and might indicate the hallmark fold of dsRNA viruses.

Structural comparison of CP has led to detection of relationships among viruses that infect organisms widely separated in evolution (Bamford et al., 2005; Benson, Bamford, Bamford, & Burnett, 2004; Krupovic & Bamford, 2008; Riffel et al., 2002). Icosahedral viruses are currently grouped in four separate lineages, the PRD1-, HK97-, and BTV-like viruses, and the picornavirus-like family (Abrescia, Bamford, Grimes, & Stuart, 2012; Bamford et al., 2005). Although many viruses are not included in these four lineages, the number of folds that satisfy the assembly constraints for a viable viral shell is nonetheless thought to be limited. Whereas *Reoviridae* $T=1$ CP share a recognizably similar fold, supporting a possible common ancestor, those of picobirna-, partiti- and totiviruses, as well as chrysoviruses, have distinct structures, and matching of any domain using traditional methods appears unlikely. The L-A virus fold is nonetheless grouped within the BTV-like lineage, although at the furthest distance in the structure-based phylogenetic tree. Using a tool such as the Structure Homology Program (Bamford et al., 2005; Benson et al., 2004; Krupovic & Bamford, 2008; Riffel et al., 2002), positioning of numerous SSE is similar between L-A and BTV, allowing L-A to be sensibly placed in the structure-based phylogeny of the BTV-like lineage.

The similarity of the conserved fold of chrysovirus and L-A CP could improve understanding of the ancestral fold and reveal structural evolutionary relationships of the dsRNA virus lineage. Conservation of the chrysovirus fold throughout evolution might be due to its intracellular transmission in simple eukaryotes, which probably prevents the high mutation rates of extracellularly transmitted viruses in complex eukaryotes. A succession of divergent evolutionary events in the ancestral chrysovirus fold, by insertions in preferential sites of the preserved α-helical core, might have led to the more complex structures observed today. Partiti- and picobirnavirus CP share an overall structural resemblance that differs from other dsRNA viruses and could be an example of convergent evolution to a similar $T=1$ capsid with the same functional requirements, but distinct quaternary structure.

6. dsRNA ORGANIZATION WITHIN THE CAPSID

Fungal dsRNA viruses have spacious capsids in comparison to the replicative cores of complex eukaryotic dsRNA viruses (Table 4.1). dsRNA in the interior of reovirus cores is very compact, with an average genome density of ~40 bp/100 nm^3 and a spacing among dsRNA strands of 25–30 Å (e.g., the core of BTV and cypoviruses) (Dryden et al., 1993; Gouet et al., 1999; Pesavento, Lawton, Estes, & Venkataram Prasad, 2001; Shaw et al., 1996). L-A virus dsRNA is loosely packed (genome density of ~20 bp/100 nm^3) with 40–45 Å spacing between filaments (Castón et al., 1997). In PcV and CnCV1, considering the capsid diameter based on 3DR (and thus, the volume available in the capsid interior) and average genome size of ~3200 bp (each segment separately encapsidated in a similar particle), the packed dsRNA would have an interstrand spacing of ~40 Å, which closely matches the relatively low density of L-A virus. The looser packing of the dsRNA would probably improve template motion in the more spacious transcriptional and replicative active particle. This greater spacing would apply to fungal and related viruses that only package a single-genomic dsRNA segment per particle and have 1–2 copies of the RdRp complex. Members of the *Reoviridae* family, whose genome consists of 10–12 segments of dsRNA packaged within the same particle, have 12 RdRp complexes around which the RNA is densely coiled.

The organization of packaged chrysovirus RNA has been analyzed in detail from 3D cryo-EM data. Although PcV and CnCV1 capsids are spacious, their dsRNA shows internal order. There are numerous interactions between the inner surface of the protein shell and the underlying genome-associated density in the chrysovirus full capsids (Fig. 4.6A); as a result, there is almost no space between the two layers (Luque et al., 2010). There are at

Table 4.1 Genome packaging densities in dsRNA viruses[a]

	dsRNA features				Inner capsid features			
Virus family	**No. of segments**	**Size[b] (kbp)**	**MW[c] (MDa)**	**Packed dsRNA spacing (nm)[d]**	***T***	**f^e (nm)**	**$_ir^f$ (nm)**	**dsRNA density (bp/100 nm^3)[g]**
Herpes simplex virus	1	~152	103.7	2.6	–	~60	43.0	46
Reoviridae								
Orthoreovirus	10	~23.5	16.0	2.6	2	~60	24.5	38
Rotavirus	11	~18.5	12.6	2.5–3.0	2	~52	23.5	34
Orbivirus	10	~19.2	13.1	3.0	2	~52	22.0	43
Aquareovirus	11	~23.6	16.0		2	~60	23.0	46
Phytoreovirus	12	~25.7	17.5		2	~57	26.0	35
Cypovirus	10	~31.4	21.4	2.5	2	~58	24.0	54
Cystoviridae, phage f6	3	~13.4	9.1		2	~50	20.0	40
Totiviridae, L-A	1	~4.6	3.1	3.6–4.0	2	~43	17.0	22
Partitiviridae, PsV-S	1 (2)	~1.7 (3.3)	1.2 (2.2)		2	~35	12.0	23
Chrysoviridae, PcV	1 (4)[h]	~3.2 (12.6)	2.2 (8.6)	4.0*	1	~40	16.0	19

[a]Data reviewed in Castón et al. (2003).
[b]Data from *Virus taxonomy: classification and nomenclature of viruses*: Ninth report of the International Committee on Taxonomy of Viruses (2012).
[c]MW were calculated assuming a mass of 682 Da/bp.
[d]Measurements were obtained mainly from cryo-EM and 3DR studies and represent average Bragg spacings. *Denotes estimated values. Interstrand spacings are calculated by introducing the factor (2/√3), if strands exhibit quasi-hexagonal packing.
[e]Outer diameter.
[f]Inner radius.
[g]Densities when volume of a perfect sphere is assumed and any other internal components are ignored.
[h]PcV dsRNA features: the genome is formed by four dsRNA molecules, but a mean value was calculated for each column as there is one dsRNA molecule/particle.

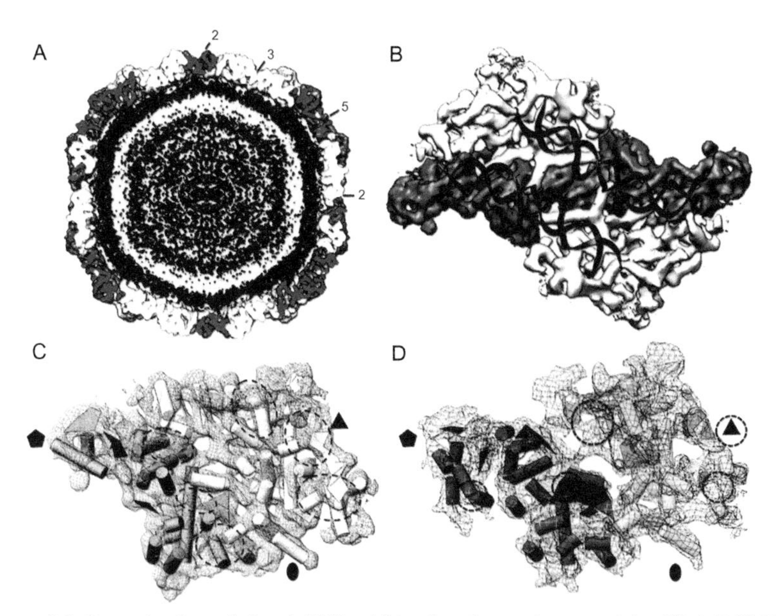

Figure 4.6 Organization of the dsRNA within the chrysovirus particle. (A) A PcV 50-Å-thick slab. Capsid shell coloring is the same as in Fig. 4.2B, contoured at a lower contour threshold to highlight the locations of the dsRNA densities. dsRNA (green) is seen as approximately four concentric layers. There are numerous contacts between the inner surface of the capsid and the outer surface of the nearest dsRNA layer. (B) Close-up view down a twofold axis from inside, showing two adjacent structural subunits. The phosphate backbone is traced as a red ribbon for the two dsRNA A-form strands. (C) PcV capsid asymmetric unit (shown as wire frames, viewed from the inside) with the SSE as in Fig. 4.2B (cyan and yellow for half-proteins A and B, respectively). dsRNA interacts with six-defined areas of the capsid inner surface (dashed ovals). (D) Segmented CnCV1 CP monomer. Protein halves A (purple) and B (green) are indicated. The CnCV1 asymmetric unit viewed from inside is shown as wire frames, with a-helices (cylinders) and b-sheets (planks). dsRNA interacts with five defined areas of the capsid inner surface (circles). Black symbols indicate icosahedral symmetry axes. *Reproduced/adapted with permission from American Society for Microbiology [from* Journal of Virology, 84, *7256–7266, 2010, http://dx.doi.org/10.1128/JVI.00432-10 and* Journal of Virology, 86, *8314–8318, 2012, http://dx.doi.org/10.1128/JVI.00802-12].* (See Page 7 in Color Section at the back of the book.)

least 360 (for PcV) and 260 (for CnCV1) interacting areas in all (Fig. 4.6C and D). In PcV, the interactions of each CP half with the RNA show overall similarities (e.g., three interacting areas in each half); this is consistent with the hypothetical structural duplication, although some differences reflect structural changes in the protein halves. Due to the numerous icosahedral positions that interact with the genome, the outermost RNA layer conforms an icosahedral cage. The bulk of genome density appears diffuse toward the center of the core, as two or three concentric layers; this density weakening suggests greater RNA disorder at greater distance from the capsid shell (Fig. 4.6A).

RNA/protein contacts appear to form tracks on the inside of the CP shell, resembling those described in the X-ray structure of BTV (Gouet et al., 1999). In PcV, this layer appears to be made of tube-like densities with dimensions of dsRNA helices (A-form helices). These dsRNA strands reinforce the connectivity between two monomers from adjacent pentamers (Fig. 4.6B) and might act as a scaffold that stabilizes the capsid. Alternatively, this internal order might be necessary for RNA synthesis (Lawton et al., 2000). dsRNA organization into a dodecahedral shell beneath the protein capsid has also been described for rotaviruses (Lawton et al., 1997; Prasad et al., 1996).

7. CONCLUSIONS AND PROSPECTS

3D Cryo-EM structures of the PcV and CnCV1 capsids at 8-Å resolution indicate that the chrysovirus capsid protein (60 copies/particle) is formed by a repeated helical core, indicative of gene duplication. The arrangement of many of these putative α-helices is conserved in the totivirus L-A capsid protein, suggesting a shared motif. This basic repeated domain could provide insight into the structural evolutionary relationships among dsRNA viruses. The 120-subunit $T=1$ capsid appears to be a conserved architecture that optimizes dsRNA replication and organization.

Although there are many functional and structural studies of members of the family *Reoviridae*, it is clear that deeper characterization of many fungal (and protozoan) dsRNA viruses is needed to compare the function and evolution of these viruses. Recent studies of fungal partitiviruses identified the first $T=1$ capsid formed by a quasi-symmetric CP dimers (Ochoa et al., 2008; Pan et al., 2009). Another notable example is infectious myonecrosis virus (IMNV), a shrimp pathogen with a totivirus-like 120-subunit capsid. The IMNV capsid surface has fibers at the fivefold axis, which are probably involved in cell entry, as this virus is transmitted extracellularly (Tang et al., 2008). Additional studies are needed to solve the atomic structure of the chrysovirus CP, which will allow us to determine whether their conserved core is a hallmark fold preserved in dsRNA viruses.

ACKNOWLEDGMENTS

We thank C. Mark for editorial assistance. This work was supported by grants from the Spanish Ministry of Science and Innovation (BFU2011-25902 to J. R. C.) and the Kentucky Science & Engineering Foundation (to S. A. G.).

REFERENCES

Abrescia, N. G., Bamford, D. H., Grimes, J. M., & Stuart, D. I. (2012). Structure unifies the viral universe. *Annual Review of Biochemistry*, *81*, 795–822.

Abrescia, N. G., Cockburn, J. J., Grimes, J. M., Sutton, G. C., Diprose, J. M., Butcher, S. J., et al. (2004). Insights into assembly from structural analysis of bacteriophage PRD1. *Nature*, *432*, 68–74.

Baker, M. L., Jiang, W., Rixon, F. J., & Chiu, W. (2005). Common ancestry of herpesviruses and tailed DNA bacteriophages. *Journal of Virology*, *79*, 14967–14970.

Baker, T. S., Olson, N. H., & Fuller, S. D. (1999). Adding the third dimension to virus life cycles: Three-dimensional reconstruction of icosahedral viruses from cryo-electron micrographs. *Microbiology and Molecular Biology Reviews*, *63*, 862–922.

Bamford, D. H., Grimes, J. M., & Stuart, D. I. (2005). What does structure tell us about virus evolution? *Current Opinion in Structural Biology*, *15*, 655–663.

Benson, S. D., Bamford, J. K., Bamford, D. H., & Burnett, R. M. (2004). Does common architecture reveal a viral lineage spanning all three domains of life? *Molecular Cell*, *16*, 673–685.

Buck, K. W., & Girvan, R. F. (1977). Comparison of the biophysical and biochemical properties of Penicillium cyaneo-fulvum virus and Penicillium chrysogenum virus. *The Journal of General Virology*, *34*, 145–154.

Cardone, G., Purdy, J. G., Cheng, N., Craven, R. C., & Steven, A. C. (2009). Visualization of a missing link in retrovirus capsid assembly. *Nature*, *457*, 694–698.

Caspar, D. L. D., & Klug, A. (1962). Physical principles in the construction of regular viruses. *Cold Spring Harbor Symposium on Quantitative Biology*, *27*, 1–24.

Castón, J. R., Ghabrial, S. A., Jiang, D., Rivas, G., Alfonso, C., Roca, R., et al. (2003). Three-dimensional structure of *Penicillium chrysogenum virus*: A double-stranded RNA virus with a genuine T=1 capsid. *Journal of Molecular Biology*, *331*, 417–431.

Castón, J. R., Luque, D., Trus, B. L., Rivas, G., Alfonso, C., Gonzalez, J. M., et al. (2006). Three-dimensional structure and stoichiometry of Helminthosporium victoriae190S totivirus. *Virology*, *347*, 323–332.

Castón, J. R., Trus, B. L., Booy, F. P., Wickner, R. B., Wall, J. S., & Steven, A. C. (1997). Structure of L-A virus: A specialized compartment for the transcription and replication of double-stranded RNA. *The Journal of Cell Biology*, *138*, 975–985.

Cheng, R. H., Castón, J. R., Wang, G.-J., Gu, F., Smith, T. J., Baker, T. S., et al. (1994). Fungal virus capsids, cytoplasmic compartments for the replication of double-stranded RNA, formed as icosahedral shells of asymmetric gag dimers. *Journal of Molecular Biology*, *244*, 255–258.

Cheng, L., Zhu, J., Hui, W. H., Zhang, X., Honig, B., Fang, Q., et al. (2010). Backbone model of an aquareovirus virion by cryo-electron microscopy and bioinformatics. *Journal of Molecular Biology*, *397*, 852–863.

Coulibaly, F., Chevalier, C., Gutsche, I., Pous, J., Navaza, J., Bressanelli, S., et al. (2005). The birnavirus crystal structure reveals structural relationships among icosahedral viruses. *Cell*, *120*, 761–772.

Dokland, T. (2000). Freedom and restraint: Themes in virus capsid assembly. *Structure Folding Design*, *8*, R157–R162.

Dryden, K., Wang, G., Yeager, M., Nibert, M., Coombs, K., Furlong, D., et al. (1993). Early steps in reovirus infection are associated with dramatic changes in supramolecular structure and protein conformation: Analysis of virions and subviral particles by cryoelectron microscopy and image reconstruction. *The Journal of Cell Biology*, *122*, 1023–1041.

Duquerroy, S., Da Costa, B., Henry, C., Vigouroux, A., Libersou, S., Lepault, J., et al. (2009). The picobirnavirus crystal structure provides functional insights into virion assembly and cell entry. *The EMBO Journal*, *28*, 1655–1665.

Gertsman, I., Gan, L., Guttman, M., Lee, K., Speir, J. A., Duda, R. L., et al. (2009). An unexpected twist in viral capsid maturation. *Nature*, *458*, 646–650.

Ghabrial, S. A., & Castón, J. R. (2011). Family Chrysoviridae. In: A. M. Q. King, M. J. Adams, E. B. Carstens & E. J. Lefkowitz (Eds.), *Virus taxonomy. Ninth report of the international committee on taxonomy of viruses* (pp. 509–513), Amsterdam: Elsevier/Academic Press.

Ghabrial, S. A., & Nibert, M. L. (2009). *Victorivirus*, a new genus of fungal viruses in the family *Totiviridae*. *Archives of Virology*, *154*, 373–379.

Gómez-Blanco, J., Luque, D., González, J. M., Carrascosa, J. L., Alfonso, C., Trus, B. L., et al. (2012). *Cryphonectria nitschkei* virus 1 structure shows that the capsid protein of chrysoviruses is a duplicated helix-rich fold conserved in fungal double-stranded RNA viruses. *Journal of Virology*, *86*(15), 8314–8318. http://dx.doi.org/10.1128/JVI.00802-12.

Gouet, P., Diprose, J. M., Grimes, J. M., Malby, R., Burroughs, J. N., Zientara, S., et al. (1999). The highly ordered double-stranded RNA genome of bluetongue virus revealed by crystallography. *Cell*, *97*, 481–490.

Grimes, J. M., Burroughs, J. N., Gouet, P., Diprose, J. M., Malby, R., Zientara, S., et al. (1998). The atomic structure of the bluetongue virus core. *Nature*, *395*, 470–478.

Harrison, S. C. (2007). Principles of virus structure. In D. M. Knipe, P. M. Howley, D. E. Griffin, R. A. Lamb, M. A. Martin, B. Roizman & S. E. Strauss (Eds.), *Fields virology* (pp. 59–98). Philadelphia: Lippincott Williams & Wilkins.

Hill, C. L., Booth, T. F., Prasad, B. V., Grimes, J. M., Mertens, P. P., Sutton, G. C., et al. (1999). The structure of a cypovirus and the functional organization of dsRNA viruses. *Nature Structural Biology*, *6*, 565–568.

Huiskonen, J. T., de Haas, F., Bubeck, D., Bamford, D. H., Fuller, S. D., & Butcher, S. J. (2006). Structure of the bacteriophage phi6 nucleocapsid suggests a mechanism for sequential RNA packaging. *Structure*, *14*, 1039–1048.

Jaalinoja, H. T., Huiskonen, J. T., & Butcher, S. J. (2007). Electron cryomicroscopy comparison of the architectures of the enveloped bacteriophages phi6 and phi8. *Structure*, *15*, 157–167.

Jiang, D., & Ghabrial, S. A. (2004). Molecular characterization of Penicillium chrysogenum virus: Reconsideration of the taxonomy of the genus Chrysovirus. *The Journal of General Virology*, *85*, 2111–2121.

Krupovic, M., & Bamford, D. H. (2008). Virus evolution: How far does the double beta-barrel viral lineage extend? *Nature Reviews. Microbiology*, *6*, 941–948.

Lawton, J. A., Estes, M. K., & Prasad, B. V. (1997). Three-dimensional visualization of mRNA release from actively transcribing rotavirus particles. *Nature Structural Biology*, *4*, 118–121.

Lawton, J. A., Estes, M. K., & Prasad, B. V. (2000). Mechanism of genome transcription in segmented dsRNA viruses. *Advances in Virus Research*, *55*, 185–229.

Li, Z., Baker, M. L., Jiang, W., Estes, M. K., & Prasad, B. V. (2009). Rotavirus architecture at subnanometer resolution. *Journal of Virology*, *83*, 1754–1766.

Liddington, R. C., Yan, Y., Moulai, J., Sahli, R., Benjamin, T. L., & Harrison, S. C. (1991). Structure of simian virus 40 at 3.8-A resolution. *Nature*, *354*, 278–284.

Lomonossoff, G. P., & Johnson, J. E. (1991). The synthesis and structure of comovirus capsids. *Progress in Biophysics and Molecular Biology*, *55*, 107–137.

Lu, G., Zhou, Z. H., Baker, M. L., Jakana, J., Cai, D., Wei, X., et al. (1998). Structure of double-shelled rice dwarf virus. *Journal of Virology*, *72*, 8541–8549.

Luque, D., Gonzalez, J. M., Garriga, D., Ghabrial, S. A., Havens, W. M., Trus, B., et al. (2010). The T=1 capsid protein of Penicillium chrysogenum virus is formed by a repeated helix-rich core indicative of gene duplication. *Journal of Virology*, *84*, 7256–7266.

Luque, D., Rivas, G., Alfonso, C., Carrascosa, J. L., Rodriguez, J. F., & Caston, J. R. (2009). Infectious bursal disease virus is an icosahedral polyploid dsRNA virus. *Proceedings of the National Academy of Sciences of the United States of America*, *106*, 2148–2152.
Luque, D., Saugar, I., Rejas, M. T., Carrascosa, J. L., Rodriguez, J. F., & Caston, J. R. (2009). Infectious Bursal disease virus: Ribonucleoprotein complexes of a double-stranded RNA virus. *Journal of Molecular Biology*, *386*, 891–901.
McClain, B., Settembre, E., Temple, B. R., Bellamy, A. R., & Harrison, S. C. (2010). X-ray crystal structure of the rotavirus inner capsid particle at 3.8 A resolution. *Journal of Molecular Biology*, *397*, 587–599.
Naitow, H., Tang, J., Canady, M., Wickner, R. B., & Johnson, J. E. (2002). L-A virus at 3.4 A resolution reveals particle architecture and mRNA decapping mechanism. *Nature Structural Biology*, *9*, 725–728.
Nakagawa, A., Miyazaki, N., Taka, J., Naitow, H., Ogawa, A., Fujimoto, Z., et al. (2003). The atomic structure of rice dwarf virus reveals the self-assembly mechanism of component proteins. *Structure*, *11*, 1227–1238.
Nandhagopal, N., Simpson, A. A., Gurnon, J. R., Yan, X., Baker, T. S., Graves, M. V., et al. (2002). The structure and evolution of the major capsid protein of a large, lipid-containing DNA virus. *Proceedings of the National Academy of Sciences of the United States of America*, *99*, 14758–14763.
Ochoa, W. F., Havens, W. M., Sinkovits, R. S., Nibert, M. L., Ghabrial, S. A., & Baker, T. S. (2008). Partitivirus structure reveals a 120-subunit, helix-rich capsid with distinctive surface arches formed by quasisymmetric coat-protein dimers. *Structure*, *16*, 776–786.
Pan, J., Dong, L., Lin, L., Ochoa, W. F., Sinkovits, R. S., Havens, W. M., et al. (2009). Atomic structure reveals the unique capsid organization of a dsRNA virus. *Proceedings of the National Academy of Sciences of the United States of America*, *106*, 4225–4230.
Patton, J. T., & Spencer, E. (2000). Genome replication and packaging of segmented double-stranded RNA viruses. *Virology*, *277*, 217–225.
Pesavento, J. B., Lawton, J. A., Estes, M. E., & Venkataram Prasad, B. V. (2001). The reversible condensation and expansion of the rotavirus genome. *Proceedings of the National Academy of Sciences of the United States of America*, *98*, 1381–1386.
Prasad, B. V., Rothnagel, R., Zeng, C. Q., Jakana, J., Lawton, J. A., Chiu, W., et al. (1996). Visualization of ordered genomic RNA and localization of transcriptional complexes in rotavirus. *Nature*, *382*, 471–473.
Rayment, I., Baker, T. S., Caspar, D. L., & Murakami, W. T. (1982). Polyoma virus capsid structure at 22.5 A resolution. *Nature*, *295*, 110–115.
Reinisch, K. M. (2002). The dsRNA Viridae and their catalytic capsids. *Nature Structural Biology*, *9*, 714–716.
Reinisch, K. M., Nibert, M. L., & Harrison, S. C. (2000). Structure of the Reovirus core at 3.6 Å resolution. *Nature*, *404*, 960–967.
Riffel, N., Harlos, K., Iourin, O., Rao, Z., Kingsman, A., Stuart, D., et al. (2002). Atomic resolution structure of Moloney murine leukemia virus matrix protein and its relationship to other retroviral matrix proteins. *Structure*, *10*, 1627–1636.
Roberts, M. M., White, J. L., Grutter, M. G., & Burnett, R. M. (1986). Three-dimensional structure of the adenovirus major coat protein hexon. *Science*, *232*, 1148–1151.
Saugar, I., Luque, D., Ona, A., Rodriguez, J. F., Carrascosa, J. L., Trus, B. L., et al. (2005). Structural polymorphism of the major capsid protein of a double-stranded RNA virus: An amphipathic alpha helix as a molecular switch. *Structure*, *13*, 1007–1017.
Shaw, A. L., Samal, S. K., Subramanian, K., & Prasad, B. V. (1996). The structure of aquareovirus shows how the different geometries of the two layers of the capsid are reconciled to provide symmetrical interactions and stabilization. *Structure*, *4*, 957–967.

Steven, A. C., Heymann, J. B., Cheng, N., Trus, B. L., & Conway, J. F. (2005). Virus maturation: Dynamics and mechanism of a stabilizing structural transition that leads to infectivity. *Current Opinion in Structural Biology, 15*, 227–236.

Tang, J., Ochoa, W. F., Sinkovits, R. S., Poulos, B. T., Ghabrial, S. A., Lightner, D. V., et al. (2008). Infectious myonecrosis virus has a totivirus-like, 120-subunit capsid, but with fiber complexes at the fivefold axes. *Proceedings of the National Academy of Sciences of the United States of America, 105*, 17526–17531.

Trask, S. D., McDonald, S. M., & Patton, J. T. (2012). Structural insights into the coupling of virion assembly and rotavirus replication. *Nature Reviews. Microbiology, 10*, 165–177.

Vancini, R., Paredes, A., Ribeiro, M., Blackburn, K., Ferreira, D., Kononchik, J. P., Jr., et al. (2012). Espirito Santo virus: A new birnavirus that replicates in insect cells. *Journal of Virology, 86*, 2390–2399.

Wickner, R. (1996). Double-stranded RNA viruses of Saccharomyces cerevisiae. *Microbiological Reviews, 60*, 250–265.

Wood, H. A., & Bozarth, R. F. (1972). Properties of viruslike particles of Penicillium chrysogenum: One double-stranded RNA molecule per particle. *Virology, 47*, 604–609.

Yu, X., Jin, L., & Zhou, Z. H. (2008). 3.88 A structure of cytoplasmic polyhedrosis virus by cryo-electron microscopy. *Nature, 453*, 415–419.

Zhang, X., Walker, S. B., Chipman, P. R., Nibert, M. L., & Baker, T. S. (2003). Reovirus polymerase lambda 3 localized by cryo-electron microscopy of virions at a resolution of 7.6 A. *Nature Structural Biology, 10*, 1011–1018.

CHAPTER FIVE

Hypovirus Molecular Biology: From Koch's Postulates to Host Self-Recognition Genes that Restrict Virus Transmission

Angus L. Dawe* and Donald L. Nuss†,1

*Department of Biology, New Mexico State University, Las Cruces, New Mexico, USA
†Institute for Bioscience and Biotechnology Research, University of Maryland, Rockville, Maryland, USA
[1]Corresponding author: e-mail address: dnuss@umd.edu

Contents

1. Introduction 110
2. Molecular Characterization of Hypovirulence-Associated dsRNAs 110
3. The *Hypoviridae* 113
4. Technical Challenges of Mycovirus Research: Development of the Hypovirus Experimental System and Completion of Koch's Postulates 115
5. Hypovirus Translation and Gene Expression 117
6. Impact of Hypovirus Infection on the Fungal Host 121
   6.1 Host signaling pathways and gene expression 122
   6.2 Host physiology 126
7. Fungal Antiviral Defense Mechanisms 128
   7.1 RNA silencing antiviral defense response in *C. parasitica* 129
   7.2 Restriction of hypovirus transmission by vic 132
8. Concluding Remarks 138
References 140

Abstract

The idea that viruses can be used to control fungal diseases has been a driving force in mycovirus research since the earliest days. Viruses in the family *Hypoviridae* associated with reduced virulence (hypovirulence) of the chestnut blight fungus, *Cryphonectria parasitica*, have held a prominent place in this research. This has been due in part to the severity of the chestnut blight epidemics in North America and Europe and early reports of hypovirulence-mediated mitigation of disease in European forests and successful application for control of chestnut blight in chestnut orchards. A more recent contributing factor has been the development of a hypovirus/*C. parasitica* experimental system that has overcome many of the challenges associated with mycovirus research, stemming primarily from the exclusive intracellular lifestyle shared by all mycoviruses. This chapter will focus on hypovirus molecular biology with an emphasis on the

Advances in Virus Research, Volume 86
ISSN 0065-3527
http://dx.doi.org/10.1016/B978-0-12-394315-6.00005-2

development of the hypovirus/*C. parasitica* experimental system and its contributions to fundamental and practical advances in mycovirology and the broader understanding of virus–host interactions and fungal pathogenesis.

1. INTRODUCTION

Hypoviruses and their host, the chestnut blight fungus *Cryphonectria parasitica*, have received considerable attention since the first report of chestnut blight in North America in 1903 and the 1977 description of double-stranded (ds) RNAs associated with a transmissible form of biological control of chestnut blight, termed hypovirulence. This level of interest increased further with demonstration of the virus-like properties of hypovirulence-associated dsRNA, the elucidation of the genome organization of prototypic *C. parasitica* hypovirus CHV-1/EP713, and development of a hypovirus reverse genetics system, the first for a mycovirus. The latter development allowed completion of Koch's postulates and marked the transition of mycovirus research from descriptive studies to hypothesis-driven experimental approaches. It also provided the means for testing the prediction that fungal viruses, like animal and plant viruses, have considerable utility for probing host functions and for engineering hypoviruses for enhanced biological control potential. Advances in fungal genomics have complemented the power of the hypovirus reverse genetics system leading to the discoveries that RNA silencing serves as an antiviral defense response in fungi and contributes to viral RNA recombination. The recent identification of vegetative incompatibility (*vic*) genes that regulate fungal self-recognition and restrict hypovirus transmission illustrates the successful application of modern techniques to solve long-standing problems in mycovirus research that are of fundamental interest and practical value. This chapter will focus on the major advances in the development of hypovirus molecular biology.

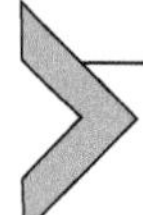

2. MOLECULAR CHARACTERIZATION OF HYPOVIRULENCE-ASSOCIATED dsRNAs

The report by Day, Dodds, Elliston, Jaynes, and Aganostakis (1977) correlating the presence of dsRNA with hypovirulence of *C. parasitica* provided the first insights into the molecular nature of hypovirulence-associated genetic elements. However, subsequent efforts to identify virus-like particles expected to be associated with the hypovirulence-associated dsRNAs were unsuccessful. Instead, the dsRNAs were shown to fractionate with club-like pleiomorphic membrane vesicles (Dodds, 1980), an association later supported by ultrastructure (Newhouse, Hoch, & MacDonald, 1983) and biochemical (Fahima, Kazmierczak, Hansen, Pfeiffer, & Van Alfen, 1993; Hansen, Van Alfen, Gillies, & Powell, 1985) analyses.

Since most mycoviruses have genomes consisting of dsRNA (Buck, 1986), the hypovirulence-associated dsRNAs were naturally thought to constitute viral genomic RNA. However, early direct analysis of the isolated dsRNAs revealed unexpected structural properties. Unrelated dsRNAs isolated from two different hypovirulent *C. parasitica* strains were found to have an ~40 nucleotide stretch of polyadenylic acid (poly A) at the 3′-terminus of one strand that was base paired to a stretch of polyuridylic acid (poly U) at the 5′-terminus of the complementary strand (Hiremath, L'Hostis, Ghabrial, & Rhoads, 1986; Tartaglia, Paul, Fulbright, & Nuss, 1986). These 3′-poly A:5′-poly U terminal structures distinguished the hypovirulence-associated dsRNAs from classical mycovirus genomic dsRNAs, which lack such structures, and were more analogous to the terminal structures found at one end of the ds replicative form RNA of single-stranded, positive-sense RNA viruses of plants and animals (Luria, Darnell, Baltimore, & Campbell, 1981; Matthews, 1981). This led Tartaglia et al. (1986) to speculate that hypovirulence-associated dsRNAs were really replicative forms of single-stranded RNA virus-like elements.

The early direct analysis of isolated hypovirulence-associated dsRNAs provided valuable landmarks for the subsequent cDNA cloning, sequence analysis, and assembly efforts that led to the elucidation of the genome organization and expression strategy of the prototypic hypovirulence-associated dsRNA isolated from a hypovirulent *C. parasitica* strain designated EP713 (Shapira, Choi, & Nuss, 1991). The nucleotide sequence derived for the 12.7-kbp dsRNA contained two large contiguous open reading frames, designated ORF A and ORF B, on the strand that contained the 3′-terminal poly A (Fig. 5.1). The two ORFs were separated by an UAAUG pentanucleotide in which the UAA was shown to serve as a stop codon for ORF A and the AUG represents the first potential initiation codon for ORF B.

The strategy used for expression of the proteins encoded by both ORFs was found to involve autoproteolytic processing events (Fig. 5.1). Two polypeptides, p29 and p40, were identified as being derived from the ORF A-encoded polyprotein, p69, as a result of autocatalytic cleavage by a papain-like protease domain within p29 (Choi, Pawlyk, & Nuss, 1991).

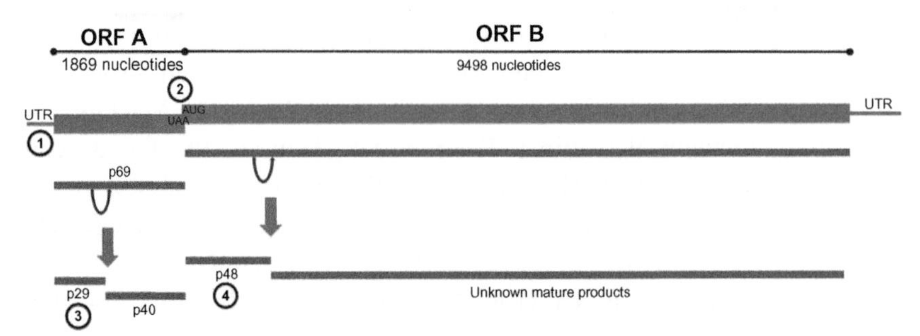

Figure 5.1 A graphical representation of the genome organization of CHV-1/EP713. Open reading frame (ORF) A results in protein product p69, which is autocatalytic, yielding the mature products p29 and p40. ORFB produces a much larger product from which p48 is released, again by autocatalysis, but the remainder of the mature products are presently unknown. Numbers in circles refer to features specifically highlighted in this review. Structure of the 5′ untranslated region (1), the importance of the pentanucleotide boundary between ORFA and ORFB (2), and the role of p48 in viral replication (4) are all discussed in Section 5. The function of p29 (3) as a suppressor of RNA silencing is highlighted in Section 7.1.2. (For color version of this figure, the reader is referred to the online version of this chapter.)

A related papain-like protease domain located in the N-terminal portion of ORF B autocatalytically releases protein p48 from the large predicted polyprotein that contains the RNA-dependent RNA polymerase (RDRP) and helicase domains required for RNA replication (Shapira & Nuss, 1991). The expression of large polyprotein precursors from an RNA containing a 3′-poly A tail is characteristic of viruses within the picornavirus supergroup (Knowles et al., 2011). In this regard, sequence alignment analyses performed by Koonin, Choi, Nuss, Shapira, and Carrington (1991) identified five domains within the ORF A and ORF B coding domains that showed significant sequence similarity with conserved motifs found in proteins encoded by members of the plant potyviruses, suggesting a common ancestry with this group of single-stranded positive-strand RNA picorna-like plant viruses. The completion of the nucleotide sequence and determination of the virus-like genome organization and expression strategy for a hypovirulence-associated RNA had two important outcomes. It led the International Committee on Taxonomy of Viruses (ICTV) to recognize this and related RNA genetic elements as viruses and to establish a new taxonomic family, the *Hypoviridae* (Hillman, Fullbright, Nuss, & Van Alfen, 1995), to accommodate them. It also provided the technical means and impetus for developing the first mycovirus reverse genetics system.

3. THE *HYPOVIRIDAE*

The *Hypoviridae* was the first virus taxonomic family established for which members do not encode a capsid protein or produce true virions (two additional nonvirion-producing virus families, the *Narnaviridae* and the *Endornaviridae*, have subsequently been added). Members of this family, like all fungal viruses, lack an extracellular phase to their life cycle. Infections cannot be initiated by inoculation of hyphae with infected cell extract or enriched fractions. Instead, these viruses are transmitted via cytoplasmic mixing following fusion (anastomosis) between vegetatively compatible fungal strains (to be discussed in a subsequent section), or to a variable extent in asexual spores. The absence of a discrete virus particle and an exclusively intracellular lifestyle makes it difficult to precisely define the genomic RNA. This complication, coupled with the original description of dsRNAs in hypovirulent *C. parasitica* strains (Day et al., 1977), resulted in the initial grouping with dsRNA virus. This erroneous grouping has been rectified in the 9th ITCV report (Nuss & Hillman, 2011) where members of the family *Hypoviridae* are grouped with the single-stranded positive-strand RNA viruses in recognition of the fact that the hypovirus coding strand transcript can initiate an infection when introduced by electroporation into fungal cell wall-free spheroplasts (to be discussed in a later section).

The family *Hypoviridae* contains a single genus, *Hypovirus*, with four species distinguished by differences in genome structure and sequence relatedness. The designations used for members of this family include CHV for *C. parasitica* hypovirus, a number indicating species relatedness and, following a backslash, the fungal host from which the virus was isolated, for example, the designation of the prototypic hypovirus isolated from *C. parasitica* strain EP713 and described in Fig. 5.1 is CHV-1/EP713.

Complete nucleotide sequences have now been reported for six members of the *Hypoviridae* distributed among the four species that comprise the family. As described earlier in Fig. 5.1, the genome organization of members of the type species CHV-1, CHV-1/EP713 (Shapira et al., 1991), CHV-1/Euro7 (Chen & Nuss, 1999), and CHV-1/EP721 (Lin et al., 2007) consists of two contiguous ORFs with a UAAUG junction encoded on a 12.7-Kb, poly A-tailed RNA. The genome organization for hypovirus CHV-2-NB58 (Hillman, Halpeern, & Brown, 1994), the type member of species CHV-2, is similar to that of the CHV-1 species except that the ORF A coding domain lacks the p29 papain-like catalytic or cleavage sites and instead

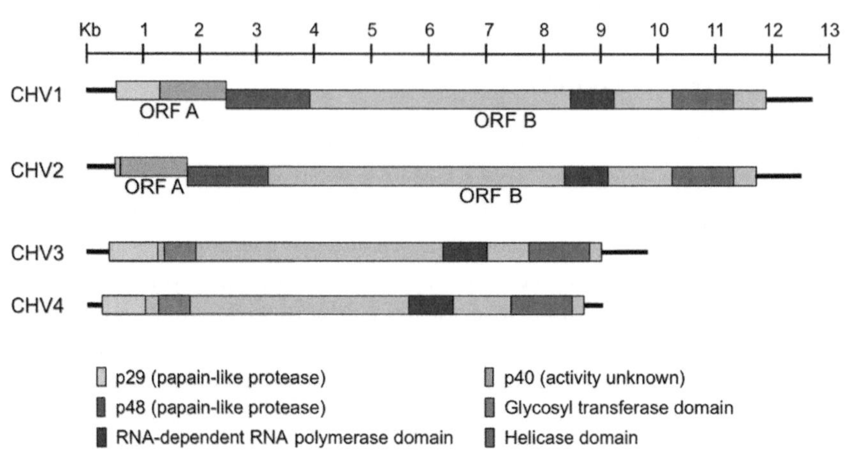

Figure 5.2 A comparison of the genome organizations of the *Hypoviridae*. Colored regions indicate stretches of sequence conservation between family members with functional identification in the key extrapolated from the known or predicted activities in CHV-1/EP713 described in the text. Black lines indicate 5′ and 3′ untranslated sequences. Poly-A tails present on the plus strand are not shown. Scale markers refer to kb RNA. *Adapted from Linder-Basso, Dynek, and Hillman (2005).* (See Page 7 in Color Section at the back of the book.)

encodes a 50-kDa protein (Fig. 5.2). ORF B of CHV-2/NB58 is organized like that of the CHV-1 species with a p52 ortholog of CHV-1 p48 and conserved helicase and RDRP domains similarly placed.

Members of species CHV-3 and CHV-4 have genomes that are smaller than those of CHV-1 and CHV-2, 9.8 and 9.1 Kb compared to 12.7 and 12.5 Kb, respectively and that contain a single rather than two ORFs. Both of the type members for species CHV-3, CHV-3/GH2 (Smart et al., 1999) and for species CHV-4, CHV-4/SR2 (Linder-Basso et al., 2005) contain putative papain-like protease domains similar to CHV-1 p29 within the N-terminal portion of the single ORF. However, functionality has been indicated only for the CHV-3 ortholog (Smart et al., 1999). The CHV-3 and CHV-4 species also contain a domain that is homologous to UDP-glucose/sterol glucosyltransferase (UGT), of yet unknown function, that is lacking in CHV-1 and CHV-2. Interestingly, the CHV-3 and CHV-4 RDRP domains were found to be more closely related to the RDRP of a virus isolated from the fungal host *Fusarium graminicola* than to the RDRP sequences of CHV-1 and CHV-2 hypoviruses. These combined observations led Linder-Basso et al. (2005) to suggest that the divergence of the CHV-1 and CHV-2 lineage from the CHV-3 and CHV-4 lineage was not a recent event. Interestingly, species CHV-2, CHV-3, and CHV-4 were all originally identified as infecting

North American *C. parasitica* strains. Species CHV-1 predominates in Europe and in Asia, the geographic origin of *C. parasitica*, but is rarely found in North America, even though European CHV-1 isolates have been introduced into several locations (Linder-Basso et al., 2005).

Two viruses closely related to hypovirus species CHV-3 and CHV-4 were recently identified infecting fungal species other than *C. parasitica*. A virus named Sclerotinia sclerotiorum hypovirus 1 (SsHV1/SZ-150) was identified in *Sclerotinia sclerotiorum* (Xie et al., 2011) and a related virus named Valsa ceratosperma hypovirus 1 (VcHV1) was recently identified infecting *Valsa ceratosperma* (Yaegshi, Kanematsu, & Ito, 2012). *S. sclerotiorum* is an economically important cosmopolitan, necrotrophic pathogen that infects over 450 plant species (Boland & Hall, 1994). *V. ceratosperma* is the causative agent of valsa canker disease of apple and pear (Kobayashi, 1970; Sakurma, 1990) and, like *C. parasitica*, forms cankers on woody plants. Both *C. parasitica* and *V. ceratosperma* are classified within the Ascomycota, Sordairomycetes, Diaporthales (Gryzenhout, Myburg, Wingfield, & Wingfield, 2006), while *S. sclerotiorum* is placed within the Leotiomycetes, a sister lineage of the Sordariomycetes (Amselem et al., 2011).

The similarities between the genomes of SsHV1 and VcHV1 and the hypovirus species CHV-3 and CHV-4 include size (9.1–10.4 kb), a single large open reading frame, and a significant level of shared sequence identity and similarity in conserved protease, UTG, polymerase, and helicase domains that are arranged in the same order. In addition, all four viruses contain significant stretches of nucleotide sequence identity in both the 5′- and 3′- noncoding terminal regions. These comparisons support the suggestion by Linder-Basso et al. (2005) of two lineages within the Hypoviridae and have led to the proposal to establish a new genus within the family (Xie et al., 2011; Yaegshi et al., 2012). Concerted efforts to identify additional hypoviruses in fungi other than *C. parasitica* are likely to be productive and the family *Hypoviridae* is very likely to see an expansion in the near future.

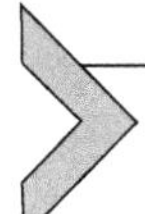

4. TECHNICAL CHALLENGES OF MYCOVIRUS RESEARCH: DEVELOPMENT OF THE HYPOVIRUS EXPERIMENTAL SYSTEM AND COMPLETION OF KOCH'S POSTULATES

Mycoviruses were discovered much more recently than viruses of plants and animals, with the first report by Hollings (1962) involving La France disease of cultivated mushroom *Agaricus bisporis*. Mycoviruses also generally have much less socio-economic impact than their plant and animal virus counterparts and often replicate without causing symptoms. However, it is the technical challenges for experimental design that have hampered the relative pace of mycovirus research. These challenges stem from the exclusive intracellular lifestyle, difficulty in initiating infections by an extracellular route, and fungal hosts that are not amenable to genetic manipulation. As a result, it is generally not experimentally feasible to demonstrate a cause and effect relationship between the presence of a mycovirus and symptoms, even when they may be of fundamental interest or potential practical application.

A solution to these basic technical challenges was provided for the hypoviruses in the form of a reverse genetics system. This key advancement was facilitated by the development of a robust DNA-mediated genetic transformation protocol for *C. parasitica* (Churchill, Ciufetti, Hansen, Van Etten, & Van Alfen, 1990). This involved the introduction of plasmid DNA containing a selectable antibiotic marker gene into cell wall-free spheroplasts, regeneration of the cell wall under selective conditions, and recovery of fungal cells containing the chromosomally integrated plasmid DNA. Inspired by the report by Racaniello and Baltimore (1981) of the launching of replicating poliovirus RNA from a cDNA copy in transformed mammalian cells, Choi and Nuss (1992) transformed virus-free *C. parasitica* spheroplasts with a plasmid containing the full-length cDNA copy of CHV-1/EP713 RNA based on the then-recently determined complete nucleotide sequence (Shapira, Choi & Nuss, 1991). The resulting transformants contained cytoplasmically replicating viral RNA that was precisely trimmed of any vector sequences (Chen, Craven, Choi, & Nuss, 1994) and were phenotypically indistinguishable from the original CHV-1/EP713-infected strain. Chen, Choi, and Nuss (1994) subsequently developed a transfection protocol in which hypovirus infection could be routinely initiated by electroporation of *in vitro*-synthesized coding strand transcripts directly into fungal spheroplasts. This protocol allowed efficient initiation of infection without the use of a selectable marker and demonstrates that the hypovirus coding strand is infectious, the hallmark of a positive-sense, single-stranded RNA virus.

Both the transformation and the transfection protocols allowed the completion of Koch's postulates, conclusively demonstrating that hypoviruses are responsible for hypovirulence and the full range of associated traits that include loss of pigmentation, reduced asexual sporulation, and female infertility. The hypovirulent strains containing the integrated viral cDNA, so-called transgenic hypovirulent strains, exhibited a novel property. While hypovirus RNA is normally excluded from ascospores resulting from a sexual cross (mating), the chromosomally integrated hypovirus cDNA copy present in the transgenic hypovirulent strain is transmitted to ascospore progeny and the viral RNA is subsequently launched in the resulting germinated fungal colony (Chen, Choi, & Nuss, 1993). Virus transmission from transgenic hypovirulent strains to ascospores has been confirmed under field conditions (Anagnostakis, Chen, Geletka, & Nuss, 1998; Root et al., 2005), and testing of the contribution of this form of transmission to the spread of virus through natural populations continues to be investigated. Both protocols have also been used to extend host range to fungal species taxonomically related to *C. parasitica* (Chen, Chen, Bowman, & Nuss, 1996; Chen, Choi, & Nuss, 1994; Sasaki et al., 2002; van Heerden et al., 2001) and the transfection protocol, in particular, has found extensive utility for experimental designs involving genetic manipulation of hypoviruses for a wide variety of studies as indicated in subsequent sections of this review (also reviewed in Dawe & Nuss, 2001; Nuss, 2005, 2011).

Recent advances in *C. parasitica* genomics and transformation efficiency have added significant additional capabilities to the hypovirus/*C. parasitica* experimental system. The Department of Energy Joint Genome Institute recently made the *C. parasitica* genome sequence publically available (http://genome.jgi-psf.org/Crypa2/Crypa2.home.html). The efficiency for gene disruption in the haploid *C. parasitica* genome was recently enhanced from ~5% to ~85% by creating a strain deficient in the *ku80* gene that encodes a key component of the nonhomologous end-joining DNA repair pathway (Lan et al., 2008). Thus, the hypovirus/*C. parasitica* experimental system provides the opportunity for facile genetic manipulation of both of the genomes of a eukaryotic virus and its host. In this regard, it is important to note that *C. parasitica* also harbors viruses of four other families including the *Reoviridae*, *Narnaviridae*, *Partitiviridae*, and *Chrysoviridae* (reviewed in Hillman & Suzuki, 2004), and thus also provides experimental capabilities for a wide range of fundamental mycovirus research.

5. HYPOVIRUS TRANSLATION AND GENE EXPRESSION

As described above, the *Hypoviridae* represent a diverse and growing collection of viruses. Besides presenting the noted experimental difficulties, the variations in sequence and genomic arrangement have also made it equally challenging to discern functional elements by comparing sequences or motifs within them. With the determination of the nucleotide sequence, Shapira, Choi and Nuss (1991) presented the organization of the elements, but progress in their functional characterization remains limited. Clearly, however, two principle processes must be accomplished—viral RNA translation and replication.

Early work on the membrane vesicles that appeared to proliferate in hypovirus-infected mycelium identified first that dsRNA was associated with these structures (Hansen et al., 1985). Subsequently, Fahima et al. (1993) conclusively demonstrated that preparations of vesicles from a hypovirus-infected fungal strain had associated RDRP activity, which was not the case for equivalent vesicles recovered from a nonhypovirus infected strain. Furthermore, the products of this reaction were found to be 80% positive strand. This supported the contention of Tartaglia et al. (1986) who had suggested that the observed dsRNA elements represent the replicative intermediates of what is, instead, a single, positive, strand of RNA. As noted above, this was confirmed by Chen, Choi & Nuss (1991).

It is well known that the untranslated regions (UTRs) of many positive-strand RNA viruses fold into complex structures that may regulate translation and RNA replication including, for example, coxsackievirus B3 (Bailey & Tapprich, 2007), rotavirus (Li et al., 2010), citrus tristeza virus (Lopez et al., 1998), and turnip crinkle virus (McCormack et al., 2008). In the *Hypoviridae*, the well-characterized organization of CHV-1/EP713 is known to include, at the 5′ end, a 495 nucleotide UTR while, at the 3′ end, an 851 nucleotide UTR is present (Shapira, Choi, et al., 1991). Rae, Hillman, Tartaglia, and Nuss (1989) found that a small portion of 5′ noncoding sequence could inhibit hypovirus translation *in vitro*, which indicated a possible function for the 5′ UTR. However, there is not a great deal of sequence similarity between the UTRs of the less well-characterized members of the *Hypoviridae*. CHV-4 has a relatively short 5′ UTR compared to CHV-1, -2, and -3, of only about 193 nucleotides long. The length of the 5′ UTRs of other species generally ranges from 369 to 487 nucleotides (Hillman & Suzuki, 2004).

Recently, Mu, Romero, Hanley, and Dawe (2011) applied a nuclease mapping approach coupled with Mfold structure prediction software (Zuker, 2003) to develop validated, constrained models of the 5′ UTR for the hypoviruses CHV-1/EP713 and CHV-1/Euro7, two hypovirus strains that exhibit variation in virulence attenuation despite >90% sequence identity (Chen & Nuss, 1999). This new analysis has revealed highly structured regions in the 5′ UTR of both viruses that may provide functional clues worthy of further investigation (Fig. 5.3). It has not been empirically determined whether hypoviruses have a 5′ cap structure to facilitate translation, although the fact that uncapped RNA produced *in vitro* is capable of

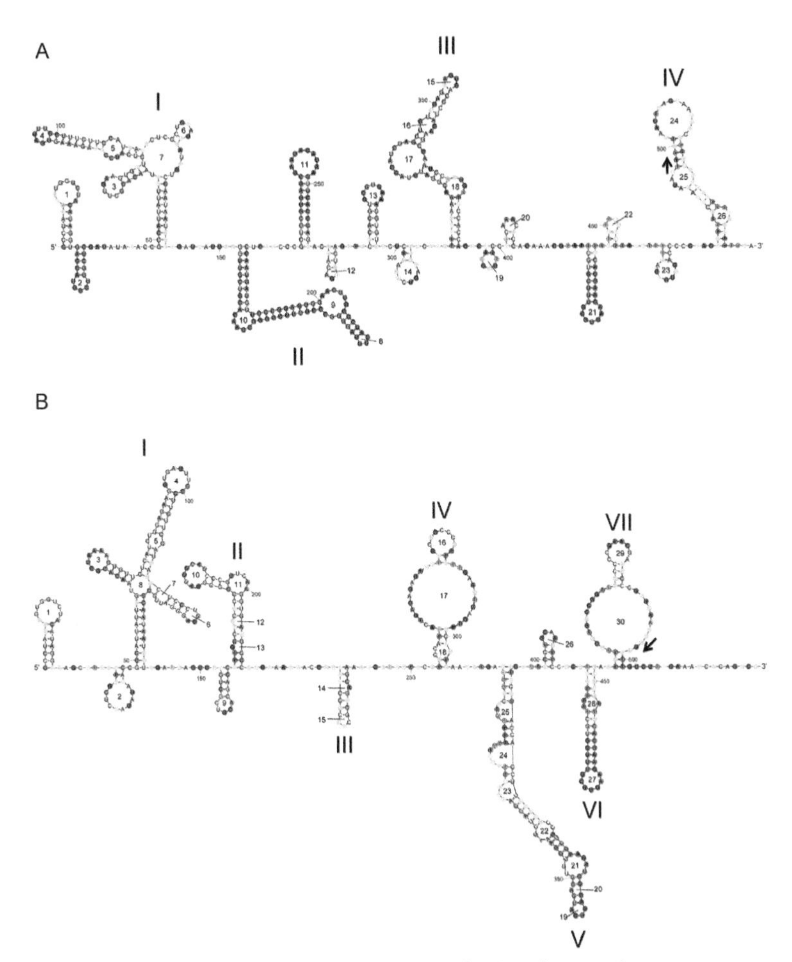

Figure 5.3 Mfold constrained predicted structure for the 5′ UTRs of CHV-1/EP713 (panel A) and CHV-1/Euro7 (panel B). The prediction was generated using the first 675 nucleotides of the genomes (nt 1–540 shown). The structure was annotated based on *P-num* and each base is colored according to its *P-num* value (Zuker & Jacobson, 1998). Colors are ranged from red to black representing unusually well-determined to poorly determined. The arrow indicates the start codon of ORF A. Each stem-loop structure was numbered sequentially; structures containing more than one stem-loop were defined as domains and are labeled with Roman numerals. *From Mu et al. (2011).* (See Page 8 in Color Section at the back of the book.)

initiating infection (Chen, Choi, & Nuss, 1994) would appear to contradict this possibility. In this regard, hypoviruses may possess a 5′ cap-independent translation mechanism first identified in picornaviruses (Pelletier, Flynn, Kaplan, Racaniello, & Sonenberg, 1988). In this instance, poliovirus, has a 750 nucleotide long 5′ UTR with seven or eight upstream AUGs (Pelletier et al., 1988). Similarly, Hepatitis C virus is also capable of cap-independent translation and contains an ~340 nucleotide long 5′ UTR with three or four upstream AUGs (Tsukiyama-Kohara, Iizuka, Kohara, & Nomoto, 1992). The CHV-1 group of isolates within the *Hypoviridae* have approximately 495 nucleotide with seven AUG triplets present upstream of the authentic start codon. The structural complexity observed by Mu et al. (2011) may support that translation in at least the CHV-1 group is cap-independent and mediated by an internal ribosome entry site (IRES), as initially proposed by Suzuki, Geletka, and Nuss (2000). Unfortunately, as noted by Baird, Turcotte, Korneluk, and Holcik (2006), it is not possible to identify an IRES based solely on secondary RNA structure, and further clarification of these initial studies will require more targeted work.

The roles that the various viral protein products play in replication and translation of the viral genome are only partially understood. As noted above, the initial product of ORFA, p69, autocatalytically releases the mature products p29 and p40. The former has been associated with host symptoms and, although the specific action remains a mystery, the host symptom determinants that reside in this protein have been mapped to cysteine residues conserved with the potyvirus-encoded HC-Pro (Craven, Pawlyk, Choi, & Nuss, 1993; Suzuki, Chen, & Nuss, 1999). However, the role of p29 also extends into RNA accumulation. Suzuki et al. (2000) demonstrated using engineered RNAs lacking various portions of p29 that amino acids 1–24 were absolutely required for virus replication. In subsequent work, it was revealed that p29 activity could function *in trans* to restore replication competency (Suzuki, Maruyama, Moriyama, & Nuss, 2003). More recently, a role for p29 in suppressing the host antiviral response has also been observed (Segers, Zhang, Deng, Sun, & Nuss, 2007; described below) confirming p29 as a multifunctional protein of major significance for viral replication and survival.

The second product from ORFA, p40, was found to be dispensable for CHV1-EP713 replication given the ability of modified constructs to retain replication competence in the absence of any portion of the p40 coding domain (Suzuki & Nuss, 2002). However, the authors did show that p40 does have a role to play in viral RNA accumulation, and that RNA accumulation correlated with the severity of host symptoms (Suzuki & Nuss, 2002).

The junction of the two open reading frames of CHV1-EP713 was predicted to be UAAUG, where the UAA portion serves as the termination codon of ORF A and the AUG portion is the 5′-proximal translation initiation codon of ORF B (Shapira, Choi, et al., 1991). Suzuki et al. (2000)

set out to exploit the potential of the CHV1-EP713 infectious cDNA clone as a vehicle for heterologous or inappropriate gene expression, and in so doing revealed unexpected flexibility in the nature of this sequence, while also demonstrating that the viral genome was capable of recognizing and eliminating most foreign insertions within a few weeks posttransfection. More surprising was the stable replication of a modified virus in which the two-ORF genetic configuration was abolished. This result suggested that the UAAUG pentanucleotide is altogether dispensable for virus replication.

A role related to viral genome replication has, however, been found for the third characterized protein product of the CHV1-EP713 genome, p48. This 48-kDa protein contains a papain-like protease domain and is autocatalytically released from the N-terminal portion of ORFB (Shapira & Nuss, 1991). Deng and Nuss (2008) found that p48 contributes to the hypovirus-associated phenotypes of reduced pigmentation and sporulation. This was similar to the requirement found by Craven et al. (1993) for p29—a viral clone lacking p29 was replication competent but caused less severe fungal symptoms. However, deletion of p48 from the CHV1-EP713 cDNA clone resulted in a failure to initiate viral replication, suggesting that p48 is not dispensable for viral RNA propagation. Most interestingly, however, p48 can be provided *in trans* by transforming fungal spheroplasts that already express p48 with the Δp48 hypoviral construct, at which time hypoviral replication and accumulation proceed normally. Similarly, passing the Δp48 hypovirus to a previously uninfected, p48 expressing, strain by anastomosis resulted in successful transmission. Subsequent transfer of the replicating Δp48 hypovirus was then possible, even to a strain not expressing p48. This indicated that, while critical for initiation of replication, p48 is not required for the maintenance of replication.

Therefore, while the complexities of mature viral protein products remain largely a mystery, it is clear that certain elements within the hypovirus genome play specific roles in replication and determination of host phenotype. Subsequent analyses will no doubt reveal further relationships of this kind, but the unique attributes of the *C. parasitica*—hypovirus infection system provide an ideal environment in which to pursue these important questions.

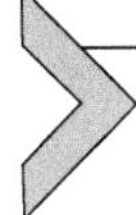

6. IMPACT OF HYPOVIRUS INFECTION ON THE FUNGAL HOST

The best-studied member of the *Hypoviridae*, CHV-1/EP713, causes reproducible, distinct, debilitation of its host colony (reviewed by Dawe & Nuss, 2001). There is heterogeneity in natural isolates of the fungus, but generally the uninfected vegetative colonies are characterized by an orange or orange/brown pigmentation and copious production of conidia (asexual spores) on the colony surface. Hypovirulent strains, however, will have a more erratic colony margin (quantified by Golinski, Boecklen, & Dawe, 2008) that grows slower and fails to generate pigment, while the numbers of conidia are greatly reduced and mating is compromised. This section will summarize recent progress toward understanding the impact of the hypovirus on host signaling pathways that control these developmental processes and the resulting effects on host physiology and metabolism.

6.1. Host signaling pathways and gene expression

Rapid developments in monitoring gene expression over the past two decades have provided new approaches for examining host responses to hypovirus infection. Initially, technology dictated that these analyses would be based on single genes such as laccase (Choi, Larson, & Nuss, 1992; Rigling, Heiniger, & Hohl, 1989; Rigling & Van Alfen, 1991). Further studies by Larson, Choi, and Nuss (1992) demonstrated that the suppression of laccase gene expression in the hypovirus-infected mycelium was due to the perturbation of a calcium/calmodulin/inositol trisphosphate-dependent positive regulatory pathway. Attempts to provide a more global approach utilized two-dimensional electrophoresis methods (Powell & Van Alfen, 1987a) or differentially hybridized host-genome libraries (Powell & Van Alfen, 1987b). These studies led to the identification of Vir2, a protein whose elimination from the fungal host led to a strain that mimicked aspects of the hypovirus-infected phenotype (Zhang, Churchill, Kazmierczak, Kim, & Van Alfen, 1993) and was later identified as a pheromone precursor (Zhang, Baasiri, & Van Alfen, 1998).

With the development of the transformation protocol (Churchill et al., 1990) and, subsequently, the cDNA hypovirus clones for transformation or transfection (Chen, Choi, et al., 1994; Choi & Nuss, 1992) described above, it was then possible to begin to address the changes that might be happening to the host as a result of hypovirus proliferation in identical strains in which the only difference was the presence or absence of the replicating RNA. Related to the pheromone signaling pathway, heterotrimeric G proteins have long been known as key effectors that transduce an extracellular signal, detected by a membrane-bound receptor, to downstream components, which may eventually lead to changes in gene expression as in the canonical system *Saccharomyces cerevisae* (see Jones & Bennett, 2011, for a recent review). Choi, Chen, and Nuss (1995) observed that a Gα ortholog, termed *cpg-1* was reduced in expression in the presence of hypovirus. Subsequently, gene deletion experiments revealed two Gα subunits with different roles, but that only *cpg-1* was affected by CHV-1/EP713. Interestingly, the phenotype of the *Δcpg-1* fungal strain (containing no hypovirus) exhibited traits that could be characterized as a more severe manifestation of those seen in hypovirus-infected colonies, namely slower growth, loss of pigmentation, no sporulation, and complete avirulence.

Clearly intact signaling pathways that allowed the fungal pathogen to monitor and respond to its environment were crucial for virulence and hypovirus-mediated disruption of them might provide for an explanation of the fungal debilitation. This was further supported following the development of a spotted cDNA microarray system that was based on a collection of expressed sequences developed by Dawe et al. (2003). Approximately 2200 genes (now known, following completion of the *C. parasitica* genome, to be 20% of the total predicted open reading frames) were represented on the microarray, which allowed a comparative analysis of large-scale gene expression between isogenic hypovirus-infected and uninfected mycelium. Altered transcript abundance was identified for 13.4% of the 2200 unique cDNAs, which represented a broad spectrum of biological functions, including stress responses, carbon metabolism, and transcriptional regulation. This is consistent with hypovirus infection resulting in a persistent reprogramming of a significant portion of the *C. parasitica* transcriptome (Allen, Dawe, & Nuss, 2003).

Establishment of the microarray methodology then permitted assessment of the differences between hypovirus-infected mycelium and cultures that lacked components of the G protein-signaling system. It was observed that more than one-half of all the transcripts changed by hypovirus infection were also changed in at least one G protein-mutant strain, with one third being changed in both (Dawe, Segers, Allen, McMains, & Nuss, 2004). Significantly, 95% of the cochanged genes were coordinately altered (i.e., either up- or downregulated, but the same direction in each data set). Comparing the levels of G proteins present also showed that these affects were posttranscriptional: mRNA for the G proteins was not reduced, but the protein levels for both Gα (CPG-1) and Gβ (CPGB-1) were (Dawe et al., 2004).

The significant, but not complete, overlap of the microarray data also demonstrated that, while G protein signaling can be concluded to be a contributor to the hypovirus-infected phenotype, it does not fully explain all the changes. Chen, Gao, Choi, and Nuss (1996) had noted that cAMP levels were elevated, as could be predicted since the Gα_i family of proteins (of which CPG-1, by sequence similarity, is a member) act to inhibit adenylyl cyclase (Kaziro, Itoh, Kozasa, Nakafuku, & Satoh, 1991). As noted above, Larson et al. (1992) linked reduced laccase expression to modulation of calcium signaling. Clearly, the effects of hypovirus infection on the host colony are the product of disruption of major signaling pathways that, in turn, may affect many other components through extensive cross talk. This leaves open two other major questions: one, if genes are being regulated in response to hypovirus replication, what are the transcription factors that control these responses? Two, is there a uniform mechanism among other mycoviruses for inducing the common phenotypic changes observed in infected colonies, principally that of reduced virulence?

To address the first question, recent studies have focused on transcription factors *Cpst12* (Deng, Allen, & Nuss, 2007) and *pro1* (Sun, Choi, & Nuss, 2009a), which were both identified as downregulated by hypovirus infection in the earliest microarray study (Allen et al., 2003). *Cpst12* was also of interest due to its relationship to similarly named genes from a collection of other filamentous fungi that are related to phenotypes that include virulence (Deng, Allen, & Nuss, 2007). Pertinently, the originator of the family name, STE-12p from *S. cerevisiae,* is activated in response to pheromone stimulation in a G protein-dependent manner and is crucial for mating (Fields & Herskowitz, 1987). Although dispensable for vegetative growth, a *ΔCpst12* strain was severely reduced for virulence and was also female infertile, another trait characteristic of hypovirus-infected mycelium (Deng, Allen, & Nuss, 2007). In contrast, a *Δpro1* strain retained full virulence while being female-sterile (Sun et al., 2009a). Microarray analysis showed considerable overlap between the genes altered in expression by the absence of *Cpst12* and the presence of hypovirus (Deng, Allen, Hillman, & Nuss, 2007) while there appeared to be a role for *pro1* in maintenance of viral replication, since virus-free sectors arose at greater frequency from *pro1*-deficient colonies (Sun et al., 2009a).

Addressing the second issue of a correlation between viral mechanisms, Allen and Nuss (2004a) compared the transcriptional changes between colonies containing CHV-1/EP713 and CHV-1/Euro7. Phenotypically, the impact of the CHV-1/Euro7 virus results in vegetative colony morphology and development as well as a level of fungal virulence, that is part way between that seen for CHV-1/EP713-infected and wild-type colonies. Therefore, mycelium containing CHV-1/EP713 exhibits traits considered "severe" while the effects of CHV-1/Euro7 may be referred to as "mild"

(Chen & Nuss, 1999). Intriguingly, while approximately one-half the number of genes was altered in expression in the context of CHV-1/Euro7 compared to CHV-1/EP713 (166 compared to 295), one-half of these (80, or 3.6% of the genomes coding potential) represented the same transcripts, suggesting significant overlap in the relative impacts. Many of these genes were noted in functional groups that included aspects of carbohydrate, lipid, and nucleic acid metabolism as well as general stress response (Allen & Nuss, 2004a). A similar transcriptional analysis of two unrelated viruses, members of the *Reoviridae*, was conducted by Deng, Allen, et al. (2007). In contrast to the CHV1 group, the reoviruses in question, MyRV1-Cp9B21 or MyRV2-CpC18, do not alter pigment production or conidiation. However, they do cause a severe reduction in virulence (Enebak, MacDonald, & Hillman, 1994; Hillman, & Suzuki, 2004). The transcriptional data supported the hypothesis that hypovirus and reovirus infections perturb common and specific pathways to cause hypovirulence and further supported the contention that modification of G protein signaling was a crucial step in the infection phenotype.

Besides G protein signaling, work by Turina, Zhang, and Van Alfen (2006) has also pointed to the relevance of the class of MAPKs (mitogen-activated protein kinases). MAPKs are a conserved group of signaling molecules that mediate responses to a diverse range of stimuli including osmotic stress and heat shock (Pearson et al., 2001). A member of this class, Cpkk1, was shown to be altered in its phosphorylation state (Turina et al., 2006) and, subsequently, to be required for virulence (Rostagno, Prodi, & Turina, 2010). Even more recently, protein secretion, likely an important component of virulence, has also been shown to be affected. Secreted aspartyl proteases are an important class of proteins that have associations with the virulence of other fungi including *Candida albicans* (De Bernardis, Sullivan, & Cassone, 2001) and *Batrachochytrium dendrobatidis* (Joneson, Stajich, Shiu, & Rosenblum, 2011). The fact that Jacob-Wilk, Moretti, Turina, Kazmierczak, and Van Alfen (2012) have demonstrated that their expression varies in the presence of CHV-1 suggests an important connection. Furthermore, a second study (Kazmierczak, McCabe, Turina, Jacob-Wilk, & Van Alfen, 2012) has provided further evidence that protein secretion is significantly compromised by CHV-1.

Together, these data have illustrated the pleiotropic effects of the presence of replicating RNA on the fungal host. These affects are specific in some cases, presumably directed at host factors important for viral maintenance and replication, but likely also reflect generic host responses to stress. Recent development of an additional technical tool by Willyerd, Kemp, and Dawe (2009), a controlled expression system using a repressible promoter, presents the intriguing possibility of controlling the initiation of hypovirus replication thus opening up to analysis the suite of changes that occur in the host as the hypovirus genome accumulates. Coupled with the recent completion of the *C. parasitica* genome (http://genome.jgi-psf.org/Crypa2/Crypa2.home.html), this experimental system is now poised to exploit the rapidly evolving development of cost-effective next-generation sequencing technologies and provide unique insights into the mycovirus–host interaction.

6.2. Host physiology

The first microarray studies had shown conclusively that a wide variety of host processes were impacted by hypovirus infection. This included elements of primary metabolism and energy generation pathways, such as glycolysis, given the changes that were observed in the expression of genes (Allen et al., 2003). In fact, clues to the changes related to host physiology were also noted some time earlier. Polashock and Hillman (1994) first observed a mitochondrial connection when they isolated a small dsRNA element from a moderately hypovirulent strain of *C. parasitica*. This element was not related to the *Hypoviridae*, but was found to reside in the mitochondrion. In a similar development, Monteiro-Vitorello, Bell, Fulbright, and Bertrand (1995) mutagenized virulent strain EP155 with ultraviolet light, then selected for slow-growing mutants. These mutants were then further screened for mitochondrial dysfunction by selecting for cyanide-resistant respiratory activity, indicating the presence of alternative oxidase that is induced when mitochondrial respiration fails. Both nuclear and mitochondrial respiration-defective mutants were found this way, with the mitochondrial ones of particular interest, since they were causative of hypovirulence and cytoplasmically transmissible between strains.

Returning to these same strains nearly a decade later, Allen and Nuss (2004b) were able to apply the then-recently developed microarray system. Dramatically, a collection of 70 genes, representing almost half of the number of detected changes in the CHV-1/EP713-infected culture, were found to be also changed in the context of the mitochondrial mutant *mit2*. Adding to the correlation, all of these 70 genes except two were coordinately changed (up or down) in expression (Allen & Nuss, 2004b). This firmly established mitochondrial dysfunction as a key physiological abnormality that was part of the consequences of hypovirus infection.

However, while the data above conclusively tied the hypovirulence-associated traits to perturbed basic cellular physiological functions, it was not until more recently that modified micromeasurement techniques and high-resolution gas analysis were used to directly assess the changes in physiological output. Using methods originally developed for the analysis of small biomass quantities such as small groups of *Caenorhabditis elegans* nematodes (Van Voorhies & Ward, 1999) and individual *Drosophila melanogaster* fruit flies (Melvin, Van Voorhies, & Ballard, 2007; Van Voorhies, Khazaeli, & Curtsinger, 2004), Dawe, Van Voorhies, Lau, Ulanov, and Li (2009) measured CO_2 production and O_2 consumption of small portions of differently aged fungal mycelium. They found significant physiological variations that occurred as a result of aging and differentiation in both infected and uninfected mycelium but that the presence of the hypovirus affected the extent of these changes. The most revealing measurement was the respiratory quotient (RQ), the ratio of CO_2 production to O_2 consumption, which has been extensively used to estimate the metabolic substrates used during oxidative phosphorylation (Gessaman & Nagy, 1988; Walsberg & Wolf, 1995) and where values above or below 1.0 indicate that the cell is utilizing metabolic processes, such as fermentative metabolism (>1.0) or substrates such as lipids or proteins (<1.0). A value of 1.0 indicates that carbohydrates are the main substrates being oxidized. As can be seen in Fig. 5.4, the metabolism of uninfected mycelium shifts from fermentative on the periphery through complete carbohydrate oxidation (correlating with the productive zone of the colony where biomass is increasing most rapidly) before finally transitioning into metabolic modes likely based on utilizing lipids and/or proteins in the aged colony interior. Interestingly, colonies infected with the hypovirus CHV-1/EP713 did not appear to transition into the metabolic pathways characteristic of older mycelium, suggesting a failure of age-dependent developmental transitions. This may support the contention of McCabe and Van Alfen (2001) that the reduced pigmentation and asexual sporulation in a hypovirus-infected colony can be likened to retaining the mycelium in a juvenile state.

Metabolomics analysis of hypovirus-infected mycelium further supported the contentions of major shifts in cellular physiology. Of 165 positively identified compounds, approximately one third were significantly altered compared to uninfected mycelium (Dawe et al., 2009). The identities of these compounds, including many precursors/intermediates in carbohydrate and lipid metabolism, were consistent with the slower growth, the altered gene expression, and the measured physiological changes. Further

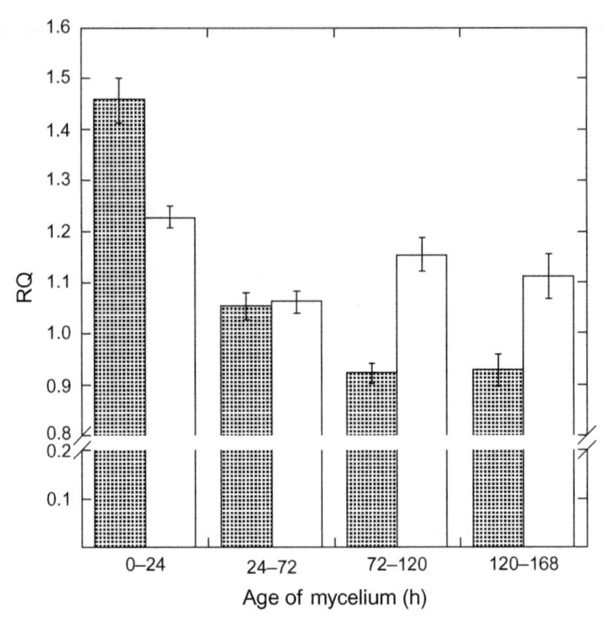

Figure 5.4 Hypovirus CHV-1/EP713 influences age-related changes in primary metabolic pathways. Mycelial plugs were removed sequentially from outer to inner regions from potato dextrose agar (solid medium) cultures of strains EP155 (shaded bars) and EP155 infected with CHV-1/EP713 (open bars) such that differently aged fungal tissues were tested from the same colony. RQ values were calculated and used to estimate the metabolic substrates used during oxidative phosphorylation as described in the text. Error bars represent the standard error of each dataset. *From Dawe et al. (2009)*

studies will, no doubt, be needed to clarify more specifically the nature of these changes, but it is clear that replication of this mycovirus results in transcriptional reprogramming, in turn, leading to basic changes in primary cellular metabolism. Whether these changes reflect a direct manipulation in order to, for instance, provide more suitable membrane surface material on which the hypovirus-encoded RDRP might replicate the viral genome, or represent a more general response, remains to be elucidated.

7. FUNGAL ANTIVIRAL DEFENSE MECHANISMS

Plants and animals employ innate immunity responses to virus infection; the interferon response predominates in animals (reviewed in Fensteri & Sen, 2009; Grandvaux, tenOever, Servant, & Hiscott, 2002) and the RNA silencing response is the primary defense in plants (Ding, 2010). Antiviral defense is less well studied in fungi, but two clear lines of defense against hypovirus infection have been demonstrated for *C. parasitica*. These include an inducible RNA silencing pathway at the cellular level and the non-self-recognition system, termed vegetative incompatibility, at the population level.

7.1. RNA silencing antiviral defense response in *C. parasitica*

RNA silencing is a conserved RNA-mediated, sequence-specific eukaryotic gene silencing mechanism involved in numerous biological process in plants and animals that include developmental and metabolic regulation, RNA stability and processing, chromosomal dynamics, and host defense (Carthew & Sontheimer, 2009; Ding & Voinnet, 2007; Ghildiyal & Zamore, 2009). RNA silencing pathways are also found in fungi. In fact, the model fungus *Neurospora crassa* was one of the first organisms used to study RNA silencing (reviewed by Li, Chang, et al., 2010) leading to the identification of the first RNA silencing pathway gene, *qde-1*, encoding the RDRP QDE-1 (Cogoni & Macino, 1999). However, the role for RNA silencing in fungi does not appear to be as extensive as shown for plants and animals. There is currently no evidence that RNA silencing contributes to developmental or metabolic regulation in fungi. Functions are currently limited to heterochromatin formation, silencing of unpaired DNA during meiosis, silencing of transposons, and other repetitive DNA and antiviral defense (reviewed in Dang, Yang, Xue, & Li, 2011).

7.1.1 Components of the C. parasitica *RNA silencing antiviral defense pathway*

The evidence that RNA silencing serves as an antiviral defense response in fungi comes primarily from studies with *C. parasitica* in which the disabling of the RNA silencing pathway by disruption of genes encoding specific core RNA silencing components, Dicer, or Argonaute, was shown to dramatically increased susceptibility to mycovirus infection (Segers et al., 2007; Sun, Choi, & Nuss, 2009b). In general terms, the roles of these conserved ribonucleases in the antiviral defense response include recognition of viral ds or structured RNA by Dicer and use of the associated RNase III-type activity to process these RNAs into small RNAs of 21–24 nts in length, termed virus-derived small (vs) RNAs. The Argonaute protein then facilitates the incorporation of the vsRNAs into an effector complex where one strand of the vsRNA is removed. The remaining strand then guides the effector complex to the cognate viral RNA, which is cleaved, or sliced, by the Argonaute-associated RNase H-like activity (reviewed in Ding, 2010).

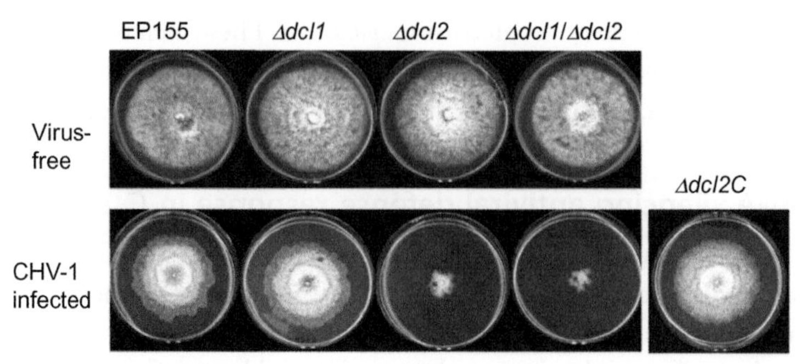

Figure 5.5 Effect of hypovirus infection on colony morphology and growth characteristics of *C. parasitica* Dicer gene disruption mutant strains. Top row: Disruption of Dicer genes *dcl1*, *dcl2*, or both *dcl1* and *dcl2* resulted in no observable phenotypic changes in the absence of virus infection. Bottom row: Hypovirus CHV-1/EP713-infected Dicer disruption mutant strain *Δdcl1* and a complemented *dcl2* mutant strain, *Δdcl2-2C*, exhibited symptoms identical to CHV-1/EP713 infected *C. parasitica* wild-type strain EP155. In sharp contrast, CHV-1/EP713 infection of the *Δdcl2* and *Δdcl1/Δdcl2* double dicer mutant strain resulted in a severe debilitation phenotype. *From Segers et al. (2007).* (See Page 9 in Color Section at the back of the book.)

Although the *C. parasitica* genome contains two Dicer genes and four Argonaute genes, only Dicer gene *dcl2* and Argonaute gene *agl2*, are required for the antiviral defense response (Segers et al., 2007; Sun, Choi, & Nuss, 2009b). As shown in Fig. 5.5, no observable phenotypic consequences resulted from disruption of either or both Dicer genes in the absence of virus infection. A similar lack of phenotypic change was observed for mutant strains in which the individual Argonaute genes, *agl1–agl4*, were disrupted (Sun et al., 2009b). A different result was observed when hypovirus CHV-1/EP713 infection was initiated in the Dicer and Argonaute mutant strains. The *dcl2* and *agl2* mutant strains became severely debilitated (Fig. 5.5 and Sun et al., 2009b). Moreover, viral RNA accumulation, particularly the viral positive-sense single-stranded RNA fraction, very significantly increased in the *dcl2* and *agl2* mutant strains (Fig. 5.6). The production of hypovirus-derived vsRNA was also demonstrated and shown to be dependent on *dcl2* and independent of *dcl1* (Zhang, Segers, Sun, Deng, & Nuss, 2008).

7.1.2 Induction and suppression of the RNA silencing antiviral defense pathway

Examination of the Dicer and Argonaute gene family expression levels revealed that *dcl2*, but not *dcl1*, transcript accumulation increased 10- to 15-fold following infection by CHV-1/EP713 or by the unrelated mycoreovirus MyRV1-9B21 (Zhang et al., 2008). Transcript accumulation for *agl2*

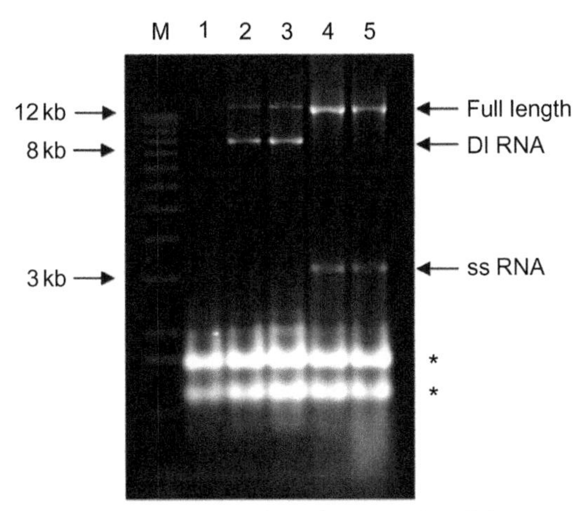

Figure 5.6 Agarose gel (1%) analysis of total RNA isolated from virus-free wild-type *C. parasitica* strain EP155 (lane 1), hypovirus CHV-1/EP713-infected strain EP155 (lane 2), CHV-1/EP713-infected Argonaute mutant strain *Δagl1* (lane 3), CHV-1/EP713-infected Argonaute mutant strain *Δagl2* (lane 4), and CHV-1/EP713-infected Dicer mutant strain *Δdcl2* (lane 5). The lane marked "M" contains 1-kb DNA size markers. The migration positions of replicative double-stranded (ds)RNAs corresponding to full-length CHV-1/EP713 and related defective interfering (DI) dsRNAs, and full-length CHV-1/EP713 single-stranded (ss)RNA are indicated by arrows at the right. The asterisks indicate the migration positions of *C. parasitica* ribosomal RNAs. *From Sun et al. (2009b).*

increased only modestly in response to hypovirus infection (Sun et al., 2009b), however, *agl2* was found to be required for *dcl2* transcript induction in response to virus infection. Interestingly, the hypovirus-encoded suppressor of RNA silencing, p29, was found to have a profound effect on the transcriptional induction of both *dcl2* and *agl2*. The accumulation of *dcl2* transcripts increased by 30- to 40-fold in response to infection by a CHV-1/EP713 virus lacking p29 (Δp29) compared to the 10- to 15-fold observed after wild-type virus infection. Accumulation of *agl2* transcripts increased over 14-fold in response to Δp29 compared to approximately two-fold in response to wild-type virus. Moreover, this difference in *agl2* transcript accumulation levels in response to wild-type and Δp29 viruses was dependent on the *agl2* promoter region. Thus, p29 appears to suppress RNA silencing, at least in part, by suppressing the transcriptional activation of the RNA silencing pathway in response to virus infection.

7.1.3 Role of RNA silencing in viral RNA recombination

It was shown during the very earliest molecular characterizations of hypovirus RNAs that this group of viruses tends to generate defective interfering (DI) RNAs at a high frequency (Hillman, Foglia, & Yuan, 2000; Shapira,

Choi, Hillman, & Nuss, 1991). Thus, it was of considerable interest to observe the absence of viral DI RNAs in extracts prepared from CHV-1/EP713-infected *Δdcl2* (Zhang & Nuss, 2008) and *Δagl2* (Sun et al., 2009b) mutant strains (Fig. 5.6). Not only did DI RNAs fail to form when the *Δdcl2* or *Δagl2* mutant strains were infected with CHV-1/EP713 by transfection with *in vitro*-synthesized transcripts, but also the DI RNAs present in CHV-1/EP713-infected wild-type strain EP155 disappeared when viral RNA was transferred to *Δdcl2* by anastomosis and reappeared when the reverse transfer was performed (Zhang & Nuss, 2008).

The apparent requirement of the RNA silencing pathway for generation of DI RNA suggested that the RNA silencing pathway also contributed to the instability of nonviral nucleotide sequences previously observed for hypovirus-based recombinant viral vectors (Suzuki et al., 2000). This was tested by transfecting several CHV-1/EP713-based vector viruses containing the enhanced green fluorescence protein coding domain into wild-type and Dicer and Argonaute mutant strains. As previously observed, the EGFP expression disappeared in the transfected wild-type strain within two colony transfers. In contrast, EGFP expression and vector virus integrity were retained in the *Δdcl2* and *Δagl2* mutant strains for prolonged culturing and transfers (Sun et al., 2009b; Zhang & Nuss, 2008). In this regard, we now routinely use the *Δdcl2*-infected strain as an experimental tool because of the decreased recombination and increased accumulation of CHV-1/EP713 RNA and proteins (X. Zhang & D. L. Nuss, unpublished observation). A role for RNA silencing in viral RNA recombination was recently reported for brome mosaic virus (Dzianott, Sztuba-Solinska, & Bujarski, 2012). These combined observations are the first to demonstrate the contribution of an RNA silencing pathway to viral RNA recombination and are of potentially broad significance with possible applications for stable production of biologicals, including vaccines, with viral RNA-based vectors.

7.2. Restriction of hypovirus transmission by vic

7.2.1 The C. parasitica *vic genetic system*

While the RNA silencing pathway serves as antiviral defense at the cellular level, the *C. parasitica* vic non-self-recognition system provides antiviral defense at the population level by restricting hypovirus transmission. Incompatible interactions result in localized programmed cell death that restricts exchange of cellular contents (Leslie & Zeller, 1996), the principle mode of mycovirus transmission (Buck, 1986). In this regard, Caten (1972) predicted that "fungal vegetative incompatibility will markedly reduce the spread of suppressive, cytoplasmic genetic elements, including viruses, from strain to strain in nature." This prediction was confirmed for *C. parasitica* by studies conducted at the Connecticut Agricultural Experiment Station in the early 1980s as reviewed by Anagnostakis (1982a). Transmission of the hypovirulence phenotype by anastomosis of a virulent strain with a hypovirulent strain (conversion of a virulent strain to hypovirulence) was shown to be restricted when the two strains belonged to different vic groups (Anagnostakis, 1983). Results of subsequent field studies conducted both in Europe and North America generally agreed that hypovirus transmission and biological control are more effective when *C. parasitica* populations have a low level of vic diversity (Anagnostakis, Hau, & Kranz, 1986; Heiniger & Rigling, 1994; Milgroom & Cortesi, 2004, Robin, Anziani, & Cortesi, 2000; Robin, Capdeville, Martin, Traver, & Colinas, 2009). The vic diversity is generally low in those regions of Europe where hypovirulence was first discovered and spreads naturally, while diversity is high in North America where instances of natural hypovirulence are limited and efforts to establish sustained hypovirulence by artificial introduction has been largely unsuccessful (reviewed in Anagnostakis, 1982a; Milgroom & Cortesi, 2004).

A series of genetic studies revealed that the vic system in *C. parasitica* is controlled by at least six genetic *vic* loci with only two alleles at each locus (Anagnostakis, 1982b; Cortesi & Milgroom, 1998; Huber, 1996). A significant advance was recently reported in the molecular identification of genes at four of the six *vic* loci using a comparative genomics approach (Choi et al., 2012). This approach was predicated on the prediction that *C. parasitica vic* loci, like the characterized corresponding loci in the model filamentous fungi *Neurospora crassa* (Glass & Dementhon, 2006) and *Podospora anserine* (Paoletti, Saupe, & Clave, 2007), would exhibit a level of allelic nucleotide sequence polymorphism and enabled by a number of genetic and genomic advances. These included a collection of 64 *C. parasitica* strains that represent all genotypes possible across the six *vic* loci assembled by Cortesi and Milgroom (1998), a genetic linkage map containing nucleotide markers linked to five of the six *vic* loci (Kubisiak & Milgroom, 2006) and a reference *C. parasitica* genome sequence assembly generated by the Department of Energy/Joint Genome Institute.

7.2.2 Identification and characterization of C. parasitica *vic genes*

C. parasitica strains that contain different alleles at one or more *vic* loci undergo an incompatible reaction when paired on artificial media resulting in a line of demarcation, or barrage, along the contact zone, due to cell death

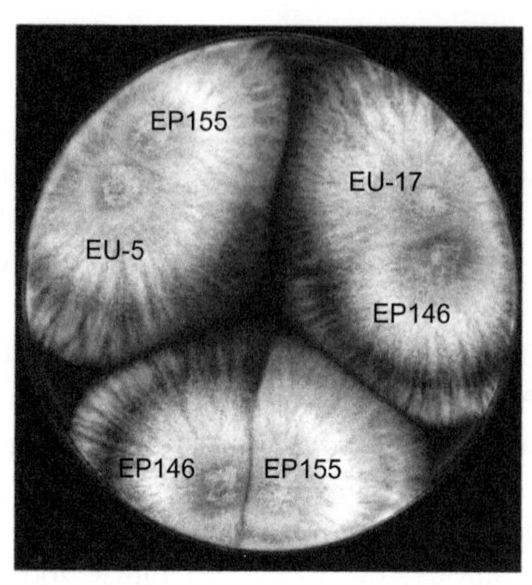

Figure 5.7 Vegetative incompatibility (*vic*) genotyping of *C. parasitica* strain EP146. Strains of *C. parasitica* that differ at one or more *vic* loci form barrage lines (a line of dead cells) when colonies of the two strains merge as observed for the pairing of strains EP155 and EP146 (bottom of the culture plate). Cortesi and Milgroom (1998) established a collection of 64 *C. parasitica* tester strains that represent all possible genotypes arising from the six genetically determined *vic* loci. Strain EP155, the strain used for generation of the reference genome sequence (http://genome.jgi-psf.org/Crypa2/Crypa2.home.html) was previously reported (Cortesi & Milgroom, 1998) to give a compatible reaction (fusion of hyphae with no barrage formation) with tester strain EU-5 (*vic* genotype *2211-22*) as shown at the left of the culture plate. Strain EP146, that was resequenced by Choi et al. (2012), was found to be compatible with tester strain EU-17, as shown at the right side of the culture plate, and thus to have the *vic* genotype *2112-11. From Choi et al. (2012).* (See Page 9 in Color Section at the back of the book.)

(Fig. 5.7). In contrast, strains that have the same alleles at all *vic* loci fuse upon contact. Thus, it is possible to determine the *vic* genotype of a strain of interest by pairing with the set of 64 genotyped tester strains to identify which tester fails to form a barrage. As indicated in Fig. 5.7, the reference strain EP155 is compatible with tester strain EU-5 that has the *vic* genotype *2211-22*, where each number refers to the allele present at the respective *vic* loci, that is, *vic1-2*, *vic2-2*, *vic3-1*, *vic4-1*, *vic6-2*, and *vic7-2*. The strain chosen for resequencing for the comparative sequence analysis, strain EP146, was found to be compatible with tester strain EU17 that has the *vic* genotype *2112-11*. Thus, the two strains contain different alleles at *vic2*, *vic4*, *vic6,* and *vic7*.

The EP146 genome was sequenced to over 11 × coverage using Roche 454 high-throughput sequencing technology and the individual reads were mapped by sequence homology to the scaffolds comprising the EP155

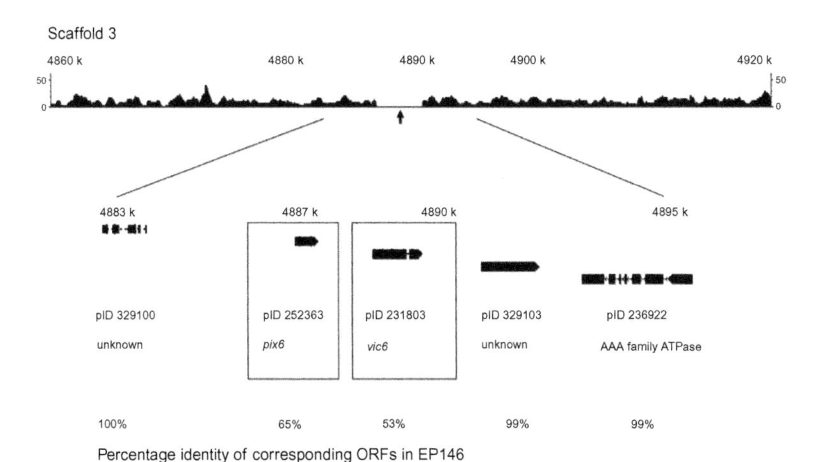

Figure 5.8 The candidate *C. parasitica vic6* locus: the region of sequence polymorphism for *C. parasitica* strains EP155 and EP146 and machine-annotated ORFs located near the genetic linkage map-predicted *vic6* locus (Kubisiak & Milgroom, 2006). The ~3.7-Kbp region of polymorphism between the EP155 and EP146 genome sequences on Scaffold 3 in the region mapped for the *vic6* locus is indicated by the absence of 454 sequence reads generated from EP146 DNA that match the corresponding EP155 reference genome sequence (arrow). The ORFs located within a 13.4-Kbp region containing the polymorphic locus, with accompanying protein ID numbers, are shown below the 454 sequence read track. The boxed candidate *vic6* gene (pID 231803), which contains a HET domain, and adjacent ORF *pix6* (pID 252363), showed a greater degree of sequence polymorphism than surrounding ORFs, which showed near perfect matches. *From Choi et al. (2012).*

reference genome assembly. Using *vic*-linked sequence markers generated by Kubisiak and Milgroom (2006) as a guide, the reference genome sequence was scanned for candidate *vic* gene polymeric regions that appeared as gaps in the density plot of matched 454 sequence reads leading to the identification of candidate loci for *vic2*, *vic4*, *vic6*, and *vic7* (see Fig. 5.8 for the *vic 6* candidate).

The polymorphic candidate *vic7* locus contained a gene that encodes a protein with a HET domain that is conserved in 9 of the 10 incompatibility genes previously characterized for *N. crassa* and *P. anserina* (Paoletti et al., 2007). The corresponding alleles in EP155 and EP146, designated *vic7-2* and *vic7-1*, respectively, are 87% identical with significant polymorphism limited to a small portion of the C-terminal domain. The candidate *vic6* locus also contained an ORF that encodes a HET domain-containing protein. In this case, the alleles for EP155 and EP146, designated *vic6-2* and *vic6-1*, respectively, shared only 53% overall identity and exhibited significant polymorphism on both sides of the conserved HET domain (Choi et al., 2012). The *vic6* locus also contained a small ORF that was subsequently given the allelic designations *pix6-2* and *pix6-1* in strains EP155 and EP146, respectively. This candidate gene showed polymorphism only in the C-terminal half of the predicted amino acid sequence (Choi et al., 2012) and will be further discussed in a later section.

The candidate *vic4* locus was found to be idiomorphic rather than polymorphic. That is, the allele in EP155, *vic4-1*, that encodes a predicted 359aa protein kinase c-like protein, was completely replaced in EP146, *vic4-2*, by a larger ORF specifying a 1628 aa protein containing NACHT-NTPase and WD repeat domains characteristic of a number of other fungal incompatibility genes (reviewed in Chevanne, Saupe, Clave, & Paoletti, 2010).

The polymorphic candidate *vic2* locus contained two open reading frames that encode proteins lacking any sequence motifs characteristic of fungal incompatibility genes. One ORF encodes a member of the patatin-like phospholipase family and the other encodes a protein related to the fungal plasmid membrane SNARE Sec9 protein. The allelic designations for these two genes are *vic2-2* and *vic2a-2* for EP155 and *vic2-1* and *vic2a-1* in EP146.

The functional role of the polymorphic *vic* alleles *vic2-2*, *vic6-2,* and *vic7-2* was examined by gene disruption analysis. The candidate *vic4* alleles were not examined because heteroallelism at the *vic4* locus was shown not to restrict hypovirus transmission, even though it does result in barrage formation (Cortesi, McCulloch, Song, Lin, & Milgroom, 2001). Interestingly, the *vic2-2*, *vic6-2*, and *vic7-2* disruption mutants exhibited an increase in virus transmission without an accompanying loss of barrage formation. Thus, as observed for heteroallelism at *vic4*, barrage formation and restriction of virus transmission are not necessarily tightly coupled. Interestingly, the increase in virus transmission for the *vic* gene disruption mutants was asymmetrical. A dramatic increase (from 5% to 100%) in transmission was observed for the *vic2-2* and *vic6-2* disruption mutants, but only when the mutant strain served as the recipient in the pairing and not when it served as the donor strain. Virus transmission for the *vic7* locus is naturally asymmetric in that virus transmission is not inhibited when the recipient contains the *vic7-2* allele (Cortesi et al., 2001) but is restricted when the recipient contains the *vic7-1* allele. One interpretation of the asymmetry in virus transmission rates observed for the *vic* gene disruption mutants is that the disruption of only one of the alleles for a heteroallelic pair causes a delay in PCD to an extent that virus transmission increases, especially when the mutant serves as the recipient, but not to the extent that barrage formation is prevented. This is consistent with the observations of Biella, Smith, Aist, Cortesi, and Milgroom (2002) showing a negative correlation between cell death and hypovirus transmission: virus transmission occurred at a higher frequency into *vic* strains, for example, *vic7*, that exhibited delayed PCD.

Although barrage formation was not eliminated by disruption of *vic2-2*, *vic6-2*, and *vic7-2*, the observed increases in virus transmission was consistent with a role of these candidate *vic* alleles in vic. A definitive role of the *vic6* alleles in vic was confirmed by independently disrupting both alleles. As was observed for the *vic6-2* mutant, disruption of the *vic6-1* allele failed to abolish barrage formation with a heteroallelic strain. However, when the two mutant strains were paired, no barrage formed, virus transmission occurred 100% in both directions and stable heterokaryons were readily formed. On the surface, a mechanism involving allelic interactions, that is, interactions between *vic6-1* and *vic6-2*, would appear to underlie the triggering of PCD at the *vic6* locus. However, the actual mechanism was shown to involve nonallelic interactions.

The indication that PCD at the *vic6* locus was triggered by interactions other than between *vic6-1* and *vic6-2* came from the observation that replacement of a disrupted *vic6-2* allele with the intact *vic6-1* allele resulted in abnormal growth characteristics and colony morphology. It was subsequently shown that disruption of the ORF adjacent to the *vic6* allele, *pix6*, also resulted in increased virus transmission without eliminating barrage formation. However, the increase in virus transmission occurred when the *pix6* mutant was the donor strain rather than the recipient strain as observed for the *vic6* mutants. As observed for the pairing of *Δvic6-1* and *Δvic6-2* mutants, the paired *Δpix6-1* and *Δpix6-2* mutants failed to form barrages and showed no resistance to virus transmission. Importantly, the disruption of *pix6-2* and *vic6-2* in the same strain eliminated barrage formation and resistance to virus transmission when paired with a *pix6-1*, *vic6-1* strain.

A model for interactions between the linked genes at the *vic6* locus is shown in Fig. 5.9. The limited set of possible interactions between the tightly linked polymorphic *vic6* and *pix6* alleles when two *vic6* heteroallelic strains are paired include two allelic interactions (*pix6-1* with *pix6-2* and *vic6-1* with *vic6-2*) and two nonallelic interactions (*pix6-1* with *vic6-2* and *pix6-2* with *vic6-1*). Since barrage formation was eliminated when one, but not both, allelic interaction was disrupted, potential *pix6* or *vic6* allelic interactions alone are clearly insufficient to cause barrage formation. In contrast, barrage formation was observed for every mutant pairing in which one

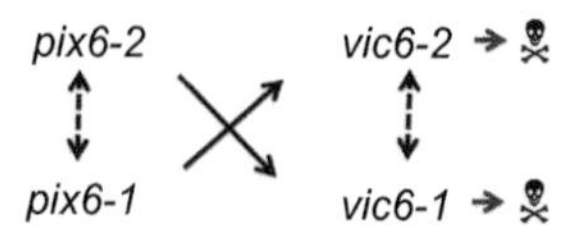

Figure 5.9 Model for interactions between the linked genes *pix6* and *vic6* at the *vic6* locus. Nonallelic interactions leading to programmed cell death as supported by data presented in Choi et al. (2012) are indicated by solid arrows. Arrow direction reflects how the *pix6* gene product potentially acts in *trans* with the *cis*-acting *vic6* gene product to effect asymmetric virus transmission. No evidence was obtained in that study for potential allelic interactions (dashed double arrows). *Modified from Choi et al. (2012).* (For color version of this figure, the reader is referred to the online version of this chapter.)

of the possible nonallelic interactions remained intact. However, the increase in virus transmission observed when one nonallelic interaction was disrupted indicates that the strength of the incompatible reaction is diminished. The fact that the *vic 2* and *vic7* also contain multiple tightly linked polymorphic genes (Choi et al., 2012) suggests the possibility of nonallelic interactions involved in triggering PCD at other *vic* loci.

The observation by Biella et al. (2002) that virus transmission was correlated with a delay in PCD may also hint at potential virus-mediated interruption of host signaling pathways related to cell death. It is highly unlikely that the protein products of the *vic* alleles are themselves also directly acting to mediate the cellular changes necessary to cause PCD. As has been noted in related fungi such as *N. crassa* and *P. anserina* (Saupe, 2000), it is expected that downstream signaling components including transcription factors will be found to be required to propagate the changes triggered by the mismatched *vic* alleles. Such a signaling pathway would then be a target for virus-mediated interference, on the premise that there would be a selective advantage for viruses that could inhibit or delay PCD, thus allowing transfer even between incompatible strains. With the advent of the genome sequence data, there are now opportunities to explore this hypothesis fully.

8. CONCLUDING REMARKS

The primary contribution of the hypovirus reverse genetics system to mycovirus research was in providing the means for going beyond correlative evidence linking mycovirus infection with altered host phenotype to a conclusive demonstration of cause and effect, that is, completion of Koch's postulates. However, with increasing advances in genomics and related technologies, the hypovirus/*C. parasitica* experimental system has added capabilities to address emerging research areas and provide solutions to long-standing problems in mycovirus research.

Investigation of the RNA silencing pathways in *C. parasitica* uncovered new insights into the induction and suppression of RNA silencing as an antiviral defense response and a role for RNA silencing in viral RNA recombination. These findings have implications for the regulation of the RNA silencing antiviral defense response in higher eukaryotes, the emergence of new viruses, and the development of more effective RNA virus vectors. The identification of genes involved in the *C. parasitica* vic system, long known to restrict hypovirus transmission, illustrates the successful application to a classic problem. An understanding of the molecular genetics underlying this fungal non-self-recognition system, including genes involved in downstream signaling leading to programmed cell death, holds promise for modulating the incompatible reaction to enhance virus transmission. The system is also poised to further exploit transcriptomic, proteomic, and metabolomic approaches to gain a comprehensive view of the relationship between virus-mediated changes in host gene expression and phenotype. It may be possible to use such knowledge to more precisely engineer hypoviruses that effectively retain the ability to reduce fungal virulence without negatively affecting ecological fitness and ability of the infected hypovirulent strains to spread and persist in the ecosystem.

While considerable progress has been made in advancing hypovirus molecular biology, translation of the new knowledge into enhanced biological control potential has lagged. This is partly due to the complexity and time constraints associated with studies conducted in the field. It is also related to the limited resources available for funding of such studies. However, there is reason to be optimistic that the pace of progress may increase. The identified *vic* genes provide powerful new tools for population genetic studies to fully understand the contribution of vic diversity to hypovirus transmission and biological control in forest ecosystems. The development of chestnut trees with increased blight resistance through breeding backcross programs involving resistant Asian chestnut trees (Burnham, Rutter, & French, 1986) provides opportunities to test the effectiveness of combining disease resistance and hypovirulence for reforestation. The potential for expanding hypovirulence to important crop pathogenic fungi, through either initiating hypovirus infection or using endogenous mycoviruses, increases with advances in fungal genetic manipulations and mycovirus molecular biology. Irrespective of the pace of this translational research, it is anticipated that the hypovirus/*C. parasitica* experimental system will continue to make major contribution to the advancement of mycovirus research.

REFERENCES

Allen, T. D., Dawe, A. L., & Nuss, D. L. (2003). Use of cDNA microarrays to monitor transcriptional responses of the chestnut blight fungus *Cryphonetria parasitica* to infection by virulence-attenuating hypoviruses. *Eukaryotic Cell*, *2*, 1253–1265.

Allen, T., & Nuss, D. (2004a). Specific and common alterations in host gene transcript accumulation following infection of the chestnut blight fungus by mild and severe hypoviruses. *Journal of Virology*, *78*, 4145–4155.

Allen, T. D., & Nuss, D. L. (2004b). Linkage between mitochondrial hypovirulence and viral hypovirulence in the chestnut blight fungus revealed by cDNA microarray analysis. *Eukaryotic Cell*, *3*, 1227–1232.

Amselem, J., Cuomo, C. A., van Kan, J. A. L., Viaud, M., Benito, E. P., et al. (2011). Genomic analysis of the Necrotrophic fungal pathogens *Sclerotinia sclerotiorum* and *Botrytis cinerea*. *PLoS Genetics*, 7, 21002230.

Anagnostakis, S. L. (1982a). Biological control of chestnut blight. *Science*, *215*, 466–471.

Anagnostakis, S. L. (1982b). Genetic analysis of *Endothia parasitica*: Linkage map of four single genes and three vegetative compatibility types. *Genetics*, *102*, 25–28.

Anagnostakis, S. L. (1983). Conversion to curative morphology in *Endothia parasitica* and its restriction by vegetative compatibility. *Mycologia*, *79*, 23–37.

Anagnostakis, S. L., Chen, B., Geletka, L. M., & Nuss, D. L. (1998). Hypovirus transmission to ascospore progeny by field-released transgenic hypovirulent strains of *Cryphonectria parasitica*. *Phytopathology*, *88*, 598–604.

Anagnostakis, S. L., Hau, B., & Kranz, J. (1986). Diversity of vegetative compatibility groups of *Cryphonectria parasitica* in Connecticut and Europe. *Plant Disease*, *70*, 536–538.

Bailey, J. M., & Tapprich, W. E. (2007). Structure of the 5′ nontranslated region of the coxsackievirus B3 genome: Chemical modification and comparative sequence analysis. *Journal of Virology*, *81*, 650–668.

Baird, S. D., Turcotte, M., Korneluk, R. G., & Holcik, M. (2006). Searching for IRES. *RNA*, *12*, 1755–1785.

Biella, S., Smith, M. L., Aist, J. R., Cortesi, P., & Milgroom, M. G. (2002). Programmed cell death correlates with virus transmission in a filamentous fungus. *Proceedings. Biological Sciences*, *269*, 2269–2276.

Boland, G. J., & Hall, R. (1994). Index of plant hosts of *Scleerotinia sclerotiorum*. *Canadian Journal of Plant Pathology*, *16*, 93–108.

Buck, K. W. (1986). Fungal virology—An overview. In K. W. Buck (Ed.), *Fungal virology* (pp. 2–84). Boca Raton: CRC Press.

Burnham, C. R., Rutter, P. A., & French, D. W. (1986). Breeding blight-resistant chestnuts. *Plant Breeding Reviews*, *4*, 347–397.

Carthew, R. W., & Sontheimer, E. J. (2009). Origins and mechanisms of miRNAs and siRNAs. *Cell*, *136*, 642–655.

Caten, C. E. (1972). Vegetative incompatibility and cytoplasmic infection in fungi. *Journal of General Microbiology*, *72*, 221–229.

Chen, B., Chen, C.-H., Bowman, B. H., & Nuss, D. L. (1996). Phenotypic changes associated with wild-type and mutant hypovirus RNA transfection of plant pathogenic fungi phylogenetically related to *Cryphonectria parasitica*. *Phytopathology*, *86*, 301–310.

Chen, B., Choi, G. H., & Nuss, D. L. (1993). Mitotic stability and nuclear inheritance of integrated viral cDNA in engineered hypovirulent strains of the chestnut blight fungus. *The EMBO Journal*, *12*, 2991–2998.

Chen, B., Choi, G. H., & Nuss, D. L. (1994). Attenuation of fungal virulence by synthetic infectious hypovirus transcripts. *Science*, *264*, 1762–1764.

Chen, B., Craven, M. G., Choi, G. H., & Nuss, D. L. (1994). cDNA-derived hypovirus RNA in transformed chestnut blight fungus is spliced and trimmed of vector nucleotides. *Virology*, *202*, 441–448.

Chen, B., Gao, S., Choi, G. H., & Nuss, D. L. (1996). Extensive alteration of fungal gene transcript accumulation and elevation of G-protein-regulated cAMP levels by a virulence-attenuating hypovirus. *Proceedings of the National Academy of Sciences of the United States of America*, *93*, 7996–8000.

Chen, B., & Nuss, D. L. (1999). Infectious cDNA clone of hypovirus CHV1-Euro7: A comparative approach to investigate virus-mediated hypovirulence of the chestnut blight fungus *Cryphonectria parasitica*. *Journal of Virology*, *73*, 985–992.

Chevanne, D., Saupe, S. J., Clave, C., & Paoletti, M. (2010). WD repeat instability and diversification of the *Podospora anserine* hnwd non-self recognition gene family. *BMC Evolutionary Biology*, *10*, 134–146.

Choi, G. H., Chen, B., & Nuss, D. L. (1995). Virus-mediated or transgenic suppression of a G-protein alpha subunit and attenuation of fungal virulence. *Proceedings of the National Academy of Sciences of the United States of America*, *92*, 305–309.

Choi, G. H., Dawe, A. L., Churbanov, A., Smith, M. L., Milgroom, M. G., & Nuss, D. L. (2012). Molecular characterization of vegetative incompatibility genes that restrict hypovirus transmission in the chestnut blight fungus *Cryphonectria parasitica*. *Genetics*, *190*, 113–127.

Choi, G. H., Larson, T. G., & Nuss, D. L. (1992). Molecular analysis of the laccase gene from the chestnut blight fungus and selective suppression of its expression in an isogenic hypovirulent strain. *Molecular Plant-Microbe Interactions*, *5*, 119–128.

Choi, G. H., & Nuss, D. L. (1992). Hypovirulence of chestnut blight fungus conferred by an infectious cDNA. *Science*, *257*, 800–803.

Choi, G. H., Pawlyk, D. M., & Nuss, D. L. (1991). The autocatalytic protease p29 encoded by a hypovirulence-associated virus of the chestnut blight fungus resembles the potyvirus-encoded protease HC-Pro. *Virology*, *183*, 747–752.

Churchill, A. C. L., Ciufetti, L. M., Hansen, H. D., Van Etten, H. D., & Van Alfen, N. K. (1990). Transformation of the fungal pathogen *Cryphonectria parasitica* with a variety of heterologous plasmids. *Current Genetics*, *17*, 25–31.

Cogoni, C., & Macino, G. (1999). Gene silencing in *Neurospora crassa* by a protein homologus to RNA-dependent RNA polymerase. *Nature*, *399*, 166–169.

Cortesi, P., McCulloch, C. E., Song, H., Lin, H., & Milgroom, M. G. (2001). Genetic control of horizontal virus transmission in the chestnut blight fungus, *Cryphonectria parasitica*. *Genetics*, *159*, 107–118.

Cortesi, P., & Milgroom, M. G. (1998). Genetic of vegetative incompatibility in *Cryphonectria parasitica*. *Applied and Environmental Microbiology*, *64*, 2988–2994.

Craven, M. G., Pawlyk, D. M., Choi, G. H., & Nuss, D. L. (1993). Papain-like protease p29 as a symptom determinant encoded by a hypovirulence-associated virus of the chestnut blight fungus. *Journal of Virology*, *67*, 6513–6521.

Dang, Y., Yang, Q., Xue, Z., & Li, Y. (2011). RNA interference in fungi: Pathways, functions and applications. *Eukaryotic Cell*, *10*, 1148–1155.

Dawe, A. L., McMains, V. C., Panglao, M., Kasahara, S., Chen, B., & Nuss, D. L. (2003). An ordered collection of expressed sequences from *Cryphonectria parasitica* and evidence of genomic microsynteny with *Neurospora crassa* and *Magnaporthe grisea*. *Microbiology*, *149*, 2373–2384.

Dawe, A. L., & Nuss, D. L. (2001). Hypoviruses and chestnut blight: Exploiting viruses to understand and modulate fungal pathogenesis. *Annual Review of Genetics*, *35*, 1–29.

Dawe, A. L., Segers, G. C., Allen, T. D., McMains, V. C., & Nuss, D. L. (2004). Microarray analysis of *Cryphonectria parasitica* Gα- and Gβγ signalling pathways reveals extensive modulation by hypovirus infection. *Microbiology*, *150*, 4033–4043.

Dawe, A. L., Van Voorhies, W. A., Lau, T. A., Ulanov, A. V., & Li, Z. (2009). Major impacts on the primary metabolism of the plant pathogen *Cryphonectria parasitica* by the virulence-attenuating virus CHV1-EP713. *Microbiology*, *155*, 3913–3921.

Day, P. R., Dodds, J. A., Elliston, J. E., Jaynes, R. A., & Aganostakis, S. L. (1977). Double-stranded RNA in *Endothia parasitica*. *Phytopathology*, *67*, 1393–1396.

De Bernardis, F., Sullivan, P. A., & Cassone, A. (2001). Aspartyl proteinases of *Candida albicans* and their role in pathogenicity. *Medical Mycology*, *39*, 303–313.

Deng, F., Allen, T. D., Hillman, B. I., & Nuss, D. L. (2007). Comparative analysis of alterations in host phenotype and transcript accumulation following hypovirus and mycoreovirus infections of the chestnut blight fungus *Cryphonectria parasitica*. *Eukaryotic Cell*, *6*, 1286–1298.

Deng, F., Allen, T. D., & Nuss, D. L. (2007). Ste12 transcription factor homologue CpST12 is down-regulated by hypovirus infection and required for virulence and female fertility of the chestnut blight fungus *Cryphonectria parasitica*. *Eukaryotic Cell*, *6*, 235–244.

Deng, F., & Nuss, D. L. (2008). Hypovirus papain-like protease p48 is required for initiation but not for maintenance of virus RNA propagation in the chestnut blight fungus Cryphonectria parasitica. *Journal of Virology*, *82*, 6369–6378.

Ding, S. W. (2010). RNA based antiviral immunity. *Nature Reviews. Immunology*, *10*, 632–644.

Ding, S. W., & Voinnet, O. (2007). Antiviral immunity directed by small RNAs. *Cell*, *130*, 413–426.

Dodds, J. A. (1980). Association of type 1 viral-like dsRNA wth club-shaped particles in hypovirulent strains of *Endothia parasitica*. *Virology*, *107*, 1–12.

Dzianott, A., Sztuba-Solinska, J., & Bujarski, J. J. (2012). Mutations in the antiviral RNAi defense pathway modify Brome mosaic virus RNA recombination profiles. *Molecular Plant-Microbe Interactions*, *25*, 97–106.

Enebak, S., MacDonald, W., & Hillman, B. (1994). Effect of dsRNA associated with isolates of *Cryphonectria parasitica* from the central Appalachians and their relatedness to other dsRNAs from North America and Europe. *Phytopathology*, *84*, 528–534.

Fahima, T., Kazmierczak, P., Hansen, D. R., Pfeiffer, P., & Van Alfen, N. K. (1993). Membrane-associated replication of the unencapsidated double-stranded RNA of the fungus *Cryphonectria parasitica*. *Virology*, *195*, 81–89.

Fensteri, V., & Sen, G. C. (2009). Interferons and viral infections. *Biofactors*, *35*, 14–20.

Fields, S., & Herskowitz, I. (1987). Regulation by the yeast mating-type locus of STE12, a gene required for cell-type-specific expression. *Molecular and Cellular Biology*, 7, 3818–3821.

Gessaman, J. A., & Nagy, K. A. (1988). Energy metabolism: Errors in gas-exchange conversion factors. *Physiological Zoology*, *61*, 507–513.

Ghildiyal, M., & Zamore, P. D. (2009). Small silencing RNAs: An expanding universe. *Nature Reviews. Genetics*, *10*, 94–108.

Glass, N. L., & Dementhon, K. (2006). Non-self recognition and programmed cell death in filamentous fungi. *Current Opinion in Microbiology*, *9*, 553–558.

Golinski, M. R., Boecklen, W. J., & Dawe, A. L. (2008). Two-dimensional fractal growth properties of the filamentous fungus *Cryphonectria parasitica*: The effects of hypovirus infection. *Journal of Basic Microbiology*, *48*, 426–429.

Grandvaux, N., tenOever, B. R., Servant, M. J., & Hiscott, J. (2002). The interferon antiviral response: From viral infection to evasion. *Current Opinion in Infectious Diseases*, *15*, 259–267.

Gryzenhout, M., Myburg, H., Wingfield, B. D., & Wingfield, M. J. (2006). Cryphonectriaceae (Diaporthales), a new family including *Cryphonectria*, *Chrysoporthe*, *Endothia* and allied genera. *Mycologia*, *98*, 239–249.

Hansen, D. R., Van Alfen, N. K., Gillies, K., & Powell, W. A. (1985). Naked dsRNA associated with hypovirulence of *Endothia parasitica* is packaged in fungal vesicles. *The Journal of General Virology*, *66*, 2605–2614.

Heiniger, U., & Rigling, D. (1994). Biological control of chestnut blight in Europe. *Annual Review of Phytopathology*, *32*, 581–599.

Hillman, B. I., Foglia, R., & Yuan, W. (2000). Satellite and defective RNAs of *Cryphonectria hypovirus* 3-Grand Haven 2, a virus species in the family *Hypoviridae* with a single open reading frame. *Virology*, *276*, 181–189.

Hillman, B. I., Fullbright, D. W., Nuss, D. L., & Van Alfen, N. K. (1995). Hypoviridae. In F. A. Murphy (Ed.), *Virus taxonomy* (pp. 261–264). New York: Springer Verlag.

Hillman, B. I., Halpeern, B. T., & Brown, M. P. (1994). A viral dsRNA element of the chestnut blight fungus with a distinct genetic organization. *Virology*, *201*, 241–250.

Hillman, B. I., & Suzuki, N. (2004). Viruses of the chestnut blight fungus. *Advances in Virus Research*, *63*, 423–473.

Hiremath, S., L'Hostis, B., Ghabrial, S., & Rhoads, R. E. (1986). Terminal structure of hypovirulence-associated dsRNAs in the chestnut blight fungus *Endothia parasitica*. *Nucleic Acids Research*, *14*, 9877–9896.

Hollings, M. (1962). Viruses associated with die-back disease of cultivated mushroom. *Nature*, *196*, 962–965.

Huber, D. H. (1996). Genetic analysis of vegetative incompatibility polymorphisms and horizontal transmission in the chestnut blight fungus *Cryphonectria parasitica*. Ph. D Thesis, Michigan State University, East Lansing, MI.

Jacob-Wilk, D., Moretti, M., Turina, M., Kazmierczak, P., & Van Alfen, N. K. (2012). Differential expression of the putative Kex2 processed and secreted aspartic proteinase gene family of *Cryphonectria parasitica*. *Fungal Biology*, *116*, 363–378.

Jones, S. K., Jr., & Bennett, R. J. (2011). Fungal mating pheromones: Choreographing the dating game. *Fungal Genetics and Biology*, *48*, 668–676.

Joneson, S., Stajich, J. E., Shiu, S. H., & Rosenblum, E. B. (2011). Genomic transition to pathogenicity in chytrid fungi. *PLoS Pathogens*, *7*, e1002338.

Kaziro, Y., Itoh, H., Kozasa, T., Nakafuku, M., & Satoh, T. (1991). Structure and function of signal-transducing GTP-binding proteins. *Annual Review of Biochemistry*, *60*, 349–400.

Kazmierczak, P., McCabe, P., Turina, M., Jacob-Wilk, D., & Van Alfen, N. K. (2012). The mycovirus CHV1 disrupts secretion of a developmentally regulated protein in *Cryphonectria parasitica*. *Journal of Virology*, *86*, 6067–6074.

Knowles, N. J., Hovi, T., Hyypiä, T., King, A. M. Q., Lindberg, M., Pallansch, M. A., et al. (2011). Picornaviridae. In A. M. Q. King, M. J. Adams, E. B. Carstens & E. J. Lefkowitz (Eds.), *Virus taxonomy: Classification and nomenclature of viruses: Ninth report of the international committee on taxonomy of viruses* (pp. 855–880). San Diego: Elsevier.

Kobayashi, T. (1970). Taxonomic studies of Japanese Diaporthaceae with special reference to their life-histories. *Bulletin of the Government Forest Experiment Station*, *226*, 1–242.

Koonin, E. V., Choi, G. H., Nuss, D. L., Shapira, R., & Carrington, J. C. (1991). Evidence for common ancestry for a chestnut blight hypovirulence-associated double-stranded RNA and a group of positive-strand RNa plant viruses. *Proceedings of the National Academy of Sciences of the United States of America*, *88*, 10647–10651.

Kubisiak, T. L., & Milgroom, M. G. (2006). Markers linked to vegetative incompatibility (*vic*) genes and a region of high heterogeneity and reduced recombination near the mating type locus (*MAT*) in *Cryphonectria parasitica*. *Fungal Genetics and Biology*, *43*, 453–463.

Lan, X., Yao, Z., Zhou, Y., Shang, J., Lin, H., Nuss, D. L., et al. (2008). Deletion of the *cpku80* gene in the chestnut blight fungus, *Cryphonectria parasitica*, enhances gene disruption efficiency. *Current Genetics*, *53*, 59–66.

Larson, T. G., Choi, G. H., & Nuss, D. L. (1992). Regulatory pathways governing modulation of fungal gene expression by a virulence-attenuating mycovirus. *The EMBO Journal*, *11*, 4539–4548.

Leslie, J. F., & Zeller, K. A. (1996). Heterokaryon incompatibility in fungi: More than just another way to die. *Journal of Genetics*, *75*, 415–424.

Li, L., Chang, S.-s., & Liu, Y. (2010). RNA interference pathways in filamentous fungi. *Cellular and Molecular Life Sciences*, *67*, 3849–3863.

Li, W., Manktelow, E., von Kirchbach, J. C., Gog, J. R., Desselberger, U., & Lever, A. M. (2010). Genomic analysis of codon, sequence and structural conservation with selective biochemical-structure mapping reveals highly conserved and dynamic structures in rotavirus RNAs with potential cis-acting functions. *Nucleic Acids Research*, *38*, 7718–7735.

Lin, H., Lan, X., Liao, H., Parsley, T. B., Nuss, D. L., & Chen, B. (2007). Genome sequence, full-length infectious cDNA clone, and mapping of viral double-stranded RNA accumulation determinant of hypovirus CHV1-EP721. *Journal of Virology*, *81*, 1813–1820.

Linder-Basso, D., Dynek, J. N., & Hillman, B. I. (2005). Genome analysis of *Cryphonectria hypovirus* 4, the most common hypovirus species in North America. *Virology*, *337*, 192–203.

Lopez, C., Ayllon, M. A., Navas-Castillo, J., Guerri, J., Moreno, P., & Flores, R. (1998). Molecular variability of the 5′- and 3′-terminal regions of citrus tristeza virus RNA. *Phytopathology*, *88*, 685–691.

Luria, S. E., Darnell, J. E., Jr., Baltimore, D., & Campbell, A. (1981). *General virology* (3rd ed.). New York: Wiley, pp. 303–42.

Matthews, R. E. F. (1981). *Plant virology* (2nd ed.). New York: Academic Press, pp. 190–250.

McCabe, P. M., & Van Alfen, N. K. (2001). Molecular basis of symptom expression by the *Cryphonectria* hypovirus. In S. M. Tavantzis (Ed.), *dsRNA genetic elements: Concepts and applications in agriculture, forestry, and medicine* (pp. 125–144). Boca Raton, FL: CRC Press.

McCormack, J. C., Yuan, X. F., Yingling, Y. G., Kasprzak, W., Zamora, R. E., Shapiro, B. A., et al. (2008). Structural domains within the 3′ untranslated region of Turnip crinkle virus. *Journal of Virology*, *82*, 8706–8720.

Melvin, R. G., Van Voorhies, W. A., & Ballard, J. W. (2007). Working harder to stay alive: Metabolic rate increases with age in Drosophila simulans but does not correlate with life span. *Journal of Insect Physiology*, *53*, 1300–1306.

Milgroom, M. G., & Cortesi, P. (2004). Biological control of chestnut blight with hypovirulence: A critical analysis. *Annual Review of Phytopathology*, *42*, 311–338.

Monteiro-Vitorello, C., Bell, J., Fulbright, D., & Bertrand, H. (1995). A cytoplasmically transmissible hypovirulence phenotype associated with mitochondrial DNA mutations in the chestnut blight fungus *Cryphonectria parasitica*. *Proceedings of the National Academy of Sciences of the United States of America*, *92*, 5935–5939.

Mu, R., Romero, T. A., Hanley, K. A., & Dawe, A. L. (2011). Conserved and variable structural elements in the 5′ untranslated region of two hypoviruses from the filamentous fungus *Cryphonectria parasitica*. *Virus Research*, *161*, 203–208.

Newhouse, J. R., Hoch, H. C., & MacDonald, W. L. (1983). The ultrastrucutre of *Endothia parasitica*. Comparison of a virulent and a hypovirulent isolate. *Canadian Journal of Botany*, *61*, 389–399.

Nuss, D. L. (2005). Hypovirulence: Mycoviruses at the fungal-plant interface. *Nature Reviews. Microbiology*, *3*, 632–642.

Nuss, D. L. (2011). Mycoviruses, RNA silencing, and viral RNA recombination. *Advances in Virus Research*, *80*, 25–48.

Nuss, D. L., & Hillman, B. I. (2011). Hypoviridae. In A. M. Q. King, M. J. Adams, E. B. Carstens & E. J. Lefkowitz (Eds.), *Virus taxonomy* (pp. 1029–1034). Oxford: Elsevier.

Paoletti, M., Saupe, S. J., & Clave, C. (2007). Genesis of a fungal non-self recognition repertoire. *PLoS One*, *2*, e283.

Pearson, G., Robinson, F., Beers Gibson, T., Xu, B. E., Karandikar, M., Berman, K., et al. (2001). Mitogen-activated protein (MAP) kinase pathways: Regulation and physiological functions. *Endocrine Reviews*, *22*, 153–183.

Pelletier, J., Flynn, M. E., Kaplan, G., Racaniello, V., & Sonenberg, N. (1988). Mutational analysis of upstream AUG codons of poliovirus RNA. *Journal of Virology, 62*, 4486–4492.
Polashock, J., & Hillman, B. (1994). A small mitochondrial double-stranded (ds) RNA element associated with a hypovirulent strain of the chestnut blight fungus and ancestrally related to yeast cytoplasmic T and W dsRNAs. *PNAS, 91*, 8680–8684.
Powell, W. A., Jr., & Van Alfen, N. K. (1987a). Two nonhomologus viruses of *Cryphonectria (Endothia) parasitica* reduce accumulation of specific virulence-associated polypeptides. *Journal of Bacteriology, 169*, 5324–5326.
Powell, W. A., & Van Alfen, N. K. (1987b). Differential accumulation of poly(A)+ RNA between virulent and double-stranded RNA-induced hypovirulent strains of *Cryphonectria (Endothia) parasitica. Molecular and Cellular Biology*, 7, 3688–3693.
Racaniello, V. R., & Baltimore, D. (1981). Cloned poliovirus complementary DNA is infectious in mammalian cells. *Science, 214*, 916–919.
Rae, B. P., Hillman, B. I., Tartaglia, J., & Nuss, D. L. (1989). Characterization of double-stranded RNA genetic elements associated with biological control of chestnut blight: Organization of terminal domains and identification of gene products. *The EMBO Journal, 8*, 657–663.
Rigling, D., Heiniger, U., & Hohl, H. R. (1989). Reduction of laccase activity in dsRNA-containing hypovirulent strains of *Cryphonectria (Endothia) parasitica. Phytopathology, 79*, 219–223.
Rigling, D., & Van Alfen, N. K. (1991). Regulation of laccase biosynthesis in the plant-pathogenic fungus *Cryphonectria parasitica* by double-stranded RNA. *Journal of Bacteriology, 173*, 8000–8003.
Robin, C., Anziani, C., & Cortesi, P. (2000). Relationship between biological control, incidence of hypovirulence, and diversity of vegetative compatibility types of *Cryphonectria parasitica* in France. *Phytopathology, 90*, 730–737.
Robin, C., Capdeville, X., Martin, M., Traver, C., & Colinas, C. (2009). *Cryphonectria parasitica* vegetative compatibility type analysis of populations in south-western France and northern Spain. *Plant Pathology, 58*, 527–535.
Root, C., Balbalian, C., Beirman, R., Geletka, L., Anagnostakis, S., Double, M., et al. (2005). Multi-seasonal field release and spermatization trials of transgenic hypovirulent strains of *Cryphonectria parasitica* containing cDNA copies of hypovirus CHV1-EP713. *Forest Pathology, 35*, 1–21.
Rostagno, L., Prodi, A., & Turina, M. (2010). Cpkk1, MAPKK of Cryphonectria parasitica, is necessary for virulence on chestnut. *Phytopathology, 100*, 1100–1110.
Sakurma, T. (1990). Valsa cankeer. In A. L. Jones & H. S. Aldwinclde (Eds.), *Compendium of apple and pear diseases* (pp. 39–40). St. Paul: APS Press.
Sasaki, A., Onoue, M., Kanematsu, S., Suzuki, K., Miyanishi, M., Suzuki, N., et al. (2002). Extending chestnut blight hypovirus range within diaporthales by biolistic delivery of Viral cDNA. *Molecular Plant-Microbe Interactions, 15*, 780–789.
Saupe, S. J. (2000). Molecular genetics of heterokaryon incompatibility in filamentous ascomycetes. *Microbiology and Molecular Biology Reviews, 64*, 489–502.
Segers, G. C., Zhang, X., Deng, F., Sun, Q., & Nuss, D. L. (2007). Evidence that RNA silencing functions as an antiviral defense mechanism in fungi. *Proceedings of the National Academy of Sciences of the United States America, 104*, 12902–12906.
Shapira, R., Choi, G. H., Hillman, B. I., & Nuss, D. L. (1991). The contribution of defective RNAs to the complexity of virus-encoded double-stranded RNA populations present in hypovirulent strains of the chestnut blight fungus *Cryphonectria parasitica. The EMBO Journal, 10*, 741–746.
Shapira, R., Choi, G. H., & Nuss, D. L. (1991). Virus-like genetic organization and expression strategy for double-stranded RNA genetic element associated with biological control of chestnut blight. *The EMBO Journal, 10*, 731–739.

Shapira, R., & Nuss, D. L. (1991). Gene expression by a hypovirulence-associated virus of the chestnut blight fungus involves two papain-like protease activities: Essential residues and cleavage site requirements for p48 autoproteolysis. *The Journal of Biological Chemistry, 266*, 19419–19425.

Smart, C. D., Yuan, W., Foglia, R., Nuss, D. L., Fulbright, D. W., & Hillman, B. I. (1999). *Cryphonectria hypovirus* 3, a virus species in the family *Hypoviridae* with a single open reading frame. *Virology, 265*, 66–73.

Sun, Q., Choi, G. H., & Nuss, D. L. (2009a). Hypovirus-responsive transcription factor gene *pro1* of the chestnut blight fungus *Cryphonectria parasitica* is required for female fertility, asexual spore development, and stable maintenance of hypovirus infection. *Eukaryotic Cell, 8*, 262–270.

Sun, Q., Choi, G. H., & Nuss, D. L. (2009b). A single Argonaute gene is required for induction of RNA silencing antiviral defense and promotes viral RNA recombination. *Proceedings of the National Academy of Sciences of the United States of America, 106*, 17927–17932.

Suzuki, N., Chen, B., & Nuss, D. L. (1999). Mapping of a hypovirus p29 protease symptom determinant domain with sequence similarity to potyvirus HC-Pro protease. *Journal of Virology, 73*, 9478–9484.

Suzuki, N., Geletka, L. M., & Nuss, D. L. (2000). Essential and dispensible virus-encoded replication elements revealed by efforts to develop hypoviruses as gene expression vectors. *Journal of Virology, 74*, 7568–7577.

Suzuki, N., Maruyama, K., Moriyama, M., & Nuss, D. L. (2003). Hypovirus papain-like protease p29 functions in trans to enhance viral double-stranded RNA accumulation and vertical transmission. *Journal of Virology, 77*, 11697–11707.

Suzuki, N., & Nuss, D. L. (2002). Contribution of protein p40 to hypovirus-mediated modulation of fungal host phenotype and viral RNA accumulation. *Journal of Virology, 76*, 7747–7759.

Tartaglia, J., Paul, C. P., Fulbright, D. W., & Nuss, D. L. (1986). Structural properties of double-stranded RNAs associated with biological control of chestnut blight fungus. *Proceedings of the National Academy of Sciences of the United States of America, 83*, 9109–9113.

Tsukiyama-Kohara, K., Iizuka, N., Kohara, M., & Nomoto, A. (1992). Internal ribosome entry site within hepatitis C virus RNA. *Journal of Virology, 66*, 1476–1483.

Turina, M., Zhang, L., & Van Alfen, N. K. (2006). Effect of *Cryphonectria* hypovirus 1 (CHV1) infection on Cpkk1, a mitogen-activated protein kinase kinase of the filamentous fungus *Cryphonectria parasitica*. *Fungal Genetics and Biology, 43*, 764–774.

van Heerden, S. W., Geletka, L. M., Preisig, O., Nuss, D. L., Wingfield, B. D., & Wingfield, M. J. (2001). Characterization of South African *Cryphonectria cubensis* isolates infected with a *C. parasitica* hypovirus. *Phytopathology, 91*, 628–632.

Van Voorhies, W. A., Khazaeli, A. A., & Curtsinger, J. W. (2004). Testing the "rate of living" model: Further evidence that longevity and metabolic rate are not inversely correlated in *Drosophila melanogaster*. *Journal of Applied Physiology, 97*, 1915–1922.

Van Voorhies, W. A., & Ward, S. (1999). Genetic and environmental conditions that increase longevity in *Caenorhabditis elegans* decrease metabolic rate. *Proceedings of the National Academy of Sciences of the United States of America, 96*, 11399–11403.

Walsberg, G., & Wolf, B. (1995). Variation in the respiratory quotient of birds and implications for indirect calorimetry using measurements of carbon dioxide production. *The Journal of Experimental Biology, 198*, 213–219.

Willyerd, K. L., Kemp, A. M., & Dawe, A. L. (2009). Controlled gene expression in the plant pathogen *Cryphonectria parasitica* by use of a copper responsive element. *Applied and Environmental Microbiology, 75*, 5417–5420.

Xie, J., Xiao, X., Fu, Y., Liu, H., Cheng, J., Ghabrial, S. A., et al. (2011). A novel mycovirus closely related to hypoviruses that infect the plant pathogenic fungus *Sclerotinia sclerotiorum*. *Virology*, *418*, 49–56.

Yaegshi, H., Kanematsu, S., & Ito, T. (2012). Molecular characterization of a new hypovirus infecting a phytopathogenic fungus, *Valsa ceratosperma*. *Virus Research*, *165*, 143–150.

Zhang, L., Baasiri, R. A., & Van Alfen, N. K. (1998). Viral repression of fungal pheromone precursor gene expression. *Molecular and Cellular Biology*, *18*, 953–959.

Zhang, L., Churchill, A. C., Kazmierczak, P., Kim, D. H., & Van Alfen, N. K. (1993). Hypovirulence-associated traits induced by a mycovirus of *Cryphonectria parasitica* are mimicked by targeted inactivation of a host gene. *Molecular and Cellular Biology*, *13*, 7782–7792.

Zhang, X., & Nuss, D. L. (2008). A host dicer is required for defective viral RNA production and recombinant virus vector RNA instability for a positive sense RNA virus. *Proceedings of the National Academy of Sciences of the United States of America*, *105*, 16749–16754.

Zhang, X., Segers, G., Sun, Q., Deng, F., & Nuss, D. L. (2008). Characterization of hypovirus-derived small RNAs generated in the chestnut blight fungus by an inducible DCL-2 dependent pathway. *Journal of Virology*, *82*, 2613–2619.

Zuker, M. (2003). Mfold web server for nucleic acid folding and hybridization prediction. *Nucleic Acids Research*, *31*, 3406–3415.

Zuker, M., & Jacobson, A. B. (1998). Using reliability information to annotate RNA secondary structures. *RNA*, *4*, 669–679.

[illegible] Y., [illegible] H., Chang, [illegible], S. A., et al. (2011). [illegible] [illegible] [illegible] [illegible]

[illegible] (2011). Molecular mechanisms of [illegible] following a photoperiodic [illegible] [illegible]. *Plant Science*, [illegible].

Chang, [illegible] R. A., & Van Allen, M. K. [illegible]. Viral suppression of floral phenotype [illegible] expression. *Molecular Plant-Microbe* [illegible]

Zhang, [illegible] & Van Allen, M. K. [illegible] [illegible] by [illegible] of a host gene. *Molecular and Cellular Biology*, [illegible].

Zhang, [illegible] (2008). [illegible] [illegible] RNA [illegible]

[illegible]

CHAPTER SIX

The Family *Narnaviridae*: Simplest of RNA Viruses

Bradley I. Hillman[1] and Guohong Cai
Department of Plant Biology and Pathology, Rutgers University, New Brunswick, New Jersey, USA
[1]Corresponding author: e-mail address: hillman@aesop.rutgers.edu

Contents

1. Introduction 150
2. Discovery and Early Work on the *Narnaviridae* with Emphasis on Selected Systems 151
3. Early Work Leading to Description of Mitoviruses 152
   3.1 *Rhizoctonia* M2 virus (=Thanatephorus cucumeris mitovirus) 155
   3.2 Examining host effects of *Narnaviridae* 156
4. Relationships Among Members of the *Narnaviridae* 158
   4.1 Relationships of members of the *Narnaviridae* to RNA viruses infecting other taxa, and host considerations 158
   4.2 RNA silencing and *Narnaviridae* 159
   4.3 Relationships among mitoviruses from different taxa 160
5. Population Biology of Members of the *Narnaviridae* 164
6. Other Genome Features and Molecular Manipulation of Members of the *Narnaviridae* 166
   6.1 Latent infection by mitoviruses 168
   6.2 Satellite and defective RNAs of mitoviruses 168
7. Conclusions and Future Directions 169
References 170

Abstract

Members of the virus family *Narnaviridae* contain the simplest genomes of any RNA virus, ranging from 2.3 to 3.6 kb and encoding only a single polypeptide that has an RNA-dependent RNA polymerase domain. The family is subdivided into two genera based on subcellular location: members of the genus *Narnavirus* have been found in the yeast *Saccharomyces cerevisiae* and in the oomycete *Phytophthora infestans* and are confined to the cytosol, while members of the genus *Mitovirus* have been found only in filamentous fungi and are found in mitochondria. None identified thus far encodes a capsid protein; like several other RNA viruses of lower eukaryotes, their genomes are confined within lipid vesicles. As more family members are discovered, their importance as genetic elements is becoming evident. The unique association of the genus *Mitovirus* with mitochondria renders them potentially valuable tools to study biology of lower eukaryotes.

Advances in Virus Research, Volume 86
ISSN 0065-3527
http://dx.doi.org/10.1016/B978-0-12-394315-6.00006-4

1. INTRODUCTION

Members of the virus family *Narnaviridae* are the simplest of known RNA viruses, consisting of a single molecule of positive-sense RNA that may be as small as 2.3 kb and encoding only an RNA-dependent RNA polymerase (RdRp) to direct their own replication. They encode no protein capsid and no virus particles other than lipid vesicles are known to be associated with infection (Fig. 6.1). The family *Narnaviridae* became the second virus family with members lacking a protein capsid (Van Regenmortel et al., 2000), following the acceptance of the family *Hypoviridae* by the International Committee for the Taxonomy of Viruses (ICTV) as the first such family in 1995 (Murphy et al., 1995). The two genera of the family, the genus *Narnavirus* and the genus *Mitovirus*, share basic genome organization properties. The single feature distinguishing the two genera is the site of

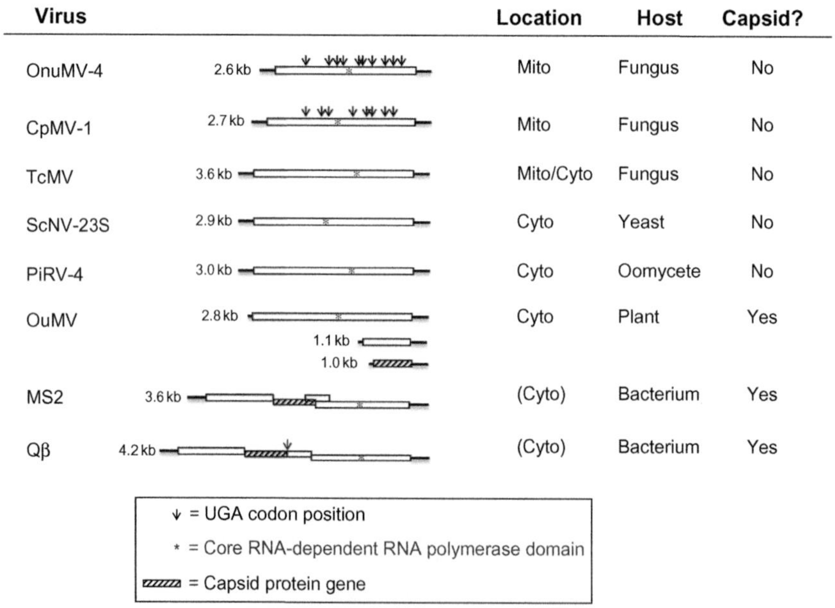

Virus		Location	Host	Capsid?
OnuMV-4	2.6 kb	Mito	Fungus	No
CpMV-1	2.7 kb	Mito	Fungus	No
TcMV	3.6 kb	Mito/Cyto	Fungus	No
ScNV-23S	2.9 kb	Cyto	Yeast	No
PiRV-4	3.0 kb	Cyto	Oomycete	No
OuMV	2.8 kb; 1.1 kb; 1.0 kb	Cyto	Plant	Yes
MS2	3.6 kb	(Cyto)	Bacterium	Yes
Qβ	4.2 kb	(Cyto)	Bacterium	Yes

↓ = UGA codon position

* = Core RNA-dependent RNA polymerase domain

▨ = Capsid protein gene

Figure 6.1 Genome organizations of selected viruses discussed in this review. Virus name abbreviations: OnuMV-4, Ophiostoma novo-ulmi mitovirus 4; CpMV-1, Cryphonectria parasitica mitovirus 1; TcMV, Thanatephorus cucumeris mitovirus; ScNV-23S, Saccharomyces cerevisiae narnavirus-23S; PiRV-4, Phytophthora infestans RNA virus 4; OuMV, Ourmia melon virus; MS2, Enterobacteria phage MS2; Qβ, Enterobacteria phage Qβ. (For color version of this figure, the reader is referred to the online version of this chapter.)

translation of the RdRp, and presumably of replication: members of the genus *Narnavirus* are found in the cytosol (the cytoplasm exclusive of organelles) and the genus *Mitovirus* are confined to mitochondria.

In this chapter, we use the term mitovirus to denote members of the genus *Mitovirus* and the term narnavirus to refer to members of the genus *Narnavirus*. For the treatment of different aspects of the family *Narnaviridae*, the reader is directed to some recent and some older reviews on different aspects of fungal viruses: Buck, Brasier, Paoletti, and Crawford (2003), Ghabrial and Suzuki (2009), Hillman and Suzuki (2004), Milgroom and Hillman (2011), Nuss (2005), and Pearson, Beever, Boine, and Arthur (2009). In particular, the recent chapter by Milgroom and Hillman (2011) covers aspects of ecology and evolution of members of the *Narnaviridae* in some detail. Further, the paper by Xie and Ghabrial (2012), while not a review, covers various aspects of mitovirus biology in some depth.

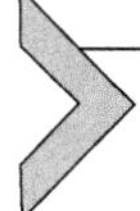

2. DISCOVERY AND EARLY WORK ON THE *NARNAVIRIDAE* WITH EMPHASIS ON SELECTED SYSTEMS

The discovery and description of viruses in the family *Narnaviridae* took two main tracks, in large part because the two genera in the family have different biological, molecular, and host properties. The two original members of the genus *Narnavirus* were identified in the yeast *Saccharomyces cerevisiae* as aberrantly segregating cytoplasmic elements that were found in a large proportion of commercial yeast strains and were induced under conditions of high stress (Garvik & Haber, 1978; Kadowaki & Halvorson, 1971; Wesolowski & Wickner, 1984), reviewed in Wickner (2001). The elements were named 20S and 23S RNA identifying the sedimentation coefficients of the ssRNAs through sucrose gradients and were associated with double-stranded (ds) RNAs designated W and T, respectively, that were also upregulated under certain high-stress conditions (Wickner, 2001). The RNA elements later came to be named Saccharomyces cerevisiae narnavirus 20S (ScNV-20S) and ScNV-23S (Hillman & Esteban, 2011).

Research on these RNAs has proceeded on several fronts for the past 40+ years, as noted in several sections below, but neither ScNV-20S nor ScNV-23S causes any quantifiable phenotypic change in the yeast host, so interest in these from virological and economic perspectives has been somewhat limited. The recent identification of a similar and related virus in the plant pathogenic oomycete *Phytophthora infestans* (Cai, Myers,

Fry, & Hillman, 2012) and the recent characterization of plant-infecting ourmiaviruses as the closest relatives of the narnaviruses (Rastgou et al., 2009) may spark renewed interest in these viruses, as their presence in *S. cerevisiae* provides unique and obvious advantages for their study.

3. EARLY WORK LEADING TO DESCRIPTION OF MITOVIRUSES

The initial description of mitoviruses was accomplished through research on two related fungal pathosystems that are often confused by those with only passing familiarity in plant pathology. One is the Dutch elm disease fungus, *Ophiostoma ulmi* and *Ophiostoma novo-ulmi* (Brasier, 1986b), and the other is the chestnut blight fungus, *Cryphonectria parasitica* (Anagnostakis, 1987). Both are ascomycetes that caused worldwide epidemics, although the biology of the two diseases is quite different. Dutch elm disease is caused by a vascular pathogen that is vectored by gallery-forming bark beetles and results in wilt symptoms (Brasier, 1986b), while chestnut blight is a canker disease that is spread without an obligate vector (Anagnostakis, 1987).

The occurrence and impact of mitoviruses in these two related systems are also quite different; mitoviruses of *O. novo-ulmi* were identified by studying the epidemiology and biology of the disease and by studying many fungal isolates from around the world (Brasier, 1979, 1986a; Hong, Cole, Brasier, & Buck, 1998a,1998b; Hong, Dover, Cole, Brasier, & Buck, 1999; Rogers, Buck, & Brasier, 1986a,1986b, 1987, 1988). Astoundingly, a single debilitated isolate of *O. novo-ulmi*, named Ld, was found to have 12 dsRNAs. Many years of work went into describing the dsRNAs in molecular terms and associating their presence or absence with the diseased phenotype of the fungus (Brasier, 1986a; Doherty, Coutts, Brasier, & Buck, 2006; Hong et al., 1998a,1998b, 1999; Rogers et al., 1986a,1986b, 1987, 1988). The dsRNAs in the Ld isolate all turned out to represent the genomes or genome derivatives of mitoviruses, and these genomes have become the scaffold for the phylogenetic tree of the genus (Fig. 6.2). Why so many different mitoviruses infected this single isolate remains an open question and will undoubtedly inform future mitovirus research.

In contrast, the single mitovirus identified thus far from *C. parasitica* was characterized through a screening and bioinformatics approach. In a project initiated by Dr. Peter Bedker, many isolates of *C. parasitica* from recovering American chestnut trees in New Jersey, USA, were screened for the presence of dsRNA, indicative of virus infection and possible association with

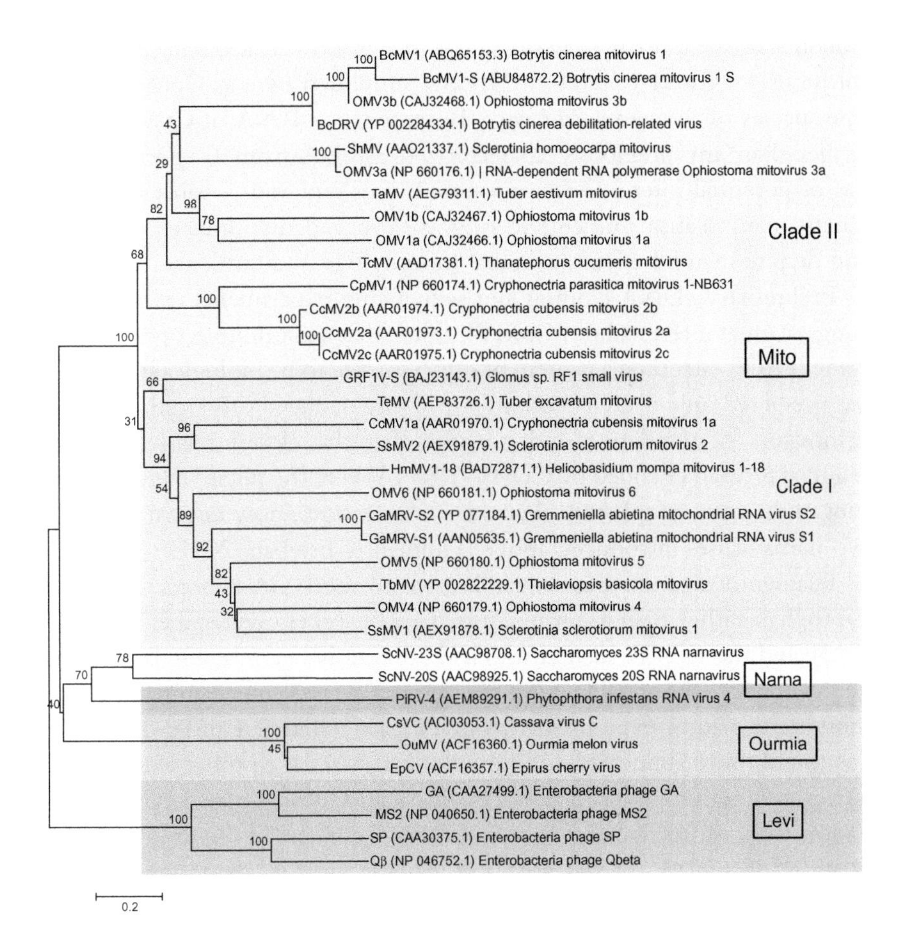

Figure 6.2 Neighbor-joining tree of viruses discussed in this review based on the alignments of full-length RdRp protein sequences. The sequences were aligned using the MUSCLE program (Edgar, 2004) implemented in MEGA5 (Tamura et al., 2011) with default settings. Numbers above the branches are bootstrap supports in percentiles in 1000 replicates. Mito, genus *Mitovirus*, Narna, genus *Narnavirus*, Ourmia, genus *Ourmiavirus*, Levi, family *Leviviridae*. The two distinct clades of the genus *Mitovirus* are shaded differently, as are the host types for the genus *Narnavirus*. (See Page 10 in Color Section at the back of the book.)

transmissible hypovirulence (Chung, Bedker, & Hillman, 1994). Most of the isolates of interest had aberrant colony morphologies and contained dsRNAs of 12–13 kb that became the basis of the description of *Cryphonectria hypovirus 2*, a species in the family *Hypoviridae* (Nuss & Hillman, 2011). One isolate, designated NB631, contained a much smaller dsRNA of ~3 kb that seemed to have little effect on colony morphology and on fungal virulence

(Polashock & Hillman, 1994). This virus is now called Cryphonectria mitovirus 1/NB631 (CpMV-1/NB631, simplified here as CpMV-1), the type species of the genus *Mitovirus*. Because the dsRNA of CpMV-1 was smaller than any previously characterized dsRNA from *C. parasitica* and was of potential interest both as a novel virus and as a possible target for genetic manipulation and engineering, we proceeded with its characterization despite its mild effect on fungal morphology and virulence.

Preliminary cDNA cloning and sequencing experiments revealed two things of great interest about CpMV-1: first, that a continuous open reading frame (ORF) encoding a protein of sufficient size to be biologically relevant was predicted only if UGA codons within the deduced ORF did not act as terminators of translation and, second, that the closest relatives of the deduced protein encoded by CpMV-1 RNA were the putative RdRp proteins of the two cytoplasmic elements of *S. cerevisiae*, now known to be the two narnaviruses introduced above (Hillman & Esteban, 2011).

Filamentous ascomycetes were known to use UGA codons to encode tryptophan rather than as terminators (Osawa, Jukes, Watenabe, & Muto, 1992), and the *O. novo-ulmi* viruses had previously been shown to be localized to the mitochondrial fraction (Rogers et al., 1987), so we initiated mitochondrial isolation experiments to investigate whether CpMV-1 dsRNA was found more abundantly in mitochondria, similar to those experiments carried out by Buck and colleagues to investigate *O. novo-ulmi* dsRNA localization. Subcellular fractionation experiments confirmed the mitochondrial location of CpMV-1 dsRNA, and genetic crosses demonstrated that CpMV-1 dsRNA was inherited in approximately 50% of single ascospore progeny when the maternal parent in matings was infected, but no transmission to ascospore progeny was apparent when the infected partner in crosses was the male (Polashock & Hillman, 1994). These results were consistent with the mitochondrial location of CpMV-1 dsRNA and the demonstrated maternal inheritance of *C. parasitica* mitochondria (Milgroom & Lipari, 1993). Why there was only ~50% inheritance to ascospore progeny is unknown. It could suggest that only a subset of mitochondria in an infected parent isolate contained dsRNA, thus leading to segregation of fungal infection in single ascospore progeny as a consequence of segregation of the infected mitochondrial population. It is also possible that dsRNA accumulation was simply suppressed and undetectable in *C. parasitica* ascospore progeny identified as dsRNA free (Polashock, Bedker, & Hillman, 1997), a possibility that must be considered in light of more recent results from other mitovirus systems in which latent or inapparent infections lead to difficulty or inability to detect dsRNA in some latently infected fungal isolates (e.g., Park, Chen, & Punja, 2006).

As CpMV-1 affects fungal virulence to only a minor degree (Polashock & Hillman, 1994), the virus is never strongly considered as a biological control agent for chestnut blight. This property of mild virulence alone would seem to increase the potential fitness of CpMV-1 in fungal populations in natural settings (Bryner & Rigling, 2011; Milgroom & Hillman, 2011; Taylor, Jarosz, Lenski, & Fulbright, 1998). Such effects have not been investigated experimentally in mitoviruses as they have been recently in *Cryphonectria* hypoviruses (Bryner & Rigling, 2012). In addition, the high level of conidial and ascospore transmission would favor ecological persistence of CpMV-1. Therefore, it is especially surprising that only a single mitovirus-containing isolate of *C. parasitica* has been identified in chestnut trees, as the transmission and biological properties described above for CpMV-1would suggest that it should be successful in natural settings. We proceeded with laboratory-based experiments to examine the biology of CpMV-1 as described later in this review.

3.1. *Rhizoctonia* M2 virus (=Thanatephorus cucumeris mitovirus)

The M2 virus identified in *Rhizoctonia solani* (teleomorph = either *Thanatephorus* or *Ceratobasidium*), a ubiquitous soilborne plant pathogen, is unique among members of the *Narnaviridae* described thus far in that it is found in both the cytosolic fraction and the mitochondrial fraction of the fungus (Jian, Lakshman, & Tavantzis, 1997; Lakshman, Jian, & Tavantzis, 1998). For consistency with GenBank and with other recent literature, the *R. solani* M2 virus will be called Thanatephorus cucumeris mitovirus (TcMV), although that name is not yet approved by ICTV and it continues to be called M2 in several recent papers. The phylogeny of TcMV dsRNA places it within the genus *Mitovirus*, but the coding sequence of the single large ORF of TcMV contains no UGA codons and therefore is predicted to encode the same protein whether translated in the cytoplasm or in the mitochondria (Lakshman et al., 1998). Whether the mitochondrial-resident and cytosolic-resident TcMV RNAs have distinct and possibly different roles in the viral infection cycle is unknown.

The story of the TcMV-*R. solani* system is complex, and many details remain unresolved. TcMV presence is associated with reduced virulence of *Rhizoctonia* (Jian et al., 1997). The TcMV RNA is slightly but not substantially larger than most mitovirus genomes, 3.6 kb encoding a protein deduced from the single large ORF of 754 amino acids (Lakshman et al., 1998).

A portion of the deduced TcMV amino acid sequence that is amino-proximal to the core RdRp domain shows intriguing amino acid sequence similarity to the pentafunctional polypeptide AROM from *S. cerevisiae* and was hypothesized to interfere with the shikimate pathway (Lakshman et al., 1998). Following up on the initial studies using the biological, biochemical, and molecular tools available, Liu et al. showed that TcMV replication was associated with upregulation of the fungal quinic acid pathway and downregulation of the shikimate pathway and that addition of quinic acid in culture resulted in reduced virulence and increase in TcMV mRNA and protein (Liu, Lakshman, & Tavantzis, 2003a, 2003b).

The complexity of the system and difficulty in teasing it apart come from two main features of the system: First, a genomic DNA fragment containing a full-length copy of TcMV was identified in a virus-free isolate of *R. solani* (Lakshman et al., 1998), so effects of the viral versus fungal copies of the encoded protein, designated pA, could not be separated in infected cultures. Second, the viral RdRp is the same polypeptide that is thought to interfere with the fungal quinate and shikimate pathways. Thus, the function of this protein in viral RNA replication cannot be separated experimentally in any reasonable way from possible function in regulation of these host biochemical pathways. In this regard, it is notable that the TcMV amino acid sequence that has similarity to the AROM protein is within a region that is relatively well conserved among the mitoviruses that are most closely related to TcMV. Indeed, several of the specific amino acid residues predicted to be homologues in the alignment of TcMV and the *S. cerevisiae* AroE domain in Fig. 7 of Lakshman et al. (1998) are the same residues predicted to be homologues in domain A of the RdRp alignment of TcMV and CpMV-1 in Fig. 3 of the same paper. Homology between the TcMV deduced pA protein and the most closely related viruses such as *O. novo-ulmi* mitovirus 3A (OnuMV3A) and Botrytis cinerea mitovirus 1 (BcMV-1) extends through the entire AroD domain identified in Fig 7 of Lakshman et al. (1998). The complete sequences of many mitoviruses and many AROM proteins, including the *R. solani* AROM protein (Lakshman, Liu, Mishra, & Tavantzis, 2006), are now available, and it would be of interest to examine the extent of the relationship among these.

3.2. Examining host effects of *Narnaviridae*

The TcMV-*R. solani* system highlights the importance of having a tractable host system to get the most meaningful information about a fungal virus. Work on TcMV and other *Rhizoctonia* viruses is greatly complicated by the limitations of working with the host fungus: *Rhizoctonia* is the asexual form (anamorph) of two sexual forms (teleomorphs) *Thanatephorus* and *Ceratobasidium*, and the biology and relationships of fungi in the complex remain relatively poorly understood (Gonzalez, Carling, Kuninaga, Vilgalys, & Cubeta, 2001). Laboratory-based classical genetics experiments are not readily approachable with *R. solani* and its teleomorphs. Further, the fungus is recalcitrant to molecular transformation in the laboratory, effectively precluding gene introductions and knock-outs. Thus, many virus/host questions that are approachable with other fungal viruses and many of the deeper questions about TcMV biology have not been with *Rhizoctonia* as the experimental host system.

Even with a more tractable host system, examining the effects of members of the *Narnaviridae* on their hosts can be complicated by the difficulty of generating isogenic, virus-free isolates with which to perform Koch's postulates. This is true with many fungal viruses, but mitoviruses present particular problems by nature of their residence in mitochondria. Our lab was unable to cure the original *C. parasitica* isolate that was infected with CpMV-1of its virus using either single conidial isolation or various chemical and hyphal tipping methods, so we looked to develop virus-free isolates from single ascospores for such studies (Polashock et al., 1997; Polashock & Hillman, 1994). As noted above, a subset of single ascospore progeny was dsRNA free, and we were able to identify dsRNA-free ascospore progeny that were vegetatively compatible with dsRNA-containing isolates. Attempts to introduce CpMV-1 into virus-free ascospore progeny by anastomosis were successful but were consistently associated with apparent recombination of mitochondrial DNA (Polashock et al., 1997). These experiments resulted in CpMV-1 infected and uninfected isolates of *C. parasitica* that were isogenic at the nuclear DNA level, but not at the level of mitochondrial DNA; thus, the specific effects of CpMV-1 on fungal phenotype could not be separated from possible effects due to mitochondrial recombination.

Park and colleagues also were unable to generate virus-free isolates of *Chalara elegans* (*Thielaviopsis basicola*) (Park et al., 2006), although they were able to demonstrate that a virus-suppressed isolate was more virulent, had more abundant mitochondria, and more regularly shaped mitochondria than the isogenic isolate in which virus was replicating actively. Similarly, Wu and colleagues were able to generate fungal isolates that they thought to be free of BcMV-1 and demonstrate that the presumably virus-free isolates were significantly more virulent on leaves of oilseed rape than isogenic isolates containing virus (Wu et al., 2007) but later found that the cured isolates contained minor quantities of BcMV RNA (Wu, Wang, & Ding,

2010). They also demonstrated that mitochondria of the isolate that was actively infected with BcMV-1 were deformed relative to their virus-free counterparts (Wu, Wang, et al., 2010). These results underscore the difficulties in working with and drawing firm conclusions from mitovirus-infected fungi.

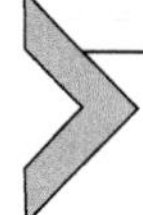

4. RELATIONSHIPS AMONG MEMBERS OF THE *NARNAVIRIDAE*

4.1. Relationships of members of the *Narnaviridae* to RNA viruses infecting other taxa, and host considerations

The closest relatives of members of the family *Narnaviridae* are members of the genus *Ourmiavirus* of plant-infecting viruses. In fact, the ourmiaviruses are phylogenetically more closely related to the narnaviruses than are the mitoviruses, a confounding factor for the current taxonomy of the family *Narnaviridae* (Cai et al., 2012; Rastgou et al., 2009). Like many plant viruses, the ourmiaviruses have multipartite genomes with their RdRp, coat protein, and movement protein genes on separate segments of positive-sense RNA (Fig. 6.1). Like the narnaviruses, ourmiaviruses replicate in cytoplasm, although their capsid proteins also have been shown to localize to the nucleolus (Crivelli, Ciuffo, Genre, Masenga, & Turina, 2011). The ourmiaviruses therefore represent one of the most interesting and experimentally promising links between fungal viruses and viruses of higher plants or animals.

Mitoviruses have not been found to date outside fungi, but mitovirus sequences have been found in plant mitochondria and in plant mitochondrial sequences embedded in plant nuclear genomes (Hong et al., 1998a; Marienfeld, Unseld, Brandt, & Brennicke, 1997; Marienfeld, Unseld, & Brennicke, 1999). Interestingly, in *Brassica* and other plant genera, the mitovirus-related sequences embedded in the mitochondrial DNA represent the GDD-containing core RdRp sequences of mitoviruses. Whether these sequences are functional in plants is currently unknown. Could these sequences represent failed attempts at mitovirus invasion of plants, or utilitarian acquisitions of heterologous viral sequences by plants? Experiments directed at the function of these sequences in plants will rely on effective methods for genetic manipulation of mitochondria.

The other close relatives of the family *Narnaviridae* are the ssRNA bacteriophages in the family *Leviviridae* (Esteban, Rodriguez-Cousino, & Esteban, 1992; Polashock & Hillman, 1994). The family *Leviviridae* is divided into two genera: *Levivirus*, which contains bacteriophages MS2 and GA, and *Allolevivirus*, which contains bacteriophages Qβ and SP (Fig. 6.1). Like the ourmiaviruses, members of the family *Leviviridae* such as Qβ or M2 are encapsidated in simple protein capsids (van Duin & Olsthoorn, 2011). Given the large and growing number of bacteria that are found to be endosymbiotic in plants or in fungi (Kobayashi & Crouch, 2009), the conjecture that the narnaviruses and their immediate relatives evolved by reduction from leviviruses is a natural one. Interestingly, members of the *Allolevivirus* genus such as Qβ contain a UGA codon that terminates translation of their capsid protein and is read through as Trp at a rate of ~6% to generate an extended capsid protein (van Duin & Olsthoorn, 2011). The effect of a UGA translation terminator that is occasionally read through in a regulatory fashion to produce a fusion protein differs from the UGA codons in mitoviruses, of course, which are constitutively read as Trp using fungal mitochondrial codon usage.

4.2. RNA silencing and *Narnaviridae*

RNA silencing, a form of RNA interference, has been shown to be an important line of defense against virus attack in fungi and other eukaryotes (Nuss, 2011, for review). Viruses that infect eukaryotes with RNA silencing defense systems presumably must have mechanisms to avoid or suppress such systems (reviewed in Wu, Zhang, Li, Jiang, & Ghabrial, 2010), and Cryphonectria hypovirus 1 has been shown to be able to overcome RNA silencing using the same type of silencing suppressor that is functional in plant-infecting potyviruses (Segers, Zhang, Deng, Sun, & Nuss, 2007). To our knowledge, nothing has been published about RNA silencing in the context of any member of the family *Narnaviridae*. Indeed, the role and targeting of mitochondria in the RNA interference pathway are just beginning to be investigated in earnest (Bandiera et al., 2011; Huang et al., 2011), so whether mitoviruses would need to escape an RNA silencing defense pathway is an open question.

Interestingly, the Dicer and Argonaute genes required for an RNA silencing system have been lost from *S. cerevisiae* (Drinnenberg et al., 2009), and the absence of these genes allows replication and maintenance of yeast totiviruses such as L-A (Drinnenberg, Fink, & Bartel, 2011). Reconstitution of the silencing system by the introduction of those genes from a closely related species, *S. castellii*, resulted in *S. cerevisiae* strains that were largely resistant to L-A. Whether this resistance extends to the two yeast narnaviruses, ScNV-20S and ScNV-23S, was not reported, but it may be notable that these viruses have been identified thus far only in yeast species lacking the competent silencing pathway. Thus, there is no evidence that members of the family *Narnaviridae* have specific mechanisms to overcome RNA silencing systems.

4.3. Relationships among mitoviruses from different taxa

As more fungal virus sequences become available, an emerging picture of the family *Narnaviridae* is coming into view. Mitoviruses appear to be among the most common of fungal viruses, occurring in many of the fungal taxa that have been examined for presence of dsRNA viruses: a minimum of 25 distinct complete mitovirus sequences have been characterized from at least 14 fungal species representing at least 11 genera (Table 6.1). Mitoviruses have not been found in nonfungal hosts.

Narnaviruses are decidedly less common than mitoviruses and, to date, have not been identified in filamentous fungi. The two confirmed narnaviruses are from *S. cerevisiae*, and a probable narnavirus was recently identified in the oomycete *P. infestans* (Cai et al., 2012). Whether the presence of mitoviruses only in filamentous fungi and narnaviruses only in nonfungal hosts has biological significance pertaining to host type is not known, given the small sample size and lack of critical work that could address such questions.

Table 6.1 Members of the virus family *Narnaviridae* and their closest relatives

Virus name	Host	Accession #	Virus family	Virus genus
Botrytis cinerea mitovirus 1	Ascomycete	ABQ65153.3	*Narnaviridae*	*Mitovirus*
Botrytis cinerea mitovirus 1S	Ascomycete	ABU84872.2	*Narnaviridae*	*Mitovirus*
Botrytis cinerea mitovirus debilitation-related virus	Ascomycete	YP_002284334.1	*Narnaviridae*	*Mitovirus*
Cryphonectria cubensis mitovirus 1a	Ascomycete	AAR01970.1	*Narnaviridae*	*Mitovirus*
Cryphonectria cubensis mitovirus 2a	Ascomycete	AAR01973.1	*Narnaviridae*	*Mitovirus*

Table 6.1 Members of the virus family *Narnaviridae* and their closest relatives—cont'd

Virus name	Host	Accession #	Virus family	Virus genus
Cryphonectria cubensis mitovirus 2b	Ascomycete	AAR01974.1	*Narnaviridae*	*Mitovirus*
Cryphonectria cubensis mitovirus 2c	Ascomycete	AAR01975.1	*Narnaviridae*	*Mitovirus*
Cryphonectria parasitica mitovirus 1-NB631	Ascomycete	NP_660174.1	*Narnaviridae*	*Mitovirus*
Glomus sp. RF1 small virus	Ascomycete	BAJ23143.1	*Narnaviridae*	*Mitovirus*
Gremmeniella abietina mitochondrial RNA virus S2	Ascomycete	YP_077184.1	*Narnaviridae*	*Mitovirus*
Gremmeniella mitovirus S1	Ascomycete	AAN05635.1	*Narnaviridae*	*Mitovirus*
Helicobasidium mompa mitovirus 1–18	Basidiomycete	BAD72871.1	*Narnaviridae*	*Mitovirus*
Rhizoctonia solani virus M2	Basidiomycete	AAD17381.1	*Narnaviridae*	*Mitovirus*
Ophiostoma mitovirus 1a	Ascomycete	CAJ32466.1	*Narnaviridae*	*Mitovirus*
Ophiostoma mitovirus 1b	Ascomycete	CAJ32467.1	*Narnaviridae*	*Mitovirus*
Ophiostoma mitovirus 3a	Ascomycete	NP_660176.1	*Narnaviridae*	*Mitovirus*
Ophiostoma mitovirus 3b	Ascomycete	CAJ32468.1	*Narnaviridae*	*Mitovirus*
Ophiostoma mitovirus 4	Ascomycete	NP_660179.1	*Narnaviridae*	*Mitovirus*
Ophiostoma mitovirus 5	Ascomycete	NP_660180.1	*Narnaviridae*	*Mitovirus*

Continued

Table 6.1 Members of the virus family *Narnaviridae* and their closest relatives—cont'd

Virus name	Host	Accession #	Virus family	Virus genus
Ophiostoma mitovirus 6	Ascomycete	NP_660181.1	*Narnaviridae*	*Mitovirus*
Sclerotinia homoeocarpa mitovirus	Ascomycete	AAO21337.1	*Narnaviridae*	*Mitovirus*
Sclerotinia sclerotiorum mitovirus 1/KL-1	Ascomycete	JQ013377	*Narnaviridae*	*Mitovirus*
Sclerotinia sclerotiorum mitovirus 2/KL-1	Ascomycete	JQ013378	*Narnaviridae*	*Mitovirus*
Thielaviopsis basicola mitovirus	Ascomycete	YP_002822229.1	*Narnaviridae*	*Mitovirus*
Tuber aestivum mitovirus	Ascomycete	AEG79311.1	*Narnaviridae*	*Mitovirus*
Tuber excavatum mitovirus	Ascomycete	JN222389	*Narnaviridae*	*Mitovirus*
Saccharomyces 20S RNA virus	Yeast	AAC98925.1	*Narnaviridae*	*Narnavirus*
Saccharomyces 23S RNA virus	Yeast	AAC98708.1	*Narnaviridae*	*Narnavirus*
Phytophthora infestans RNA 4	Oomycete	AEM89291.1	*Narnaviridae*	*Unassigned*
Cassava virus C	Plant	ACI03053.1	Unassigned	*Ourmiavirus*
Epirus cherry virus	Plant	ACF1357.1	Unassigned	*Ourmiavirus*
Ourmia melon virus	Plant	ACF16360.1	Unassigned	*Ourmiavirus*
Enterobacteria phage MS2	Bacterium	NP040650.1	*Leviviridae*	*Levivirus*
Enterobacteria phage GA	Bacterium	CAA27499.1	*Leviviridae*	*Levivirus*
Enterobacteria phage Qβ	Bacterium	NP046752.1	*Leviviridae*	*Allolevivirus*
Enterobacteria phage SP	Bacterium	CAA30375.1	*Leviviridae*	*Allolevivirus*

Mitoviruses have been identified in both ascomycetes and basidiomycetes, but not in lineages of fungi outside the dikarya. Only three basidiomycetes are represented as mitovirus hosts and that includes *Thanatephorus* and *Ceratobasidium* as independent hosts of TcMV (Charlton & Cubeta, 2007). Plant pathogens dominate the list of mitovirus hosts: all except for two ectomycorrhizal fungi in the genus *Tuber* (three if including two partial mitovirus sequences from ESTs; Menotta, Amicucci, Sisti, Gioacchini, & Stocchi, 2004) are plant pathogens. Thus far, only the sequences of the mitoviruses associated with ectomycorrhizal hosts have been reported (Stielow, Bratek, Klenk, Winter, & Menzel, 2012; Stielow, Klenk, Winter, & Menzel, 2011), with no significant biological study yet done.

Based on sequence relationships, the mitoviruses characterized to date fall into two ancestral clades (Doherty et al., 2006; Stielow et al., 2012; Xie & Ghabrial, 2012; Fig. 6.2). When more than one mitovirus has been characterized from a single fungal host species or even isolate, it is often the case that both ancestral clades are represented by the different viruses (Botella, Tuomivirta, Vervuurt, Diez, & Hantula, 2012; Doherty et al., 2006), supporting the hypothesis of multiple independent entries of viruses into host species. The two viruses infecting different ectomycorrhizal fungal species (*Tuber excavatum* and *T. aestivum*) are found in different mitovirus clades. The two mitoviruses identified in basidiomycetes, Helicobasidium mompa mitovirus 1 (Osaki, Nakamura, Nomura, Matsumoto, & Yoshida, 2005) and TcMV1 (Lakshman et al., 1998b), also are in different phylogenetic clades. The idea of codivergence of viruses and their fungal hosts was examined in an interesting recent study using more recently available statistical methods of cophylogenetic analysis (Goker, Scheuner, Klenk, Stielow, & Menzel, 2011), but the dataset for members of the *Narnaviridae* and their hosts was relatively small and selective, preventing strong conclusions to be made.

An unusual feature of mitovirus comparative genomics is the unexpected number of very closely related sequences identified in unrelated fungal taxa. The first of these identified was the virus conspecific with OnuMV-3A-Ld in *Sclerotinia homoeocarpa*, which was given the name OnuMV-3A-Sh12B (Deng, Xu, & Boland, 2003). In that case, the two genomes were 92.4% identical at the nucleotide level, meaning that they were strains of a single virus. Interestingly, OnuMV-3A-Ld does not appear to be associated with reduced virulence of *O. novo-ulmi*, whereas OnuMV-3A-Sh12B was isolated from a hypovirulent isolate of *S. homoeocarpa*.

A parallel but slightly different situation was found to occur with the Botrytis cinerea debilitation-associate mitovirus, BcMV-1, and the

O. novo-ulmi virus, OnuMV-3B. The two viral genomes share 95% nucleotide sequence identity, although the BcMV-1 genome contains an additional 472 nt because of a substantially longer 5′ noncoding sequence and a slightly longer 3′-noncoding sequence (Wu, Wang, et al., 2010). As with the *S. homoeocarpa*/OnuMV-3A-Ld example, BcMV-1 caused hypovirulence in its *Botrytis* host, but OnuMV-3B was not associated with hypovirulence of *O. novo-ulmi*. The level of sequence similarity in these examples, especially in the coding region, is higher than has been found with other fungal virus genera. It could be that the mitochondrial localization of replication of these RNAs uniquely affects their evolution.

Interestingly, both OnuMV-3A-Sh12B and BcMV-1 were found to be associated with accessory RNAs: a satellite RNA unrelated to the genomic RNA of OnuMV-3A-Sh12B was characterized and found not to be associated with virulence or genomic RNA replication competence (Deng & Boland, 2004). A presumably defective RNA associate with BcMV-1 was also not found to have an effect on genomic RNA replication.

There is a notable difference between two mitovirus sequences in the database and others characterized to date: *S. sclerotiorum* mitovirus 1/KL-1 (SsMV-1/KL1) and 2a (CcMV-2a) are the only two mitoviruses that have poly (A) tails and they each have lower AU contents than other mitoviruses, 51% in the case of CcMV-2a and 53% in the case of SsMV-1/KL-1 (Xie & Ghabrial, 2012). Although these features are suggestive of a possible cytosolic rather than a mitochondrial subcellular location, the presence of 6 UGA codons in SsMV-1/KL-1 and 15 UGA codons in CcMV-2a indicates that both are translated in mitochondria (van Heerden, 2004; Xie & Ghabrial, 2012). Phylogenetically, the two mitoviruses with low AU contents and poly(A) tails are each other's closest relatives and are within clade I, but they are not basal to that clade (Xie & Ghabrial, 2012; Fig. 6.2). With only two such examples characterized to date, whether the low AU content and poly(A) tail of these elements represent vestigial features inherited from the progenitor from which other clade I elements evolved, or features that were gained subsequent to entry of the progenitor clade I elements into fungal mitochondria is a matter of speculation.

5. POPULATION BIOLOGY OF MEMBERS OF THE *NARNAVIRIDAE*

Population-level studies of members of the *Narnaviridae* are confined to a few examples. The yeast-infecting narnaviruses ScNV-20S and ScNV-23S have been studied more than many other fungal viruses in large part because of the economic importance of the host. In general, these studies have focused on industrial yeasts, with wild yeasts included as lesser components of the studies. Lopez, Gil, Vicente Carbonell, and Navarro (2002) examined a total of 160 industrial and wild yeast isolates representing two species of *Saccharomyces*, *S. cerevisiae* and *S. diastaticus*, and found that 27/160 contained only ScNV-20S, 4/160 contained only ScNV-23S, 15/160 contained both viruses, and the remaining 114 isolates were virus-free. A similar ratio of ScNV-20S and ScNV-23S was identified in 70 isolates of *S. cerevisiae* by Nakayashiki, Kurtzman, Edskes, and Wickner (2005). As these viruses have been shown to be stress induced (Esteban et al., 1992; Matsumoto, Fishel, & Wickner, 1990), Lopez et al. (2002) discussed the possibility of using these viruses as molecular probes in industrial settings, and this was further examined using natural strains (Maqueda, Zamora, Rodriguez-Cousino, & Ramirez, 2010). The idea was also examined by Esteban and colleagues, who developed an engineered form of ScNV-23S for use as a molecular tag to follow commercial yeast strains used in the wine industry (Esteban & Rodriguez-Cousino, 2008).

Cubeta and colleagues have pursued studies on TcMV (called the M2 virus in these papers) in the context of its fungal host, *R. solani*, from a population standpoint. They found that TcMV was common and recombining in field populations of *R. solani* (Charlton, Carbone, Tavantzis, & Cubeta, 2006). They demonstrated that the transmission of TcMV between somatically incompatible field isolates of *R. solani* anastomosis group 3 could occur but that such transmission events were rare and recipient isolates appeared to lose virus when maintained in culture (Charlton & Cubeta, 2007). Closely related isolates of TcMV were identified in both *Thanatephorus* and *Ceratobasidium*, of interest because the two are thought to represent non-interbreeding populations within the *Rhizoctonia* species complex (Charlton, Carbone, Tavantzis, & Cubeta, 2008). In that study, no evidence for coevolution of TcMV with its hosts and no evidence for association of virus with geographic origin were found (Charlton et al., 2008). As noted above in discussing host effects of TcMV, the complex biology of *R. solani* renders it difficult to study in the context of virus populations as well.

A recent study by Botella et al. (2012) followed up at the population level studies on mitoviruses that were previously identified in the ascomycete *Gremmeniella abietina* var. *abietina*, which causes a canker disease of conifers. Work on this system began with initial characterization of one mitovirus species (Tuomivirta & Hantula, 2003) and subsequent characterization of another (Tuomivirta & Hantula, 2005). The recent population study

(Botella et al., 2012) examined a total of 353 fungal isolates from Canada, Spain, Switzerland, Turkey, and USA for the mitoviruses, finding 60 isolates from Spain and Finland but none from the other regions bearing mitoviruses. The Spanish population was found to harbor both mitovirus species, whereas the Finnish population harbored only one. This study provides a very nice initiation point to examine mitovirus populations in the context of nuclear and mitochondrial haplotype in populations of this fungus and could provide a system through which to examine questions of mitovirus transmission posed elsewhere in this review.

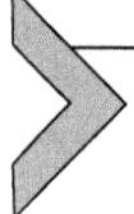

6. OTHER GENOME FEATURES AND MOLECULAR MANIPULATION OF MEMBERS OF THE *NARNAVIRIDAE*

As genome sizes of members of the *Narnaviridae* are very small, they are excellent candidates for potential molecular manipulation, including development and deployment of infectious RNA from cDNA clones. And indeed, this has been accomplished with the two yeast narnaviruses—both ScNV-23S and ScNV-20S RNAs have been launched from full-length clones and shown to replicate autonomously (Esteban & Fujimura, 2003; Esteban, Vega, & Fujimura, 2005). In those instances, experiments typical of those done with infectious cDNA clones of RNA viruses were completed: besides demonstration of autonomous replication, requirements for *cis*-acting element in terminal sequences and specific RdRp catalytic residues were examined. The experiments with ScNV-23S were especially significant, as it had not been found in strains devoid of ScNV-20S, and there was speculation that ScNV-23S might require ScNV-20S for replication, which turned out not to be true (Esteban et al., 2005). As both the ScNV-20S and ScNV-23S RNAs encode only an RdRp, of course, broader examination of viral elements through reverse genetics is somewhat limited.

A reverse genetics system for the related ourmiaviruses of plants has also been developed and used to investigate replication and subcellular localization of these related viruses (Crivelli et al., 2011). An important next step is to develop a heterologous system for replication, for example, ourmiavirus replication in a yeast system as discussed by Crivelli and colleagues (Crivelli et al., 2011; M. Turina, personal communication).

The mitoviruses present a separate problem in the context of reverse genetics, and it has not been solved as far as we are aware. Though their small size makes them easy to clone, their mitochondrial subcellular location is difficult to reach. Mitochondrial transformation and transfection have been very challenging, and the repeatability that is needed for useful reverse genetics studies of mitoviruses is not yet available.

Assuming that one begins with full-length cDNA clones, experiments to deliver infectious RNA to mitochondria could be approached in several ways: (1) deliver RNA transcripts that were generated *in vitro* to fungal protoplasts and allow pooled protoplasts to develop and merge into colonies, and then screen such colonies subsequently for replicating RNA (e.g., Chen, Choi, & Nuss, 1994); (2) transform the uninfected fungal host with a plasmid bearing a full-length cDNA copy of viral RNA, which is launched from the nucleus and exported to the cytoplasm (Choi & Nuss, 1992; Esteban & Fujimura, 2003); or (3) transform mitochondria of the uninfected host with a plasmid bearing a full-length cDNA copy of viral RNA, which is then transcribed, translated, and replicated within the transformed mitochondria. Infectivity by the first two methods, in which positive-sense full-length RNA is effectively delivered either directly or indirectly to the fungal cytoplasm, relies on the RNA being imported from cytoplasm into mitochondria. The third method is particularly challenging technically, as mitochondrial transformation or transfection is very difficult and has been shown to be successful in only a limited number of cases (Lightowlers, 2011; Mileshina, Ibrahim, et al., 2011; Mileshina, Koulintchenko, Konstantinov, & Dietrich, 2011; Yu et al., 2012). A variety of methods for mitochondrial transformation and transfection have been attempted, with systems designed for both DNA and RNA importation to mitochondria, but results have been inconsistent and difficult to reproduce (Lightowlers, 2011).

Few labs have reported attempting *in vitro* delivery of mitovirus constructs to fungi to initiate infection. Conversations with Dr. Ken Buck some years ago indicated that his lab was working on a system for transfection of *O. novo-ulmi* mitoviruses, but no results from those studies have yet been published. Our lab has tried and been unsuccessful at the first and second of the above methods, that is, delivering full-length positive-sense RNA copies of the CpMV-1 genome launched from the *C. parasitica* nucleus following transformation as a cDNA clone, or directly to the cytoplasm as full-length transcripts. We have not made serious attempts at the third method, that is, transformation or transfection of *C. parasitica* mitochondria. We also attempted to make a cytoplasmic replicon of CpMV-1 by mutating its nine UGA codons, which would encode Trp in mitochondria but would be termination codons in the cytoplasm, to UGG, which would encode Trp in cytoplasmic translation, but RNA replication was not achieved with that construct.

Besides the single ORF, members of the *Narnaviridae* present only the termini of their RNAs to examine using reverse genetic approaches. Terminal sequences of mitovirus RNAs have been predicted to fold into stem-loop structures (e.g., Hong et al., 1998a, 1998b, 1999; Lakshman et al., 1998; Park et al., 2006; Tuomivirta & Hantula, 2003; Wu, Wang, et al., 2010; Wu, Zhang, et al., 2010; Xie & Ghabrial, 2012), and some but not others are predicted to form panhandle structures due to terminal inverse complementarity of nucleotide sequences (e.g., Hong et al., 1998a,1998b, 1999; Lakshman et al., 1998; Osaki et al., 2005). Lengths of 5′- and 3′-nontranslated sequences are variable and have been summarized in Xie and Ghabrial (2012) and Stielow et al. (2012).

6.1. Latent infection by mitoviruses

The observation of possible mitovirus latency originated during the early research with *O. novo-ulmi* mitoviruses (Rogers et al., 1986a). This phenomenon of great reduction in viral RNA accumulation in some mitovirus infections, generally viewed as latent infections, is being observed with increasing consistency in other mitovirus systems as well (Melzer, Deng, & Boland, 2005; Park et al., 2006; Wu, Wang, et al., 2010). In such latent infections, dsRNA is not detectable by gel electrophoresis but is detectable by RT-PCR or Q-PCR-based methods. The basis for such latent infections is unknown, but this brings into focus the question of how many mitochondria within a given cell are mitovirus infected and how that population may expand or contract in a given mycelium. As noted above, all single conidial isolates but only ~50% of single ascospore isolates from CpMV-1-infected isolates used as females in matings were infected with CpMV-1, suggesting that only a subset of mitochondria may be virus infected (Polashock et al., 1997; Polashock & Hillman, 1994). How the proportion of infected mitochondria within a mycelium affects overall RNA accumulation requires more study, and these experiments are difficult.

6.2. Satellite and defective RNAs of mitoviruses

Satellite and defective RNAs are subviral RNAs that may rely on a replication-competent parent virus for their maintenance and may or may not affect virus symptoms in the host. Both types of subviral RNAs are common in fungal viruses, as they are in plant viruses (e.g., ;Hillman, Foglia, & Yuan, 2000 Shapira, Choi, Hillman, & Nuss, 1991). Both types of RNAs have also been identified in the *Narnaviridae*. Defective RNAs have been identified in

O. novo-ulmi strain Ld; segments 7 and 10 of the original hypovirulent isolate were shown not to encode functional proteins, but they are not known to affect virus-associated symptoms (Cole, Muller, Hong, Brasier, & Buck, 1998; Hong et al., 1998b). A defective RNA was also identified in the *B. cinerea* isolate CanBc-1c-78, named BcMV-1S (Wu, Wang, et al., 2010; Wu et al., 2007). BcMV-1S was found to be a simple deletion mutant of BcMV-1 that fits the criteria of a defective interfering RNA by being derived from and reducing accumulation of the parent viral RNA (Wu, Wang, et al., 2010). Examining OnuMV-3A infection of *S. homoeocarpa*, Deng and Boland (2004) identified a satellite RNA population that was heterogeneous in sequence and appeared not to affect accumulation of parent virus replication to a significant extent. These results are interesting in the context of considerations of RNA silencing and mitoviruses introduced earlier. Nuss and colleagues found that Dicer and Argonaute genes of *C. parasitica* were required for RNA recombination and hypovirus DI RNA accumulation (Sun, Choi, & Nuss, 2009; Zhang & Nuss, 2008). Mitoviruses with DI RNAs could thus provide systems through which to investigate some aspects of RNA silencing.

7. CONCLUSIONS AND FUTURE DIRECTIONS

Members of the family *Narnaviridae* are challenging and important subjects of future study from several perspectives. As the "minimalists" of RNA viruses containing no genes other than an RdRp, requirements for replication can potentially be addressed directly. In yeast, this is being addressed (Esteban, Vega, & Fujimura, 2008) and in the near future should allow for direct comparisons with nonfungal viruses that have exploited yeast as a vehicle to examine genes required for replication (for reviews, see Janda & Ahlquist, 1993; Nagy & Pogany, 2006, 2012).

The mitochondrial localization of mitoviruses in simple haploid hosts offers a major opportunity for study. Many of the fungal hosts of mitoviruses have accessible classical genetics systems and molecular transformation systems available, and are completely sequenced. The challenge of mitochondrial transformation and transfection thus also presents the opportunity to use mitoviruses as molecular probes of mitochondrial function and biology. A question that still requires work is clearer definition of the level of restriction of mitoviruses within mitochondria. In this context, *Rhizoctonia* TcMV and its relatives (Lakshman et al., 1998a) could be important, as such viruses could represent a transition between the presumably purely mitochondrial lifestyle of mitoviruses and the purely cytosolic lifestyle of narnaviruses.

From the standpoint of investigating virus evolution, the mitoviruses could also represent important and unique models. RNA viruses generally are known to have high rates of nucleotide substitution (e.g., Holmes, 2003), but the acquisition of mitoviruses by unrelated fungal species, which could be a relatively common phenomenon based on numbers from the small sample size to date, and the apparent subsequent lack of substantial evolution at the nucleotide level following acquisition (Deng et al., 2003; Wu, Wang, et al., 2010) deserve experimental scrutiny. Studies such as those by Botella et al. (2012) have begun to examine nucleotide substitution rates in mitoviruses at the population level, an important step. The availability of increasingly informative labeling and imaging technologies will allow for investigation of the cellular/molecular mechanisms of horizontal transmission between divergent species of viruses that are confined to mitochondria, but the experiments themselves will still be difficult and require the most tractable experimental systems.

The natural history of members of the *Narnaviridae* is also poorly understood. Unfortunately, the work that is required for studies of fungal viruses in natural settings is substantial, and most scientists who are equipped to do those studies are not equipped for the difficult and resource-consuming studies that are required to push the field of virus ecology forward. As with other studies of viruses in populations, the use of deep sequencing of environmental samples is changing the way these studies are approached (Diemer & Stedman, 2012). One such recent study on grapevine plants (Al Rwahnih, Daubert, Urbez-Torres, Cordero, & Rowhani, 2011) revealed a large number of presumed mycovirus sequences including mitovirus sequences, although the latter were not examined in detail. We anticipate that the rapid increase in metagenomic analysis of environmental samples will provide valuable insights to drive these studies.

REFERENCES

Al Rwahnih, M., Daubert, S., Urbez-Torres, J. R., Cordero, F., & Rowhani, A. (2011). Deep sequencing evidence from single grapevine plants reveals a virome dominated by mycoviruses. *Archives of Virology*, *156*, 397–403.

Anagnostakis, S. L. (1987). Chestnut blight: The classical problem of an introduced pathogen. *Mycologia*, *79*, 23–37.

Bandiera, S., Ruberg, S., Girard, M., Cagnard, N., Hanein, S., Chretien, D., et al. (2011). Nuclear outsourcing of RNA interference components to human mitochondria. *PLoS One*, *6*, e20746.

Botella, L., Tuomivirta, T. T., Vervuurt, S., Diez, J. J., & Hantula, J. (2012). Occurrence of two different species of mitoviruses in the European race of *Gremmeniella abietina* var. abietina, both hosted by the genetically unique Spanish population. *Fungal Genetics and Biology*, *116*, 872–882.

Brasier, C. M. (1979). A cytoplasmically transmitted disease of *Ceratocystis ulmi*. *Nature*, *305*, 220–223.
Brasier, C. M. (1986a). The d-factor in *Ceratocystis ulmi*: Its biological characteristics and implications for Dutch elm disease. In K. W. Buck (Ed.), *Fungal virology* (pp. 177–208). Boca Raton, FL: CRC Press.
Brasier, C. M. (1986b). The population biology of Dutch elm disease: Its principal features and some implications for other host-pathogen systems. In D. S. Ingram & P. H. Williams (Eds.), *Advances in plant pathology* (pp. 55–118). New York: Academic Press.
Bryner, S. F., & Rigling, D. (2011). Temperature-dependent genotype-by-genotype interaction between a pathogenic fungus and its hyperparasitic virus. *The American Naturalist*, *177*, 65–74.
Bryner, S. F., & Rigling, D. (2012). Virulence not only costs but also benefits the transmission of a fungal virus. *Evolution*, *66*, 2540–2550.
Buck, K. W., Brasier, C. M., Paoletti, M., & Crawford, L. J. (2003). Virus transmission and gene flow between two species of Dutch elm disease fungi, *Ophiostoma ulmi* and *O. novo-ulmi*: Deleterious viruses as selective agents for gene introgression. In R. S. Hails, J. E. Beringer & H. C. J. Godfray (Eds.), *Genes in the environment* (pp. 26–45). Oxford: Blackwell Publishing.
Cai, G., Myers, K., Fry, W. E., & Hillman, B. I. (2012). A member of the virus family *Narnaviridae* from the plant pathogenic oomycete *Phytophthora infestans*. *Archives of Virology*, *157*, 165–169.
Charlton, N. D., Carbone, I., Tavantzis, S. M., & Cubeta, M. A. (2006). Analysis of genetic diversity and evolutionary history of the M2 dsRNA of *Rhizoctonia solani* AG-3. *Inoculum*, *57*, 14.
Charlton, N. D., Carbone, I., Tavantzis, S. M., & Cubeta, M. A. (2008). Phylogenetic relatedness of the M2 double-stranded RNA in *Rhizoctonia* fungi. *Mycologia*, *100*, 555–564.
Charlton, N. D., & Cubeta, M. A. (2007). Transmission of the M2 double-stranded RNA in *Rhizoctonia solani* anastomosis group 3 (AG-3). *Mycologia*, *99*, 859–867.
Chen, B., Choi, G. H., & Nuss, D. L. (1994). Attenuation of fungal virulence by synthetic infectious hypovirus transcripts. *Science*, *264*, 1762–1764.
Choi, G. H., & Nuss, D. L. (1992). Hypovirulence of chestnut blight fungus conferred by an infectious viral cDNA. *Science*, *257*, 800–803.
Chung, P. H., Bedker, P. J., & Hillman, B. I. (1994). Diversity of *Cryphonectria parasitica* hypovirulence-associated double-stranded RNAs within a chestnut population in New Jersey. *Phytopathology*, *84*, 984–990.
Cole, T. E., Muller, B. M., Hong, Y., Brasier, C. M., & Buck, K. W. (1998). Complexity of virus-like double-stranded RNA elements in a diseased isolate of the Dutch elm disease fungus, *Ophiostoma novo-ulmi*. *Journal of Phytopathology*, *146*, 593–598.
Crivelli, G., Ciuffo, M., Genre, A., Masenga, V., & Turina, M. (2011). Reverse genetic analysis of Ourmiaviruses reveals the nucleolar localization of the coat protein in *Nicotiana benthamiana* and unusual requirements for virion formation. *Journal of Virology*, *85*, 5091–5104.
Deng, F., & Boland, G. J. (2004). A satellite RNA of *Ophiostoma novo-ulmi mitovirus 3a* in hypovirulent Isolates of *Sclerotinia homoeocarpa*. *Phytopathology*, *94*, 917–923.
Deng, F., Xu, R., & Boland, G. J. (2003). Hypovirulence-associated dsRNA from *Sclerotinia homeocarpon* is conspecific with *Ophiostoma novo-ulmi* mitovirus 3a-Ld. *Phytopathology*, *93*, 1407–1414.
Diemer, G. S., & Stedman, K. M. (2012). A novel virus genome discovered in an extreme environment suggests recombination between unrelated groups of RNA and DNA viruses. *Biology Direct*, 7, 13.
Doherty, M., Coutts, R. H., Brasier, C. M., & Buck, K. W. (2006). Sequence of RNA-dependent RNA polymerase genes provides evidence for three more distinct mitoviruses in *Ophiostoma novo-ulmi* isolate Ld. *Virus Genes*, *33*, 41–44.

Drinnenberg, I. A., Fink, G. R., & Bartel, D. P. (2011). Compatibility with killer explains the rise of RNAi-deficient fungi. *Science*, *333*, 1592.

Drinnenberg, I. A., Weinberg, D. E., Xie, K. T., Mower, J. P., Wolfe, K. H., Fink, G. R., et al. (2009). RNAi in budding yeast. *Science*, *326*, 544–550.

Edgar, R. C. (2004). MUSCLE: Multiple sequence alignment with high accuracy and high throughput. *Nucleic Acids Research*, *32*, 1792–1797.

Esteban, R., & Fujimura, T. (2003). Launching the yeast 23S RNA Narnavirus shows 5' and 3' cis-acting signals for replication. *Proceedings of the National Academy of Sciences of the United States of America*, *100*, 2568–2573.

Esteban, R., & Rodriguez-Cousino, N. (2008). 23S RNA-derived replicon as a 'molecular tag' for monitoring inoculated wine yeast strains. *Yeast*, *25*, 359–369.

Esteban, L. M., Rodriguez-Cousino, N., & Esteban, R. (1992). T double-stranded RNA (dsRNA) sequence reveals that T and W dsRNAs form a new RNA family in *Saccharomyces cerevisiae*. *The Journal of Biological Chemistry*, *267*, 10874–10881.

Esteban, R., Vega, L., & Fujimura, T. (2005). Launching of the yeast 20 s RNA narnavirus by expressing the genomic or antigenomic viral RNA in vivo. *The Journal of Biological Chemistry*, *280*, 33725–33734.

Esteban, R., Vega, L., & Fujimura, T. (2008). 20S RNA Narnavirus defies the antiviral activity of SKI1/XRN1 in *Saccharomyces cerevisiae*. *The Journal of Biological Chemistry*, *283*, 25812–25820.

Garvik, B., & Haber, J. E. (1978). New cytoplasmic genetic element that controls 20S RNA synthesis during sporulation in yeast. *Journal of Bacteriology*, *134*, 261–269.

Ghabrial, S. A., & Suzuki, N. (2009). Viruses of plant pathogenic fungi. *Annual Review of Phytopathology*, *47*, 353–384.

Goker, M., Scheuner, C., Klenk, H. P., Stielow, J. B., & Menzel, W. (2011). Codivergence of mycoviruses with their hosts. *PLoS One*, *6*, e22252.

Gonzalez, D., Carling, D. E., Kuninaga, S., Vilgalys, R., & Cubeta, M. A. (2001). Ribosomal DNA systematics of *Ceratobasidium* and *Thanatephorus* with *Rhizoctonia* anamorphs. *Mycologia*, *93*, 1138–1150.

Hillman, B. I., & Esteban, R. (2011). Family Narnaviridae. In A. M. Q. King, M. J. Adams, E. B. Castens & E. J. Lefkowitz (Eds.), *Virus taxonomy: Ninth report of the international committee for the taxonomy of viruses* (pp. 1025–1030). New York: Elsevier.

Hillman, B. I., Foglia, R., & Yuan, W. (2000). Satellite and defective RNAs of *Cryphonectria hypovirus* 3-Grand Haven 2, a virus species in the family *Hypoviridae* with a single open reading frame. *Virology*, *276*, 181–189.

Hillman, B. I., & Suzuki, N. (2004). Viruses of the chestnut blight fungus, *Cryphonectria parasitica*. *Advances in Virus Research*, *65*, 423–472.

Holmes, E. C. (2003). Error thresholds and the constraints to RNA virus evolution. *Trends in Microbiology*, *11*, 543–546.

Hong, Y., Cole, T. E., Brasier, C. M., & Buck, K. W. (1998a). Evolutionary relationships among putative RNA-dependent RNA polymerases encoded by a mitochondrial virus-like RNA in the Dutch elm disease fungus, *Ophiostoma novo-ulmi*, by other viruses and virus-like RNAs and by the *Arabidopsis* mitochondrial genome. *Virology*, *246*, 158–169.

Hong, Y., Cole, T. E., Brasier, C. M., & Buck, K. W. (1998b). Novel structure of two virus-like RNA elements from a diseased isolate of the Dutch elm disease fungus, *Ophiostoma novo-ulmi*. *Virology*, *242*, 80–89.

Hong, Y., Dover, S. L., Cole, T. E., Brasier, C. M., & Buck, K. W. (1999). Multiple mitochondrial viruses in an isolate of the Dutch elm disease fungus, *Ophiostoma novo-ulmi*. *Virology*, *258*, 118–127.

Huang, L., Mollet, S., Souquere, S., Le Roy, F., Ernoult-Lange, M., Pierron, G., et al. (2011). Mitochondria associate with P-bodies and modulate microRNA-mediated RNA interference. *The Journal of Biological Chemistry*, *286*, 24219–24230.

Janda, M., & Ahlquist, P. (1993). RNA-dependent replication, transcription, and persistence of brome mosaic virus RNA replicons in *S. cerevisiae*. *Cell*, *72*, 961–970.
Jian, J., Lakshman, D. K., & Tavantzis, J. (1997). Association of distinct double-stranded RNAs with enhanced or diminished virulence in *Rhizoctonia solani* infecting potato. *Molecular Plant-Microbe Interactions*, *10*, 1002–1009.
Kadowaki, K., & Halvorson, H. O. (1971). Appearance of a new species of ribonucleic acid during sporulation in *Saccharomyces cerevisiae*. *Journal of Bacteriology*, *105*, 826–830.
Kobayashi, D. Y., & Crouch, J. A. (2009). Bacterial/fungal interactions: From pathogens to mutualistic endosymbionts. *Annual Review of Phytopathology*, *47*, 63–82.
Lakshman, D. K., Jian, J., & Tavantzis, S. M. (1998). A double-stranded RNA element from a hypovirulent strain of *Rhizoctonia solani* occurs in DNA form and is genetically related to the pentafunctional AROM protein of the shikimate pathway. *Proceedings of the National Academy of Sciences of the United States of America*, *95*, 6425–6429.
Lakshman, D. K., Liu, C., Mishra, P. K., & Tavantzis, S. (2006). Characterization of the arom gene in *Rhizoctonia solani*, and transcription patterns under stable and induced hypovirulence conditions. *Current Genetics*, *49*, 166–177.
Lightowlers, R. N. (2011). Mitochondrial transformation: Time for concerted action. *EMBO Reports*, *12*, 480–481.
Liu, C., Lakshman, D. K., & Tavantzis, S. M. (2003a). Expression of a hypovirulence-causing double-stranded RNA is associated with up-regulation of quinic acid pathway and down-regulation of shikimic acid pathway in *Rhizoctonia solani*. *Current Genetics*, *42*, 284.
Liu, C., Lakshman, D. K., & Tavantzis, S. M. (2003b). Quinic acid induces hypovirulence and expression of a hypovirulence-associated double-stranded RNA in *Rhizoctonia solani*. *Current Genetics*, *43*, 103.
Lopez, V., Gil, R., Vicente Carbonell, J., & Navarro, A. (2002). Occurrence of 20S RNA and 23S RNA replicons in industrial yeast strains and their variation under nutritional stress conditions. *Yeast*, *19*, 545–552.
Maqueda, M., Zamora, E., Rodriguez-Cousino, N., & Ramirez, M. (2010). Wine yeast molecular typing using a simplified method for simultaneously extracting mtDNA, nuclear DNA and virus dsRNA. *Food Microbiology*, *27*, 205–209.
Marienfeld, J. R., Unseld, M., Brandt, P., & Brennicke, A. (1997). Viral nucleic acid sequence transfer between fungi and plants. *Trends in Genetics*, *13*, 260–261.
Marienfeld, J., Unseld, M., & Brennicke, A. (1999). The mitochondrial genome of Arabidopsis is composed of both native and immigrant information. *Trends in Plant Science*, *4*, 495–502.
Matsumoto, Y., Fishel, R., & Wickner, R. B. (1990). Circular single-stranded RNA replicon in *Saccharomyces cerevisiae*. *Proceedings of the National Academy of Sciences of the United States of America*, *87*, 7628–7632.
Melzer, M. S., Deng, F., & Boland, G. J. (2005). Asymptomatic infection and distribution of *Ophiostoma* mitovirus 3A (OMV3A) in populations of *Sclerotinia homoeocarpa*. *Journal of Plant Pathology*, *27*, 610–615.
Menotta, M., Amicucci, A., Sisti, D., Gioacchini, A. M., & Stocchi, V. (2004). Differential gene expression during pre-symbiotic interaction between *Tuber borchii* Vittad. and *Tilia americana* L. *Current Genetics*, *46*, 158–165.
Mileshina, D., Ibrahim, N., Boesch, P., Lightowlers, R. N., Dietrich, A., & Weber-Lotfi, F. (2011). Mitochondrial transfection for studying organellar DNA repair, genome maintenance and aging. *Mechanisms of Aging and Development*, *132*, 412–423.
Mileshina, D., Koulintchenko, M., Konstantinov, Y., & Dietrich, A. (2011). Transfection of plant mitochondria and in organello gene integration. *Nucleic Acids Research*, *39*, e115.
Milgroom, M. G., & Hillman, B. I. (2011). The ecology and evolution of fungal viruses. In C. J. Hurst (Ed.), *Studies in virus ecology* (pp. 221–257). New York: John Wiley & Sons, Inc.

Milgroom, M. G., & Lipari, S. E. (1993). Maternal inheritance and diversity of mitochondrial DNA in the chestnut blight fungus, *Cryphonectria parasitica*. *Phytopathology*, *83*, 563–567.
Murphy, F. A., Fauquet, C. M., Bishop, D. H. L., Ghabrial, S. A., Jarvis, A. W., Martelli, G. P., et al. (1995). *Virus taxonomy. Classification and nomenclature of viruses. Sixth report of the international committee on taxonomy of viruses*, Wien, New York: Springer.
Nagy, P. D., & Pogany, J. (2006). Yeast as a model host to dissect functions of viral and host factors in tombusvirus replication. *Virology*, *344*, 211–220.
Nagy, P. D., & Pogany, J. (2012). The dependence of viral RNA replication on co-opted host factors. *Nature Reviews. Microbiology*, *10*, 137–149.
Nakayashiki, T., Kurtzman, C. P., Edskes, H. K., & Wickner, R. B. (2005). Yeast prions [URE3] and [PSI+] are diseases. *Proceedings of the National Academy of Sciences of the United States of America*, *102*, 10575–10580.
Nuss, D. L. (2005). Hypovirulence: Mycoviruses at the fungal-plant interface. *Nature Reviews. Microbiology*, *3*, 632–642.
Nuss, D. L. (2011). Mycoviruses, RNA silencing, and viral RNA recombination. *Advances in Virus Research*, *80*, 25–48.
Nuss, D. L., & Hillman, B. I. (2011). Family Hypoviridae. In A. M. Q. King, M. J. Adams, E. B. Castens & E. J. Lefkowitz (Eds.), *Virus taxonomy: Ninth report of the international committee for the taxonomy of viruses* (pp. 999–1003). New York: Elsevier.
Osaki, H., Nakamura, H., Nomura, K., Matsumoto, N., & Yoshida, K. (2005). Nucleotide sequence of a mitochondrial RNA virus from the plant pathogenic fungus, *Helicobasidium mompa* Tanaka. *Virus Research*, *107*, 39–46.
Osawa, S., Jukes, T. H., Watenabe, K., & Muto, A. (1992). Recent evidence for evolution of the genetic code. *Microbiological Reviews*, *56*, 229–264.
Park, Y., Chen, X., & Punja, Z. K. (2006). Molecular and biological characterization of a mitovirus in *Chalara elegans* (*Thielaviopsis basicola*). *Phytopathology*, *96*, 468–479.
Pearson, M. N., Beever, R. E., Boine, B., & Arthur, K. (2009). Mycoviruses of filamentous fungi and their relevance to plant pathology. *Molecular Plant Pathology*, *10*, 115–128.
Polashock, J. J., Bedker, P. J., & Hillman, B. I. (1997). Movement of a small mitochondrial double-stranded RNA element of *Cryphonectria parasitica*: Ascospore inheritance and implications for mitochondrial recombination. *Molecular & General Genetics*, *256*, 566–571.
Polashock, J. J., & Hillman, B. I. (1994). A small mitochondrial double-stranded (ds) RNA element associated with a hypovirulent strain of the chestnut blight fungus and ancestrally related to yeast cytoplasmic T and W dsRNAs. *Proceedings of the National Academy of Sciences of the United States of America*, *91*, 8680–8684.
Rastgou, M., Habibi, M. K., Izadpanah, K., Masenga, V., Milne, R. G., Wolf, Y. I., et al. (2009). Molecular characterization of the plant virus genus Ourmiavirus and evidence of inter-kingdom reassortment of viral genome segments as its possible route of origin. *The Journal of General Virology*, *90*, 2525–2535.
Rogers, H. J., Buck, K. W., & Brasier, C. M. (1986a). The D2-factor in *Ophiostoma ulmi*: Expression and latency. In J. A. Bailey (Ed.), *Biology and molecular biology of plant-pathogen interactions* (pp. 393–400). Berlin: Springer-Verlag.
Rogers, H. J., Buck, K. W., & Brasier, C. M. (1986b). Transmission of double-stranded RNA and a disease factor in *Ophiostoma ulmi*. *Plant Pathology*, *35*, 277–287.
Rogers, H. J., Buck, K. W., & Brasier, C. M. (1987). A mitochondrial target for the double-stranded RNAs in diseased isolates of the fungus that causes dutch elm disease. *Nature*, *329*, 558–560.
Rogers, H. J., Buck, K. W., & Brasier, C. M. (1988). dsRNA and disease factors of the aggressive subgroup of *Ophiostoma ulmi*. In Y. Koltin & M. J. Leibowitz (Eds.), *Viruses of fungi and simple eukaryotes* (pp. 327–351). New York: Marcel Dekker.

Segers, G. C., Zhang, X., Deng, F., Sun, Q., & Nuss, D. L. (2007). Evidence that RNA silencing functions as an antiviral defense mechanism in fungi. *Proceedings of the National Academy of Sciences of the United States of America, 104*, 12902–12906.

Shapira, R., Choi, G. H., Hillman, B. I., & Nuss, D. L. (1991). The contribution of defective RNAs to the complexity of viral-encoded double-stranded RNA populations present in hypovirulent strains of the chestnut blight fungus *Cryphonectria parasitica*. *The EMBO Journal, 10*, 741–746.

Stielow, J. B., Bratek, Z., Klenk, H. P., Winter, S., & Menzel, W. (2012). A novel mitovirus from the hypogeous ectomycorrhizal fungus *Tuber excavatum*. *Archives of Virology, 157*, 787–790.

Stielow, B., Klenk, H. P., Winter, S., & Menzel, W. (2011). A novel *Tuber aestivum* (Vittad.) mitovirus. *Archives of Virology, 156*, 1107–1110.

Sun, Q., Choi, G. H., & Nuss, D. L. (2009). A single Argonaute gene is required for induction of RNA silencing antiviral defense and promotes viral RNA recombination. *Proceedings of the National Academy of Sciences of the United States of America, 106*, 17927–17932.

Tamura, K., Peterson, D., Peterson, N., Stecher, G., Nei, M., & Kumar, S. (2011). MEGA5: Molecular evolutionary genetics analysis using maximum likelihood, evolutionary distance, and maximum parsimony methods. *Molecular Biology and Evolution, 28*, 2731–2739. http://dx.doi.org/10.1093/molbev/msr1121.

Taylor, D. R., Jarosz, A. M., Lenski, R. E., & Fulbright, D. W. (1998). The acquisition of hypovirulence in host-pathogen systems with three trophic levels. *The American Naturalist, 151*, 343–355.

Tuomivirta, T. T., & Hantula, J. (2003). *Gremmeniella abietina* mitochondrial RNA virus S1 is phylogenetically related to the members of the genus *Mitovirus*. *Archives of Virology, 148*, 2429–2436.

Tuomivirta, T. T., & Hantula, J. (2005). Three unrelated viruses occur in a single isolate of *Gremmeniella abietina* var. *abietina* type A. *Virus Research, 110*, 31–39.

van Duin, J., & Olsthoorn, R. C. M. (2011). Family Leviviridae. In A. M. Q. King, M. J. Adams, E. B. Cartens & E. J. Lefkowitz (Eds.), *Virus taxonomy: Ninth report of the international committee for the taxonomy of viruses*. New York: Elsevier.

van Heerden, S. W. (2004). *Studies on* Cryphonectria cubensis *in South Africa with special reference to mycovirus infection*. Pretoria: Faculty of Natural and Agricultural Science. University of Pretoria, 132 p.

Van Regenmortel, M. H. V., Fauquet, C. M., Bishop, D. H. L., Carstens, E., Estes, M., Lemon, S., MsGeoch, D., Wickner, R. B., Mayo, M. A., Pringle, C. R., & Maniloff, J. (2000). *Virus taxonomy. Seventh report of the international committee for the taxonomy of viruses*. New York: Academic Press.

Wesolowski, M., & Wickner, R. B. (1984). Two new double-stranded RNA molecules showing non-Mendelian inheritance and heat inducibility in *Saccharomyces cerevisiae*. *Molecular and Cellular Biology, 4*, 181–187.

Wickner, R. B. (2001). Viruses of yeasts, fungi, and parasitic microorganisms. In D. M. Knipe & P. M. Howley (Eds.), *Fields virology* (pp. 629–658). (4th ed.). Philadelphia: Lippencott Williams & Wilkens.

Wu, Q., Wang, X., & Ding, S. W. (2010). Viral suppressors of RNA-based viral immunity: Host targets. *Cell Host & Microbe, 8*, 12–15.

Wu, M., Zhang, L., Li, G., Jiang, D., & Ghabrial, S. A. (2010). Genome characterization of a debilitation-associated mitovirus infecting the phytopathogenic fungus *Botrytis cinerea*. *Virology, 406*, 117–126.

Wu, M. D., Zhang, L., Li, G. Q., Jiang, D. H., Hou, M. S., & Huang, H. C. (2007). Hypovirulence and double-stranded RNA in *Botrytis cinerea*. *Phytopathology, 97*, 1590–1599.

Xie, J., & Ghabrial, S. A. (2012). Molecular characterization of two mitoviruses co-infecting a hypovirulent isolate of the plant pathogenic fungus *Sclerotinia sclerotiorum*. *Virology*, *428*, 77–85.

Yu, H., Koilkonda, R. D., Chou, T. H., Porciatti, V., Ozdemir, S. S., Chiodo, V., et al. (2012). Gene delivery to mitochondria by targeting modified adenoassociated virus suppresses Leber's hereditary optic neuropathy in a mouse model. *Proceedings of the National Academy of Sciences of the United States of America*, *109*, E1238–E1247.

Zhang, X., & Nuss, D. L. (2008). A host dicer is required for defective viral RNA production and recombinant virus vector RNA instability for a positive sense RNA virus. *Proceedings of the National Academy of Sciences of the United States of America*, *105*, 16749–16754.

CHAPTER SEVEN

Viruses of the White Root Rot Fungus, *Rosellinia necatrix*

Hideki Kondo*, **Satoko Kanematsu**† **and Nobuhiro Suzuki***[,1]
*Institute of Plant Science and Resources (IPSR), Okayama University, Chuou, Kurashiki, Okayama, Japan
†Apple Research Station, Institute of Fruit Tree Science, NARO, Morioka, Iwate, Japan
[1]Corresponding author: e-mail address: nsuzuki@rib.okayama-u.ac.jp

Contents

1. Introduction 178
   1.1 General properties of the white root rot fungus, *Rosellinia necatrix* 178
   1.2 Population structure and life cycle of *R. necatrix* in fruit tree orchards 179
   1.3 Extensive virus search 183
   1.4 Epidemiological analysis of viruses in *R. necatrix* 185
2. Viruses of *R. necatrix* 186
   2.1 *Megabirnaviridae* 186
   2.2 *Partitiviridae* 190
   2.3 *Reoviridae* 194
   2.4 *Quadriviridae* (new family) 196
   2.5 *Totiviridae* 198
   2.6 Unclassified viruses 198
3. Expansion of Host Ranges of *R. necatrix*-Infecting Viruses 199
   3.1 Techniques for investigation of experimental host ranges 199
   3.2 Member of the class Sordariomycetes, *C. parasitica*, as a host of viruses infecting *R. necatrix* 202
   3.3 Other fungi as experimental hosts of viruses naturally infecting *R. necatrix* 203
4. Conclusions and Prospects 203
Acknowledgments 206
References 206

Abstract

Rosellinia necatrix is a filamentous ascomycete that is pathogenic to a wide range of perennial plants worldwide. An extensive search for double-stranded RNA of a large collection of field isolates led to the detection of a variety of viruses. Since the first identification of a reovirus in this fungus in 2002, several novel viruses have been molecularly characterized that include members of at least five virus families. While some cause phenotypic alterations, many others show latent infections. Viruses attenuating the virulence of a host fungus to its plant hosts attract much attention as agents for virocontrol (biological control using viruses) of the fungus, one of which is currently being tested in experimental fields. Like the *Cryphonectria parasitica*/viruses, the

Advances in Virus Research, Volume 86
ISSN 0065-3527
http://dx.doi.org/10.1016/B978-0-12-394315-6.00007-6

R. necatrix/viruses have emerged as an amenable system for studying virus/host and virus/virus interactions. Several techniques have recently been developed that enhance the investigation of virus etiology, replication, and symptom induction in this mycovirus/fungal host system.

1. INTRODUCTION

1.1. General properties of the white root rot fungus, *Rosellinia necatrix*

The genus *Rosellinia* (subphylum: Pezizomycotina; class: Sordariomycetes; subclass: Xylariomycetidae; order: Xylariales; family: Xylariaceae) was created by de Notaris in 1844 and 116 species have been reported within this genus (according to *Index Fungorum* in April 2012; http://www.indexfungorum.org/). However, only some species are known as phytopathogenic fungi, among which *Rosellinia necatrix* Prill. is the best studied root rot pathogen. *R. necatrix* is a soil-borne fungus that is widespread throughout the world, including temperate, subtropical, and tropical areas. *R. necatrix* can invade the roots of a wide range of woody and herbaceous plants. Over 400 (437) host plant species are collectively listed with references in a database provided by the USDA-ARS (http://nt.ars-grin.gov/fungaldatabases/). The number of host species of this fungus is considered to be increasing further with increasing reports of the disease on new hosts (reviewed in Pliego, Lopez-Herrera, Ramos, & Cazorla, 2012; ten Hoopen & Krauss, 2006).

Since the first reports of disease in vineyards in Germany and France about 135 years ago, perennial crops, especially fruit trees (including apple in the Middle East and USA, grape in Europe, and avocado in Spain), in land-intensive orchards have been seriously damaged by *R. necatrix*. In Japan, serious yield and quality losses have been reported in apple and Japanese pear orchards, as well as in expensive table grapes grown in glasshouses. An extensive survey in 2007 in the Nagano prefecture, which is one of the main fruit tree production areas in Japan, revealed that 3.2% and 6% of apple and Japanese pear fields in orchards, respectively, were infested with *R. necatrix* (Eguchi, 2010).

R. necatrix usually occurs repeatedly in particular spots in infested orchards and seldom disperses to neighboring orchards, unlike air-borne fungal pathogens. When mycelial inoculum of the dormancy stage encounters roots, mycelial masses and strands start to propagate on the roots, penetrate from the lenticels of epidermal cells and junctions between epidermal cells, and then invade the epidermis and xylem of the roots, forming fan-shaped mycelial strands (Pliego et al., 2009). The symptoms in the upper parts of the plants cannot be recognized in early stages of root infection. Therefore, diagnosis of this disease is difficult and laborious because infected roots in the soil can only be detected by digging up the soil and observing the roots by eye, trapping mycelia around the infected roots by the bait twig method (Eguchi, Kondo, & Yamagishi, 2009), or by using molecular tools (Schena, Nigro, & Ippolito, 2002; Shishido, Kubota, & Nakamura, 2012).

The control of white root rot in infested fields is very difficult; nevertheless, many attempts at control, including cultural, chemical, and biological, have been made (reviewed in ten Hoopen & Krauss, 2006). Breeding strategies for disease resistance seem to be quite difficult, owing to the wide host range of *R. necatrix*. In Japan, drenching with the fungicide fluazinam (50–200 l/tree) or hot water is practical in commercial orchards to control white root rot. These methods are sometimes effective, but not always, and recurrence occurs due to the limitation of the suppressive period of mycelial growth. In order to protect fruit trees from *R. necatrix*, sustainable management systems with minimal impact on the environment are strongly recommended. To meet these demands, we exploited the mycovirus, which attenuates the virulence of the host fungus, to control white root rot in fruit tree orchards.

1.2. Population structure and life cycle of *R. necatrix* in fruit tree orchards

Mycoviruses infecting filamentous fungi are generally transmitted either horizontally or intracellularly through mycelial fusion between vegetatively compatible individuals of a single species or vertically through spores. No extracellular phases of mycoviruses have been reported (Ghabrial & Suzuki, 2009; Pearson, Beever, Boine, & Arthur, 2009). Therefore, programmed cell death caused by mycelial incompatibility hampers mycovirus transmission between individuals belonging to dissimilar vegetative compatibility groups (Choi et al., 2012; Cortesi, McCulloch, Song, Lin, & Milgroom, 2001). The limiting factors for the biocontrol of chestnut blight at a population level are reviewed by Milgroom and Cortesi (2004), and the population structure of the host fungi is one of the crucial factors with regard to the spread of biocontrol agents (hypoviruses) to the *Cryphonectria parasitica* population.

R. necatrix isolates differing in mycelial compatibility groups (MCGs) produce dark demarcation lines at the colony junction when paired on oatmeal agar plates (Ikeda et al., 2011). Hyphal reactions of *R. necatrix* between and within MCGs were observed under light and electron microscopy (Inoue, Kanematsu, Park, & Ikeda, 2011). Interestingly, incompatible cell death of *R. necatrix* between hyphae of different MCGs was triggered without hyphal fusion, in contrast to the observations with *Neurospora crassa* (Glass, Jacobson, & Shiu, 2000). The clonality of each MCG was confirmed by genetic markers such as: telomere fingerprinting (Aimi, Kano, Yotsutani, & Morinaga, 2002), intersimple sequence repeat-PCR and universally primed-PCR (Ikeda, Nakamura, Arakawa, & Matsumoto, 2004), and amplified fragment length polymorphism (AFLP) (Kanematsu, 2013). Therefore, isolates belonging to identical MCGs are considered to represent the same genotypes.

Nakamura and Matsumoto (2006) surveyed the population structures of *R. necatrix* in Japanese fruit tree orchards based on MCG and collated distribution patterns of mycovirus and MCG (see later). Figure 7.1 illustrates the typical pattern of population structures of *R. necatrix* in Japanese pear orchards. In general, neighboring trees are colonized by the same MCGs, and the disease occurs in patches (Fig. 7.1A). These results agree with the widely accepted epidemiological observation that *R. necatrix* spreads to neighboring trees by mycelial development along root contact in the orchard (Pliego et al., 2012; ten Hoopen & Krauss, 2006). Some fields rarely showed high diversity, even though each MCG was considered to expand among trees also through root contact in the field (Fig. 7.1B). Such a random distribution pattern of MCGs considered to reflect field history; the land was leveled by bulldozer about 30 years ago, and the farmer initially buried large amount of wood debris to improve drainage. In addition, the random distribution pattern of the MCGs may be enhanced by agricultural practices, such as removal of diseased roots, disinfection of infested spots by fungicide, and replanting of young trees. Populations from different fields consist of distinct MCGs, suggesting a high genetic diversity in overall populations of *R. necatrix*. This particular MCG distribution pattern, involving clonal development with several MCGs in a field in spite of high variation in the total population, coincided with the report of avocado orchards in southern Spain (Perez Jimenez, Jimenez Diaz, & Lopez Herrera, 2002).

The reproductive strategy of *R. necatrix* certainly affects the population diversity and structure of this fungus. Asexual spores, conidia (3–4 μm × 2 μm), are produced on the tip of synnemata which develop on severely diseased roots under scattering light (Nakamura, Ikeda, Arakawa, & Matsumoto, 2002). Conidia germinate only slightly, but never

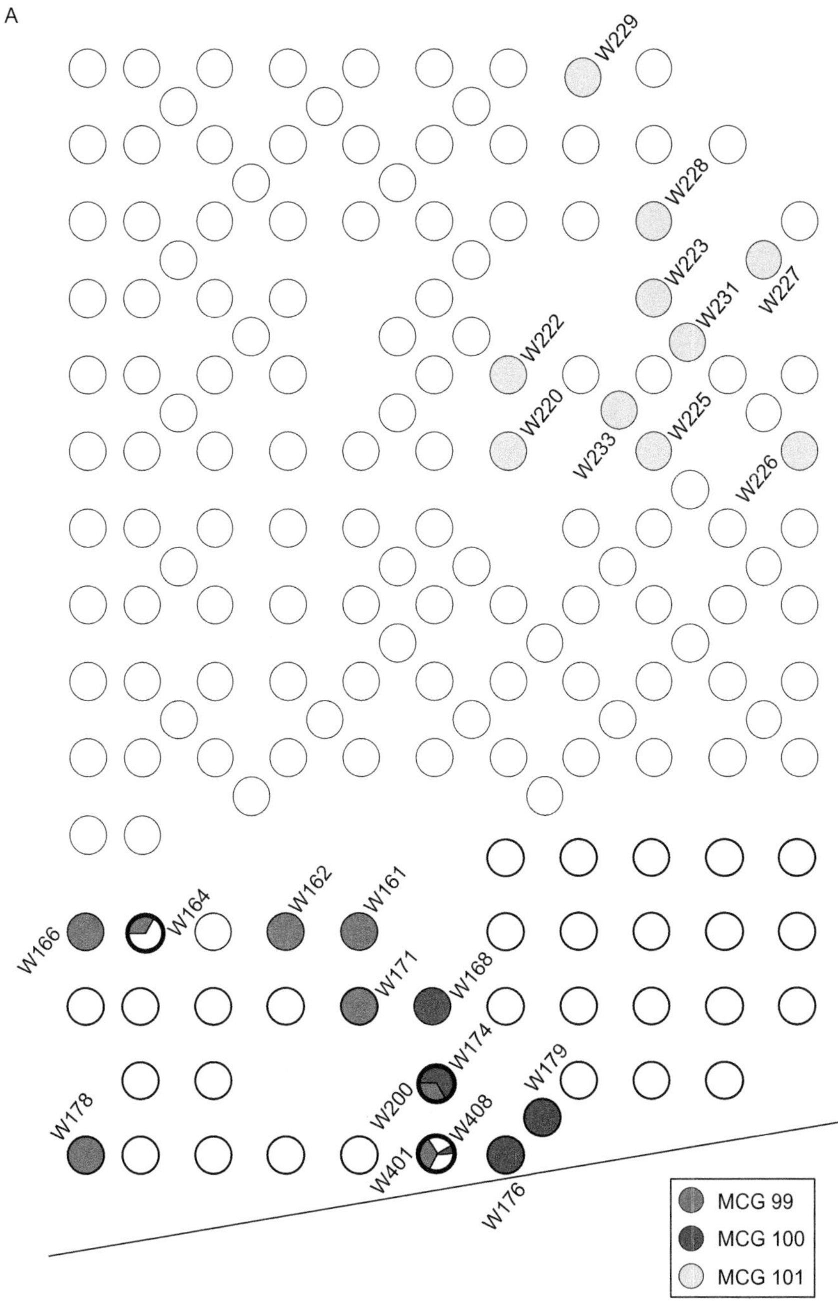

Figure 7.1

(Continued)

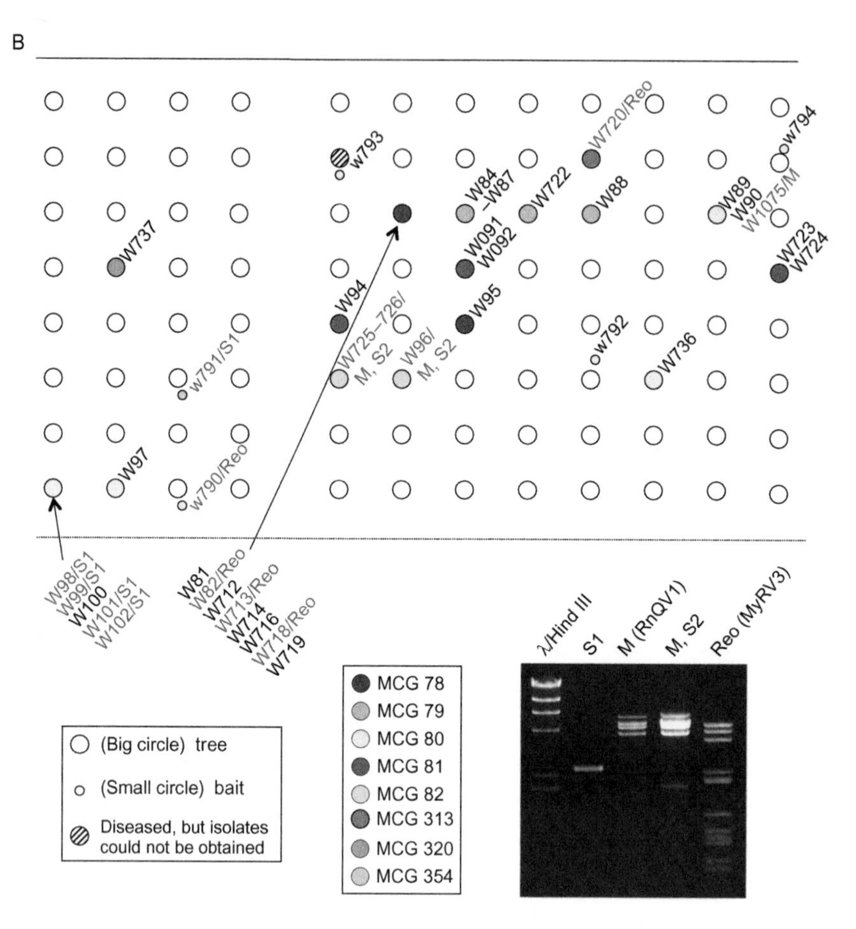

Figure 7.1—Cont'd Population structures of *R. necatrix*. (A) Typical low levels of genetic diversity measured by MCG shown in a Japanese pear orchard in Chiba Prefecture, Japan. dsRNA was detected in some of the isolates obtained from this orchard. (B) The population structure of *R. necatrix* in a Japanese pear orchard in Saga Prefecture, Japan. The field shows relatively complex MCGs possibly reflecting field history and the activities of farmers. Each large circle refers to a tree. Smaller circles indicate fungal isolates obtained by the bait twig method, which involved inserting cut twigs into the soil around the trunk as bait for *R. necatrix*. MCGs of fungal isolates are shown by different colors. The dsRNA detected in each isolate was categorized into S, M, or Reo, as indicated following the slash. The dsRNA profile revealed by agarose gel electrophoresis was also shown. The M dsRNA in this field was coined Rosellinia necatrix quadrivirus 1 (RnQV1), and the reo was identified as Rosellinia necatrix mycoreovirus 3 (MyRV3) (see Fig. 7.2). (See Page 11 in Color Section at the back of the book.)

develop into mycelia in all of the tested media *in vitro*. Following conidial production, perithecia develop underneath synnemata, and sexual spores (ascospore; 30–50 μm × 5–8 μm) are discharged from the ostioles of the mature perithecia. Therefore, conidia are considered to act as spermatia for sexual development (Nakamura, Uetake, Arakawa, Okabe, & Matsumoto, 2000). Most of the *R. necatrix* isolates from a single ascospore retain pathogenicity and can propagate on infected roots (Ikeda et al., 2004), whereas the role of conidia as a propagule is doubtful.

R. necatrix has a heterothallic life cycle (Ikeda et al., 2011). Most of ascospore progenies originating from perithecium represent different MCGs (Ikeda et al., 2011). Siblings from a single ascospore could be distinguished by telomere fingerprinting (Aimi et al., 2002) and AFLP analysis (Kanematsu, 2013). Ascospore discharge in fruit tree orchards seems to be rare, because farmers do not usually leave dead roots long enough for stromatal development in order to prevent the spread of disease to neighboring trees (Nakamura et al., 2000; Perez Jimenez et al., 2002). The role of ascospores is yet to be elucidated in relation to the epidemiology of *R. necatrix* (Pliego et al., 2012; ten Hoopen & Krauss, 2006). However, a high diversity of MCGs in the overall populations of *R. necatrix* in nature may suggest the following steps of population development in fruit tree orchards: (1) ascospores are introduced into the orchard and infect roots, (2) MCGs clonally develop by extending mycelia on infected roots, and most of them may fail to establish infection at this stage or at the previous stage in the field, and (3) mycelial aggregates within the pseudosclerotia are produced in the root debris and persist for years in the soil (Ogata & Ochiai, 1994) to reinfect roots of the newly planted trees. It is noteworthy that each isolate from a single ascospore is cured of mycovirus infection as in many other ascomycetous fungi (Ikeda et al., 2004). Therefore, field isolates of *R. necatrix* (e.g., Arakawa, Nakamura, Uetake, & Matsumoto, 2002; Ikeda, Nakamura, & Matsumoto, 2005) are likely to acquire mycoviruses during clonal development in the soil.

1.3. Extensive virus search

No information regarding mycovirus infection in *R. necatrix* was available in a project led by Matsumoto (1998) in the beginning of *R. necatrix* virocontrol (biological control using viruses). In order to find mycoviruses in *R. necatrix*, fungal cultures were extensively isolated from all over Japan and more than 1000 isolates have been collected to date. Many of the isolates were obtained from diseased roots. All the trees in some fields were surveyed by digging up the soil around the trunk carefully to detect infected tissues. The bait twig method was also applied to isolate *R. necatrix* from the soil in infested fields. Briefly, mulberry twigs were buried for a period of weeks to months in the soil as the bait, and then the fungal isolates were obtained from the mycelia developed on the twigs (Eguchi et al., 2009). Ascospore progenies were also isolated from perithecia produced on diseased roots via the enhancement of stromatal development, as described by Nakamura et al. (2000) and Ikeda et al. (2011).

To facilitate RNA virus detection from the collection, double-stranded (ds) RNAs were used as criterion to confirm RNA virus infection, since the genomes of many mycoviruses could be detected in their dsRNA form. dsRNA was extracted from *R. necatrix* mycelia following the procedure of Arakawa et al. (2002), including total nucleic acid digestion with S1 nuclease and DNase. This procedure could minimize both sample size and extraction scale and was therefore suitable for examining many samples simultaneously. The investigation identified the presence of relevant varieties of dsRNAs in *R. necatrix* at a frequency of about 20% (53 out of 254 MCGs) (Arakawa et al., 2002; Ikeda et al., 2004). The dsRNAs were categorized in the following size ranges: L, more than 10 kbp; M, from 3.5 to 10 kbp; S, less than 3 kbp; and Reo, 1 to 4 kbp with 11 to 12 segments (genome of a mycoreovirus). Variability of the dsRNAs was expected because most of the dsRNAs did not hybridize with the probes prepared from several dsRNAs by Northern hybridization (Ikeda et al., 2004). Some of *R. necatrix* isolates were thought to harbor a mixed infection of viruses (e.g., strain W8; Sasaki, Miyanishi, Ozaki, Onoue, & Yoshida, 2005) because they possessed a combination of different dsRNAs categorized in different size groups.

In order to identify the viruses that could reduce the virulence of *R. necatrix*, strains with abnormal, debilitating colonies were mainly selected. Several methods were applied for etiological analysis: curing viruses by hyphal tipping, transmitting viruses to virus-free isogenic strains through hyphal fusion, and transfection with purified virus particles via protoplasts. During these processes, methods for protoplast preparation and transformation methods were developed for *R. necatrix* (Aimi, Taguchi, Tanaka, Kitamoto, & Morinaga, 2005; Kanematsu et al., 2004; Kano, Kurita, Kanematsu, & Morinaga, 2011; Pliego et al., 2009; Shimizu, Ito, & Kanematsu, 2012) and can now be reliably applied for the above purpose. Strains tagged with antibiotic resistance genes and fluorescent proteins are useful for the analysis of virus transmission between individual mycelia (Kanematsu et al., 2004; Yaegashi, Sawahata, Ito, & Kanematsu, 2011).

The virulence of virus-infected and virus-free isogenic strains were compared by inoculating several plants with mycelia: apple fruits (Sasaki et al., 2007; Sasaki, Kanematsu, Onoue, Oyama, & Yoshida, 2006), roots of potted plants including lupines (Uetake, Nakamura, Arakawa, Okabe, & Matsumoto, 2001), apple seedlings (Kanematsu et al., 2004), and apple root stocks (*Malus prunifolia* var. ringo) (Chiba et al., 2009). Most of the viruses were associated with latent infections, although some specific viruses reduced the virulence of *R. necatrix* as described below.

1.4. Epidemiological analysis of viruses in *R. necatrix*

The diversity of dsRNA distribution in *R. necatrix* is distinct not only in different MCGs but also within MCGs isolated from the same trees (Arakawa et al., 2002; Ikeda et al., 2005). Figure 7.1B shows an example of the mosaic distribution of dsRNAs in *R. necatrix* isolates belonging to different as well as the same MCGs in an orchard. This field yielded fungal isolates with M- and S-sized dsRNAs in addition to mycoreoviruses. The M-sized dsRNAs in this orchard were characterized in detail and were proposed to belong to a new virus species (see quadrivirus below). Most of the dsRNAs were localized in a single MCG, but some dsRNAs were detected from different MCGs (Arakawa et al., 2002; Ikeda et al., 2005, for example, mycoreoviruses were detected from the three MCGs in Fig. 7.1B). The mechanism of virus infection and uneven distribution in *R. necatrix* population are yet to be elucidated, although some clues have been provided in recent research.

Yaegashi et al. (2011) developed the colony-print immunoassay method to detect the distribution of dsRNAs in a mycelial colony *in vitro*. The method showed relatively uneven distribution of a mycoreovirus (MyRV3) and a quadrivirus (Rosellinia necatrix quadrivirus 1, RnQV1) when compared to a partitivirus (Rosellinia necatrix partitivirus 1, RnPV1), a megabirnavirus (Rosellinia necatrix megabirnavirus 1, RnMBV1) in a colony of *R. necatrix*. This may contribute to the mosaic distribution of dsRNAs in the same MCGs around a tree in an orchard.

In addition, dynamic mycovirus infection was confirmed in an apple orchard in the Nagano prefecture. Two *R. necatrix* strains (W563; virus-free MCG139, NW10; N10 virus (tentative name)-infected, MCG442) were introduced in an apple orchard. These strains were retrieved from infested apple roots and the genetic identity was confirmed with W563 or NW10.

The N10 virus was transmitted to W563, even though the transmission of the N10 virus was not confirmed between the two strains *in vitro*. More surprisingly, at least six novel RNA viruses (three partitiviruses, one victorivirus, and two unclassified new viruses; see later) were found to be infecting these retrieved isolates from unknown sources (Yaegashi et al., 2013). It would be of considerable interest to elucidate the mechanism of virus infection in these strains in the orchard.

2. VIRUSES OF *R. necatrix*

Like other fungi, such as *C. parasitica* (Hillman & Suzuki, 2004) and *Sclerotinia sclerotiorum* (Chapter 8), *R. necatrix* hosts a number of viruses possibly belonging to a total of five tentative and definitive families to date. Notably, sequence analyses of those viruses suggested that they might all be new species. As is the case for viruses from other fungal hosts, most *R. necatrix* viruses detected to date have dsRNA genomes with only a few exceptions assumed to have single-stranded (ss) RNA genomes. However, it is worth noting that field isolates were screened for viruses based on the presence or absence of dsRNAs, which may have masked possible DNA viruses in this fungus (as discussed in Section 1.3 of this chapter). The occurrence of geminivirus-like DNA viruses in *S. sclerotiorum* (Yu et al., 2010) may encourage rescreening of the fungal isolates for DNA viruses. Below are the viruses of *R. necatrix* whose molecular characteristics have been revealed. It is believed that these are only some of the viruses that *R. necatrix* can host in nature.

2.1. *Megabirnaviridae*

RnMBV1 was isolated from a dsRNA-positive fungal strain, W779, originally collected from a bait twig buried in a Japanese pear orchard infested with *R. necatrix* in Ibaraki prefecture (Fig. 7.1B) (Ikeda et al., 2004). The biological and molecular attributes of RnMBV1 places it in a distinct virus family for which the name "*Megabirnaviridae*" was proposed (Figs. 7.2 and 7.3) (Chiba et al., 2009). Megabirnaviruses were named based on the large-sized (*mega*), bisegmented nature of their ds*RNA* genome, as for the families *Birnaviridae* and *Picobirnaviridae*. However, it should be worth noting that these families show different gene organizations and no significant sequence similarities (Delmas, 2012; Delmas, Mundt, Vakharia, & Wu, 2012). A phylogenetic tree generated based on an alignment of RNA-dependent RNA polymerase (RdRp) sequences places RnMBV1 into a clade that is distinct from other mycoviruses (Fig. 7.3), which accommodates

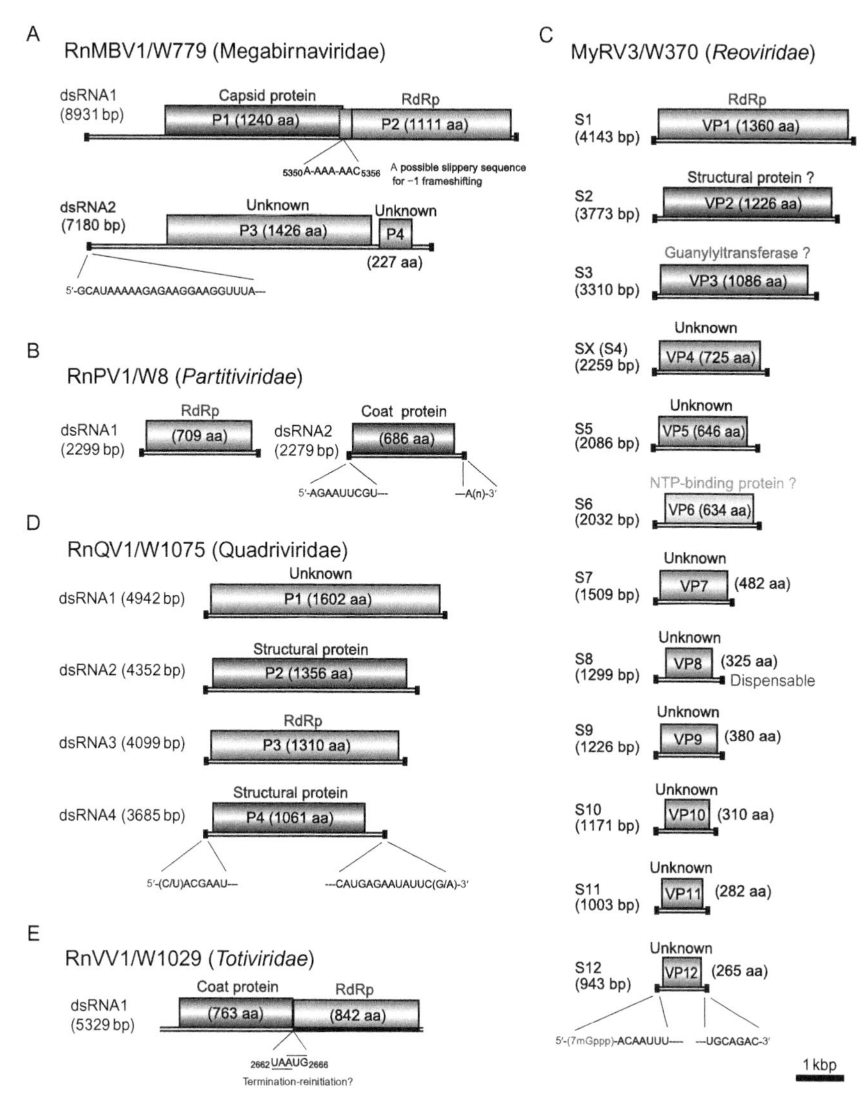

Figure 7.2 Genome organization of representative *R. necatrix* viruses. Schematic representation of the genomic organization of five *R. necatrix* viruses: (A) RnMBV1 (megabirnavirus); (B) RnPV1 (partitivirus); (C) MyRV3 (mycoreovirus); (D) RnQV1 (quadrivirus); (E) RnVV1 (victorivirus). Solid thick lines and open boxes denote genomic dsRNAs and open reading frames (ORFs), respectively. Sizes of the genome segments and encoded proteins are shown in parentheses. Assigned functions are shown within or above ORF boxes. The conserved terminal stretches among segments are detailed. (See Page 12 in Color Section at the back of the book.)

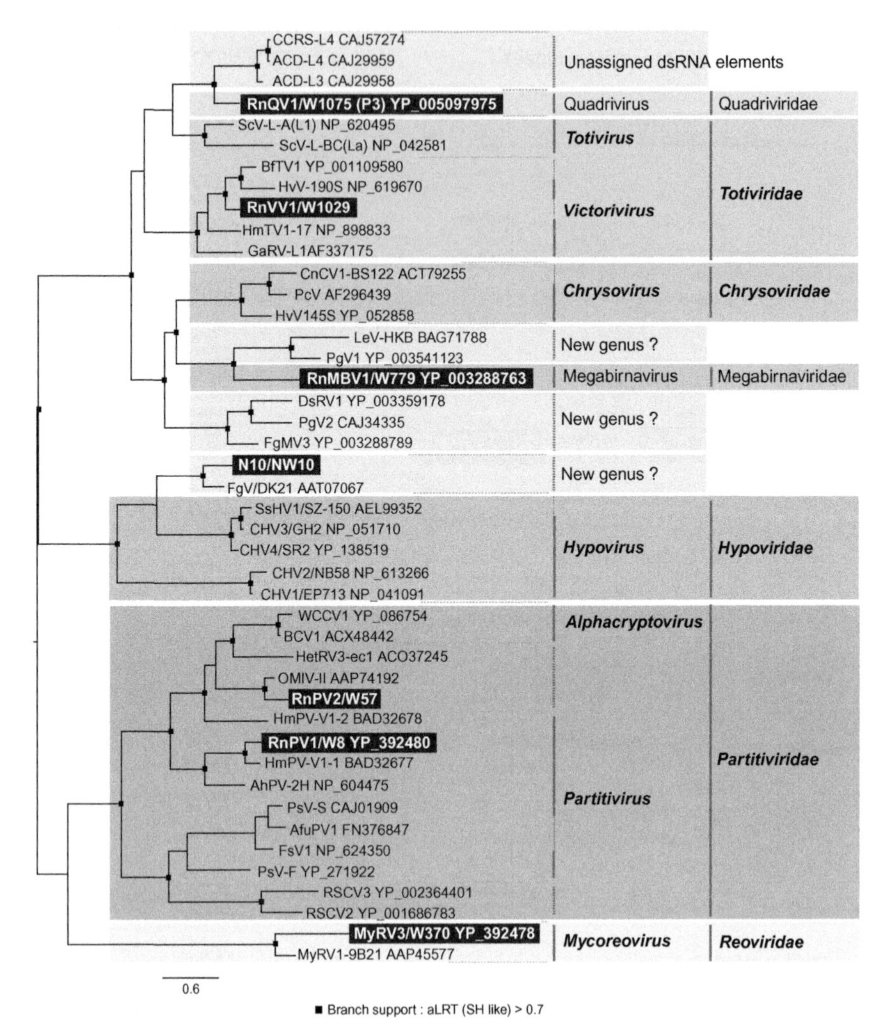

Figure 7.3 Phylogenetic tree of viruses of *R. necatrix*. A dendrogram was generated by the ML method based on the RdRp sequences of fungal viruses including *R. necatrix* viruses that are discussed in this chapter. Viruses highlighted in black are those infecting *R. necatrix*: RnMBV1, RnPV1, RnPV2, RnVV1, RnQV1, MyRV3, and the N10 virus (tentative name). Other mycoviruses representative of major taxonomical groups include: Saccharomyces cerevisiae virus L-A (ScV-L-A, the genus *Totivirus*) (Icho & Wickner, 1989), Helminthosporium victoriae virus 190S (HvV190S, the genus *Victorivirus*) (Huang & Ghabrial, 1996), Cryphonectria hypovirus 1 (CHV1, the family *Hypoviridae*) (Shapira, Choi, & Nuss, 1991), White clover cryptic virus 1 (WCCV1, the genus *Alphacryptvirus*) (Boccardo & Candresse, 2005), AhPV (the genus *Partitivirus*) (Oh & Hillman, 1995), Penicillium chrysogenum virus (PCV, the genus *Chrysovirus*) (Jiang & Ghabrial, 2004), and Mycoreovirus 1 (MyRV1, the genus *Mycoreovirus*) (Suzuki, Supyani, Maruyama, & Hillman, 2004). (See Page 13 in Color Section at the back of the book.)

two virus-like dsRNA elements from two separate basidiomycetes, *Lentinula edodes* (LeV-HKB) and *Phlebiopsis gigantea* (PgV1).

RnMBV1 forms rigid spherical particles of approximately ~50 nm in diameter. Infectious virions can be purified using either sucrose gradient or cesium sulfate equilibrium gradient centrifugation (Salaipeth et al., 2013). Two genomic dsRNA segments, termed dsRNA1 and dsRNA2, are encapsidated by the major capsid protein of 135 kDa, possibly separately. As shown in Fig. 7.2A, each segment possesses two open reading frames (ORFs), making a collective total of four ORFs, entitled ORF1 to ORF4. ORF1 codes for the major capsid protein of 135 kDa (P1), while the ORF2-encoded protein (P2) has typical RdRp motifs. Little is known about the functional roles of ORF3- or ORF4-encoded protein products (P3 and P4), and these ORFs show no significant sequence similarities with *e*-values <0.1 to known proteins in the databases. The dispensability of ORF3 and ORF4 may be revealed by the analysis of rearranged dsRNA2 generated *in vitro*.

During repeated transfection tests, mutant strains of RnMBV1 (RnMBV1/R) were obtained. The viral genome of RnMBV1/R retained two dsRNA segments, dsRNA1 and the newly emerging dsRNA3, but not dsRNA2. Analyses of dsRNA3 revealed that it originated from dsRNA1 by almost complete duplication of ORF2, in a tandemly arranged nature. Purified virions of RnMBV1 and RnMBV1/R showed similar infectivity of *R. necatrix* when tested by transfection of host protoplasts (detailed data will be reported elsewhere by Kanematsu et al.).

Other features of RnMBV1 that distinguish it from other RNA viruses include an extremely long (~1.6 kb) 5′-untranslated region (UTR) of the genome segments, which show very high levels of sequence similarity between the two segments (Chiba et al., 2009). The RnMBV1 5′-UTR represents one of the longest UTRs found in RNA virus genomes, which is longer than those of aphthoviruses or cardioviruses within the family *Picornaviridae* (Knowles et al., 2012). The failure to detect subgenomic RNA (Salaipeth et al., 2013) poses a major question about genome expression strategy: how are the upstream (ORF1 and ORF3) and downstream ORFs (ORF2 and ORF4) of the segments translated from genome segment-sized mRNAs? The extremely long 5′-UTR contains 28 small ORFs that precede ORF1, suggesting that ORF1 and ORF3 might be expressed via a noncanonical translation strategy. The most likely mechanism is internal ribosome entry site (IRES)-mediated translation, as is the case for picornaviruses. If this is the case, the responsible IRES sequences should be identified, but these are often difficult to predict based on sequence comparisons. A clue to the expression of ORF2 comes from the sequence features closest to ORF2. ORF2 is in −1 frame with respect to ORF1. A slippery sequence, 5′-A-AAA-AAC-3′, and a predicted stem-loop structure are found immediately upstream and downstream of the termination codon of ORF1. These suggest that RnMBV1 ORF2 is expressed from full-length dsRNA1 mRNA as a fusion product with capsid protein via −1 ribosomal frameshifting. The fact that a minor amount of a large protein of 250 kDa is detected in purified virion preparations supports this notion. Similar sequence features are not detected between ORF3 and ORF4. It is an open question whether ORF3 is expressed. A reporter system was developed in which a single firefly (*Photinus pyralis*) luciferase gene was used in *C. parasitica* filamentous fungus (Guo et al., 2009). A duel reporter system for the investigation of possible sequence elements allowing non-canonical translation is now being developed using two luciferase genes from a sea pansy *Renilla reniformis* and a firefly (Salaipeth et al., 2013). This technology should be helpful for exploring the mechanism underlying the translation of RnMBV1 ORFs.

A virus etiology was established using two approaches. For the first approach, an isogenic virus-free strain W1015 was obtained by protoplast regeneration. This strain manifested a greater growth rate and virulence to apple rootstock relative to the original virus-infected strain (W779) and yet showed a reduced virulence after receiving the virus again via anastomosis with strain W779. Second, two other virus-free strains, W97 and W370T1, were transfected with RnMBV1-purified particles (see Section 3.1 for the detailed method), which were applied for other virus/host systems (Hillman, Supyani, Kondo, & Suzuki, 2004; Sasaki et al., 2007, 2006). Accordingly, the two strains could be converted into hypovirulent strains with reduced growth rates (Fig. 7.4). It is important to note that the two strains are vegetatively incompatible with each other and with the original strain W1015 (Ikeda et al., 2005). This represents compelling evidence for the ability of RnMBV1 to confer hypovirulence to the host.

2.2. *Partitiviridae*

As in many other filamentous fungi (Ghabrial, Ochao, Baker, & Nibert, 2008; Vainio, Korhonen, Tuomivirta, & Hantula, 2010), partitiviruses are omnipresent in this fungus. There are at least six partitivirus species termed Rosellinia necatrix partitivirus 1–6 (RnPV1 to RnPV6) and more are likely to be identified. The best characterized among them is RnPV1-W8, which

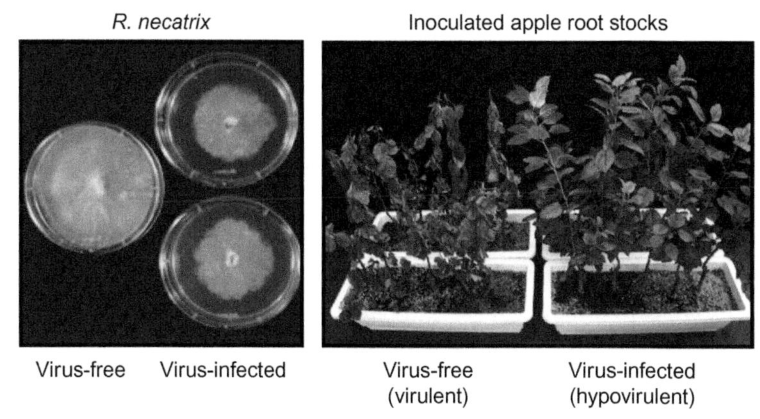

Figure 7.4 Colony morphology and virulence of RnMBV1-infected fungal strains. The left panel shows virus-free (left) and RnMBV1-infected colonies cultured on potato dextrose agar (PDA) for 10 days. The right panel shows apple root stocks (10 each) inoculated in the soil with either a virus-free, virulent fungal strain (left) or a RnMBV1-infected hypovirulent fungal strain (right). Inoculated plants were photographed 4 weeks after inoculation. *Adapted from Chiba et al. (2009).* (For color version of this figure, the reader is referred to the online version of this chapter.)

was isolated from a hypovirulent fungal strain, W8 (Sasaki et al., 2005), that was coinfected with Rosellinia necatrix megabirnavirus 2 (RnMBV2) (Sasaki, 2013b). As shown in Fig. 7.2B, RnPV1 has a typical partitivirus genome organization: the RdRp-encoding dsRNA1 is 2299 nt long and the coat protein (CP)-encoding dsRNA2 is 2279 nt in length, excluding an interrupted poly(A) tract. The genome segments of RnPV2-W57 (RnPV2) are smaller than the counterparts of RnPV1. Other partitiviruses (RnPV3 to RnPV6) were isolated from virulent strains (W1029, W1030, W1031, and W113) of *R. necatrix*. RnPV3 to RnPV6 are less characterized biologically and molecularly compared to RnPV1-W8 or RnPV2-W57. Their attributes will be published elsewhere.

The transfection of strain W97 with purified RnPV1 particles resulted in no overt phenotypic effects, suggesting that RnPV1 alone is not responsible for the hypovirulence in W8 (Sasaki et al., 2006). Either mixed infection with RnMBV2 or single infection by RnMBV2 may cause attenuated virulence of the host fungus. RnPV1-W8 is unable to induce macroscopic symptoms alone (Sasaki et al., 2006), but it remains unknown whether it contributes to the hypovirulence of strain W8 that was coinfected with RnMBV2. Chiba et al. (2013a) applied hyphal tipping to obtain isogenic virus-free fungal strains from strain W57. A biological comparison between virus-free and the original virus-infected strains showed no noticeable association between RnPV2-W57 and host phenotypic alterations in *R. necatrix*. In most cases, partitiviruses show latent infections, as in the case of plant-infecting members of the family *Partitiviridae*. Some partitiviruses, however, cause phenotypic effects occasionally, as does the Heterobasidion RNA virus 3 (HetRV3) in *Heterobasidion* spp. (Vainio et al., 2010), and a partitivirus naturally infecting *Aspergillus fumigatus* alters colony morphology (Bhatti et al., 2011). Interestingly, RnPV1 induces enhancement of pigmentation in an experimental host, *C. parasitica* (Kanematsu, Sasaki, Onoue, Oikawa, & Ito, 2010). Phenotypic changes induced by a partitivirus may depend on culture conditions and host fungi.

The RnPV2 CP sequence was recently brought to the attention of not only virologists but also botanists and geneticists. Surprisingly, a blast search of the RnPV2 CP sequence yields hits with a plant gene product, auxin indole-3-acetic acid (IAA) Leucine resistant 2 (ILR2), in addition to its plant homologues (partitivirus CP-like sequences, PCLSs) and other partitivirus CP sequences (Chiba et al., 2011) (Fig. 7.5). The plant gene was reported to regulate the homeostasis of phytohormone auxin in *Arabidopsis thaliana*. This sequence, and other homologous sequences, is believed to have been transferred from ancient partitiviruses that possibly infected plants. The direction of horizontal transfer is assumed to be from "virus to plant" based on the PCLS detection profiles in which some PCLS′, such as PCLS1, which is homologous to AtILR2 (AtPCLS1), are found on orthologous chromosomal positions of the *Arabidopsis*-related plants and nonorthologous chromosomal positions of distantly related plants (*Mimulus guttatus*). However, they are not found in relatively distantly related plants in the family, such as *Brassica rapa* and *Brassica napus*. Further support for this notion comes from a patchy pattern of PCLS′ in a phylogenetic tree where extant virus CP sequences and PCLSs are grouped together (Fig. 7.5). This finding led the authors to detect a similar horizontal transfer of positive-sense and negative-sense RNA virus genomes into plant genomes (Chiba et al., 2011). Independently, Jiang's group showed that the CP sequence of a partitivirus from *S. sclerotiorum* was the most closely related to ILR2 and horizontally transferred PCLSs in plant chromosomes.

Intriguingly, similar findings were reported earlier for animal and fungi Belyi, Levine, & Skalka, 2010; Horie et al., 2010; Liu, Fu, Li, et al., 2011, Liu, Fu, Xie, et al., 2011; Taylor & Bruenn, 2009; Taylor, Leach, & Bruenn, 2010). As the number of available genome sequences increase, more horizontal gene transfer examples will be unraveled. Unlike retroviruses or

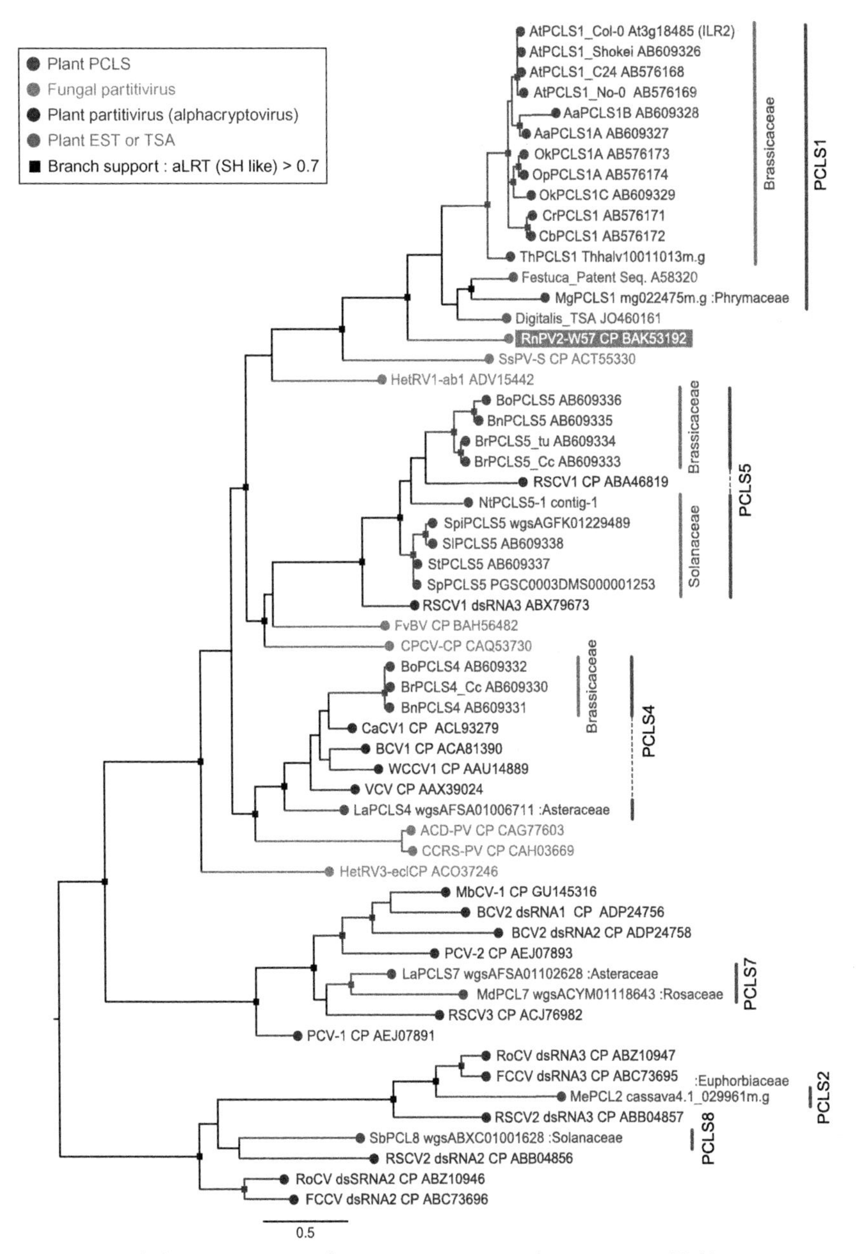

Figure 7.5 Phylogenetic tree of partitivirus CPs and partitivirus CP-like sequences (PCLSs). An alignment of CP sequences of representative partitiviruses and PCLSs integrated into plant genomes was analyzed by the ML method. The accession numbers are shown next to the sequence names in the figure. Part of this analysis was reported by Chiba et al. (2011). Newly analyzed sequences in the tree includes Heterobasidion RNA

pararetroviruses, dsRNA viruses do not code for integrases. In bornavirus-infected cells, cDNAs are detectable, suggesting reverse-transcriptase (RT) activities. Horie et al. (2010) suspect that transposon-derived RT may be responsible for the reverse transcription of bornavirus RNA. How these RNA viral sequences are reverse transcribed and integrated into host chromosomes is an open question. Investigation of the possible functional roles of these integrated nonretro RNA viral sequences will be an interesting challenge.

RnPV2 has another interesting feature. In addition to dsRNA1 and dsRNA2, RnPV2 harbors defective interfering dsRNA1 (DI-dsRNA1) at an approximate ratio of 1:12–15 (dsRNA1:DI-dsRNA1), which could be generated by a single internal deletion event accompanied by nucleotide substitutions that possess an interrupted ORF potentially encoding a half-sized, defective RdRp ORF. This is a very rare case for a partitivirus. Virion transfection with RnPV2 leads to the infrequent elimination of DI-dsRNA1 with a relatively higher accumulation of dsRNA1 and enhanced replication. Furthermore, DI-dsRNA1 plays a role in modulating the expression of symptoms in its experimental host, *C. parasitica* (Chiba et al., 2013).

2.3. *Reoviridae*

The presence of reoviruses (Mycoreovirus 1 (MyRV1) and Mycoreovirus 2 (MyRV2)) in a fungal host, *C. parasitica*, was suggested for the first time by Enebak, Hillman, and Macdonald (1994), but their genome sequences were not fully determined until 2004 (Hillman et al., 2004; Suzuki et al., 2004). Another reovirus (later named Mycoreovirus 3, MyRV3) was detected in *R. necatrix*, which had 12 genome segments (S1–S12) in a hypovirulent strain, W370 (Fig. 7.2C). MyRV3-W370 was shown to reduce mycelial

virus 1 (HetRV1-ab1, putative partitivirus), pepper cryptic virus 1 (PCV-1, putative partitivirus), pepper cryptic virus 2 (PCV-2, putative partitivirus), Mulberry cryptic virus 1 (MbCV-1 putative partitivirus), *Digitalis purpurea* transcriptome shotgun assembly (TSA) contig00181, *Festuca pratensis* patent sequence (WO9637613) that share sequence identity with the *F. pratensis* ESTs (Chiba et al., 2011), *Thellungiella halophila* gene Thhalv10011013m.g, the genome shotgun (WGS) sequence AGFK01229489 of *Solanum pimpinellifolium* (SpiPCLS5), and *Lactuca sativa* WGS sequences AFSA01006711 (LaPCLS4) and AFSA01102628 (LaPCLS7). (See Page 14 in Color Section at the back of the book.)

growth rate and confer hypovirulence to the host fungus (Kanematsu et al., 2004). Unlike RnMBV1, this virus was unstably maintained and tended to be lost during the subculture of infected fungal strains. This phenomenon may be related to the uneven distribution in a single colony (Yaegashi et al., 2011). It is anticipated that MyRV3 is relatively widely distributed among *R. necatrix* isolates (Arakawa et al., 2002).

MyRV3 was investigated at the molecular level by Osaki's research group (Osaki et al., 2002; Wei, Osaki, Iwanami, Matsumoto, & Ohtsu, 2003, 2004). Like MyRV1 and MyRV2, which infect *C. parasitica*, MyRV3 is placed in the genus *Mycoreovirus* within the *Spinareovirinae* ("turreted") group of the family *Reoviridae* (Attoui, 2012). Functional analyses of MyRV3 proteins are hampered by the lack of reverse genetics, which has been developed for some animal reoviruses (Boyce, Celma, & Roy, 2008; Kobayashi et al., 2007; Komoto, Sasaki, & Taniguchi, 2006; Matsuo, Celma, & Roy, 2010) and the unavailability of MyRV3 mutants, except for a variant lacking segment 8 (S8) (discussed below). However, bioinformatic analysis assigned functions to a few MyRV3 genome segments. Sequence motifs common to the *Spinareovirinae* are found in VP1 (RdRp) (Wei et al., 2004), VP3 (guanylyltransferase) (Spear, Sisterson, & Stenger, 2012; Supyani, Hillman, & Suzuki, 2007), and VP6 (NTP-binding) (Nibert & Kim, 2004; Spear et al., 2012; Suzuki et al., 2004). Although no rearranged segments have been reported for MyRV3, some functional insights are gained from MyRV1 counterparts for which rearrangement mutants are available. MyRV1 rearrangements occur spontaneously and are also induced by a multifunctional protein p29, encoded by Cryphonectria hypovirus 1-EP713 (CHV1-EP713) and contribute to the identification of functional roles of genome segments (Eusebio-Cope, Sun, Hillman, & Suzuki, 2010; Sun & Suzuki, 2008; Tanaka, Eusebio-Cope, Sun, & Suzuki, 2012; Tanaka, Sun, Tsutani, & Suzuki, 2011). For example, the MyRV1 S10-coded VP10 was shown to contribute to the suppression of aerial hyphae caused by the virus, while the S4-coded VP4 was shown to be involved in efficient virus transmission through conidia and symptom induction. By analogy, the MyRV3 counterparts of these viral proteins are likely to be dispensable for replication and exert similar activities in the virus replication cycle.

Despite the close relationship between MyRV3 and MyRV1, they differ in the number of genome segments: 11 for MyRV1 versus 12 for MyRV3. In this regard, it is of note that MyRV3 lacking S8, generated during subculturing in the laboratory, is replication competent. This is a rare type of gene alteration and it is different from intragenic rearrangements commonly found across reovirus genera (Tanaka et al., 2012). Even after intergenic rearrangements, reovirus genome segments retained the terminal sequence domains necessary for genome packaging and/or replication. Furthermore, MyRV1 has no counterpart of MyRV3 S8 (Suzuki et al., 2004). This may suggest that MyRV1 and MyRV2 may have had a genome composed of 12 segments and lost one of them during the course of evolution. The question then is what functional role MyRV3 S8 plays. A close relationship to members of the genus *Coltivirus* within the family *Reoviridae* may provide some clues for addressing this question. No vector is reported for any mycoviruses, unlike plant and animal viruses. The genus *Mycoreovirus* is closely related phylogenetically to members of the genus *Coltivirus*, which infect mammals vectored by ticks belonging to the order Acari. Based on this and the close ecological association of host fungi with some Acari members (mites), Hillman et al. (2004) hypothesized that mycoreoviruses may be transmitted horizontally by Acari spp. in nature. It is of interest to speculate that MyRV3 S8 is somehow involved in this extracellular horizontal transfer.

2.4. *Quadriviridae* (new family)

A quadripartite virus, Rosellinia necatrix quadrivirus 1 (RnQV1), with a spherical particle structure of ~45 nm diameter, was isolated from the fungal strain W1075 in a Japanese pear orchard in the Saga prefecture (Fig. 7.1B). Like RnMBV1, this virus may belong to a new virus family for which the family "*Quadriviridae*" is proposed. An agarose gel profile of the four genomic dsRNAs of this virus is very similar to that of a member of the family *Chrysoviridae*, Helminthosporium victoriae virus 145S (HvV145S) (Lin et al., 2012), although their size ranges are different: 3.9–4.9 kbp for RnQV1 and 2.8–3.6 kbp for HvV145S. Intriguingly similar dsRNA profiles were found in multiple field isolates as reported earlier by Arakawa et al. (2002), suggesting a relatively widespread occurrence of this virus in *R. necatrix* in Japan.

The four dsRNAs, 1–4, have conserved terminal sequences and each segment shows sequence heterogeneity at the very end (5′-C/U——G/A-3′ for the plus-strand), which is rare for a viral genome. It is anticipated from sequence analysis of RACE clones and the expected nucleotide sequence complementarity that the plus-strand of each genome segment appears to be 5′-C——G-3′ or 5′-U———A-3′. Each segment has a single large ORF corresponding to 86–97% of its segment size (Fig. 7.2D). dsRNA3 encodes RdRp (P3) with low levels of sequence identity to those from other mycoviruses. The highest sequence identity was shared with RdRp encoded by the Amasya cherry disease (ACD)-associated L dsRNA3 and L dsRNA4 and the cherry chlorotic rusty spot (CCRS)-associated L dsRNA3 and L dsRNA4. Likewise, the dsRNA1-encoded P1 shows modest sequence identity to polypeptides encoded by ACD L dsRNA1 and dsRNA2, but not to any other known proteins. It should be noted that the origin of these ACD and CCRS dsRNA elements is unknown.

In addition to differences in the encoded protein sequences and the sequence heterogeneity at the extreme termini of the genome segments, the nature of multistructural proteins makes RnQV1 distinct from other mycoviruses. Purified virions comprise at least six proteins with sizes of 110, 100, 60, 40, 37, and 30 kDa. Lin et al. (2012) performed peptide mass fingerprinting and MS/MS analyses and suggested that the 100, 60, 40, and 37 kDa polypeptides are encoded by dsRNA2, while the others are likely to be derived from P4 (dsRNA4 protein). Given the coding capacity of dsRNA2 (P2: 1356 amino acids) and dsRNA4 (P4: 1061 amino acids), the structural protein components, with the exception of the 110 kDa protein, are expected to be cleavage or degradation products from P2 or P4 in infected fungal cells, during virion purification or the storage of virus solutions. The observation that the ratio of these proteins varies from preparation to preparation when examined by the intensity of Coomassie Brilliant Blue-stained bands on SDS-polyacrylamide gel, suggests proteolytic degradation during purification procedures, to some extent. RnQV1 is the first mycovirus example, excluding mycoreoviruses (Tanaka et al., 2011), that has multiple major structural proteins encoded by two segments. Recently, a similar example was found in a bisegmented dsRNA virus infecting *Botrytis porri* (Wu et al., 2012). Cryo-electron microscopy of RnQV1 will provide a better understanding of its morphology.

Elimination of RnQV1 from the original W1075 strain via hyphal tipping results in indistinguishable colony morphology (Lin et al., 2012). Thus, the effects of RnQV1 on the host fungus are believed to be minimal, if at all. Transfection with purified virus particles has been unsuccessful, which might be associated with the multicomponent pattern of virus structural proteins.

Another RnQV1 strain was recently identified by us in other *R. necatrix* isolates, such as W1118, whose RdRp showed approximately 74% amino acid sequence identity to that of RnQV1-W1075 (Lin et al., 2013). However, this new strain can be distinguished from the RnQV1-W1075 strain via agarose gel electrophoresis pattern of the genome segments and the stability of the structural proteins during virus purification. The second and fourth largest segments are smaller in size than those of the W1075 strain, while the third largest segment is larger. The structural proteins of the new strain appeared to be more resistant to proteolysis than those of RnQV1-W1075 (Lin et al., 2013), suggesting that it may be more suitable for virion structural analysis than RnQV1-W1075.

2.5. *Totiviridae*

Like partitiviruses, members of the genus *Victorivirus* are ubiquitous among filamentous fungi (Ghabrial & Suzuki, 2009). However, not many fungal isolates with totivirus-like dsRNA profiles were found (Arakawa et al., 2002), although mixed infections may have obscured its detection in *R. necatrix*. Recently, a few fungal isolates, W1028 to W1030, from the Nagano prefecture were confirmed to be coinfected by victoriviruses and partitiviruses (Yaegashi et al., 2013). Thorough molecular and biological characterization of one of the victoriviruses from W1029 is underway and these data will be reported elsewhere. Briefly, commonality and dissimilarity to other reported victoriviruses includes the presence of a sequence element facilitating translation termination/reinitiation at the junction between two ORFs (Guo et al., 2009; Li, Havens, Nibert, & Ghabrial, 2011) and sequence divergence (41% RdRp amino acid sequence identity to the closest relative, Botryotinia fuckeliana totivirus 1) (Chiba et al., 2013b) (Fig. 7.2E). This modest RdRp sequence identity supports the suggestion that the newly isolated W1029 virus is a novel victorivirus species. The effect of virus infection on host morphology was minimal, thus indicating that a distinct host–virus interaction occurs in the *Helminthosporium victoriae*/Helminthosporium victoriae virus 190S (HvV-190S) pathosystem and is known as a debilitation-conferring infection (Ghabrial & Suzuki, 2009).

2.6. Unclassified viruses

The aforementioned viruses are well characterized except the victoriviruses. It is worth noting that there are other poorly defined RNA viruses that are known to occur in some strains of *R. necatrix*. For example, partial sequence analyses showed the occurrence in two fungal strains of two viruses with undivided RNA genomes of 6 and 10–12 kbp, which are related to dsRNA elements from *P. gigantea* (Kozlakidis et al., 2009) and *Fusarium graminearum* dsRNA mycovirus-3 (FgMV3) (Yu, Kwon, Lee, Son, & Kim, 2009), and *Fusarium graminearum virus*-DK21 (FgV-DK21) (Kwon, Lim, Park, Park, & Kim, 2007) (see Fig. 7.3, N10 virus). The RdRp sequences of these viruses share approximately 30–40% identity to their closest relatives. After complete characterization, the data will be published elsewhere.

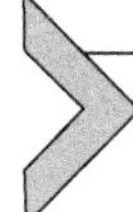

3. EXPANSION OF HOST RANGES OF *R. necatrix*-INFECTING VIRUSES

3.1. Techniques for investigation of experimental host ranges

It is generally very difficult to inoculate fungal hosts with mycoviruses, unlike bacterial, plant, and animal viruses. Therefore, the experimental host range of a mycovirus was investigated only after 1994. Despite a few rare successful examples of artificial introduction of virions (Castro, Kramer, Valdivia, Ortiz, & Castillo, 2003; el-Sherbeini & Bostian, 1987; Stanway & Buck, 1984), the mycoviruses used were long regarded as "noninfectious," "indigenous," or "heritable" (Ghabrial, 2001). However, some milestone technology developments have been made in the past few decades. In the beginning of the 1990s, Nuss and coworkers developed two reverse genetic systems for CHV1, either with transformation using infectious full-length CHV1 cDNA (Choi & Nuss, 1992) or with transfection by synthetic transcripts derived from full-length viral cDNA (Chen, Choi, & Nuss, 1994). Similar approaches are now applicable to the *Diaporthe RNA virus* (Moleleki, van Heerden, Wingfield, Wingfield, & Preisig, 2003) and two narnaviruses (Esteban & Fujimura, 2003; Esteban, Vega, & Fujimura, 2005). Unlike these viruses with RNA genomes closely related to (+)ssRNA viruses, no plasmid-based artificial infection systems are available for dsRNA mycoviruses.

In 2004, reproducible transfection systems using purified virions were established for the MyRV1/*C. parasitica* (Hillman et al., 2004) (Fig. 7.6A). This technique is basically similar to that of DNA transformation and entails mixing fungal protoplasts and purified virions in the presence of polyethylene glycol and calcium ions. This method has proved to be applicable for different viruses with dsRNA and ssDNA genomes. Examples include members of the genus *Mycoreovirus* (MyRV3) (Kanematsu et al., 2010; Sasaki et al., 2007), the genus *Partitivirus* (RnPV1) (Kanematsu et al., 2010; Sasaki et al., 2006), and the family *Megabirnaviridae* (RnMBV1) (Chiba et al., 2009) and recently found *S. sclerotiorum* hypovirulence-associated DNA virus

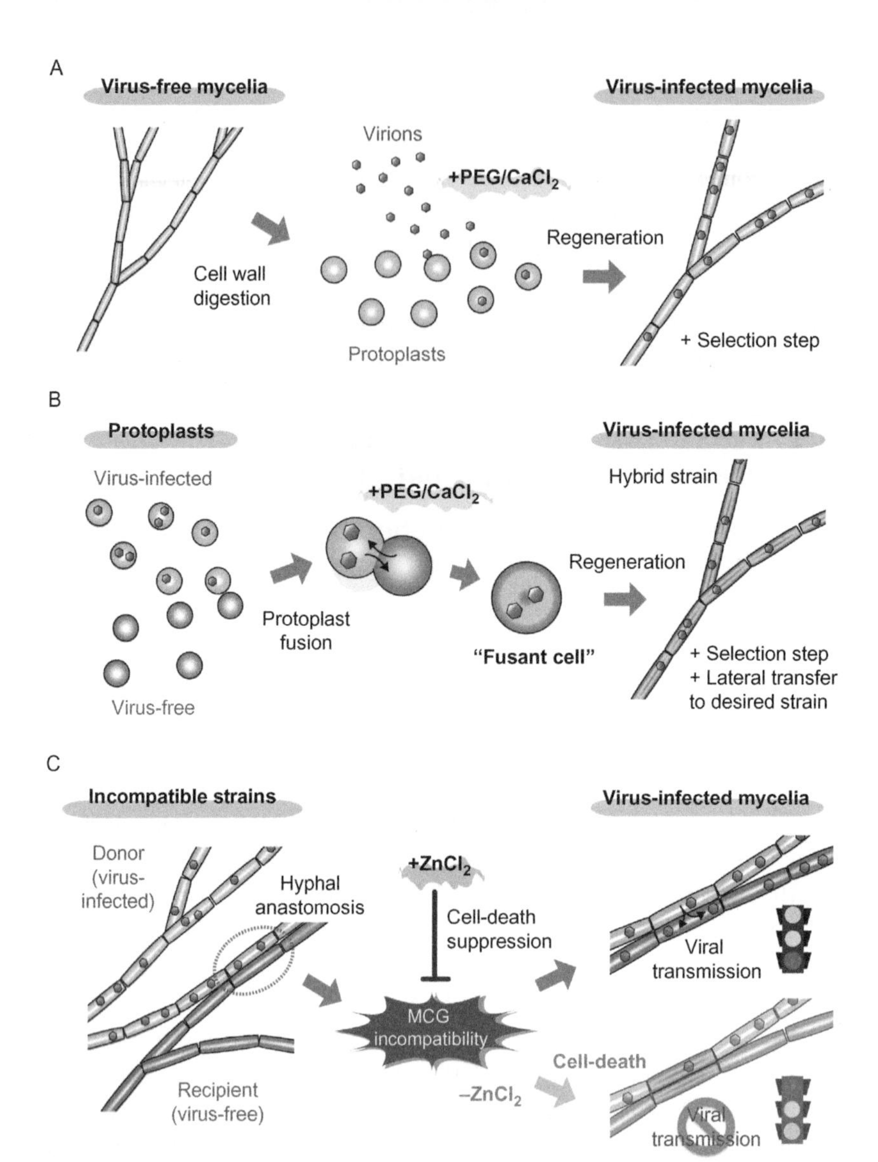

Figure 7.6 Illustration of technical advances to introduce *R. necatrix* viruses. Schematic representations of virion transfection (A), protoplast fusion (B), and anastomosis in the presence of chemicals between mycelially incompatible strains (C). (A) Procedure for the artificial introduction of RnMBV1 into the white root rot fungus. Virus-free fungal mycelia were digested with cell wall-degrading enzymes to prepare protoplasts (Kanematsu et al., 2004). The resultant protoplasts were transfected with purified virions in the presence of $CaCl_2$ and PEG. Surviving protoplasts were regenerated during which RnMBV1 were spread horizontally within a colony. (B) Intra- and interspecies protoplast fusion can be achieved with *R. necatrix* (Sasaki, 2013a). Protoplasts prepared from donor and recipient fungal strains are fused with the aid of PEG and $CaCl_2$ as in the transfection procedure. Fused cells are subject to culturing on a selection medium, for example, in the presence of a drug, which allows selective growth of the donor. (C) Coculturing of a virus-free and a virus-infected *R. necatrix* strain in a medium containing $ZnCl_2$ at an appropriate concentration enhances hyphal fusion, otherwise mycelially incompatible, resulting in lateral virus transfer. (See Page 15 in Color Section at the back of the book.)

(Yu et al., 2010). Many mycoviruses have dsRNA genomes that are encapsidated into spherical particles. These particles are expected to contain virion-associated RNA transcriptase activities assuming that dsRNAs are inactive as templates for translation. Once introduced into host fungal cells, the transcriptase is activated to deliver viral mRNAs. The transfection method may be widely used to introduce other mycoviruses and will continue to contribute to the determination of as-yet-unexplored cause–effect relationships between fungal phenotype and infecting viruses. To this regard, it is worth attempting to test for infectivity of ssRNA particles, such as those of barnaviruses, alphaflexiviruses, and gammaflexviruses.

Other potential methods to examine experimental host ranges include protoplast fusion and chemically mediated removal of mycelial incompatibility. Successful virus transfer via protoplast fusion was reported between different species in *Aspergilli* for the first time (van Diepeningen, Debets, & Hoekstra, 1998). Later, this method was used for horizontal virus transfer between diverse fungi belonging to different genera (Lee, Yu, Son, Lee, & Kim, 2011; van Diepeningen et al., 2000) (Fig. 7.6B). However, caution should be taken when determining viral effects, because protoplast fusion involves mixing all cellular genetic components including nuclei and mitochondria. An intriguing question to be addressed is how widely protoplast fusion can be used experimentally for horizontal virus transfer. Is this method applicable for organisms belonging to different families, orders, classes, or kingdoms?

After a screen of approximately 100 chemical compounds, Ikeda's group developed a novel method for allowing horizontal virus transfer between two vegetatively incompatible fungal strains of a single species using $ZnCl_2$ (Fig. 7.6C). In a different manner from the three above methods, this protocol does not require preparation of fungal protoplasts. The inclusion of an appropriate concentration (0.5–1.5 mM) of this chemical in media enhances hyphal fusion and the lateral transfer of RnMBV1 between incompatible *R. necatrix* strains, which are otherwise impaired in virus cross-strain transmission. Similar effects were confirmed with zinc vitriol by the same group. These data are published elsewhere by Ikeda and Inoue. Genes involved in mycelial incompatibility are known in some fungi, such as *N. crassa* (Glass & Dementhon, 2006; Kaneko, Dementhon, Xiang, & Glass, 2006) and *C. parasitica* (Choi et al., 2012; Smith, Gibbs, & Milgroom, 2006). There may be diverse mechanisms governing mycelial incompatibility and associated cell death. It would be interesting to test whether $ZnCl_2$ and $ZnSO_4$, the most likely zinc ions, function in other fungal species as they do in *R. necatrix*, and to explore what the molecular effects of $ZnCl_2$ are on these genes. This chemical abolishment of incompatibility barriers has so far been used between strains of single species. It would also be intriguing to examine whether the chemical is useful for overcoming interspecies barriers imposed by mechanisms different from those operating between intraspecies barriers.

3.2. Member of the class Sordariomycetes, *C. parasitica*, as a host of viruses infecting *R. necatrix*

C. parasitica is a versatile organism that could serve as a model host for studies on virus/virus and virus/host interactions based on several features. The genome sequence is publicly available (http://genome.jgi-psf.org/Crypa2/Crypa2.home.html), single-gene knock-out technology is available (Gao & Nuss, 1996; Kasahara & Nuss, 1997), multitransformation is possible (Faruk, Eusebio-Cope, Suzuki, 2008; Faruk, Izumino, Suzuki, 2008; Kasahara, Wang, & Nuss, 2000; Segers, Zhang, Deng, Sun, & Nuss, 2007), its sexual and asexual life cycles can be completed in relatively short periods of time (Hillman & Suzuki, 2004), viral RNA-based and virion-based transfection is available for some viruses and their mutants (Chen et al., 1994; Hillman et al., 2004; Lin et al., 2007; Suzuki, Chen, & Nuss, 1999; Suzuki, Maruyama, Moriyama, & Nuss, 2003), and viral cDNA-based reverse genetics has been developed for hypoviruses (Choi & Nuss, 1992). Therefore, once introduced and found to be infectious in *C. parasitica*, viruses of the *R. necatrix* family can be investigated in view of virus/host interactions using existing molecular and genetic tools that are available for the standard strain (EP155) of *C. parasitica*.

Kanematsu et al. (2010) tested some members in the Sordariomycetes class, including *C. parasitica*, for susceptibility to two viruses, a partitivirus, RnPV1-W8, and a mycoreovirus, MyRV3-W370. The authors showed that the two viruses can infect *C. parasitica* and cause phenotypic changes, including reduced virulence. Taking advantage of the availability of the infectious full-length cDNA clone of CHV1, its ability to infect and induce symptoms in *C. parasitica*-related and unrelated fungi, such as *Valsa ceratosperma* and *Diaporthe* sp. G-type was noted (Chen, Chen, Bowman, & Nuss, 1996; Sasaki et al., 2002). The report by Kanematsu et al. (2010) is different from previous ones in that transfection with virions was used. To date, *C. parasitica* has been shown to support replication of members in the *Megabirnaviridae*, *Reoviridae*, *Partitiviridae*, and *Totiviridae* families that were originally isolated from *R. necatrix*. *C. parasitica* mutants will be readily tested for the effect of infections by these viruses. It remains unknown how similar or different EP155 genes respond to infections by these transfected viruses and CHV1 (Allen, Dawe, & Nuss, 2003; Shang et al., 2008).

Table 7.1 Infectivity of viruses of *R. necatrix* to other fungi and of heterologous viruses to *R. necatrix*

	Virus			
Host	**MyRV3**	**RnPV1**	**RnMBV1**	**CHV1**
C. parasitica	+	+	+	NA
Diaporthe sp.	+	+	ND	+
V. ceratosperma	+	+	+	+
G. cingulata	−	+	ND	+
R. necatrix	NA	NA	NA	ND

Data are taken mainly from Kanematsu et al. (2010) and some are unpublished. Fungal strains used as experimental hosts are *Cryphonectria parasitica* EP155/2, *Diaporthe* sp. G-type 3a, *Valsa ceratosperma* AVC53, *Glomerella cingulata* 046-71, and *R. necatrix* W97. Viral strains used are MyRV3-RnW370, RnPV1-W8, and RnMBV1-W779, and CHV1-EP713. +: virus infected. −: virus uninfected. NA, not applicable; ND, not determined.

3.3. Other fungi as experimental hosts of viruses naturally infecting *R. necatrix*

Kanematsu et al. (2010) further showed that all the other six tested combinations (two viruses, RnPV1 and MyRV3 × 3 host fungal strains) (Table 7.1) showed successful transfections. Only one combination (*Glomerella cingulata* × MyRV3) provided an unsuccessful transfection, suggesting that *G. cingulata* is unable to support MyRV3 replication. Another interesting finding is the transference of hypovirulence by MyRV3 to the other two susceptible fungal hosts (*V. ceratosperma* and *Diaporthe* sp. G-type) as in the natural host, *R. necatrix*. This observation represents the potential of MyRV3 to serve as a virocontrol agent in other phytopathogenic fungi. Moreover, RnPV1, while exerting no overt effects on the natural host, *R. necatrix* (Sasaki et al., 2006), can infect three fungi (*G. cingulata*, *V. ceratosperma*, and *Diaporthe* sp. G-type) and alter their phenotype (Table 7.1). The effects of partitivirus infection are influenced by many factors.

4. CONCLUSIONS AND PROSPECTS

Characterization of the viruses from *R. necatrix* mentioned above has enhanced our understanding of the great diversity of mycoviruses (Figs. 7.2 and 7.3) and allowed the establishment of *R. necatrix*/mycoviruses as a useful experimental system. The mycovirus diversity is strongly supported by recent reports on the identification of an increasing number of novel viruses from all major groups of host fungi that include arbuscular mycorrhizal fungi (Aoki et al., 2009; Cai, Myers, Hillman, & Fry, 2009; Ikeda, Shimura, Kitahara, Masuta, & Ezawa, 2012; Liu et al., 2009; Urayama et al., 2010; Wu et al., 2012). *R. necatrix* viruses also provide interesting future challenges. For example, RnMBV1 dsRNA1 and dsRNA2 have extremely long 5′-UTRs of 1.6 kb. These UTRs must have sequence elements involved in a noncanonical translation mechanism given the failure of detecting subgenomic RNA. Novel *R. necatrix* viruses, such as RnMBV1 and RnQV1, can be purified to relatively large quantities. Therefore, these viruses can serve as good materials for structural analysis that may lead to discoveries of unique virion structures such as a novel T (triangulation number) $= 1$ (pseudo-$T = 2$) icosahedral architecture that was revealed for chrysoviruses (Caston et al., 2003).

C. parasitica has been established as the best phytopathogenic fungus model for the investigation of virus/virus and virus/host interactions (Dawe & Nuss, 2001; Faruk, Eusebio-Cope, et al., 2008; Faruk, Izumino, et al., 2008; Hillman & Suzuki, 2004; Jacob-Wilk, Turina, & Van Alfen, 2006; Nuss, 2011; Sun, Nuss, & Suzuki, 2006; Sun & Suzuki, 2008) and the biocontrol of a plant pathogenic fungal disease, that is, virocontrol (Heiniger & Rigling, 1994; MacDonald & Fulbright, 1991; Milgroom & Cortesi, 2004). Comparison with the *C. parasitica*/viruses will provide insights into the *R. necatrix* system. *R. necatrix* has some suitable features for such studies. For example, protoplasts of *R. necatrix* are relatively easy to prepare and are competent in transformation with DNA (Kanematsu et al., 2004; Pliego et al., 2009) and transfection with purified virion (Fig. 7.6) (Kanematsu et al., 2010; Sasaki et al., 2007, 2006). As discussed in the preceding section, this fungus can support diverse homologous and heterologous viruses, that is, members of at least four established families and one tentative family, and unassigned species (Figs. 7.2 and 7.3), which induce distinct symptoms or show asymptomatic infections. Like the characteristics of *C. parasitica*/mycoviruses, these features of the *R. necatrix*/viruses meet the criteria as an experimental fungus/virus to study virus/host and virus/virus interactions. There are some limitations with the *R. necatrix*/virus: unlike *C. parasitica*, *R. necatrix* conidia rarely germinate and develop into mycelia in the laboratory (Nakamura et al., 2002); this impedes virus elimination from virus-harboring field strains and, thus, generation of homokaryotic *R. necatrix* strains. Protoplasts, while relatively easy to prepare, are difficult to store in the competent form to be transfected or transformed. This requires researchers to make fresh protoplasts whenever transformation or transfection assays are performed. Also, the fungus produces melanin when it becomes old or is fused with another vegetatively incompatible strain. Furthermore, while the genome of *R. necatrix* is currently being sequenced, it is not yet available publicly.

White root rot is one of the most destructive diseases of perennial crops in Japan (Matsumoto, 1998). Like other soil-borne fungal pathogens, *R. necatrix* is very difficult to control. Virocontrol is an attractive measure that entails biocontrol using viruses infecting organisms that are pathogenic to important crops (in this case *R. necatrix*). One of the best studied examples is the tritrophic relationship of the chestnut/chestnut blight/hypovirus. In the system, two application methods are tested experimentally: (1) the direct application of virus-containing hypovirulent strains to the cankers, with the expectation of therapeutic effects and (2) spraying of transgenically modified fungal spores carrying CHV1-cDNA clones that can launch a viral infection, with the expectation of preventive effects (Anagnostakis, Chen, Geletka, & Nuss, 1998). To practice "virocontrol" of the white root rot fungus, a different approach from the ones used for air-borne or vector-transmissible *C. parasitica* is required, given its soil-borne nature and inability to sporulate. Fungal strains must be isolated from infested orchards, and virocontrol agents must be introduced experimentally into the strains, before placing them back into the soil. Considering clonal development of several MCGs of the fungus in an infested orchard as stated above (Fig. 7.1A), it is anticipated that a virus-containing hypovirulent fungal strain can efficiently serve as a donor strain for the spread of the virocontrol agent (mycovirus), once introduced into an orchard (Fig. 7.7). The rationale behind this idea is that mycelial incompatibility among strains governed genetically (Ikeda et al., 2011; Inoue et al., 2011) is considered not to limit horizontal virus transfer in the soil of a treated orchard. Rather, simple *R. necatrix* population structure (Fig. 7.1A) may enhance dissemination of virocontrol agents. From the practical point of view, Japanese fruits are generally costly enough to justify this made-to-order type of treatment. However, the question that remains is: which viruses meet the conditions as virocontrol agents? RnMBV1 is a strong candidate as a virocontrol agent (Fig. 7.4) because of its ability to reduce host fungal virulence stably and to be infectious as particles to any *R. necatrix* strain (Chiba et al., 2009). So far, it has been shown that the virus can be purified in large amounts and stored at −80 °C (Salaipeth et al., 2013) for long periods of time without reducing infectivity in transfection assays. These are the prerequisites that a virocontrol agent should have. Kanematsu's group is testing the effects of introduction of hypovirulent

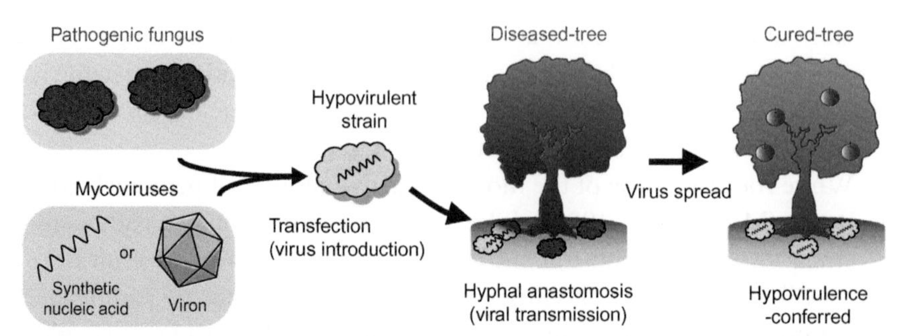

Figure 7.7 Virocontrol strategy. To implement virocontrol of the soil-borne white root rot disease, the virulent fungal strain needs to be isolated from a diseased tree, to be converted into a hypovirulent strain, and lastly to be placed back to the infested soil expecting therapeutic effects of the hypovirulent strain. The conversion of a virulent to hypovirulent fungal strain can be made by introducing a virocontrol agent such as RnMBV1 into the former strain. Near-clonal spread of *R. necatrix* in an orchard or field (see Fig. 7.2) is expected to spread a virocontrol agent efficiently via anastomosis. (For the color version of this figure, the reader is referred to the online version of this chapter.)

strains into the test plot in Tsukuba Fruit Tree Experimental Station expecting therapeutic effects. Whether RnMBV1 is able to disseminate in a field plot and control the white root rot disease will be unraveled through continuous field observations.

ACKNOWLEDGMENTS

The authors are grateful to Yomogi Inc. (to N. S.) and a project "Development of Introduction Protocols of Virocontrol Agents into Phytopathogenic Fungi" led by S. K. via the Program for Promotion of Basic and Applied Researches for Innovations in Bio-Oriented Industry (to S. K. and H. K.) for financial support during the study. We also wish to thank Drs. Naoyuki Matsumoto, Tsutomu Morinaga, Hitoshi Nakamura, Ken-ichi Ikeda, Sotaro Chiba, and Atsuko Sasaki for their fruitful discussions and for sharing unpublished data with us.

REFERENCES

Aimi, T., Kano, S., Yotsutani, Y., & Morinaga, T. (2002). Telomeric fingerprinting of the white root rot fungus, *Rosellinia necatrix*: A useful tool for strain identification. *FEMS Microbiology Letters*, *217*, 95–101.

Aimi, T., Taguchi, H., Tanaka, Y., Kitamoto, Y., & Morinaga, T. (2005). *Agrobacterium tumefaciens*-mediated genetic transformation of the white root rot ascomycete *Rosellinia necatrix*. *Mycoscience*, *46*, 27–31.

Allen, T. D., Dawe, A. L., & Nuss, D. L. (2003). Use of cDNA microarrays to monitor transcriptional responses of the chestnut blight fungus *Cryphonectria parasitica* to infection by virulence-attenuating hypoviruses. *Eukaryotic Cell*, *2*, 1253–1265.

Anagnostakis, S. L., Chen, B. S., Geletka, L. M., & Nuss, D. L. (1998). Hypovirus transmission to ascospore progeny by field-released transgenic hypovirulent strains of *Cryphonectria parasitica*. *Phytopathology*, *88*, 598–604.

Aoki, N., Moriyama, H., Kodama, M., Arie, T., Teraoka, T., & Fukuhara, T. (2009). A novel mycovirus associated with four double-stranded RNAs affects host fungal growth in *Alternaria alternata*. *Virus Research*, *140*, 179–187.

Arakawa, M., Nakamura, H., Uetake, Y., & Matsumoto, N. (2002). Presence and distribution of double-stranded RNA elements in the white root rot fungus *Rosellinia necatrix*. *Mycoscience*, *43*, 21–26.

Attoui, H., et al. (2012). Family *Reoviridae*. In A. M. Q. King, M. J. Adams, E. B. Carstens & E. J. Lefkowits (Eds.), *Virus taxonomy: Ninth report of the International Committee for the Taxonomy of Viruses* (pp. 541–637), New York: Elsevier, Academic Press.

Belyi, V. A., Levine, A. J., & Skalka, A. M. (2010). Unexpected inheritance: Multiple integrations of ancient bornavirus and ebolavirus/marburgvirus sequences in vertebrate genomes. *PLoS Pathogens*, *6*(7), e1001030.

Bhatti, M. F., Jamal, A., Petrou, M. A., Cairns, T. C., Bignell, E. M., & Coutts, R. H. (2011). The effects of dsRNA mycoviruses on growth and murine virulence of *Aspergillus fumigatus*. *Fungal Genetics and Biology*, *48*, 1071–1075.

Boccardo, G., & Candresse, T. (2005). Complete sequence of the RNA1 of an isolate of white clover cryptic virus 1, type species of the genus *Alphacryptovirus*. *Archives of Virology*, *150*, 399–402.

Boyce, M., Celma, C. C., & Roy, P. (2008). Development of reverse genetics systems for bluetongue virus: Recovery of infectious virus from synthetic RNA transcripts. *Journal of Virology*, *82*, 8339–8348.

Cai, G., Myers, K., Hillman, B. I., & Fry, W. E. (2009). A novel virus of the late blight pathogen, *Phytophthora infestans*, with two RNA segments and a supergroup 1 RNA-dependent RNA polymerase. *Virology*, *392*, 52–61.

Caston, J. R., Ghabrial, S. A., Jiang, D., Rivas, G., Alfonso, C., Roca, R., et al. (2003). Three-dimensional structure of *Penicillium chrysogenum* virus: A double-stranded RNA virus with a genuine $T=1$ capsid. *Journal of Molecular Biology*, *331*, 417–431.

Castro, M., Kramer, K., Valdivia, L., Ortiz, S., & Castillo, A. (2003). A double-stranded RNA mycovirus confers hypovirulence-associated traits to *Botrytis cinerea*. *FEMS Microbiology Letters*, *228*, 87–91.

Chen, B. S., Chen, C. H., Bowman, B. H., & Nuss, D. L. (1996). Phenotypic changes associated with wild-type and mutant hypovirus RNA transfection of plant pathogenic fungi phylogenetically related to *Cryphonectria parasitica*. *Phytopathology*, *86*, 301–310.

Chen, B., Choi, G. H., & Nuss, D. L. (1994). Attenuation of fungal virulence by synthetic infectious hypovirus transcripts. *Science*, *264*, 1762–1764.

Chiba, S., Kondo, H., Tani, A., Saisho, D., Sakamoto, W., Kanematsu, S., et al. (2011). Widespread endogenization of genome sequences of non-retroviral RNA viruses into plant genomes. *PLoS Pathogens*, 7(7), e1002146.

Chiba, S., Salaipeth, L., Lin, Y. H., Sasaki, A., Kanematsu, S., & Suzuki, N. (2009). A novel bipartite double-stranded RNA mycovirus from the white root rot fungus *Rosellinia necatrix*: Molecular and biological characterization, taxonomic considerations, and potential for biological control. *Journal of Virology*, *83*, 12801–12812.

Chiba, S., Lin, Y.-H., Kondo, H., Kanematsu, S., & Suzuki, N. (2013a). Effects of defective-interfering RNA on symptom induction by, and replication of, a novel partitivirus from a phytopathogenic fungus *Rosellinia necatrix*. *Journal of Virology*, *87*, 2330–2341.

Chiba, S., Lin, Y.-H., Kondo, H., Kanematsu, S., & Suzuki, N. (2013b). A novel victorivirus from a phytopathogenic fungus, *Rosellinia necatrix* is infectious as particles and targeted by RNA silencing. Manuscript in preparation.

Choi, G. H., Dawe, A. L., Churbanov, A., Smith, M. L., Milgroom, M. G., & Nuss, D. L. (2012). Molecular characterization of vegetative incompatibility genes that restrict hypovirus transmission in the chestnut blight fungus *Cryphonectria parasitica*. *Genetics*, *190*, 113–127.

Choi, G. H., & Nuss, D. L. (1992). Hypovirulence of chestnut blight fungus conferred by an infectious viral cDNA. *Science*, *257*, 800–803.

Cortesi, P., McCulloch, C. E., Song, H. Y., Lin, H. Q., & Milgroom, M. G. (2001). Genetic control of horizontal virus transmission in the chestnut blight fungus, *Cryphonectria parasitica*. *Genetics*, *159*, 107–118.

Dawe, A. L., & Nuss, D. L. (2001). Hypoviruses and chestnut blight: Exploiting viruses to understand and modulate fungal pathogenesis. *Annual Review of Genetics*, *35*, 1–29.

Delmas, B. (2012). Family *Picobirnaviridae*. In A. M. Q. King, M. J. Adams, E. B. Carstens & E. J. Lefkowits (Eds.), *Virus taxonomy: Ninth report of the International Committee for the Taxonomy of Viruses* (pp. 535–539), New York: Elsevier, Academic Press.

Delmas, B., Mundt, E., Vakharia, V. N., & Wu, J. L. (2012). Family *Birnaviridae*. In A. M. Q. King, M. J. Adams, E. B. Carstens & E. J. Lefkowits (Eds.), *Virus taxonomy: Ninth report of the International Committee for the Taxonomy of Viruses* (pp. 499–507), New York: Elsevier, Academic Press.

Eguchi, N. (2010). Studies on the ecology and control of major fungal diseases in Japanese pear. *Bulletin of Nagano Prefecture. Nanshin Agricultural Experiment Station*, *4*, 1–133.

Eguchi, N., Kondo, K., & Yamagishi, N. (2009). Bait twig method for soil detection of *Rosellinia necatrix*, causal agent of white root rot of Japanese pear and apple, at an early stage of tree infection. *Journal of General Plant Pathology*, *75*, 325–330.

el-Sherbeini, M., & Bostian, K. A. (1987). Viruses in fungi: Infection of yeast with the K1 and K2 killer viruses. *Proceedings of the National Academy of Sciences of the United States of America*, *84*, 4293–4297.

Enebak, S. A., Hillman, B. I., & Macdonald, W. L. (1994). A hypovirulent isolate of *Cryphonectria parasitica* with multiple, genetically unique dsRNA segments. *Molecular Plant-Microbe Interactions*, 7, 590–595.

Esteban, R., & Fujimura, T. (2003). Launching the yeast 23S RNA Narnavirus shows 5′ and 3′ cis-acting signals for replication. *Proceedings of the National Academy of Sciences of the United States of America*, *100*, 2568–2573.

Esteban, R., Vega, L., & Fujimura, T. (2005). Launching of the yeast 20 s RNA narnavirus by expressing the genomic or antigenomic viral RNA *in vivo*. *The Journal of Biological Chemistry*, *280*, 33725–33734.

Eusebio-Cope, A., Sun, L., Hillman, B. I., & Suzuki, N. (2010). *Mycoreovirus 1* S4-coded protein is dispensable for viral replication but necessary for efficient vertical transmission and normal symptom induction. *Virology*, *397*, 399–408.

Faruk, M. I., Eusebio-Cope, A., & Suzuki, N. (2008). A host factor involved in hypovirus symptom expression in the chestnut blight fungus, *Cryphonectria parasitica*. *Journal of Virology*, *82*, 740–754.

Faruk, M., Izumimoto, M., & Suzuki, N. (2008). Characterization of mutants of the chestnut blight fungus (*Cryphonectria parasitica*) with unusual hypovirus symptoms. *Journal of General Plant Pathology*, *74*, 425–433.

Gao, S., & Nuss, D. L. (1996). Distinct roles for two G protein alpha subunits in fungal virulence, morphology, and reproduction revealed by targeted gene disruption. *Proceedings of the National Academy of Sciences of the United States of America*, *93*, 14122–14127.

Ghabrial, S. A. (2001). Fungal viruses. In O. Maloy & T. Murray (Eds.), *Encyclopedia of plant pathology*, Vol. 1, (pp. 267–269). New York: John Wiley & Sons.

Ghabrial, S., Ochao, W., Baker, T., & Nibert, M. (2008). Partitiviruses: General features. In B. W. J. Mahy & M. H. V. Van Regenmortel (Eds.), (3rd ed.). *Encyclopedia of virology*, Vol. 4, (pp. 68–75). Oxford: Elsevier.

Ghabrial, S., & Suzuki, N. (2009). Viruses of plant pathogenic fungi. *Annual Review of Phytopathology*, *47*, 353–384.

Glass, N. L., & Dementhon, K. (2006). Non-self recognition and programmed cell death in filamentous fungi. *Current Opinion in Microbiology*, *9*, 553–558.

Glass, N. L., Jacobson, D. J., & Shiu, P. K. (2000). The genetics of hyphal fusion and vegetative incompatibility in filamentous ascomycete fungi. *Annual Review of Genetics, 34*, 165–186.

Guo, L. H., Sun, L., Chiba, S., Araki, H., & Suzuki, N. (2009). Coupled termination/reinitiation for translation of the downstream open reading frame B of the prototypic hypovirus CHV1-EP713. *Nucleic Acids Research, 37*, 3645–3659.

Heiniger, U., & Rigling, D. (1994). Biological control of chestnut blight in Europe. *Annual Review of Phytopathology, 32*, 581–599.

Hillman, B. I., Supyani, S., Kondo, H., & Suzuki, N. (2004). A reovirus of the fungus *Cryphonectria parasitica* that is infectious as particles and related to the coltivirus genus of animal pathogens. *Journal of Virology, 78*, 892–898.

Hillman, B. I., & Suzuki, N. (2004). Viruses of the chestnut blight fungus, *Cryphonectria parasitica*. *Advances in Virus Research, 63*, 423–472.

Horie, M., Honda, T., Suzuki, Y., Kobayashi, Y., Daito, T., Oshida, T., et al. (2010). Endogenous non-retroviral RNA virus elements in mammalian genomes. *Nature, 463*, 84–87.

Huang, S., & Ghabrial, S. A. (1996). Organization and expression of the double-stranded RNA genome of *Helminthosporium victoriae 190S virus*, a totivirus infecting a plant pathogenic filamentous fungus. *Proceedings of the National Academy of Sciences of the United States of America, 93*, 12541–12546.

Icho, T., & Wickner, R. B. (1989). The double-stranded RNA genome of yeast virus L-A encodes its own putative RNA polymerase by fusing two open reading frames. *The Journal of Biological Chemistry, 264*, 6716–6723.

Ikeda, K., Inoue, K., Nakamura, H., Hamanaka, T., Ohta, T., Kitazawa, H., et al. (2011). Genetic analysis of barrage line formation during mycelial incompatibility in *Rosellinia necatrix*. *Fungal Biology, 115*, 80–86.

Ikeda, K., Nakamura, H., Arakawa, M., & Matsumoto, N. (2004). Diversity and vertical transmission of double-stranded RNA elements in root rot pathogens of trees, Helicobasidium mompa and *Rosellinia necatrix*. *Mycological Research, 108*, 626–634.

Ikeda, K., Nakamura, H., & Matsumoto, N. (2005). Comparison between *Rosellinia necatrix* isolates from soil and diseased roots in terms of hypovirulence. *FEMS Microbiology Ecology, 54*, 307–315.

Ikeda, Y., Shimura, H., Kitahara, R., Masuta, C., & Ezawa, T. (2012). A novel virus-like double-stranded RNA in an obligate biotroph arbuscular mycorrhizal fungus: A hidden player in mycorrhizal symbiosis. *Molecular Plant-Microbe Interactions, 25*, 1005–1012.

Inoue, K., Kanematsu, S., Park, P., & Ikeda, K. (2011). Cytological analysis of mycelial incompatibility in *Rosellinia necatrix*. *Fungal Biology, 115*, 87–95.

Jacob-Wilk, D., Turina, M., & Van Alfen, N. K. (2006). Mycovirus *Cryphonectria hypovirus 1* elements cofractionate with trans-Golgi network membranes of the fungal host *Cryphonectria parasitica*. *Journal of Virology, 80*, 6588–6596.

Jiang, D., & Ghabrial, S. A. (2004). Molecular characterization of Penicillium chrysogenum virus: Reconsideration of the taxonomy of the genus *Chrysovirus*. *The Journal of General Virology, 85*, 2111–2121.

Kaneko, I., Dementhon, K., Xiang, Q., & Glass, N. L. (2006). Nonallelic interactions between het-c and a polymorphic locus, pin-c, are essential for nonself recognition and programmed cell death in *Neurospora crassa*. *Genetics, 172*, 1545–1555.

Kanematsu, S. (2013). Differentiation of *Rosellinia necatrix* mycelial incompatibility groups by amplified fragment length polymorphism. Unpublished raw data.

Kanematsu, S., Arakawa, M., Oikawa, Y., Onoue, M., Osaki, H., Nakamura, H., et al. (2004). A reovirus causes hypovirulence of *Rosellinia necatrix*. *Phytopathology, 94*, 561–568.

Kanematsu, S., Sasaki, A., Onoue, M., Oikawa, Y., & Ito, T. (2010). Extending the fungal host range of a partitivirus and a mycoreovirus from *Rosellinia necatrix* by inoculation of protoplasts with virus particles. *Phytopathology, 100*, 922–930.

Kano, S., Kurita, T., Kanematsu, S., & Morinaga, T. (2011). Agrobacterium tumefaciens-mediated transformation of the plant pathogenic fungus *Rosellinia necatrix*. *Mikrobiologiia, 80*, 86–92.

Kasahara, S., & Nuss, D. L. (1997). Targeted disruption of a fungal G-protein beta subunit gene results in increased vegetative growth but reduced virulence. *Molecular Plant-Microbe Interactions, 10*, 984–993.

Kasahara, S., Wang, P., & Nuss, D. L. (2000). Identification of *bdm-1*, a gene involved in G protein beta-subunit function and alpha-subunit accumulation. *Proceedings of the National Academy of Sciences of the United States of America, 97*, 412–417.

Knowles, N. J., Hovi, T., Hyypiä, T., King, A. M. Q., Lindberg, A. M., Pallansch, M. A., et al. (2012). Family Picornaviridae. In: A. M. Q. King, M. J. Adams, E. B. Carstens & E. J. Lefkowits (Eds.), *Virus taxonomy: Ninth report of the International Committee for the Taxonomy of Viruses* (pp. 855–880), New York: Elsevier, Academic Press.

Kobayashi, T., Antar, A. A., Boehme, K. W., Danthi, P., Eby, E. A., Guglielmi, K. M., et al. (2007). A plasmid-based reverse genetics system for animal double-stranded RNA viruses. *Cell Host & Microbe, 1*, 147–157.

Komoto, S., Sasaki, J., & Taniguchi, K. (2006). Reverse genetics system for introduction of site-specific mutations into the double-stranded RNA genome of infectious rotavirus. *Proceedings of the National Academy of Sciences of the United States of America, 103*, 4646–4651.

Kozlakidis, Z., Hacker, C. V., Bradley, D., Jamal, A., Phoon, X., Webber, J., et al. (2009). Molecular characterisation of two novel double-stranded RNA elements from *Phlebiopsis gigantea*. *Virus Genes, 39*, 132–136.

Kwon, S. J., Lim, W. S., Park, S. H., Park, M. R., & Kim, K. H. (2007). Molecular characterization of a dsRNA mycovirus, Fusarium graminearum virus-DK21, which is phylogenetically related to hypoviruses but has a genome organization and gene expression strategy resembling those of plant potex-like viruses. *Molecules and Cells, 23*, 304–315.

Lee, K. M., Yu, J., Son, M., Lee, Y. W., & Kim, K. H. (2011). Transmission of Fusarium boothii mycovirus via protoplast fusion causes hypovirulence in other phytopathogenic fungi. *PLoS One, 6*(6), e21629.

Li, H., Havens, W. M., Nibert, M. L., & Ghabrial, S. A. (2011). RNA sequence determinants of a coupled termination-reinitiation strategy for downstream open reading frame translation in *Helminthosporium victoriae virus 190S* and other victoriviruses (Family *Totiviridae*). *Journal of Virology, 85*, 7343–7352.

Lin, Y. H., Chiba, S., Tani, A., Kondo, H., Sasaki, A., Kanematsu, S., et al. (2012). A novel quadripartite dsRNA virus isolated from a phytopathogenic filamentous fungus, *Rosellinia necatrix*. *Virology, 426*, 42–50.

Lin, Y.-H., Hisano, S., Yaegashi, H., Kanematsu, S., & Suzuki, N. (2013). A second quadrivirus strain from a phytopathogenic filamentous fungus. *Rosellinia necatrix*. *Archives in Virology*, http://dx.doi.org/10.1007/s00705-012-1580-8.

Lin, H., Lan, X., Liao, H., Parsley, T. B., Nuss, D. L., & Chen, B. (2007). Genome sequence, full-length infectious cDNA clone, and mapping of viral double-stranded RNA accumulation determinant of hypovirus CHV1-EP721. *Journal of Virology, 81*, 1813–1820.

Liu, H., Fu, Y., Jiang, D., Li, G., Xie, J., Peng, Y., et al. (2009). A novel mycovirus that is related to the human pathogen hepatitis E virus and rubi-like viruses. *Journal of Virology, 83*, 1981–1991.

Liu, H., Fu, Y., Li, B., Yu, X., Xie, J., Cheng, J., et al. (2011). Widespread horizontal gene transfer from circular single-stranded DNA viruses to eukaryotic genomes. *BMC Evolutionary Biology, 11*, 276.

Liu, H., Fu, Y., Xie, J., Cheng, J., Ghabrial, S. A., Li, G., et al. (2011). Widespread endogenization of densoviruses and parvoviruses in animal and human genomes. *Journal of Virology, 85*, 9863–9876.

MacDonald, W. L., & Fulbright, D. W. (1991). Biological-control of chestnut blight-use and limitations of transmissible hypovirulence. *Plant Disease, 75*, 656–661.

Matsumoto, N. (1998). Biological control of root diseases with dsRNA based on population structure of pathogenes. *Japan Agricultural Research Quarterly, 32*, 31–35.

Matsuo, E., Celma, C. C., & Roy, P. (2010). A reverse genetics system of African horse sickness virus reveals existence of primary replication. *FEBS Letters, 584*, 3386–3391.

Milgroom, M. G., & Cortesi, P. (2004). Biological control of chestnut blight with hypovirulence: A critical analysis. *Annual Review of Phytopathology, 42*, 311–338.

Moleleki, N., van Heerden, S. W., Wingfield, M. J., Wingfield, B. D., & Preisig, O. (2003). Transfection of *Diaporthe perjuncta* with *Diaporthe RNA virus*. *Applied and Environmental Microbiology, 69*, 3952–3956.

Nakamura, H., Ikeda, K., Arakawa, M., & Matsumoto, N. (2002). Conidioma production of the white root rot fungus in axenic culture under near-ultraviolet light radiation. *Mycoscience, 43*, 251–254.

Nakamura, H., & Matsumoto, N. (2006). Mycelial compatibility group-based determination of population structures of *R. necatrix* in Japanese fruit tree orchards. Unpublished raw data.

Nakamura, H., Uetake, Y., Arakawa, M., Okabe, I., & Matsumoto, N. (2000). Observations on the teleomorph of the white root rot fungus, Rosellinia necatrix, and a related fungus, *Rosellinia aquila*. *Mycoscience, 41*, 503–507.

Nibert, M. L., & Kim, J. (2004). Conserved sequence motifs for nucleoside triphosphate binding unique to turreted *Reoviridae* members and coltiviruses. *Journal of Virology, 78*, 5528–5530.

Nuss, D. L. (2011). Mycoviruses, RNA silencing, and viral RNA recombination. *Advances in Virus Research, 80*, 25–48.

Ogata, T., & Ochiai, M. (1994). Longevity of apple white root rot fungus in infected roots in soil. *Annals of the Phytopathological Society of Japan, 60*, 354.

Oh, C.-S., & Hillman, B. I. (1995). Genome organization of a partitivirus from the filamentous ascomycete *Atkinsonella hypoxylon*. *The Journal of General Virology, 76*, 1461–1470.

Osaki, H., Wei, C. Z., Arakawa, M., Iwanami, T., Nomura, K., Matsumoto, N., et al. (2002). Nucleotide sequences of double-stranded RNA segments from a hypovirulent strain of the white root rot fungus *Rosellinia necatrix*: Possibility of the first member of the *Reoviridae* from fungus. *Virus Genes, 25*, 101–107.

Pearson, M. N., Beever, R. E., Boine, B., & Arthur, K. (2009). Mycoviruses of filamentous fungi and their relevance to plant pathology. *Molecular Plant Pathology, 10*, 115–128.

Perez Jimenez, R. M., Jimenez Diaz, R. M., & Lopez Herrera, C. J. (2002). Somatic incompatibility of *Rosellinia necatrix* on avocado plants in southern Spain. *Mycological Research, 106*, 239–244.

Pliego, C., Kanematsu, S., Ruano-Rosa, D., de Vicente, A., Lopez-Herrera, C., Cazorla, F. M., et al. (2009). GFP sheds light on the infection process of avocado roots by *Rosellinia necatrix*. *Fungal Genetics and Biology, 46*, 137–145.

Pliego, C., Lopez-Herrera, C., Ramos, C., & Cazorla, F. M. (2012). Developing tools to unravel the biological secrets of *Rosellinia necatrix*, an emergent threat to woody crops. *Molecular Plant Pathology, 13*, 226–239.

Salaipeth, L., Eusebio-Cope, A., Chiba, S., Kanematsu, S., & Suzuki, N. (2013). Biological and molecular characterization of Rosellinia necatrix megabirnavirus 1 in an experimental host *Cryphonectria parasitica*. Manuscript in preparation.

Sasaki, A. (2013a). A protoplast fusion method for mycovirus lateral transmission within a host fungal species. Unpublished raw data.

Sasaki, A. (2013b). Partial characterization of Rosellinia megabirnavirus 2. Unpublished raw data.

Sasaki, A., Kanematsu, S., Onoue, M., Oikawa, Y., Nakamura, H., & Yoshida, K. (2007). Artificial infection of *Rosellinia necatrix* with purified viral particles of a member of the genus *Mycoreovirus* reveals its uneven distribution in single colonies. *Phytopathology, 97*, 278–286.

Sasaki, A., Kanematsu, S., Onoue, M., Oyama, Y., & Yoshida, K. (2006). Infection of *Rosellinia necatrix* with purified viral particles of a member of *Partitiviridae* (RnPV1-W8). *Archives of Virology, 151*, 697–707.

Sasaki, A., Miyanishi, M., Ozaki, K., Onoue, M., & Yoshida, K. (2005). Molecular characterization of a partitivirus from the plant pathogenic ascomycete *Rosellinia necatrix*. *Archives of Virology, 150*, 1069–1083.

Sasaki, A., Onoue, M., Kanematsu, S., Suzaki, K., Miyanishi, M., Suzuki, N., et al. (2002). Extending chestnut blight hypovirus host range within diaporthales by biolistic delivery of viral cDNA. *Molecular Plant-Microbe Interactions, 15*, 780–789.

Schena, L., Nigro, F., & Ippolito, A. (2002). Identification and detection of *Rosellinia necatrix* by conventional and real-time Scorpin-PCR. *European Journal of Plant Pathology, 108*, 355–366.

Segers, G. C., Zhang, X., Deng, F., Sun, Q., & Nuss, D. L. (2007). Evidence that RNA silencing functions as an antiviral defense mechanism in fungi. *Proceedings of the National Academy of Sciences of the United States of America, 104*, 12902–12906.

Shang, J., Wu, X., Lan, X., Fan, Y., Dong, H., Deng, Y., et al. (2008). Large-scale expressed sequence tag analysis for the chestnut blight fungus *Cryphonectria parasitica*. *Fungal Genetics and Biology, 45*, 319–327.

Shapira, R., Choi, G. H., & Nuss, D. L. (1991). Virus-like genetic organization and expression strategy for a double-stranded RNA genetic element associated with biological control of chestnut blight. *The EMBO Journal, 10*, 731–739.

Shimizu, T., Ito, T., & Kanematsu, S. (2012). Transient and multivaridate system for transformation of a fungal pathogen, *Rosellinia necatrix*, using autonomously replicating vectors. *Current Genetics, 58*, 129–138.

Shishido, M., Kubota, I., & Nakamura, H. (2012). Development of real-time PCR assay using TaqMan probe for detection and quantification of *Rosellinia necatrix* in plant and soil. *Journal of General Plant Pathology, 78*, 115–120.

Smith, M. L., Gibbs, C. C., & Milgroom, M. G. (2006). Heterokaryon incompatibility function of barrage-associated vegetative incompatibility genes (vic) in *Cryphonectria parasitica*. *Mycologia, 98*, 43–50.

Spear, A., Sisterson, M. S., & Stenger, D. C. (2012). Reovirus genomes from plant-feeding insects represent a newly discovered lineage within the family *Reoviridae*. *Virus Research, 163*, 503–511.

Stanway, C. A., & Buck, K. W. (1984). Infection of protoplasts of the wheat take-all fungus, *Gaeumannomyces-graminis* var tritici, with double-stranded-RNA viruses. *The Journal of General Virology, 65*, 2061–2065.

Sun, L., Nuss, D. L., & Suzuki, N. (2006). Synergism between a mycoreovirus and a hypovirus mediated by the papain-like protease p29 of the prototypic hypovirus CHV1-EP713. *The Journal of General Virology, 87*, 3703–3714.

Sun, L., & Suzuki, N. (2008). Intragenic rearrangements of a mycoreovirus induced by the multifunctional protein p29 encoded by the prototypic hypovirus CHV1-EP713. *RNA, 14*, 2557–2571.

Supyani, S., Hillman, B. I., & Suzuki, N. (2007). Baculovirus expression of the 11 mycoreovirus-1 genome segments and identification of the guanylyltransferase-encoding segment. *The Journal of General Virology, 88*, 342–350.

Suzuki, N., Chen, B., & Nuss, D. L. (1999). Mapping of a hypovirus p29protease symptom determinant domain with sequence similarity to potyvirus HC-Pro protease. *Journal of Virology, 73*, 9478–9484.

Suzuki, N., Maruyama, K., Moriyama, M., & Nuss, D. L. (2003). Hypovirus papain-like protease p29 functions in trans to enhance viral double-stranded RNA accumulation and vertical transmission. *Journal of Virology, 77*, 11697–11707.

Suzuki, N., Supyani, S., Maruyama, K., & Hillman, B. I. (2004). Complete genome sequence of *Mycoreovirus-1*/Cp9B21, a member of a novel genus within the family *Reoviridae*, isolated from the chestnut blight fungus *Cryphonectria parasitica*. *The Journal of General Virology, 85*, 3437–3448.

Tanaka, T., Eusebio-Cope, A., Sun, L., & Suzuki, N. (2012). Mycoreovirus genome alterations: Similarities to and differences from rearrangements reported for other reoviruses. *Frontiers in Virology, 3*, 186. http://dx.doi.org/10.3389/fmicb.2012.00186.

Tanaka, T., Sun, L., Tsutani, K., & Suzuki, N. (2011). Rearrangements of mycoreovirus 1 S1, S2 and S3 induced by the multifunctional protein p29 encoded by the prototypic hypovirus Cryphonectria hypovirus 1 strain EP713. *The Journal of General Virology, 92*, 1949–1959.

Taylor, D. J., & Bruenn, J. (2009). The evolution of novel fungal genes from non-retroviral RNA viruses. *BMC Biology, 7*, 88.

Taylor, D. J., Leach, R. W., & Bruenn, J. (2010). Filoviruses are ancient and integrated into mammalian genomes. *BMC Evolutionary Biology, 10*, 193.

ten Hoopen, G. M., & Krauss, U. (2006). Biology and control of *Rosellinia bunodes*, *Rosellinia necatrix* and *Rosellinia pepo*: A review. *Crop Protection, 25*, 89–107.

Uetake, Y., Nakamura, H., Arakawa, M., Okabe, I., & Matsumoto, N. (2001). Inoculation of *Lupinus luteus* with white root rot fungus, *Rosellinia necatrix*, to estimate virulence. *Journal of General Plant Pathology, 67*, 285–287.

Urayama, S., Kato, S., Suzuki, Y., Aoki, N., Le, M. T., Arie, T., et al. (2010). Mycoviruses related to chrysovirus affect vegetative growth in the rice blast fungus *Magnaporthe oryzae*. *The Journal of General Virology, 91*, 3085–3094.

Vainio, E. J., Korhonen, K., Tuomivirta, T. T., & Hantula, J. (2010). A novel putative partitivirus of the saprotrophic fungus *Heterobasidion ecrustosum* infects pathogenic species of the *Heterobasidion annosum* complex. *Fungal Biology, 114*, 955–965.

van Diepeningen, A. D., Debets, A. J., & Hoekstra, R. F. (1998). Intra- and interspecies virus transfer in Aspergilli via protoplast fusion. *Fungal Genetics and Biology, 25*, 171–180.

van Diepeningen, A. D., Debets, A. J., Slakhorst, S. M., Fekete, C., Hornok, L., & Hoekstra, R. F. (2000). Interspecies virus transfer via protoplast fusions between *Fusarium poae* and black *Aspergillus* strains. *Fungal Genetics Newsletter, 47*, 99–100.

Wei, C. Z., Osaki, H., Iwanami, T., Matsumoto, N., & Ohtsu, Y. (2003). Molecular characterization of dsRNA segments 2 and 5 and electron microscopy of a novel reovirus from a hypovirulent isolate, W370, of the plant pathogen *Rosellinia necatrix*. *The Journal of General Virology, 84*, 2431–2437.

Wei, C. Z., Osaki, H., Iwanami, T., Matsumoto, N., & Ohtsu, Y. (2004). Complete nucleotide sequences of genome segments 1 and 3 of *Rosellinia* anti-rot virus in the family *Reoviridae*. *Archives of Virology, 149*, 773–777.

Wu, M., Jin, F., Zhang, J., Yang, L., Jiang, D., & Li, G. (2012). Characterization of a novel bipartite double-stranded RNA mycovirus conferring hypovirulence in the phytopathogenic fungus *Botrytis porri*. *Journal of Virology, 86*, 6605–6619.

Yaegashi, H., Nakamura, H., Sawahata, T., Sasaki, A., Iwanami, Y., Ito, T., et al. (2013). Appearance of mycovirus-like double-stranded RNAs in the white root rot fungus, *Rosellinia necatrix*, in an apple orchard. *FEMS Microbiology Ecology*, *83*, 49–62.

Yaegashi, H., Sawahata, T., Ito, T., & Kanematsu, S. (2011). A novel colony-print immunoassay reveals differential patterns of distribution and horizontal transmission of four unrelated mycoviruses in *Rosellinia necatrix*. *Virology*, *409*, 280–289.

Yu, J., Kwon, S. J., Lee, K. M., Son, M., & Kim, K. H. (2009). Complete nucleotide sequence of double-stranded RNA viruses from *Fusarium graminearum* strain DK3. *Archives of Virology*, *154*, 1855–1858.

Yu, X., Li, B., Fu, Y., Jiang, D., Ghabrial, S. A., Li, G., et al. (2010). A geminivirus-related DNA mycovirus that confers hypovirulence to a plant pathogenic fungus. *Proceedings of the National Academy of Sciences of the United States of America*, *107*, 8387–8392.

CHAPTER EIGHT

Viruses of the Plant Pathogenic Fungus *Sclerotinia sclerotiorum*

Daohong Jiang[*,†,1], **Yanping Fu**[†], **Li Guoqing**[*,†] **and Said A. Ghabrial**[‡]

[*]The State Key Lab of Agricultural Microbiology, Huazhong Agricultural University, Wuhan, Hubei Province, P. R. China

[†]Provincial Key Lab of Plant Pathology of Hubei Province, College of Plant Science and Technology, Huazhong Agricultural University, Wuhan, Hubei Province, P. R. China

[‡]Department of Plant Pathology, University of Kentucky, Lexington, Kentucky, USA

[1]Corresponding author: e-mail address: daohongjiang@mail.hzau.edu.cn

Contents

1. Introduction 216
2. Mycoviruses of *S. sclerotiorum* 218
   2.1 DNA virus—Sclerotinia sclerotiorum hypovirulence-associated DNA virus 1 218
   2.2 Sclerotinia RNA viruses 222
3. The Interaction Between Mycovirus and *S. sclerotiorum* 238
4. Coinfection of *S. sclerotiorum* with Multiple Mycoviruses 240
5. The Potential Use of Mycoviruses to Control Sclerotinia Diseases 241
6. Conclusions and Prospects 242

Acknowledgments 243

References 244

Abstract

Sclerotinia sclerotiorum is a notorious plant fungal pathogen with a broad host range including many important crops, such as oilseed rape, soybean, and numerous vegetable crops. Hypovirulence-associated mycoviruses have attracted much attention because of their potential as biological control agents for combating plant fungal diseases and for use in fundamental studies on fungal pathogenicity and other properties. This chapter describes several mycoviruses that were isolated from hypovirulent strains except for strain Sunf-M, which has a normal phenotype. These viruses include the geminivirus-like mycovirus Sclerotinia sclerotiorum hypovirulence-associated DNA virus 1 (SsHADV-1), Sclerotinia debilitation-associated RNA virus (SsDRV), Sclerotinia sclerotiorum RNA virus L (SsRV-L), Sclerotinia sclerotiorum hypovirus 1 (SsHV-1), Sclerotinia sclerotiorum mitoviruses 1 and 2 (SsMV-1, SsMV-2), and Sclerotinia sclerotiorum partitivirus S (SsPV-S). Unlike many other fungi, incidences of mixed infections with two or more mycoviruses in *S. sclerotiorum* are particularly high and very common. The interaction between SsDRV and *S. sclerotiorum* is likely to be unique. The significance of these mycoviruses to fungal ecology and viral evolution and the potential for biological control of Sclerotinia diseases using mycoviruses are discussed.

Advances in Virus Research, Volume 86
ISSN 0065-3527
http://dx.doi.org/10.1016/B978-0-12-394315-6.00008-8

1. INTRODUCTION

Sclerotinia sclerotiorum is an ascomycetous plant pathogenic fungus with a wide host range; it attacks more than 400 species of plant hosts in many families including Brassicaceae (Cruciferae), Fabaceae (Leguminosae), Solanaceae, Asteraceae, and Apiaceae (Umbelliferae) (Boland & Hall, 1994; Bolton, Thomma, & Nelson, 2006). With such a broad host range, *S. sclerotiorum* is widely distributed across the world. *S. sclerotiorum* produces sclerotia to over-season; when environmental conditions are fit for germination, sclerotia may germinate myceliogenically to produce hyphae that infect the lower parts of plants, such as leaves and basal stems, or germinate carpogenically to produce apothecia and release ascospores into the air (Roper et al., 2010). The discharged ascospores then infect petals and senescent leaves, and these often contact with other parts (such as pods, leaves, branches, and stems) of the plant; consequently, the plant becomes extensively infected and is killed (Steadman, 1979). Infected plants often show soft rot symptoms and are killed in a short period of time. *S. sclerotiorum* grows on lesions with flourishing aerial hyphae, and thus the diseases caused by this pathogen are often called white molds. New sclerotia are produced on lesions at the late stage of infection, which drop into the soil or attach to the debris of diseased plant residues, or mix with seeds for dormancy. *S. sclerotiorum* does not produce any conidial spores, but may produce microspores; however, these microspores cannot germinate further, and their function is not known (Bolton et al., 2006). Field symptoms, lesions, carpogenical sclerotial germination, and colony morphology are shown in Fig. 8.1.

As a typical necrotrophic fungal pathogen, *S. sclerotiorum* secretes oxalic acid and many cell wall-degrading enzymes to kill plant cells and tissues and then uptakes nutrients from the dead tissues; oxalic acid is viewed as a key pathogenicity factor of this pathogen (Cessna, Sears, Dickman, & Low, 2000; Hegedus & Rimmer, 2005; Kim, Min, & Dickman, 2008; Riou, Freyssinet, & Fevre, 1992; Williams, Kabbage, Kim, Britt, & Dickman, 2011). Recently, the genome sequence of *S. sclerotiorum* became available (Amselem et al., 2011); this progress is likely to significantly promote the understanding of *S. sclerotiorum* at the molecular level. However, the problems for controlling Sclerotinia diseases still exist. A major problem is the lack of resistant cultivars. The interaction between this pathogen and its host does not clearly follow "the Gene for Gene" hypothesis: the virulence differentiation or host-specific differentiation of *S. sclerotiorum* does not exist

Figure 8.1 Brief introduction to *Sclerotinia sclerotiorum*. (A) Symptoms caused on the basal stem of rapeseed (*Brassica napus*); white mass represents immature sclerotia; (B) Overseasoned sclerotia in a stubble of rapeseed that germinate in the spring; (C) Typical colony morphology of *S. sclerotiorum*; dark sclerotia are uniformally distributed at the colony edge; (D) Field-grown rapeseed plants that were killed prematurely by *S. sclerotiorum*; the green-looking plants possibly escaped infection. (See Page 16 in Color Section at the back of the book.)

and strains isolated from one host show similar virulence on other hosts. Although the genome sequence of *S. sclerotiorum* was released and compared with other fungal genomes, no similar *avr* genes were found. However, the host's resistance against *S. sclerotiorum* is horizontal in nature, which is mediated by QTL genes (Ronicke, Hahn, Vogler, & Friedt, 2005; Zhao & Meng, 2003) but not by major resistance genes. Recently, oxalic acid degrading enzyme genes were transformed into plants to enhance their resistance against *S. sclerotiorum* (Dong et al., 2008; Hu et al., 2003; Kesarwani, Azam, Natarajan, Mehta, & Datta, 2000); however, there is still a long way before these transgenic plants can be evaluated under field conditions. In China, fungicides are mainly used to control crop diseases caused by *S. sclerotiorum*, which leads to many negative impacts on both environmental safety and food safety; furthermore, fungicide resistant isolates have frequently been isolated from fields (Pan, Wang, & Wu, 1997). To reduce the usage of fungicides, alternative approaches need to be developed.

Hypovirulence-associated mycoviruses have attracted much interest because of their potential for exploitation as biological control agents. Hypovirus-mediated hypovirulence of the chestnut blight fungal pathogen *Cryphonectria* (*Endothia*) *parasitica* was first reported in 1965 (Grente, 1965), and then it was utilized successfully to control chestnut blight in Europe (Anagnostakis, 1982; Nuss, 1992). Hypovirulence and dsRNA elements in *S. sclerotiorum* were first reported in the early 1990s (Boland, 1992), but these dsRNA elements were not studied further. The original aim for studying mycoviruses of *S. sclerotiorum* in our lab was to screen hypovirulence-associated mycoviruses for implementation in biological control of *Sclerotinia* diseases.

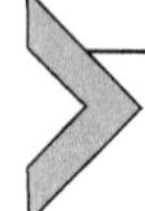

2. MYCOVIRUSES OF *S. SCLEROTIORUM*

2.1. DNA virus—Sclerotinia sclerotiorum hypovirulence-associated DNA virus 1

Sclerotinia sclerotiorum hypovirulence-associated DNA virus 1 (SsHADV-1) is the first characterized DNA virus that infects a fungus. SsHADV-1 was isolated from a hypovirulent strain, DT-8, of *S. sclerotiorum*. Strain DT-8 was originally isolated from a sclerotium produced on a diseased rapeseed plant. The strain is weakly virulent and possesses other debilitating traits, such as slow growth on PDA, production of small sclerotia, and development of abnormal colony morphology (Yu et al., 2010). Since hypovirulence and its associated traits are transmissible, strain DT-8 shows typical characteristics of virus-mediated debilitation/hypovirulence. Although strain DT-8 was once suspected of being infected by an RNA virus, no viral RNA elements were isolated from its mycelial mass. However, two small DNA elements with sizes of 2.0 and 0.3 kb were isolated, and subsequent sequencing analysis led to the discovery of a single-stranded circular DNA virus. Since this DNA virus is involved in the hypovirulence of strain DT-8, it was named SsHADV-1. The full length of SsHADV-1 viral genome is 2166 nucleotide (nt) and comprises only two putative open reading frames (ORF), one coding for a replication initiation protein (Rep) on the antisense strand and the second coding for a coat protein (CP) on the sense strand of viral DNA. There are two intergenic regions with similar sizes between the two ORFs, the large one (LIR) is 133 nt, and the small one (SIR) is 119 nt. The genome structure of SsHADV-1 is similar to those of viruses in the genus *Mastrevirus* in the Family

Geminiviridae. However, there are two distinguishing differences: first, SsHADV-1 has only two ORFs, but viruses in the genus *Mastrevirus* have four ORFs; second, SsHADV-1 has a novel nonanucleotide TAATATT↓AT at the apex of a potential stem–loop structure on the LIR. Virus particles of SsDHAV-1 were extracted from a mycelial mass of strain DT-8 and isometric particles (20–22 nm in diameter) were observed with an electron microscope; the viral coat proteins were extracted from the particles and subjected *for* N-terminal sequencing analyzed to confirm the prediction of viral ORFs. Mycoviruses are common in fungi; however, previously only dsRNA or ssRNA viruses had been identified. Viruses with DNA genomes have not been reported to infect true fungi (different from protists and oomycetes); the discovery of SsHADV-1 provided convincing evidence that fungi could be infected by DNA viruses. The hypovirulence of strain DT-8, viral particles, genomic organization, and phylogenic analysis of SsHADV-1 are shown in Fig. 8.2.

SsHADV-1 is phylogenetically closely related to geminiviruses. The Rep of SsHADV-1 shares the highest sequence identity with the Reps of geminiviruses that infect dicotyledonous plants belonging to the genus *Mastrevirus*, but it is distinct from ssDNA viruses in the families *Nanoviridae* and *Circoviridae*. The Rep of SsHADV-1 is so similar to these of geminiviruses that SsHADV-1 could be grouped as a virus in the family of *Geminiviridae* if just based on the Rep amino acid sequences. However, there is no known viral protein in the family *Geminiviridae* that is phylogenetically related to the coat protein of SsHADV-1. Thus, SsHADV-1 is distinct from geminiviruses; they are likely to share common ancestors, but SsHADV-1 acquired the coat protein gene independently.

SsHADV-1 CP shared the highest sequence similarity (37.8%) with a marine putative protein (Marine_PP) from the metagenomic data of a microbial community from Sargasso Sea (accession number: AACY024124290), and a short gene sequence possibly coding for a viral Rep was presented on the same fragment as this Marine_PP gene; thus, a marine virus is likely to exist that is closely related to SsHADV-1 (Yu et al., 2010). Recently, a DNA virus, namely Mosquito VEM GeminiFungivirus-SDBVL G, was found in mosquitoes through viral metagenomics (Ng et al., 2011). Obviously, this virus is significantly different from SsDHAV-1 since the viral genome of SDBVL G has three ORFs. However, proteins coded by those three ORFs are highly similar to the coat protein and the Rep of SsHADV-1. ORF 1 codes for coat protein showed 32% amino acid identity to the SSHADV1 coat protein; ORF2 codes for a geminivirus replication

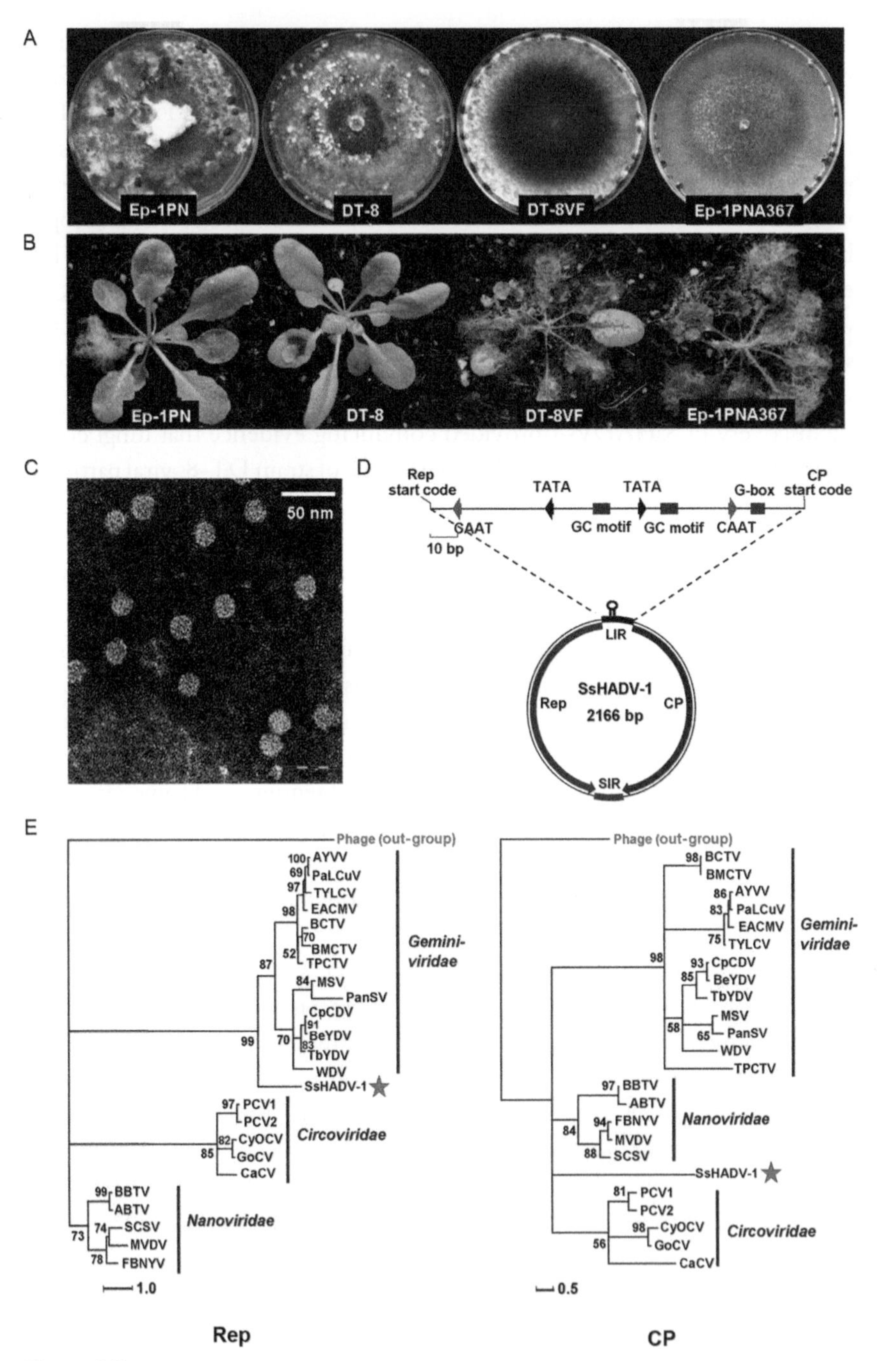

Figure 8.2

(Continued)

catalytic domain pfam 00799 showed 34% amino acid identity to SsHADV1. Moreover, the viral protein coded by ORF3 has 62% amino acid identity to the Rep of SsHADV1 (Ng et al., 2011). Although, geminiviruses are transmitted by insects, they could not replicate in insect cells; thus, SsHADV-1, SDBVL G, and the putative marine DNA virus possibly share recent ancestors since their coat proteins could be clustered phylogenetically, and together they share a recent ancestor with geminiviruses.

SsHADV-1 may represent a group of small single-stranded circular DNA viruses. DNA viral sequences similar to that of SsDHAV-1 were found in feces of wild rodents (Phan et al., 2011). Van den Brand et al. (van den Brand et al., 2012) identified DNA viral genome sequences from the feces of European badger and pine marten, and found that a virus, named European badger (Meles meles) fecal virus (MmFV), is phylogenetically closely related to SsHADV-1. MmFV shares similar characteristics of genomic organization of SsHADV-1, such as the presence of only two ORFs on the viral genome as well as LIR and SIR. These viral DNAs are likely of mushroom virus origin since mushrooms are a popular food for wild animals or, alternatively, they are derived from the mushroom genome. Previously, our

Figure 8.2 Hypovirulence-associated traits of strain DT-8 of *Sclerotinia sclerotiorum*. (A) Abnormal colony morphology of strain DT-8 grown on a PDA plate at 20 °C for 15 days; (B) Hypovirulent phenotype of strain DT-8 as exhibited by infected *Arabidopsis thaliana* plants, which were maintained at 20 °C for 4 days postinoculation. The RNA virus-infected hypovirulent strain Ep-1PN and its sexual progeny Ep-1PNA367 (virus-free) were used as controls; DT-8VF is a virus-free culture derived by hyphal tipping of strain DT-8 and showed a normal phenotype of *S. sclerotiorum*. (C) Viral particles observed under transmission electron microscopy. The particles were purified from mycelia of strain DT-8 and negatively stained with 1% uranyl acetate. Scale bars = 50 nm. (D) Genome organization of SsHADV-1. Functional ORFs (coding for CP and Rep) are displayed as thick arrows. The positions of the potential stem–loop structure, large intergenic region (LIR), and small short intergenic region (SIR) are marked; LIR is also shown in an expanded form to indicate the elements of the bidirectional promoter. (E) Phylograms of the Rep and CP of SsHADV-1 and selected circular ssDNA viruses in the families *Geminiviridae*, *Nanoviridae*, and *Circoviridae*. The NJ algorithm was performed using MEGA version 4.0, and the ML was performed using program TREE-PUZZLE version 5.2. The NJ algorithm and ML were used to independently generate tentative phylogenetic trees, and the resulting trees had similar topologies. The resulting ML tree is shown, and the numbers at the nodes represent the bootstrap values (%) inferred from 1000 puzzling steps. The scale relates branch lengths to the number of substitutions per site. The position of SsHADV-1 is indicated by a red star. *Modified from Yu et al. (2010).* (See Page 17 in Color Section at the back of the book.)

research showed that SsHADV-1 Rep gene-like sequences have been found in the genomes of many fungi, including mushrooms *Tuber melanosporum* and *Laccaria bicolor* (Liu, Fu, Xie, et al., 2011; Liu, Fu, Li, et al., 2011), suggesting that SsHADV-1 like viruses are likely widespread in fungi. Most recently, a viral DNA with similarities to SsDHAV-1 was PCR amplified from a DNA sample extracted from cassava leaves suggesting that the generated viral sequences were amplified from a mycovirus infecting endophytic fungi associated with cassava (Dayaram et al., 2012).

Unlike geminiviruses whose purified particles and viral ssDNA are not infectious, purified SsHADV-1 particles or genomic DNA preparations, extracted from either virus particles or directly from mycelial mass, are infectious when mediated by polyethylene glycol (PEG) (Yu et al., 2010). SsHADV-1 is also different from RNA mycoviruses in that it could be transmitted from strain DT-8 to other vegetatively incompatible strains of *S. sclerotiorum* with relatively high frequency. Although the mechanism for viral transmission from an infected strain to its vegetatively incompatible strains is not known, this finding indicated that SsHADV-1 or SsHADV-1-like viruses may have greater potential for biological control of plant fungal diseases.

2.2. Sclerotinia RNA viruses

2.2.1 Sclerotinia sclerotiorum debilitation-associated RNA virus

Sclerotinia sclerotiorum debilitation-associated RNA virus (SsDRV) was isolated from *S. sclerotiorum* strain Ep-1PN, which was originally isolated from a sclerotium on a diseased eggplant (*Solanum melongena*). Strain Ep-1PN grows slowly on PDA medium with frequent sectoring and excessive hyphal branching at hyphal tips resulting in abundant aerial mycelium and abnormal colony morphology (Fig. 8.2). Although strain Ep-1PN produces abundant oxalic acids, a toxin used by *S. sclerotiorum* to kill host cells and tissues, it has a very weak virulence on eggplant, rapeseed, and other hosts (Jiang, Li, Fu, Yi, & Wang, 1998; Li, Wang, Huang, & Zhou, 1996). Furthermore, strain Ep-1PN is also very weakly virulent on *Arabidopsis thaliana* (Fig. 8.2). Three dsRNA segments with estimated sizes of 7.4, 6.4, and 1.0 kbp were isolated from the mycelia of strain Ep-1PN (Fig. 8.3); current evidence indicates that the 6.4 kb dsRNA segment is associated with the hypovirulence of strain Ep-1PN since subcultures from this strain that lost the 6.4 kb dsRNA element are virulent (Li, Jiang, Wang, Zhu, & Rimmer, 1999; Liu, 2011).

The 6.4-kb dsRNA element represents the replicative form of the ssRNA genomic RNA of SsDRV. The full-length genome of SsDRV is 5419 nt,

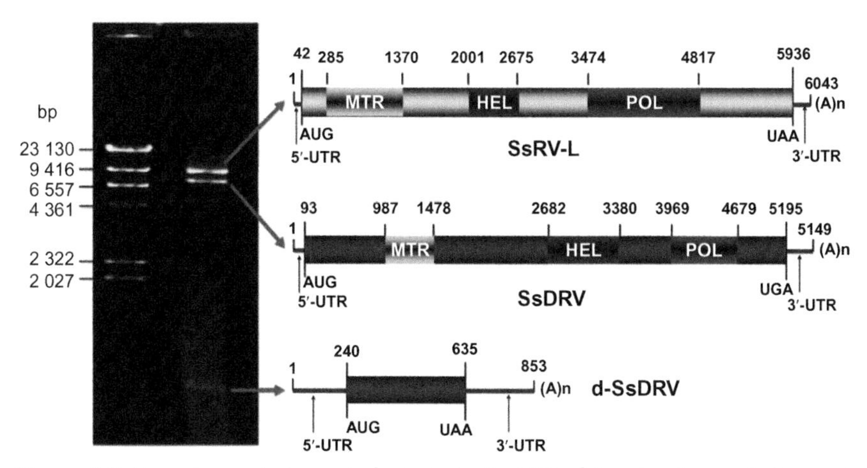

Figure 8.3 Mycoviruses in hypovirulent strain Ep-1PN of *S. sclerotiorum*. Left panel, dsRNA profile of samples extracted from hyphal mass shows three dsRNA bands. Right panel, Schematic representations of the genomic organization of mycovirus *Sclerotinia sclerotiorum* RNA virus L (SsRV-L) (Re-drew from Liu et al., 2009, Copyright © American Society for Microbiology) *Sclerotinia sclerotiorum* debilitation-associated RNA virus (SsDRV) (Re-drew from Xie et al., 2006, Copyright © Society for General Microbiology) and a defective dsRNA of SsDRV (d-SsDRV) (Re-drew from Liu, 2011, Copyright © Huazhong Agricultural University), respectively. (For color version of this figure, the reader is referred to the online version of this chapter.)

excluding the poly (A) tail at the 3′ terminus. A large ORF (nt positions 93–5195) coding for a putative protein with 1700 amino acid residues is highly similar to the viral RNA replicases of positive-stranded RNA viruses. Like some other ss(+)RNA viruses, the putative RNA replicase of SsDRV contains three conserved domains, namely, methyl transferase, helicase, and RNA-dependent RNA polymerase (RdRp) (Fig. 8.3). However, SsDRV does not produce typical virions since it lacks a gene for coat protein (Xie et al., 2006).

Based on phylogenetic analysis of the replication protein, SsDRV is closely related to plant viruses in the genus *Allexivirus* (family *Alphaflexiviridae*), such as garlic virus E, garlic virus C, garlic virus X, and shallot virus X, and to a mycovirus Botrytis virus F in the genus *Mycoflexivirus* (family *Gammaflexiviridae*) (Fig. 8.4). However, SsDRV is significantly different from viruses in the family of *Alphaflexiviridae* based on genomic organization. The genome of SsDRV is quite simple, has only one ORF, but lacks coat protein and movement protein genes. Recently, based on characteristics of SsDAV, a new genus *Sclerodarnavirus* in the family *Alphaflexividae* was established (http://www.ictvonline.org/virusTaxonomy.asp? Src=NCBI & ictv_id=20091014&bhcp=1). Although SsDRV and another virus Sclerotinia sclerotiorum RNA virus L (SsRV-L)

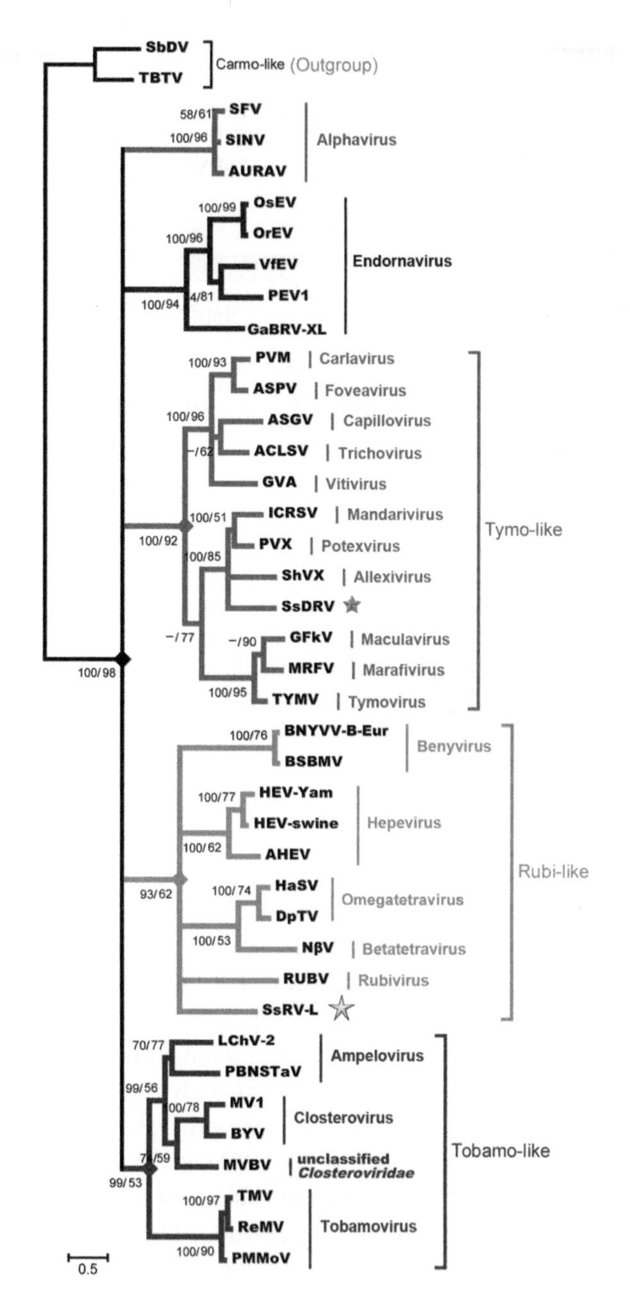

Figure 8.4

(Continued)

coinfect the hypovirulent strain Ep-1PN, SsDRV could replicate and proliferate in *S. sclerotiorum* independently. A defective RNA with a size of 1.0 kb is usually associated with SsDRV infection (Fig. 8.3).

SsDRV was the first mycovirus characterized at the molecular level from *S. sclerotiorum*. Two aspects of this mycovirus are likely to be of considerable interest. First, the fact that SsDRV is closely related to plant viruses may raise questions regarding its origin. *S. sclerotiorum* has a wide host range and shares the same plant hosts as viruses in the genus *Allexivirus S. sclerotiorum* is likely to have acquired viruses from virus-infected host plants. It might be possible that the host plants synthesize some antifungal substances (such as chitinases and glucanases) to weaken fungal cell walls thus offering viruses a chance to enter into fungal cells. Plant viruses inside fungal cells may have a unique way for evolution, such as deleting genes for movement and coat proteins since these two genes are not necessary for viruses that replicate and move in fungal cells. There is also the possibility that plants acquire viruses from fungal pathogens during infection. However, mycoviruses, like SsDRV, are quite simple in their genome structures since they code only for an RNA replicase. The question then is how these viruses acquire other genes (e.g., genes coding for coat and movement proteins) to be typical plant viruses. The second question is the potential use of SsDRV for the control of Sclerotinia diseases. This topic is discussed in detail in Section 5 of this chapter.

2.2.2 *Sclerotinia sclerotiorum* RNA virus L

SsRV-L was the first reported ss (+) RNA mycovirus related to a human ss (+) RNA virus. It was also isolated from the hypovirulent strain Ep-1PN, which is coinfected by SsDRV (Fig. 8.3). The 7.4-kb dsRNA element in strain Ep-1PN represents a replicative form of SsRV-L genomic ssRNA.

Figure 8.4 Phylogenetic analysis of the conserved motifs and flanking sequences of RdRp derived from aligned deduced amino acid sequences of SsRV-L and selected viruses. The NJ algorithm and ML were used to generate tentative phylogenetic trees independently. The NJ algorithm was performed using PAUP* 4.0b10 (Sinauer Associates, Sunderland, MA), and ML was performed using the program TREE-PUZZLE version 5.2. The resulting ML tree is shown. The first number indicated at the nodes represents the bootstrap values (%) calculated from the NJ tree inferred from 1000 bootstrap replicates, and the second number represents the quartet puzzling support values (%) inferred from 10,000 puzzling steps; a minus sign (−) indicates that a node is absent in the corresponding NJ method. Only bootstrap or quartet puzzling support values of >50% are indicated. The scale relates branch lengths to the number of substitutions per site. Yellow and red stars—positions of the two viruses coinfecting *S. sclerotiorum*. *Copyright © American Society for Microbiology, Liu et al., 2009.* (See Page 18 in Color Section at the back of the book.)

The genome of SsRV-L is 6043 nts, excluding the poly (A) tail. This viral genome has only one large ORF (nt positions 42–5936) that encodes a protein with significant similarity to the RNA replicases of "alphavirus-like" supergroup of positive-strand RNA viruses. The RNA replicase of SsRV-L contains conserved methyltransferase, helicase, and RNA-dependent RNA polymerase motifs (Fig. 8.3). Homology analysis showed that SsRV-L has significant sequence similarity to the replicases of *Hepatitis E virus* (HEV), *swine hepatitis E virus* (HEV-swine), and *Avian hepatitis E virus*, viruses infecting humans and animals. The RdRp domain of SsRV-L shares 27% identity and 43% similarity with HEV, and 28% identity and 44% similarity with HEV-swine. The RdRp conserved motifs of SsRV-L also share 23% identity and 40% similarity with plant viruses including mint virus 1, and 24% identity and 41% similarity with plum bark necrosis and stem pitting-associated virus, two viruses in the family *Closteroviridae*. The RNA helicase motifs of SsRV-L also share significant sequence similarities with the insect viruses Helicoverpa armigera stunt virus (HaSV) and Dendrolimus punctatus tetravirus (DpTV) belonging to the genus *Omegatetravirus* in the family *Tetraviridae* (Liu et al., 2009).

The superfamily of alpha-like viruses comprises three lineages, namely rubi-like, tobamo-like, and tymo-like viruses (Koonin & Dolja, 1993), and phylogenetic analysis revealed that SsRV-L belongs to the rubi-like subfamily in the "alphavirus-like" supergroup of positive-strand RNA viruses (Fig. 8.4). Viruses in both tobamo- and tymo-like subfamilies infect plants, but viruses in the rubi-like subfamily have a wide host range: they may infect plants, vertebrates, insects, and protists. The discovery of SsRV-L demonstrated that viruses in the rubi-like subfamily may also infect fungi; thus, the host range of rubi-like viruses is more diverse than once thought.

The origin of SsRV-L is not clearly known, but it is likely that SsRV-1 was derived from plant viruses, possibly from those related to tobamoviruses, as SsRV-1 is closely related phylogenetically to plant viruses in the family *Closteroviridae*. We have also suggested that HEV and the insect viruses HaSV and DpTV were likewise derived from plant viruses. Humans are likely to acquire HEV from infected plants directly, or from animals that acquired the HEV-like virus from plants, and insects are likely to acquire viruses through feeding on virus-infected plants (Liu et al., 2009).

With a few exceptions, such as mycoviruses in the families *Reoviridae* and *Chrysoviridae* as well as the recently reported Rosellinia necatrix quadrivirus 1 (RnQV1) in the family *Quadriviridae* (Lin et al., 2012), mycoviruses often have quite simple genomes. The genomic organization of SsRV-L and

SsDRV are similar to each other: they have only one ORF for RNA replicase, and both of them lack genes for coat protein and movement protein. Viruses in the family *Hypoviridae* and genus *Mitovirus* as well as unclassified mycoviruses including Diaporthe ambigua RNA virus (DaRV) (Preisig, Moleleki, Smit, Wingfield, & Wingfield, 2000) are also examples of mycoviruses that lack coat and movement proteins. This phenomenon suggests that coat protein for many mycoviruses is not necessary for infection and replication in fungal cells and that mycoviruses in fungi have a unique way of moving from one cell to other cells. The subgenomic transcription strategy for ss(+) RNA viruses is likely to be a mechanism for the regressive evolution of mycoviruses. Once the viral RNA polymerase gene could replicate in fungal cells independently and efficiently, other genes, such as coat protein gene and movement protein genes are deleted. Interestingly, many plant and animal ss (+) RNA viruses were found to replicate in yeast (Nagy, 2008).

2.2.3 Sclerotinia sclerotiorum hypovirus 1

Sclerotinia sclerotiorum hypovirus 1 (SsHV-1) was isolated from *S. sclerotiorum* strain SZ-150. Strain SZ-150 was isolated from a sclerotium obtained from a diseased rapeseed (*Brassica napus*) plant in Suizhou, P. R. China. Strain SZ-150 grows on PDA medium slowly with frequent sectoring and abnormal colony morphology and produces no, or a few, sclerotia that are abnormally distributed in the colony (Xie et al., 2011). Unlike strain Ep-1PN, the aerial hyphae in colony of strain SZ-150 is not rich, old hyphae often accumulate dark pigment. It almost lost its ability to attack detach leaves of rapeseed. Two dsRNA segments with sizes of ~9.5 and ~3.6 kb were extracted from a mycelial mass of strain SZ-150. However, further cDNA cloning and sequencing analysis revealed that the 9.5 kb dsRNA segment is actually composed of two independent dsRNA segments of very similar size but which cannot be easily separated by agarose gel electrophoresis. One of these two segments represents the replicative form of the ssRNA viral genome of SsHV-1 (Fig. 8.5).

The complete genomic sequence of SsHV1 is 10,398 nts long excluding the 3′-terminal poly (A) tail (Fig. 8.5). This genome has only one large putative ORF coding for a polyprotein. The ORF starts at AUG (542–544) and terminates at UAG (9386–9388), and the polyprotein is 2948 amino acids in length with an approximate molecular mass of 337 kDa. The polyprotein contains four conserved domains, namely papain-like protease (Prot), UDP glucose/sterol glucosyltransferase (UGT), RNA-dependent RNA polymerase (POL), and viral RNA Helicase (Hel), as identified by the Conserved Domain Architecture Retrieval Tool on the NCBI website.

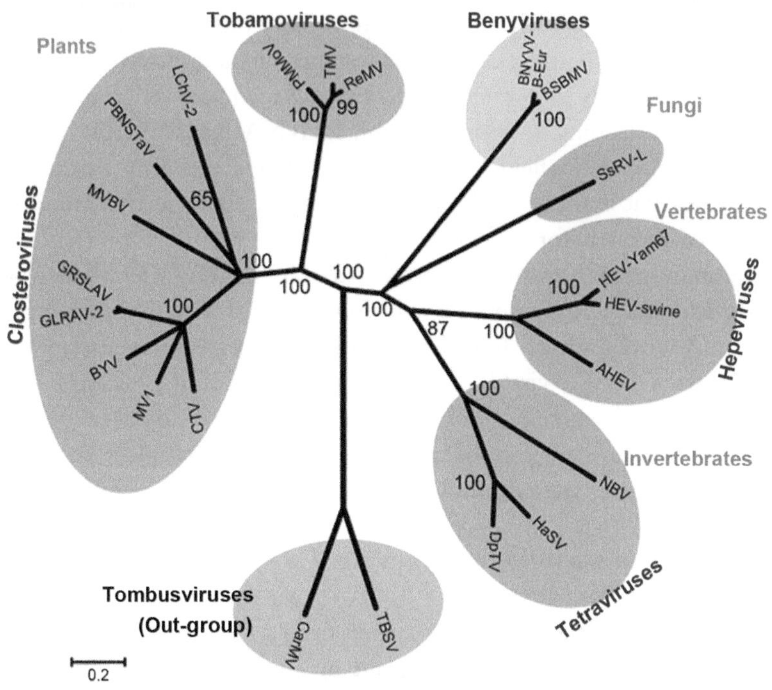

Figure 8.5 Phylogenetic analysis of the conserved amino acid sequences of RdRp domain of selected representatives of Rubi-like and Tobamo-like viruses. The NJ algorithm and ML were used to generate tentative phylogenetic trees independently. The NJ algorithm was performed using PAUP* 4.0b10, and ML was performed using the program TREE-PUZZLE5.2. The resulting NJ tree is shown. The number indicated at the nodes represents the bootstrap values (%) calculated from the NJ tree inferred from 1000 bootstrap replicate. Only bootstrap support values of >50% are indicated. The ML tree also matched the tree topology, except that the branch marked with an asterisk was not found (Not shown in Fig. 8.5). The scale relates branch lengths to the number of substitutions per site. The tree was outgroup-rooted to the Tombusviruses. The full virus names are CTV, Citrus tristeza virus (AAO12717); MV1, Mint virus 1(YP_224091); BYV, Beet yellows virus (AAC25115); GLRAV-2, Grapevine leafroll-associated virus 2 (AAC40856); GRSLAV, Grapevine rootstock stem lesion-associated virus (NP_835337); MVBV, Mint vein banding virus (AAS57939); PBNSTaV, Plum bark necrosis and stem pitting-associated virus (YP_001552324); LcHV2, Little cherry virus 2 (AAP87784); PMMoV, Pepper mild mottle virus (CAC59955); TMV, Tobacco mosaic virus (AAD44327); ReMV, Rehmannia mosaic virus (YP_001041889); BNYVV-B-Eur, Beet necrotic yellow vein virus-F2, France (CAA28795); BSBMV, Beet soil-borne mosaic virus (NP_612601); HEV-Yam67, Hepatitis E virus Yam67 (AAM66329); HEV-swine, Swine hepatitis E virus (ABB88699); AHEV, Avian hepatitis E virus (AAS45830); NβV, Nudaurelia capensis beta virus (NP_048059); HaSV, Helicoverpa armigera stunt virus (NP_049235); DpTV, Dendrolimus punctatus tetravirus (YP_025094); TBSV, Tomato bushy stunt virus (ACT67403.1); and CarMV, Carnation mottle virus (CAA26726.1) . (*from Liu, 2011*, Copyright © Huazhong Agricultural University). (For color version of this figure, the reader is referred to the online version of this chapter.)

Phylogenetic analysis of the complete amino acid sequences of SsHV-1 and related mycoviruses revealed that SsHV1/SZ-150 is most closely related to CHV3/GH2 and CHV4/SR2 in the family *Hypoviridae* (Fig. 8.5). Multiple alignment analysis showed that SsHV-1 shares a high sequence identity with that of CHV3/GH2 (58.1%) and CHV4/SR2 (48.0%). Furthermore, the 5′ UTR regions of SsHV-1 and CHV3/GH2 and CHV4/SR2 are highly conserved. SsHV-1 has longer 3′ UTR than that of these two hypoviruses; however, the extreme end of the 3′ UTR regions of the three viruses are highly similar. Although the genome size of SsHV-1 is about 500 nt longer than that of CHV3/GH2 (9.8 kb), the viral RNA organization and the conserved domains in the polyprotein of the two viruses are very similar, which suggests that SsHV1/SZ-150 also belongs to the family *Hypoviridae*, and it is most closely related to the hypoviruses CHV3/GH2 and CHV4/SR2 (Fig. 8.6).

The small dsRNA segment in strain SZ-150 is a satellite-like RNA of SsHV-1. The complete cDNA sequence is 3643 nt in length excluding the poly (A) tail, and it comprises a single large putative ORF (nt positions 318 to 2237) coding for a hypothetical protein of 638 amino acids. The first 104-nt stretch of the 5′ UTR of the satellite-like RNA is 100% identical to that of SsHV-1, and the remaining region of the 5′ UTR shares 51.5% identity with the corresponding region of SsHV-1. The satellite-encoded putative protein does not show any similarity with the polyprotein encoded by the RNA genome of SsHV-1 and lacks any conserved domains for RNA replicase. Since SsHV-1 could replicate without the small dsRNA segment, and the small dsRNA segment always accompanies SsHV-1, the small dsRNA segment was suggested to be a satellite-like RNA of SsHV-1. The size of this satellite-like RNA is larger than that of any known satellite RNAs; furthermore, it shares a high sequence identity with the 5′ UTR of SsHV1/SZ-150; these unusual properties support the hypotheses that this satellite-like RNA is a unique RNA element or represents a new type of satellite RNA.

SsHV-1 is the first reported naturally occurring hypovirus that infects a fungus other than *C. parasitica*. Recently, a hypovirus (Valsa ceratosperma hypovirus 1, VcHV1), which is closely related to SsHV-1 was isolated and characterized from the fruit tree pathogenic fungus, *Valsa ceratosperma* (Yaegashi, Kanematsu, & Ito, 2012). The genome of VcHV1 is 9543 nt including the poly (A) tail and is 855 bp shorter than that of SsHV-1; however, the sizes of their ORFs are very close and their encoded proteins are 2941 and 2948 amino acid residues, respectively. SsHV-1 has a 668 bp longer

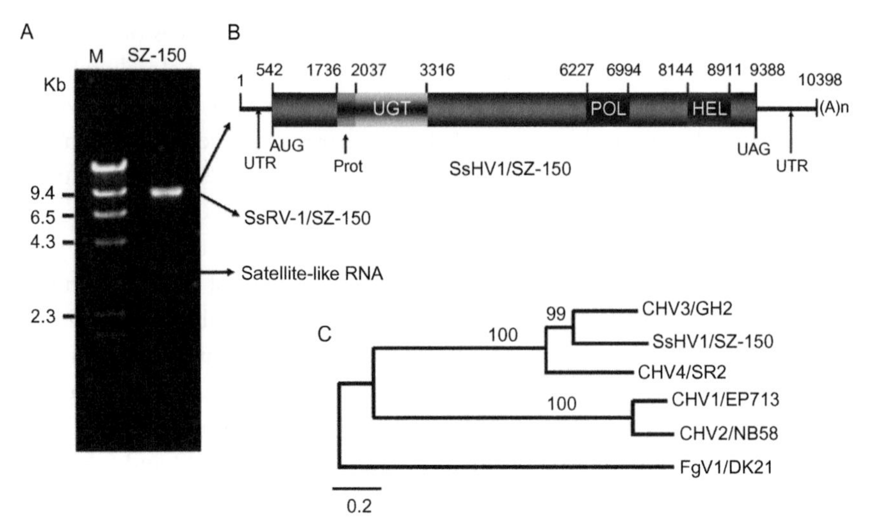

Figure 8.6 Mycoviruses of strain SZ-150 of *S. sclerotiorum*. (A) dsRNA profile of samples extracted from hyphal mass. (B) Schematic representation of the genomic organization of mycovirus SsHV1/SZ-150. Open reading frame encodes a putative viral protein with 2948 amino acids that contains four conserved domains: papain-like proteinase (Prot), UDP glucose/sterol glucosyltransferase (UGT), RNA-dependent RNA polymerase (POL), and viral RNA Helicase (HEL). (C) Phylogenetic analysis of SsHV1/SZ-150 and selected mycoviruses based on full-length amino acid sequences of the viral polyproteins. A neighbor joining an unrooted tree is shown. Bootstrap values (%) obtained with 1000 replicates are indicated on branches and branch lengths correspond to genetic distance; scale bar at lower left corresponds to a genetic distance of 0.2. Mycovirus (acronym/strain; GenBank accession) included Cryphonectria parasitica hypovirus 1 (CHV1/EP713; NP_041091.1), Cryphonectria parasitica hypovirus 2 (CHV2/NB58; NP_613266.1), Cryphonectria parasitica hypovirus 3 (CHV3/GH2; NP_051710.1), Cryphonectria parasitica hypovirus 4 (CHV4/SR2; YP_138519.1), and Fusarium graminearum virus 1 (FgV1/DK21; AAT07067.2). *Modified form Xie et al. (2011)).* (For color version of this figure, the reader is referred to the online version of this chapter.)

3′ UTR than that of VcHV1. Phylogenetically, *V. ceratosperma* is quite closely related to *C. parasitica*, the host of Cryphonectria hypovirus 1-4. Furthermore, both fungi infect trees; however, *S. sclerortiorum* is distant phylogenetically from these two fungi, and tree hosts of *S. sclerotiorum* are unknown. Thus, viruses in the genus Hypovirus 3 are likely to have a wide host range.

Strain SZ-150 is a hypovirulent strain of *S. sclerotiorum*; however, SsHV-1 alone is not the primary causal agent of hypovirulence in strain SZ-150 of *S. sclerotiorum*. Strain SZ-150 was coinfected by two viruses and a satellite-like RNA element. Since fungal strains infected only with the uncharacterized mycovirus (with similar size to SsHV1) did not show any debilitation traits and they had normal phenotypes, this uncharacterized mycovirus does not appear to contribute to the hypovirulence of strain SZ-150. On the other hand, strains infected with SsHV-1, but not the satellite-like RNA element, show a very slight change from the wild-type strain. Interestingly, CHV3/GH2 has only a slight impact on the host's virulence, growth, and conidiation (Linder-Basso, Dynek, & Hillman, 2005; Smart et al., 1999), while VcHV1 does not have any impact on the host's virulence. It seems that the satellite-like RNA element is the main contributor to the hypovirulence phenotype.

Recently, a totivirus and a hypovirus similar to SsHV1 were found in strain SNZ1-3-5 of *S. sclerotiorum*. Interestingly, when this SsHV1-like virus infects *S. sclerotiorum* alone, strong impact on host could be observed, while the totivirus alone has little or no effect. Thus, it seems that this totivirus could mitigate the negative effects caused by the SsHV1-like virus (Li, Y., et al., unpublished data). This phenomenon also suggested that SsHV1 has certain diversity at the species level; some viral strain may cause serious debilitation in their host.

2.2.4 Sclerotinia sclerotiorum mitoviruses

Two mitoviruses were isolated from a hypovirulent strain (strain KL-1) of *S. sclerotiorum* (Xie & Ghabrial, 2012). Strain KL-1 was isolated from a sclerotium on diseased lettuce in a farm at Lexington, KY, USA. The hypovirulence traits of strain KL-1 could be transmitted to strain 1980 by hyphal anastomosis using a dual culture method (Fig. 8.7). Although a single dsRNA segment of ~3.0 kp on agarose gel was extracted from the mycelial mass of strain KL-1, cDNA cloning and sequencing analysis revealed that there are actually two dsRNA segments with similar sizes; this was confirmed by electrophoresis on a 15% nondenaturing polyacrylamide gel (Fig. 8.7). The larger segment, named dsRNA1, is 2513 nt; the smaller one, named dsRNA 2, is 2421 nt excluding a poly (A) tail (Xie & Ghabrial, 2012).

Mitochondrial codon usage indicated that both dsRNAs 1 and 2 have one large ORF encoding a putative viral RNA replicase typical of mitoviruses. Thus, dsRNA1 and dsRNA2 are the viral genomes of two mitoviruses, which were named Sclerotinia sclerotiorum mitovirus 1 (SsMV1/KL-1) and SsMV2/KL-1. The 5′ UTR of SsMV1/KL-1 is 418 nt long, and its 3′ UTR is extremely short, only 16 nt long; the ORF encodes a RNA replicase of 691 amino acid (aa) residues. The 5′ UTR and 3′ UTR of SsMV2/KL-1 are 311 and 82 nt long, respectively; its ORF encodes an RNA replicase of 676 amino acid residues. The two mitoviruses are significantly different from each other both in genomic size and organization and in similarity of RNA replicase sequence. The deduced amino acid sequences of SsMV1/KL-1 and SsMV2/KL-1 are 33.3% identical, thus, they are independent mitoviruses (Xie & Ghabrial, 2012).

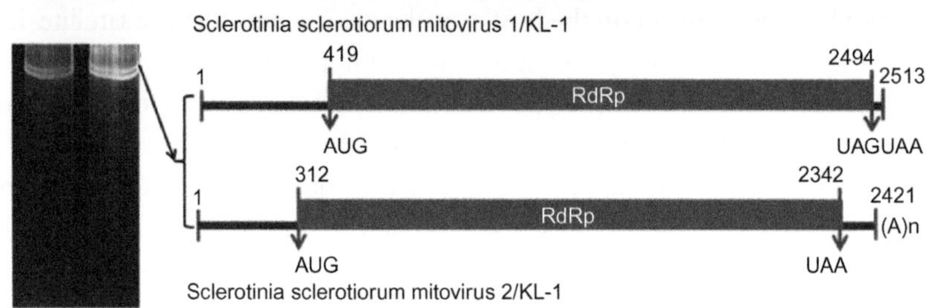

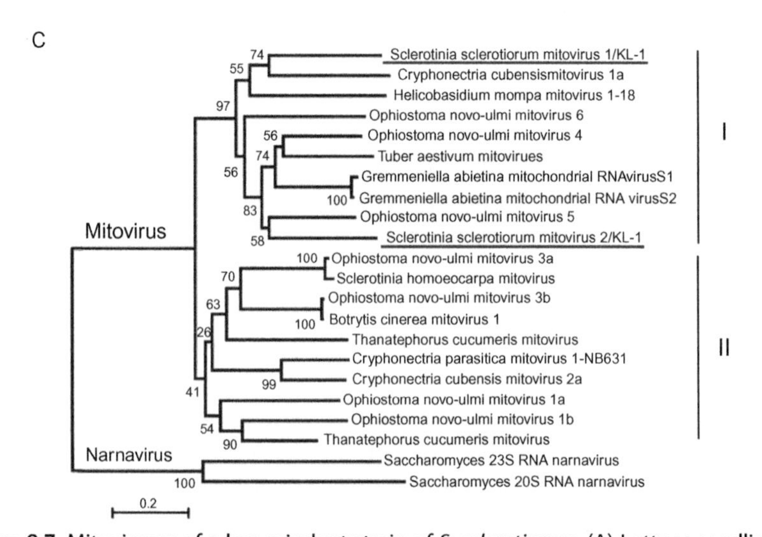

Figure 8.7 Mitoviruses of a hypovirulent strain of *S. sclerotiorum*. (A) Lettuce seedlings inoculated with the mitovirus-infected hypovirulent strain KL-1 (left panel), the virus-free virulent strain 1980^{hyg} (center panel), or strain 1980^{hyg}-converted (right panel), which was generated by hyphal anastomosis between strain KL-1 and strain 1980^{hyg}. (B) (left) Polyacrylamide gel electrophoresis of dsRNA on nondenaturing 15% polyacrylamide gel and genomic organization of SsMV1/KL-1 and SsMV2/KL-1. An ethidium bromide-stained gel showing two dsRNA segments is shown; all dsRNA samples were treated with both DNase I and S1 nuclease prior to electrophoresis. (Right) Schematic representation of the genomic organization of the mitoviruses SsMV1 (SsMV1/KL-1) and SsMV2 (SsMV2/KL-1). (C) A neighbor joining an unrooted phylogenetic tree constructed based on alignment of the respective RdRp amino acid sequences of SsMV1 (SsMV1/KL-1), SsMV2 (SsMV2/KL-1), and other mitoviruses is shown. Bootstrap values (%) obtained with 1000 replicates are indicated on branches and branch lengths correspond to genetic distance; scale bar at lower left corresponds to a genetic distance of 0.2. The results of phylogenetic analysis indicate that mitoviruses could be separated into two large clusters (I and II). *Modified from Xie and Ghabrial (2012); Virology.* (See Page 19 in Color Section at the back of the book.)

Phylogenetic analysis of SsMV1/KL-1 and SsMV2/KL-1 with other selected viruses revealed that members of the genus *Mitovirus* were divided into two clusters I and II, and that both SsMV1/KL-1 and SsMV2/KL-1 belong to cluster I, which includes Cryphonectria cubensis mitovirus 1a (CcMV1a), Helicobasidium mompa mitovirus 1-18, Ophiostoma novo-ulmi mitovirus 6 (OnuMV6), OnuMV4, Thielaviopsis basicola mitovirus, Gremmeniella abietina mitochondrial RNA virus S1, and OnuMV5 (Fig. 8.7). SsMV1/KL-1 is closely related to OnuMV5, a mitovirus that infects the plant fungal pathogen *Ophiostoma novo-ulmi*, and SsMV2/KL-1 is closely related to CcMV1a that infects fungal pathogen *Cryphonectria cubensis* (Xie & Ghabrial, 2012).

Mitoviruses vary considerably and are widespread among fungi, and they are evolving via horizontal transfer from one species to another distant species. The 3′-UTRs of mitoviruses reported are variable in length (Stielow, Klenk, Winter, & Menzel, 2011); however, the length of the 3′-UTRs of SsMV1/KL-1 is extremely short (16 nt) compared with that of other mitoviruses and SsMV2/KL-1 (82 nt). Like the mitovirus CcMV1a, the genome of SsMV2/KL-1 contains a poly (A) tail at its 3′-terminal sequence, while other reported mitoviruses do not have this structure (Xie & Ghabrial, 2012). A mitovirus, BcMV1, was isolated from strain CanBc-1 of *Botrytis cinerea*, a fungus belonging to the family *Sclerotiniaceae* (Wu, Zhang, Li, Jiang, & Ghabrial, 2010). The complete genome of BcMV1 is 2804 nt, and it belongs to cluster II phylogenetically (Fig. 8.7). Interestingly, BcMV-1 has a long 5′ UTR (477 nt), and this mitovirus also has a defective RNA, 2171-nt in length. Further study showed that this defective RNA could suppress the replication of BcMV1 and could be cotransmissible with BcMV1 through hyphal anastomosis. BcMV-1 shares 95% nt sequence identity with OnuMV3b, which suggests that this mitovirus has been horizontally transferred between *B. cinerea* and *O. novo-ulmi* or their relatives. A similar phenomenon was observed between *Sclerotinia homoeocarpa* and *O. novo-ulmi*; Sclerotinia homoeocarpa mitovirus shares 93% nt sequence identity with Ophiostoma mitovirus 3a (Deng, Xu, & Boland, 2003).

2.2.5 Sclerotinia sclerotiorum partitivirus S

Sclerotinia sclerotiorum partitivirus S (SsPV-S) was isolated from strain Sunf-M of *S. sclerotiorum*. Strain Sunf-M was isolated from a sclerotium on residues of a diseased sunflower (*Helianthus annuus*) plant in Inner Mongolia, P. R. China. Strain Sunf-M showed a normal phenotype of *S. sclerotiorum*, and it grew on PDA rapidly and formed typical colony morphology; it also had a strong virulence on its hosts. DsRNA samples were extracted from hyphal mass of strain Sunf-M, and three dsRNA bands were observed on an agarose gel. The largest dsRNA segment represents a viral genome of an unclassified mycovirus, *S. sclerotiorum* nonsegmented virus L (SsNsV-L), and the smallest band is about 2.2 kb in size, as shown in Fig. 8.8 (Liu et al., 2012).

This 2.2-kb dsRNA band comprised two dsRNA segments of a partitivirus viral genome; the two dsRNA segments are 1856 and 1783 bp in length, excluding a poly (A) tail, respectively. This virus was named SsPV (Liu et al., 2012). The 1856-bp dsRNA encodes a putative RdRp protein (YP_003082248.1) of 579 aa, while the 1783 bp encodes a putative coat protein (YP_003082249.1) of 507 aa. Phylogenetic analysis of the RdRp of SsPV and that of other selected partitiviruses showed that SsPV is closely related to Flammulina velutipes isometric virus (BAH08700.1); Ophiostoma quercus partitivirus (CAJ34337.1); Oyster mushroom isometric virus II (AAP74192.1); and a plant cryptovirus, Vicia faba partitivirus 1 (ABJ99996.1). These viruses shared the same clade with other viruses in the genus Alphacrytovirus, which were usually found in plants (Fig. 8.9).

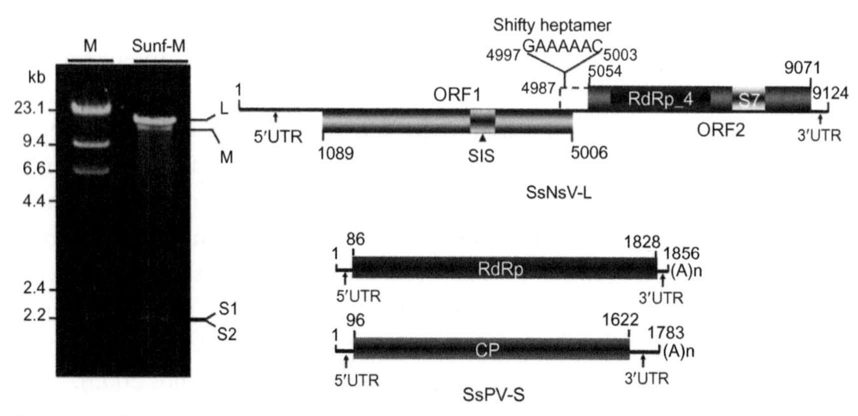

Figure 8.8 Mycoviruses in strain Sunf-M of *S. sclerotiorum*. Left, Agarose gel electrophoresis of dsRNA samples extracted from hyphal mass. Right, Schematic representations of the genomic organization of *Sclerotinia sclerotiorum* nonsegmented virus L (SsNsV-L) and Sclerotinia sclerotiorum Partitivirus S (SsPV-S), respectively. The conserved domains of deduced proteins of SsNsV-L are shown: SIS, Sugar Isomerase domain; S7, Phytoreovirus S7 protein; RdRP_4, Viral RNA-dependent RNA polymerase. *Modified from Liu et al. (2012).* (For the color version of this figure, the reader is referred to the online version of this chapter.)

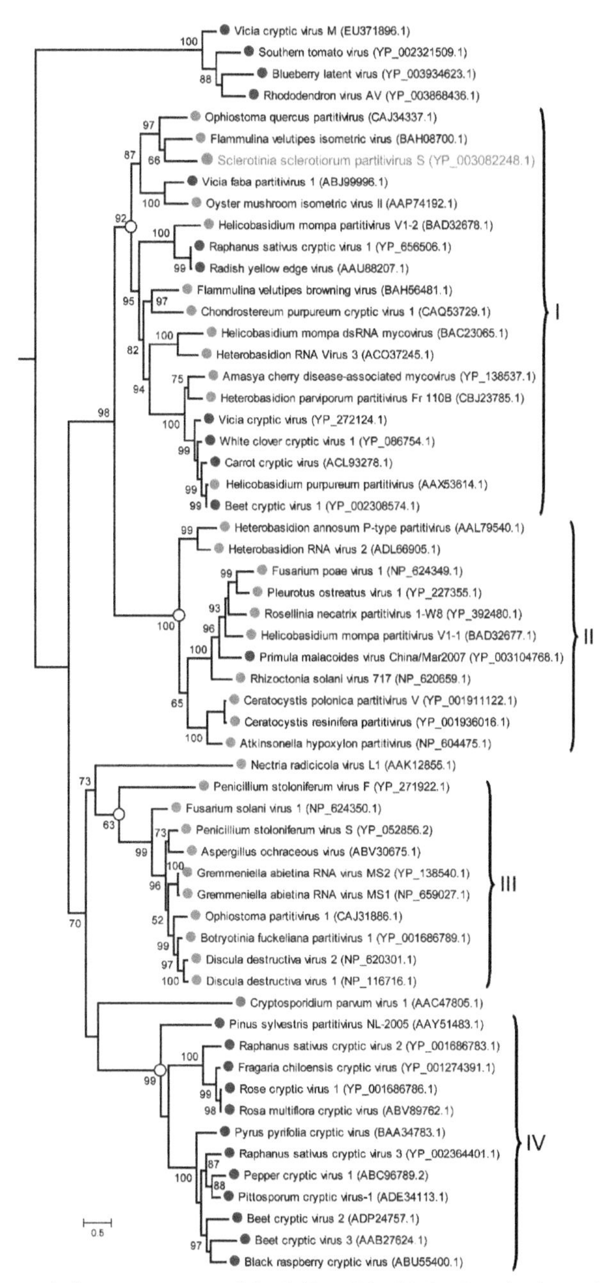

Figure 8.9 ML phylogenetic trees of the RdRps (A) of SsPV-S and other selected viruses in the family *Partitiviridae*. Viral lineages are color-coded to reflect their host range. The trees were rooted with STV-like lineages. Only *p*-values of the SH-like approximate likelihood ratios (SH-aLRT) >0.5(50%) are indicated. All scale bars correspond to 0.5 amino acid substitutions per site. *From Liu et al. (2012).* (For the color version of this figure, the reader is referred to the online version of this chapter.)

Interestingly, the coat protein encoded by the SsPV-S shows high similarity to IAA-leucine resistant 2 protein (ILR2) of *A. thaliana* (Liu et al., 2012). The amino acid similarity and identity between the coat protein and ILR2 are 43% (132/307) and 26% (82/307) with an *E*-value of <5e–16, respectively. It is possible that a horizontal gene transfer (HGT) event has occurred from the ancestor of SsPV-S to the ancestor of *Arabidopsis*. Recently, a novel partitivirus (Rosellinia necatrix partitivirus 2, RnPV 2) was identified from a tree fungal pathogen, *Rosellinia necatrix* (Chiba et al., 2011). Phylogenetic analysis revealed that the coat proteins encoded by these two partitiviruses have high similarity and identity; the similarity between the two viruses is 25% (92/361) and identity is 45% (162/361) with an *E*-value of <2e–23. Although RnPV 2 dsRNA1, which codes for an RdRp, is not currently available from the GenBank, SsPV-S and RnPV 2 CPs are related to each other, and apparently share a more recent ancestor. The discovery of RnPV 2 confirms further that an HGT event has occurred once between SsPV-S-like partitiviruses and plants, and the direction of this HGT was most likely from virus to plants (Fig. 8.10).

2.2.6 *Novel viruses to be explored from* S. sclerotiorum

Although debilitated fungal strains can be easily isolated from the field, the mechanism underlying debilitation is sometimes difficult to ascertain. *S. sclerotiorum* strain XG36-1 is an example. Strain XG36-1 was isolated from a typical lesion on the stem of rapeseed induced by *S. sclerotiorum* at a field near Xiaogan City, P. R. China. Strain XG36-1 shows debilitation traits as it grew slowly and the hyphal tips often branch excessively. Many sectors developed at the colony margin on PDA and sclerotial production of strain XG36-1 was significantly suppressed, and only a few sclerotia were produced in the colony at later stage. The cytoplasm of strain XG36-1 was destroyed and granulated; membranes of both nuclei and mitochondria were disintegrated. The amount of mitochondria in cells was significantly reduced, only a few cristae in cavity-full mitochondria were observed. The abnormal phenotypes of strain XG36-1 could be cured using protoplast regeneration since some protoplast regenerants showed the normal phenotype of the wild-type strain, and they recovered their full virulence. Like the other hypovirulent strains mentioned earlier, the sclerotia of strain XG36-1 seldom produced apothecia. All tested single ascospore offspring of strain XG36-1 showed a typical wild-type phenotype of *S. sclerotiorum*. Furthermore, the debilitation traits of XG36-1 are transmissible in dual cultures

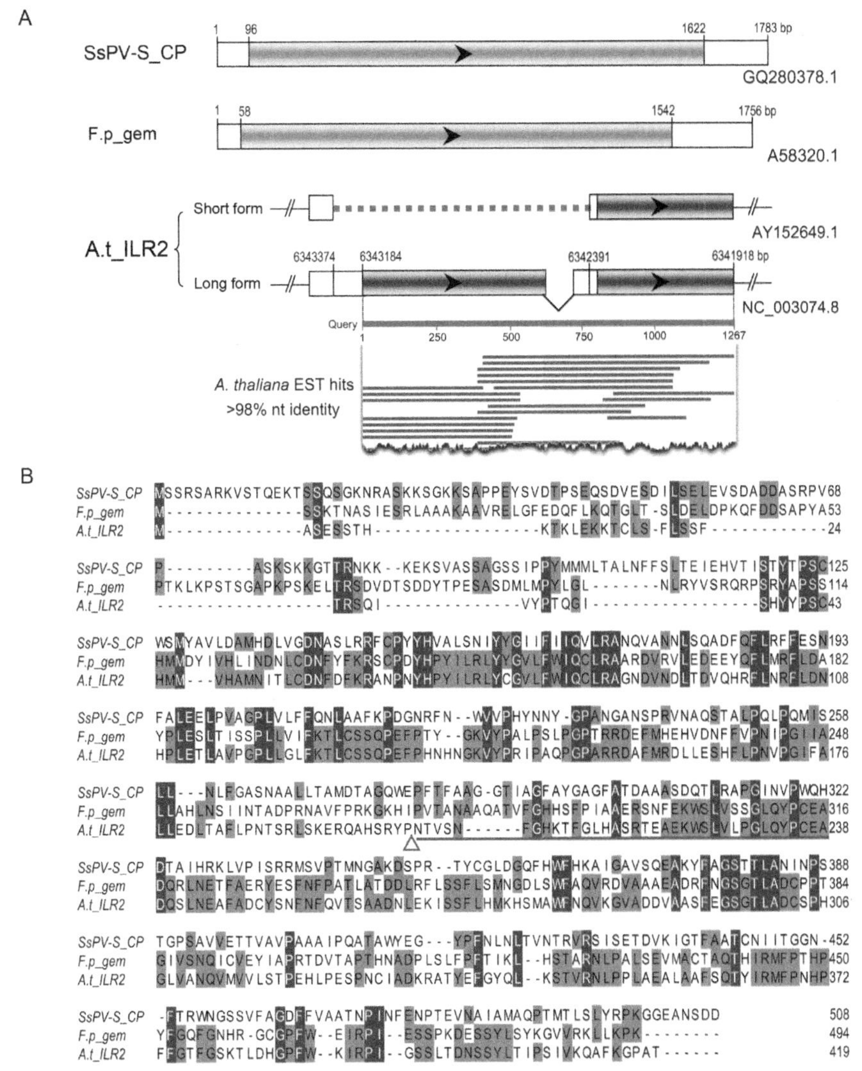

Figure 8.10 Schematic representation (A) and multiple alignment of predicted amino acid sequences (B) of ILR2 of *Arabidopsis thaliana*, gem of Festuca pratensis, and coat protein gene of SsPV-S. (A) Colored rectangular boxes with arrowheads indicate open reading frames (ORFs). The region indicated by a broken horizontal line is deleted in the Wassilewskija (Ws) and Landsberg erecta (Ler) accessions of *Arabidopsis*. The matched regions of expressed sequences and the nt identity are indicated. A red triangle in (B) indicates predicted frameshift sites in ILR2 and the region of predicted ILR2 intron is marked with a red line. Identical residues between sequences are shaded. F.p_gem, the gem of Festuca pratensis; SsPV-S_CP, coat protein gene of Sclerotinia sclerotiorum partitivirus S; A.t_ILR2, IAA-leucine resistant 2 protein of *Arabidopsis thaliana*; CP, coat protein; EST, expressed sequence tag. *Copyright © American Society for Microbiology, Liu, Fu, Xie, et al. (2011).* (See Page 20 in Color Section at the back of the book.)

with the wild-type isolate on PDA. Thus, there exists a mobile element that mediates the debilitation of strain XG36-1 (Zhang et al., 2009).

Virus-like particles could be observed in the cells of strain XG36-1 but not in the cells of the wild-type strain. The virus-like particles were isometric, with a diameter of ~40 nm, and several particles were enveloped by a single layer membrane. Virus-like particles could rarely be isolated from the mycelial mass of strain XG36-1 after ultracentrifugation. However, all attempts to extract dsRNA either directly from hyphae or from ultracentrifugation pellets were unsuccessful. Previously, spontaneously occurring hypovirulent strains derived from isolate S10 of *S. sclerotiorum* showed transmission of hypovirulence and its associated traits, but no dsRNAs segments were detected (Li, Huang, Laroche, & Acharya, 2003). Similar phenomenon appears to occur in other fungal species. Strain CanBc-3 of *B. cinerea* isolated from rapeseed field had a normal phenotype; however, a spontaneous mutant CanBC-3HV was isolated from strain CanBc-3 when it grew on PDA. Mutant CanBC-3HV showed typical hypovirulence traits: it grew on PDA slowly, sporulated sporadically, and formed abnormal colony morphology with few sclerotia. Under SEM, intrahyphal hyphae in colonies of mutant CanBC-3HV were frequently observed. Furthermore, no dsRNA segment or extra DNA segments were extracted from mutant CanBC-3HV (Cao et al., 2011).

The hypovirulence and its associated traits of strain XG36-1 and other strains are undoubtedly caused by mycoviruses or other mobile elements. It is possible that the titer of mycovirus in hyphae is too low to be extracted, or an improper approach for dsRNA extraction was applied since the mobile elements may not be RNA viruses, but other molecular parasites, such as viroid-like RNAs or DNA viruses with large genomes. New approaches are needed to isolate and characterize the properties of these mobile elements, and RNA-seq techniques and related bioinformatic tools are likely to be a powerful way.

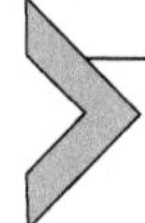

3. THE INTERACTION BETWEEN MYCOVIRUS AND *S. SCLEROTIORUM*

The *hypovirus*/*C. parasitica* system has become a good model for studying the interaction between mycoviruses and their hosts. This system has been successfully utilized in fundamental studies that led to the discovery of several important signal pathways involved in the pathogenicity and reproduction of the chestnut blight pathogen (Nuss, 2005; Parsley, Chen,

Geletka, & Nuss, 2002) and utilized to understand the antiviral mechanism in fungi (Segers, Zhang, Deng, Sun, & Nuss, 2007; Sun, Choi, & Nuss, 2009; Zhang & Nuss, 2008). However, the interaction between other mycoviruses and their hosts may not be necessarily similar since both viruses and hosts are different. Therefore, it is worthwhile to explore the interactions between mycoviruses and other fungi.

In a previous study, using the SsDRV/*S. sclerotiorum* system, 150 genes of *S. sclerotiorum* whose expressions in the virus-infected strain Ep-1PN were significantly suppressed were identified (Li et al., 2008). These genes represented a broad spectrum of biological functions including carbon and energy metabolism, protein synthesis and transport, signal transduction, and stress response. Among these genes, some are emphasized here. The first gene codes for inorganic pyrophosphatase (SS1G_04783, the EST accession number DN795959.1), which is involved in energy production and conversion; SS1G_04783 in the wild-type strain is expressed constitutively during hyphal growth, sclerotial development, sclerotial germination, and infection. However, in strain Ep-1PN, the expression of SS1G_04783 is significantly suppressed. Considering the shortage of energy, it may not be surprising that strain Ep-1PN grows slowly and is hypovirulent. The second gene is coding for cyclophilin A (SS1G_11216, the EST accession number: DN795949.1). Cyclophilin is a virulence determinant in *Magnaporthe grisea* (Viaud, Balhadère, & Talbot, 2002) and is involved in the pathogenicity of *B. cinerea* (Viaud, Brunet-Simon, Brygoo, Pradier, & Levis, 2003). Interestingly, a cyclophilin gene is required for virulence in the chestnut blight fungus, and its expression in hypovirus-infected strain of *C. parasitica* was also suppressed (Chen et al., 2007). SS1G_11216 is also a constitutively expressed gene in the wild-type strain of *S. sclerotiorum*, while during an early infection stage its expression is about threefold higher than that in the hyphal growth stage. The hyphal growth of the *S. sclerotiorum* wild-type strain was suppressed significantly when the PDA medium was amended with Cyclosporine A. The colony of wild-type strain growing on Cyclosporine A-amended PDA was similar to that of the colony of strain Ep-1PN, which developed on a cyclosporine A-free PDA plate. Thus, cyclophilin A is likely to play a critical role in the life of *S. sclerotiorum*. The third gene is coding for integrin-like protein (SS1G_14133, the EST accession number: DN795890). The expression of SS1G_14133 has two peaks: one is at the early stage of sclerotial formation, and the other is at the early stage of infection on the leaves of *A. thaliana*; at other stages, such as hyphal extension and sclerotial germination, its expression can be detected but is relatively low.

SS1G_14133 is likely to have functions involved in the sclerotial development and pathogenicity of *S. sclerotiorum* (Zhu et al., 2013).

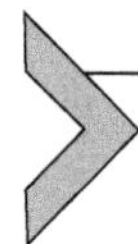

4. COINFECTION OF *S. SCLEROTIORUM* WITH MULTIPLE MYCOVIRUSES

Coinfection of fungi with two or more viruses is of common occurrence. For example, two dsRNA viruses, the totivirus HvV190S and the chrysovirus HvV145S, coinfect *Cochliobolus victoriae* (Sanderlin & Ghabrial, 1978), and two totiviruses, SsRV-1 and SsRV-2, were reported to coinfect the same strain of *Sphaeropsis sapinea* (Preisig, Wingfield, & Wingfield, 1998). Mycoviruses coinfections were also found in other fungi, such as in *Phlebiopsis gigantea* (Kozlakidis et al., 2009), *Chalara elegans* (Park, Chen, & Punja, 2006), in endophytic and entomopathogenic fungus *Tolypocladium cylindrosporum* (Herrero & Zabalgogeazcoa, 2011), and even in the oomycete pathogen *Phytophthora infestans* (Cai, Myers, Fry, & Hillman, 2011). However, the frequency of coinfections in *S. sclerotiorum* is unexpectedly very high. SsRV-L and SsDRV coinfect strain Ep-1PN of *S. sclerotiorum*; SsHV-1, and an unclassified dsRNA virus were found in strain SZ-150 and two mitoviruses (SsMV-1 and SsMV-2) in strain KL-1. SsPV-S coinfects with SsNsV-L in strain Sunf-M, and we even found that SsHADV-1 coinfects with an RNA virus (Liu, S., et al., unpublished data). Furthermore, multisegments of dsRNA were frequently extracted from *S. sclerotiorum* strains, and some of them are being subjected to sequencing analysis.

Although two or more mycoviruses may frequently occur in one strain of *S. sclerotiorum*, some of them can replicate independently. We found that both SsRV-L and SsDRV could be separated from each other, and that they can replicate in *S. sclerotiorum* independently. Through the separation of these viruses, we confirmed that it is SsDRV that causes the hypovirulence of strain Ep-1PN (Li et al., 1999; Liu, 2011). However, SsDRV is not likely to be stable in *S. sclerotiorum*, since it is very easy to isolate cultures which are SsDRV-free but contains only SsRV-L from sclerotia or hyphal tips; we also found that SsDRV could be removed easily by protoplast regeneration. This phenomenon suggested that SsRV-L could replicate in *S. sclerotiorum* efficiently. A similar phenomenon was observed in the viral coinfection system in strain SZ-150, SsHV-1 and its satellite-like RNA element were easily removed through either hyphal tip subculture or protoplast regeneration. We further found that strains infected by SsHADV-1 are easily cured. This implies that *S. sclerotiorum* exists as a special antivial mechanism against either debilitation-associated DNA virus or RNA virus selectively.

The unique property of *S. sclerotiorum* is that it does not produce any conidial spores but forms sclerotia. Sclerotia can germinate to form apothecia and produce ascospores; alternatively, it also can germinate to form infectious hyphae directly (Bolton et al., 2006). In most cases, mycoviruses are removed from sexual spores of their fungal hosts; furthermore, some asexual spores may be virus-free with high frequency. The unusually high frequency of coinfection of mycoviruses in *S. sclerotiorum* is most likely due to the special asexual reproductive fashion of *S. sclerotiorum*. Mycoviruses may be efficiently maintained in sclerotia., We have also suggested that field strains when infected by one mycovirus may be infected with other viruses easily since its viral resistance will be weakened and subsequently weaken the vegetative incompatible reaction between virus donor and virus recipient.

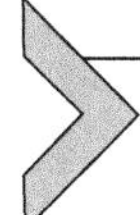

5. THE POTENTIAL USE OF MYCOVIRUSES TO CONTROL SCLEROTINIA DISEASES

Hypoviruses were utilized to control chestnut blight disease successfully in European countries (Anagnostakis, 1982). This attracted many interests in discovering new hypovirulence-associated mycoviruses from other plant fungal pathogens. However, the biocontrol of chestnut blight with hypovirus was not successful in North America, possibly due to the fact that transmission of mycoviruses under field conditions is limited by the vegetative incompatibility grouping of the fungal host and the fitness of the hypovirus-infected strains utilized (Milgroom & Cortesi, 2004). To overcome the limitation of vegetative incompatibility reactions, the cDNA genome of hypovirus was transformed into the host's genome, and the transgenic strains converted to hypovirulence could transmit the hypovirus persistently. Since there are only two mating type genes in *C. parasitica*, the hypovirus in transgenic strain could be transmitted in most strains in field (Anagnostakis, Chen, Geletka, & Nuss, 1998; Chen, Choi, & Nuss, 1994). Recently, protoplasts of several fungi have been successfully transfected with purified virus particles mediated by PEG, and through such studies, it was possible for mycoviruses to infect different vegetative groups, even extend the host range of some mycoviruses (Chiba et al., 2009; Hillman, Supyani, Kondo, & Suzuki, 2004; Kanematsu, Sasaki, Onoue, Oikawa, & Ito, 2010; Lee, Yu, Son, Lee, & Kim, 2011; Sasaki et al., 2007, 2002; Yu et al., 2010). Most recently, polymorphic genes associated with four vegetative incompatibility (vic) loci of *C. parasitica* were identified, disruption of alleles in one of two strains enhanced hypovirus transmission (Choi et al., 2012).

These approaches demonstrate that there are still many ways to overcome the obstacles for using mycoviruses to control fungal diseases.

Using hypovirulence-associated mycoviruses to control Sclerotinia diseases is likely to be very difficult. First, *S. sclerotiorum* has developed very complicated nonself-recognition systems (vegetative incompatibility); strains belonging to several compatibility groups may occur in the same field (Kohn, Stasovski, Carbone, Royer, & Anderson, 1991). This extensive diversity of vegetative incompatibility hinders the spread of mycoviruses in field populations of *S. sclerotiorum*. Second, *S. sclerotiorum*, which does not produce conidia, limits mycoviruses' vertical transmission in the field since no mycoviruses have been transmitted through ascospores so far. Third, unlike the host of *C. parasitica*, which is a fruit tree and can live in an orchard or in a forest for a long time, most hosts of *S. sclerotiorum* are herbaceous crops that are harvested shortly after growth termination; these short-lived hosts are not likely to help virus colonization of Sclerotinia isolates under field conditions. Furthermore, the homothallic property of *S. sclerotiorum* seals off the door for virus transmission through hyphal anastomosis during sexual reproduction and discourages the viral cDNA integration strategy, which has been used in the C. parasitica/hypovirus system for control of chestnut blight. Based on the reasons described earlier, it may be very difficult and impractical to control diseases caused by *S. sclerotiorum* with mycoviruses if only current methods for mycovirus delivery are used.

Theoretically, there are possible ways to solve the difficult problems for using mycoviruses to control *S. sclerotiorum* diseases. For example, create an artificial vector for mycovirus transmission. *C. minitans* is a mycoparasite of *S. sclerotiorum*; it parasitizes both hyphae and sclerotia of *S. sclerotiorum* (Li et al., 2006; Whipps & Gerlagh, 1992). In fact, we identified a totivirus (CmRV) in strain Chy-1 of *C. minitans* (Cheng et al., 2003). Hypovirulence-associated mycoviruses that infect *S. sclerotiorum* will be transfected into *C. minitans*, and following replication, this virus could then be returned to *S. sclerotiorum* from a virus-infected strain of *C. minitans* during hyphal parasitization. This strategy is likely to overcome the limitation of mycovirus transmission caused by vegetative incompatibility and to be feasible for field use since *C. minitans* could easily produce large amounts of conidial spores.

6. CONCLUSIONS AND PROSPECTS

The mycoviruses of *S. sclerotiorum* have already shown strong diversity: both DNA and RNA viruses have already been found in *S. sclerotiorum*, and other mycoviruses belonging to the families *Totiviridae*, *Chrysoviridae*, and

Reoviridae have been isolated and identified. It may not be surprising that new viruses representing new types of fungal viruses are isolated in the future.

Sclerotia are likely to play an important role in mycovirus survival. Like other ascomycetous fungi, *S. sclerotiorum* also has an unknown mechanism to rid itself of virus infection through sexual reproduction. Sclerotia, however, have two types of germinations, which may help viruses to be retained permanently in *S. sclerotiorum*. When sclerotia germinate carpogenically, this resultant virus-free fungus may infect hosts in a regular fashion; while at the same time, sclerotia also maintain a way for germinating myceliogenically, when this virus-infected hyphae fuse with virus-free hyphae, viruses may easily transmit into new individuals via hyphal anastomosis. Thus, *S. sclerotiorum* is a good fungus for studying the ecology and evolution of viruses.

Reasonably, mycoviruses are to be used for biological control of plant diseases, and successfully used to probe fungal pathogenicity and other biological properties at the molecular level (Nuss, 2005). Mycoviruses are likely to be used to explore the basic principles of virus life cycle, such as virus replication and proliferation, structure, and ecology and evolution. Pioneering work on the interaction between hypovirus and its host has been elegantly presented by Dr. Nuss research group (Lin et al., 2007; Nuss, 2005; Segers et al., 2007; Sun et al., 2009; Zhang & Nuss, 2008; see also Chapter 5). Studying mycoviruses also may contribute to better understanding of virion structure, especially at nanometer level (Caston et al., 2003, 2006; Ochoa et al., 2008; Pan et al., 2009; Tang et al., 2010, 2010). Studying mycoviruses also enhance our understanding of ecology and evolution of viruses in virosphere (Chiba et al., 2011; Goker, Scheuner, Klenk, Stielow, & Menzel, 2011; Lee et al., 2011; Liu, Fu, Li, et al., 2011; Liu, Fu, Xie, et al., 2011; Liu et al., 2010). Human beings have to face the frequent emergence of new viruses, which threaten both food production and human health. Do mycoviruses provide any clues for the emergence of novel viruses? At least, mycoviruses are likely to be used for antiviral drug screening since many of them are closely related to human and plant viruses.

Mycoviruses are viruses that infect fungi and replicate in fungi. However, the host range of mycoviruses is not likely to be restricted to fungi in the future since more and more mycoviruses are found to be closely related to other viruses that do not infect fungi.

ACKNOWLEDGMENTS

This research was supported by China National Funds for Distinguished Young Scientists (31125023), the Special Fund for Agro-scientific Research in the Public Interest (201103016), and China Agriculture Research System (nycytx-00514).

REFERENCES

Amselem, J., Cuomo, C. A., van Kan, J. A. L., Viaud, M., Benito, E. P., et al. (2011). Genomic analysis of the necrotrophic fungal pathogens *Sclerotinia sclerotiorum* and *Botrytis cinerea*. *PLoS Genetics*, 7(8), e1002230.

Anagnostakis, S. L. (1982). Biological control of chestnut blight. *Science*, *215*, 466–471.

Anagnostakis, S. L., Chen, B. S., Geletka, L. M., & Nuss, D. L. (1998). Hypovirus transmission to ascospore progeny by field-released transgenic hypovirulent strains of *Cryphonectria parasitica*. *Phytopathology*, *88*, 598–604.

Boland, G. J. (1992). Hypovirulence and double-stranded RNA in *Sclerotinia sclerotiorum*. *Canadian Journal of Plant Pathology*, *14*, 10–17.

Boland, G. J., & Hall, R. (1994). Index of plant hosts of *Sclerotinia sclerotiorum*. *Canadian Journal of Plant Pathology*, *16*, 93–100.

Bolton, M., Thomma, B. P. H. J., & Nelson, B. (2006). *Sclerotinia sclerotiorum* (Lib.) de Bary: Biology and molecular traits of a cosmopolitan pathogen. *Molecular Plant Pathology*, *7*, 1–16.

Cai, G., Myers, K., Fry, W. E., & Hillman, B. I. (2011). Co-infection of a single *Phytophthora infestans* isolate by two distinct viruses. *Phytopathology*, *101*, S25.

Cao, J. B., Zhou, Y., Zhang, L., Zhang, J., Yang, L., Qin, L. H., et al. (2011). DsRNA-free transmissible hypovirulence associated with formation of intra-hyphal hyphae in *Botrytis cinerea*. *Fungal Biology*, *115*, 660–671.

Caston, J. R., Ghabrial, S. A., Jiang, D., Rivas, G., Alfonso, C., Roca, R., et al. (2003). Three-dimensional structure of *Penicillium chrysogenum virus*: A double-stranded RNA virus with a genuine $T=1$ capsid. *Journal of Molecular Biology*, *331*, 417–431.

Caston, J. R., Luque, D. L., Trus, B. L., Rivas, G., Alfonso, C., Gonzalez, J. M., et al. (2006). Three-dimensional structure and stoichiometry of *Helminthosporium victoriae 190S totivirus*. *Virology*, *347*, 323–332.

Cessna, S. G., Sears, V. E., Dickman, M. B., & Low, P. S. (2000). Oxalic acid, a pathogenicity factor for *Sclerotinia sclerotiorum*, suppresses the oxidative burst of the host plant. *The Plant Cell*, *12*, 2191–2200.

Chen, B. S., Choi, G. H., & Nuss, D. L. (1994). Attenuation of fungal virulence by synthetic infectious hypovirus transcripts. *Science*, *264*, 1762–1764.

Chen, M., Jiang, M., Shang, J., Lan, X., Yang, F., Huang, J., et al. (2007). CYP1, a hypovirus-regulated cyclophilin, is required for virulence in the chestnut blight fungus. *Molecular Plant Pathology*, *12*, 239–246.

Cheng, J., Jiang, D., Fu, Y. P., Li, G., Peng, Y., & Ghabrial, S. A. (2003). Molecular characterization of a dsRNA totivirus infecting the sclerotial parasite *Coniothyrium minitans*. *Virus Research*, *93*, 41–50.

Chiba, S., Kondo, H., Tani, A., Saisho, D., Sakamoto, W., Kanematsu, S., et al. (2011). Widespread endogenization of genome sequences of non-retroviral RNA viruses into plant genomes. *PLoS Pathogens*, 7, e1002146.

Chiba, S., Salaipeth, L., Lin, Y. H., Sasaki, A., Kanematsu, S., & Suzuki, N. (2009). A novel bipartite double-stranded RNA mycovirus from the white root rot fungus *Rosellinia necatrix*: Molecular and biological characterization, taxonomic considerations, and potential for biological control. *Journal of Virology*, *83*, 12801–12812.

Choi, G. H., Dawe, A. L., Churbanov, A., Smith, M. L., Milgroom, M. G., & Nuss, D. L. (2012). Molecular characterization of vegetative incompatibility genes that restrict hypovirus transmission in the chestnut blight fungus *Cryphonectria parasitica*. *Genetics*, *190*, 113–127.

Dayaram, A., Jaschke, A., Hadfield, J., Baschiera, M., Dobson, R. C., Shepherd, D. N., et al. (2012). Molecular characterisation of a novel cassava associated circular ssDNA virus. *Virus Research*, *166*, 130–135.

Deng, F., Xu, R., & Boland, G. J. (2003). Hypovirulence-associated double-stranded RNA from *Sclerotinia homoeocarpa* is conspecific with *Ophiostoma novo-ulmi Mitovirus 3a-Ld. Phytopathology, 93*, 1407–1414.

Dong, X., Ji, R., Guo, X., Foster, S. J., Chen, H., Dong, C., et al. (2008). Expressing a gene encoding wheat oxalate oxidase enhances resistance to *Sclerotinia sclerotiorum* in oilseed rape (*Brassica napus*). *Planta, 228*, 331–340.

Goker, M., Scheuner, C., Klenk, H. P., Stielow, J. B., & Menzel, W. (2011). Codivergence of mycoviruses with their hosts. *PLoS One, 6*, e22252.

Grente, J. (1965). Les formes hypovirulentes d'Endothia parasitica et les espoirs de lutte contre le chancre du chantaingnier. *Comptes Rendus de l'Academie d'Agriculture de France, 51*, 1033–1037.

Hegedus, D. D., & Rimmer, S. R. (2005). Sclerotinia sclerotiorum: When "to be or not to be" a pathogen? *FEMS Microbiology Letters, 251*, 177–184.

Herrero, N., & Zabalgogeazcoa, I. (2011). Mycoviruses infecting the endophytic and entomopathogenic fungus *Tolypocladium cylindrosporium*. *Virus Research, 160*, 409–413.

Hillman, B. I., Supyani, S., Kondo, H., & Suzuki, N. (2004). A reovirus of the fungus *Cryphonectria parasitica* that is infectious as particles and related to the *Coltivirus* genus of animal pathogen. *Journal of Virology, 78*, 892–898.

Hu, X., Bidney, D. L., Yalpani, N., Duvick, J. P., Crasta, O., Folkerts, O., et al. (2003). Overexpression of a gene encoding hydrogen peroxide-generating oxalate oxidase evokes defense responses in sunflower. *Plant Physiology, 133*, 170–181.

Jiang, D., Li, G., Fu, Y., Yi, X., & Wang, D. (1998). Transmissible hypovirulent element in isolate Ep-1PN of *Sclerotinia sclerotiorum*. *Chinese Science Bulletin, 43*, 779–781.

Kanematsu, S., Sasaki, A., Onoue, M., Oikawa, Y., & Ito, T. (2010). Extending the fungal host range of a partitivirus and a mycoreovirus from *Rosellinia necatrix* by inoculation of protoplasts with virus particles. *Phytopathology, 100*, 922–930.

Kesarwani, M., Azam, M., Natarajan, K., Mehta, A., & Datta, A. (2000). Oxalate decarboxylase from Collybia velutipes. Molecular cloning and its overexpression to confer resistance to fungal infection in transgenic tobacco and tomato. *The Journal of Biological Chemistry, 275*, 7230–7238.

Kim, K. S., Min, J. Y., & Dickman, M. B. (2008). Oxalic acid is an elicitor of plant programmed cell death during *Sclerotinia sclerotiorum* disease development. *Molecular Plant-Microbe Interactions, 21*, 605–612.

Kohn, L. M., Stasovski, E., Carbone, I., Royer, J., & Anderson, J. B. (1991). Mycelial incompatibility and molecular markers identify genetic variability in populations of *Sclerotinia sclerotiorum*. *Phytopathology, 81*, 480–485.

Koonin, E. V., & Dolja, V. V. (1993). Evolution and taxonomy of positive-strand RNA viruses: Implications of comparative analysis of amino acid sequences. *Critical Reviews in Biochemistry and Molecular Biology, 28*, 375–430.

Kozlakidis, Z., Hacker, C. V., Bradley, D., Jamal, A., Phoon, X., Webber, J., et al. (2009). Molecular characterisation of two novel double-stranded RNA elements from *Phlebiopsis gigantea*. *Virus Genes, 399*, 132–136.

Lee, K. M., Yu, J., Son, M., Lee, Y. W., & Kim, K. H. (2011). Transmission of *Fusarium boothii* mycovirus via protoplast fusion causes hypovirulence in other phytopathogenic fungi. *PLoS One, 6*, e21629.

Li, H., Fu, Y., Jiang, D., Li, G., Ghabrial, S. A., & Yi, X. (2008). Down-regulation of *Sclerotinia sclerotiorum* gene expression in response to infection with Sclerotinia sclerotiorum debilitation-associated RNA virus. *Virus Research, 135*, 95–106.

Li, G., Huang, H. C., Laroche, A., & Acharya, S. N. (2003). Occurrence and characterization of hypovirulence in the tan sclerotial isolate S10 of *Sclerotinia sclerotiorum*. *Mycological Research, 107*, 1350–1360.

Li, G., Huang, H. C., Miao, H. J., Erichson, R. S., Jiang, D., & Xiao, Y. (2006). Biological control of Sclerotinia diseases of rapeseed by aerial applications of the mycoparasite *Coniothyrium minitans*. *European Journal of Plant Pathology, 114*, 345–355.

Li, G., Jiang, D., Wang, D., Zhu, B., & Rimmer, R. (1999). Double-stranded RNAs associated with the hypovirulence of *Sclerotinia sclerotiorum* strain Ep-1PN. *Progress in Natural Science, 9*, 836–841.

Li, G., Wang, D., Huang, H. C., & Zhou, Q. (1996). Polymorphisms of *Sclerotinia sclerotiorum* isolated from eggplant in Jiamusi, Heilongjiang Province. *ACTA Phytopathologica sinica, 26*, 237–242.

Lin, Y. H., Chiba, S., Yani, A., Kondo, H., Sasaki, A., Kanematsu, A., et al. (2012). A novel quadripartite dsRNA virus isolated from a phytopathogenic filamentous fungus, *Rosellinia necatrix*. *Virology, 426*, 42–50.

Lin, H., Lan, X., Liao, H., Parsley, T. B., Nuss, D. L., & Chen, B. (2007). Genome sequence, full-length infectious cDNA clone, and mapping of viral double-stranded RNA accumulation determinant of hypovirus CHV1-EP721. *Journal of Virology, 81*, 1813–1820.

Linder-Basso, D., Dynek, J. N., & Hillman, B. I. (2005). Genome analysis of *Cryphonectria hypovirus 4*, the most common hypovirus species in North America. *Virology, 337*, 192–203.

Liu, H. (2011). Endogenization and evolutionary genomics of eukaryotic non-retroviral viruses. Huazhong Agricultural University, Wuhan, Hubei Province, P. R. Chian, PhD thesis.

Liu, H., Fu, Y., Jiang, D., Li, G., Xie, J., Peng, Y., et al. (2009). A novel Mycovirus that is related to the human pathogen *Hepatitis E Virus* and Rubi-like viruses. *Journal of Virology, 83*, 1981–1991. http://dx.doi.org/10.1128/JVI.01897-08.

Liu, H., Fu, Y., Jiang, D., Li, G., Xie, J., Cheng, J., et al. (2010). Widespread horizontal gene transfer from double-stranded RNA viruses to eukaryotic nuclear genomes. *Journal of Virology, 84*, 11876–11887.

Liu, H., Fu, Y., Li, B., Yu, X., Xie, J., Cheng, J., et al. (2011). Widespread horizontal gene transfer from circular single-stranded DNA viruses to eukaryotic genomes. *BMC Evolutionary Biology, 11*, 276.

Liu, H., Fu, Y., Xie, J., Cheng, J., Ghabrial, S. A., Li, G., et al. (2012). Evolutionary genomics of mycovirus-related dsRNA viruses reveals cross-family horizontal gene transfer and evolution of diverse viral lineages. *BMC Evolutionary Biology, 12*, 91.

Liu, H., Fu, Y., Xie, J., Cheng, J., Ghabrial, S. A., Li, G., et al. (2011). Widespread horizontal gene transfer from double-stranded RNA viruses to eukaryotic nuclear genomes. *Journal of Virology, 84*, 11876–11887. http://dx.doi.org/10.1128/JVI.00955-10.

Milgroom, M. G., & Cortesi, P. (2004). Biological control of chestnut blight with hypovirulence: A critical analysis. *Annual Review of Phytopathology, 42*, 311–338.

Nagy, P. D. (2008). Yeast as a model host to explore plant virus-host interactions. *Annual Review of Phytopathology, 46*, 217–242.

Ng, T. F. F., Willner, D. L., Lim, Y. W., Schmieder, R., Chau, B., et al. (2011). Broad surveys of DNA viral diversity obtained through viral metagenomics of mosquitoes. *PLoS One, 6*, e20579.

Nuss, D. L. (1992). Biological control of chestnut blight: An example of virus-mediated attenuation of fungal pathogenesis. *Microbiological Reviews, 56*, 561–576.

Nuss, D. L. (2005). Hypovirulence: Mycoviruses at the fungal-plant interface. *Nature Reviews. Microbiology, 3*, 632–642.

Ochoa, W. F., Havens, W. M., Sinkovits, R. S., Nibert, M. L., Ghabrial, S. A., & Baker, T. S. (2008). Partitivirus structure reveals a 120-subunit, helix-rich capsid with distinctive surface arches formed by quasisymmetric coat-protein dimers. *Structure, 16*, 776–786.

Pan, J., Dong, L., Lin, L., Ochoa, W. F., Sinkovits, R. S., Havens, W. M., et al. (2009). Atomic structure reveals the unique capsid organization of a dsRNA virus. *Proceedings of the National Academy of Sciences of the United States of America, 106*, 4225–4230.
Pan, Y. L., Wang, Z. Y., & Wu, H. Z. (1997). Resistance to carbendazim in *Sclerotinia sclerotiorum*. *Chinese Journal of Oil Crop Sciences, 19*, 1767–1768.
Park, Y., Chen, X., & Punja, Z. K. (2006). Diversity, complexity and transmission of double-stranded RNA elements in *Chalara elegans* (synanam. *Thielaviopsis basicola*). *Mycological Research, 110*, 697–704.
Parsley, T. B., Chen, B., Geletka, L. M., & Nuss, D. L. (2002). Differential modulation of cellular signaling pathways by mild and severe hypovirus strains. *Eukaryotic Cell, 1*, 401–413.
Phan, T. G., Kapusinszky, B., Wang, C., Rose, R. K., Lipton, H. L., et al. (2011). The fecal viral flora of wild rodents. *PLoS Pathogens*, 7(9), e1002218. http://dx.doi.org/10.1371/journal.ppat.1002218.
Preisig, O., Moleleki, N., Smit, W., Wingfield, B. D., & Wingfield, M. J. (2000). A novel RNA mycovirus in a hypovirulent isolate of the plant pathogen *Diaporthe ambigua*. *The Journal of General Virology, 81*, 3107–3114.
Preisig, O., Wingfield, B. D., & Wingfield, M. J. (1998). Coinfection of a fungal pathogen by two distinct double-stranded RNA viruses. *Virology, 252*, 399–406.
Riou, C., Freyssinet, G., & Fevre, M. (1992). Purification and characterization of extracellular pectinolytic enzymes produced by *Sclerotinia sclerotiorum*. *Applied and Environmental Microbiology, 58*, 578–583.
Ronicke, S., Hahn, V., Vogler, A., & Friedt, W. (2005). Quantitative trait loci analysis of resistance to *Sclerotinia sclerotiorum* in sunflower. *Phytopathology, 95*, 834–839.
Roper, M., Seminara, A., Bandi, M. M., Cobb, A., Dillard, H. R., & Pringle, A. (2010). Dispersal of fungal spores on a cooperatively generated wind. *Proceedings of the National Academy of Sciences of the United States of America, 107*, 17474–17479.
Sanderlin, R. S., & Ghabrial, S. A. (1978). Physicochemical properties of two distinct types of virus-like particles from *Helminthosporium victoriae*. *Virology, 87*, 142–151.
Sasaki, A., Kanematsu, S., Onoue, M., Oikawa, Y., Nakamura, H., & Yoshida, K. (2007). Artificial infection of *Rosellinia necatrix* with purified viral particles of a member of the genus Mycoreovirus reveals its uneven distribution in single colonies. *Phytopathology, 97*, 278–286.
Sasaki, A., Onoue, M., Kanematsu, S., Suzaki, K., Miyanishi, M., Suzuki, N., et al. (2002). Extending chestnut blight hypovirus host range within diaporthales by biolistic delivery of viral cDNA. *Molecular Plant-Microbe Interactions, 15*, 780–789.
Segers, G. C., Zhang, X., Deng, F., Sun, Q., & Nuss, D. L. (2007). Evidence that RNA silencing functions as an antiviral defense mechanism in fungi. *Proceedings of the National Academy of Sciences of the United States of America, 104*, 12902–12906.
Smart, C. D., Yuan, W., Foglia, R., Nuss, D. L., Fulbright, D. W., & Hillman, B. I. (1999). *Cryphonectria hypovirus 3*, a virus species in the family hypoviridae with a single open reading frame. *Virology, 265*, 66–73.
Steadman, J. R. (1979). Control of plant diseases caused by *Sclerotinia* species. *Phytopathology, 69*, 904–907.
Stielow, B., Klenk, H. P., Winter, S., & Menzel, W. (2011). A novel *Tuber aestivum* (Vittad.) mitovirues. *Archives of Virology, 156*, 1107–1110.
Sun, Q., Choi, G. H., & Nuss, D. L. (2009). A single Argonaute gene is required for induction of RNA silencing antiviral defense and promotes viral RNA recombination. *Proceedings of the National Academy of Sciences of the United States of America, 106*, 17927–17932.
Tang, J., Havens, W. M., Ochoa, W. F., Guu, T. S., Ghabrial, S. A., Nibert, M. L., et al. (2010). Backbone trace of partitivirus capsid protein from electron cryomicroscopy and homology modeling. *Biophysical Journal, 99*, 685–694.

Tang, J., Ochoa, W. F., Li, H., Havens, W. M., Nibert, M. L., Ghabrial, S. A., et al. (2010). Structure of Fusarium poae virus 1 shows conserved and variable elements of partitivirus capsids and evolutionary relationships to picobirnavirus. *Journal of Structural Biology*, *172*, 363–371.

van den Brand, J. M. A., van Leeuwen, M., Schapendonk, C. M., Simon, J. H., Haagmans, B. L., Osterhaus, A. D. M. E., et al. (2012). Metagenomic analysis of the viral flora of pine marten and european badger feces. *Journal of Virology*, *86*, 2360–2365.

Viaud, M., Balhadère, P., & Talbot, N. J. (2002). A *Magnaporthe grisea* cyclophilin acts as a virulence determinant during plant infection. *The Plant Cell*, *14*, 917–930.

Viaud, M., Brunet-Simon, A., Brygoo, Y., Pradier, J. M., & Levis, C. (2003). Cyclophilin A and calcineurin functions investigated by gene inactivation, cyclosporin A inhibition and cDNA arrays approaches in the phytopathogenic fungus *Botrytis cinerea*. *Molecular Microbiology*, *50*(5), 1451–1465.

Whipps, J. M., & Gerlagh, M. (1992). Biology of *Coniothyrium minitans* and its potential for use in disease biocontrol. *Mycological Research*, *96*, 897–907.

Williams, B., Kabbage, M., Kim, H. J., Britt, R., & Dickman, M. B. (2011). Tipping the balance: Sclerotinia sclerotiorum secreted oxalic acid suppresses host defenses by manipulating the host redox environment. *PLoS Pathogens*, 7, e1002107.

Wu, M., Zhang, L., Li, G., Jiang, D., & Ghabrial, S. A. (2010). Genome characterization of a debilitation-associated mitovirus infecting the phytopathogenic fungus *Botrytis cinerea*. *Virology*, *406*, 117–126.

Xie, J., & Ghabrial, S. A. (2012). Molecular characterizations of two mitoviruses co-infecting a hypovirulent isolate of the plant pathogenic fungus *Sclerotinia sclerotiorum*. *Virology*, *428*, 77–85.

Xie, J., Wei, D., Jiang, D., Fu, Y., Li, G., Ghabrial, S., et al. (2006). Characterization of debilitation-associated mycovirus infecting the plant-pathogenic fungus *Sclerotinia sclerotiorum*. *The Journal of General Virology*, *87*, 241–249.

Xie, J., Xiao, X., Fu, Y., Liu, H., Cheng, J., Ghabrial, S. A., et al. (2011). A novel mycovirus closely related to hypoviruses that infects the plant pathogenic fungus Sclerotinia sclerotiorum. *Virology*, *418*, 49–56.

Yaegashi, H., Kanematsu, S., & Ito, T. (2012). Molecular characterization of a new hypovirus infecting a phytopathogenic fungus, *Valsa ceratosperma*. *Virus Research*, *165*, 143–150.

Yu, X., Li, B., Fu, Y., Jiang, D., Ghabrial, S. A., Li, G., et al. (2010). A geminivirus-related DNA mycovirus that confers hypovirulence to a plant pathogenic fungus. *Proceedings of the National Academy of Sciences of the United States of America*, *107*, 8387–8392.

Zhang, L., Fu, Y., Xie, J., Jiang, D., Li, G., & Yi, X. (2009). A novel virus that infecting hypovirulent strain XG36-1 of plant fungal pathogen *Sclerotinia sclerotiorum*. *Virology Journal*, *6*, 96.

Zhang, X., & Nuss, D. L. (2008). A host dicer is required for defective viral RNA production and recombinant virus vector RNA instability for a positive sense RNA virus. *Proceedings of the National Academy of Sciences of the United States of America*, *105*, 16749–16754.

Zhao, J., & Meng, J. (2003). Genetic analysis of loci associated with partial resistance to *Sclerotinia sclerotiorum* in rapeseed (*Brassica napus* L.). *Theoretical and Applied Genetics*, *106*, 759–764.

Zhu, W., Wei, W., Fu, Y., Cheng, J., Xie, J., Li, G., et al. (2013). A secretory protein of necrotrophic fungus *Sclerotinia sclerotiorum* that suppresses host resistance. *PLoS ONE*, *8*(1), e53901.

CHAPTER NINE

Viruses of Botrytis

Michael N. Pearson*,[1] and Andrew M. Bailey†
*School of Biological Sciences, The University of Auckland, Auckland, New Zealand
†School of Biological Sciences, University of Bristol, Woodland Road, Bristol, BS8 1UG, United Kingdom
[1]Corresponding author: e-mail address: m.pearson@auckland.ac.nz

Contents

1. Introduction 250
   1.1 The importance of Botrytis as a pathogen 250
   1.2 Viruses of Botrytis 252
2. Molecular Characterization and Phylogenetic Relationships of Botrytis Viruses 254
   2.1 Methods used to detect viruses and their merits and limitations 254
   2.2 Characterized and partially characterized Botrytis viruses and their phylogenetic relationships to other fungal and plant viruses 255
3. Effects of Viruses on Botrytis 259
   3.1 Methods used to study the effects of mycoviruses and their merits and limitations 259
   3.2 Effects of viruses on Botrytis phenotypes 260
   3.3 Effects of viruses on fungal metabolism 261
   3.4 Insights from Botrytis genome data 262
4. Virus Transmission 264
   4.1 Botrytis life cycles and possible routes of virus transmission 264
   4.2 Distribution and incidence of Botrytis viruses 265
   4.3 Vegetative compatibility groups and virus transmission 266
5. Conclusions and Future Research 267
References 269

Abstract

Botrytis cinerea (gray mold) is one of the most widespread and destructive fungal diseases of horticultural crops. Propagation and dispersal is usually by asexual conidia but the sexual stage (*Botryotinia fuckeliana* (de Bary) Whetzel) also occurs in nature. DsRNAs, indicative of virus infection, are common in *B. cinerea*, but only four viruses (Botrytis virus F (BVF), Botrytis virus X (BVX), Botrytis cinerea mitovirus 1 (BcMV1), and Botrytis porri RNA virus) have been sequenced. BVF and BVX are unusual mycoviruses being ssRNA flexous rods and have been designated the type species of the genera Mycoflexivirus and Botrexvirus (family *Betaflexivirdae*), respectively. The reported effects of viruses on *Botrytis* range from negligible to severe, with Botrytis cinerea mitovirus 1 causing hypovirulence. Little is currently known about the effects of viruses on *Botrytis* metabolism but recent complete sequencing of the *B. cinerea* genome now provides an opportunity to investigate the

Advances in Virus Research, Volume 86
ISSN 0065-3527
http://dx.doi.org/10.1016/B978-0-12-394315-6.00009-X

host–pathogen interactions at the molecular level. There is interest in the possible use of mycoviruses as biological controls for *Botrytis* because of the common problem of fungicide resistance. Unfortunately, hyphal anastomosis is the only known mechanism of horizontal virus transmission and the large number of vegetative incompatibility groups in *Botrytis* is a potential constraint on the spread of an introduced virus. Although some *Botrytis* viruses, such as BVF and BVX, are known to have international distribution, there is a distinct lack of epidemiological data and the means of spread are unknown.

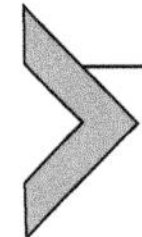

1. INTRODUCTION

1.1. The importance of Botrytis as a pathogen

Botrytis cinerea is one of the most readily recognized fungal pathogens of plants, being commonly observed in a domestic setting where it is often found causing gray mold on soft fruits such as strawberries (Elad, Williamson, Tudzynski, & Delen, 2004; Williamson, Tudzynski, Tudzynski, & van Kan, 2007). The most obvious symptom of infection is the production of abundant aerial hyphae, coupled with massive sporulation generating gray conidiospores, hence the common name of gray mold for this fungus. While it can often be a serious postharvest problem, the fungus is also a serious pathogen of growing plants; indeed, in 1968 the host range had already been recorded in excess of 235 different plant species (Macfarlane, 1968) and may now be larger still. As such, it is one of the most prevalent and widespread of the plant pathogenic fungi, being readily detectable in the majority of climates and across all habitable continents. Despite having been isolated from habitats ranging from arctic to deserts, it is primarily regarded as a pathogen associated with high humidity; hence it is common in tropical locations, in dense canopied crops in temperate regions, or within enclosed growth environments or other areas with protected cropping, such as under plastic.

In addition to growth on soft fruits, the fungus can infect many other plant tissues including petals, leaves, stems, and tubers. Infection is usually initiated by the asexual conidia landing on a suitable plant surface where, depending on the available moisture, the conidium germinates, and the germ tube either penetrates the epidermis directly, or grows into wounds or other natural openings. *B. cinerea* can produce a wide range of cell wall-degrading enzymes to assist in breaching the physical barriers imposed by plant cell walls (Urbanek & Zalewska-Sobczak, 2003), and can also produce a range of different toxins that may help establish disease by reducing the ability of the plant to fight off the infection (Choquer et al., 2007). The fungus proliferates through the dead or dying plant tissues and then undergoes sporulation, allowing transmission to other hosts.

There have been several detailed population analyses of *B. cinerea* (e.g., Staats, van Baarlen, & van Kan, 2004). They show that rather than being a single species, *B. cinerea* encompasses at least two similar but genetically isolated groups of fungi, separable by physical characteristics such as spore size and genetically based on various gene markers and repetitive elements. There are also many other species of Botrytis in addition to *B. cinerea* (Staats et al., 2004). The majority of them are host-specific, infecting a very limited range of plants, and tend to be found primarily on the monocotyledonous plants, although there are host-specific species to be found on several species of legume. These species of Botrytis have received less attention than *B. cinerea* and so, *B. cinerea* will be the main focus of this review.

It is the asexual form of the fungus that is most commonly encountered; abundant aerial mycelia emerge from fruit or other plant tissues, producing prodigious numbers of gray multinucleate conidia, while the sexual stage of the fungus, more correctly called *Botryotina fuckeliana* (de Bary) Whetzel, is rarely observed in the field. If the climatic conditions are not conducive to survival of the asexual spores or their ready infection of plant material (e.g., low temperature or low humidity), *B. cinerea* can also generate sclerotial tissues. These are multicellular tissues of 1–2 mm × 2–5 mm, which comprise matted hyphae having thickened and heavily melanized cell walls. They facilitate the survival of the fungus, either through conditions of drought/dehydration, or through freezing, thereby allowing the fungus to survive over periods that preclude successful transmission or in the absence of any suitable host stages in the field. When favorable growth conditions are restored, the sclerotia give rise to conidiophores bearing asexual conidiospores to resume the infection cycle.

The sclerotia also have a role in the sexual stage of the fungus. Under appropriate environmental conditions, the sclerotia can produce receptive apothecial hyphae, acting as the "female" partner for sexual reproduction. *B. cinerea* also produces uninucleate microconidia, which act as spermatia for the fungus. *B. cinerea* is heterothallic, having two different mating types; should a sexually compatible microconidium make contact with the apothecial hyphae, fertilization will take place allowing sexual development. This results in the production of a cup-shaped fruiting body or apothecium, with the inner surface lined with numerous asci, each containing eight uninucleate ascospores derived from meiotic divisions of the transient diploid stage of the ascus. The ascospores are forcibly ejected into the air, which, like conidia, are capable of establishing infection in susceptible plant tissues.

When infection occurs, it does not always immediately result in the production of a spreading lesion. Depending on the hosts' physiology, the fungus can enter a latent phase, giving rise to a symptomless infection. Only when the physiology of the plant material becomes suitable for growth does the *B. cinerea* start to grow and proliferate through the tissues. This can be a problem when trying to control *B. cinerea* on certain crops, for instance, infection of grapes and strawberries often occurs during flowering; however, after invading the plant tissues, the fungus may remain latent until the fruit starts to ripen. At this point, *B. cinerea* resumes its necrotrophic growth throughout the fruit giving rise to severe symptoms. Control of fruit infection therefore has to target the flowering period of the crop.

Control of infection can be achieved in a number of different ways. With the exception of tomato, there are few crops where cultivars are available that display increased resistance to *B. cinerea*, therefore other cultural control methods are often applied. Good crop hygiene including efforts to minimize the sources of infection for *B. cinerea* may be productive; hence, removal of alternative hosts, coupled with removal and incineration of crop debris, fallen fruit, or prunings from vines, is often recommended (Jaspers, Seyb, Trought, & Balasubramaniam, 2012). *B. cinerea* is primarily a problem during periods of high humidity; therefore, modification of the environment to reduce humidity can help to control the disease. This can be achieved by ensuring good ventilation of protected crops, and in crops with high canopy densities such as grapes, thinning of the canopy to increase airflow has often been recommended to reduce the severity of infection.

There are also a number of chemical control measures that are effective in control of *B. cinerea*. Several agricultural fungicides have proven effective in controlling infection; however, as with many fungicides, there are concerns that resistance may increase leading to loss of control, and given that many of the crops are consumed without further processing, there are often concerns about fungicide residues within the fruits, limiting the periods during which chemical controls can be applied to the crop. This means that there are still demands for improved crop protection, and interest in the feasibility of biological control measures is ongoing. Given the ability of mycoviruses to cause hypovirulence in other fungi, this area has also been explored in *B. cinerea*.

1.2. Viruses of Botrytis

Several authors have reported virus-like particles and dsRNAs in *B. cinerea*, only a few of which have been further studied to determine their sequence or association with virus particles (Table 9.1). Howitt, Beever, Pearson, and Forster (1995) detected dsRNAs in 143 of 200 New Zealand *B. cinerea* isolates from a range of host plants, including grape, kiwifruit, tomato,

Table 9.1 Virus particles and dsRNAs detected in *Botrytis cinerea*

Genus	Species	Particle morphology	Genome	References
Unassigned	Unassigned	Isometric 30–45 nm	dsRNA[a]	Howitt et al. (1995)
Unassigned	Unassigned	Bacilliform, 25×63 nm	dsRNA[a]	Howitt et al. (1995)
Unassigned	Unassigned	Unencapsidated (various)	dsRNA	Howitt et al. (1995)
Unassigned	Unassigned	Isometric ~40 nm	dsRNA, 8.3 kb	Vilches and Castillo (1997)
Unassigned	Unassigned	Unencapsidated	dsRNA, 2.0 kb	Vilches and Castillo (1997)
Unassigned	Unassigned	Unencapsidated	dsRNA, 1.4 kb	Vilches and Castillo (1997)
Unassigned	Unassigned	Isometric, 28 nm	dsRNA, 1.8 kb	Castro et al. (1999)
Mycoflexivirus	*Botrytis virus F*	Flexuous, ~720 nm	+ssRNA, ~6.8 kb	Howitt et al. (2001)
Unassigned	Unassigned	Isometric, 33 nm	dsRNA, 6.8 kb	Castro et al. (2003)
Botrexvirus	*Botrytis virus X*	Flexuous, ~720 nm	+ssRNA, ~6.9 kb	Howitt et al. (2006)
Mitovirus	*Botrytis cinerea* mitovirus 1[b]	Unencapsidated	dsRNA, 3.0 kb	Wu et al. (2007, 2010)
Unassigned	*Botrytis porri* RNA virus 1[b]	Isometric ~35 nm	dsRNAs, $6.2 + 5.9$ kb	Wu et al. (2012)

[a]Particles associated with dsRNA, genome not confirmed.
[b]Strain of *Ophiostoma novo-ulmimitovirus 3b*.

cucumber, French bean, blackberry, and strawberry, and from various geographic locations. The dsRNA profiles varied in the number of dsRNA species (1–8), their size (800–15000 bp), and relative concentration. Some, but not all, were associated with the presence of isometric and/or bacilliform particles. Similarly, Vilches and Castillo (1997) detected three dsRNA species in a single *B. cinerea* isolate, only one of which copurified with isometric virus particles also found in the isolate. Castro, Kramer, Valdivia, Ortiz, and Castillo (1999, 2003) detected a single dsRNA species associated with both 28 and 33 nm isometric particles and (Wu, Zhang, Li, Jiang, & Ghabrial, 2010; Wu et al., 2007) reported unencapsidated dsRNAs, which they identified as members of the genus Mitovirus. Howitt, Beever, Pearson, and Forster (2001, 2006) detected two ssRNA filamentous viruses, Botrytis virus F (BVF) and Botrytis virus X (BVX), belonging to the family *Flexiviridae* and Wu et al. (2012) describe a novel dsRNA isometric virus from *Botrytis porri*. These viruses are discussed in greater detail in the following section.

Although many of the viruses detected in Botrytis appear to have little or no obvious effect on their host (Howitt et al., 1995), hypovirulence or other adverse phenotypic effects have been reported by several authors (Castro et al., 1999, 2003; Wu et al., 2007, 2010, 2012). Given the importance of Botrytis species, especially *B. cinerea*, as pathogens of a wide range of horticultural crops and the problems of fungicide resistance and residues, the possibility of using mycoviruses as a biological control is appealing. Consequently, there is substantial interest in the study of Botrytis viruses and the question of whether there are circumstances where these may be used as a viable control measure. However, even though hypovirulent viruses have been identified in Botrytis, if these are to be used as a biological control agent there is still the critical issue of virus spread from donor isolates to the target population. Given that *B. cinerea* populations contain at least 66 compatibility groups (Beever & Weeds, 2004), this potentially presents a major obstacle to the successful use of mycoviruses to control this fungus.

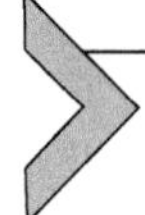

2. MOLECULAR CHARACTERIZATION AND PHYLOGENETIC RELATIONSHIPS OF BOTRYTIS VIRUSES

2.1. Methods used to detect viruses and their merits and limitations

In common with many other mycoviruses, most of the Botrytis viruses were originally detected by screening fungal isolates for dsRNAs using methods based on that originally described by Morris and Dodds (1979). This method has proven to be extremely useful for detecting new RNA mycoviruses as it is robust and does not require any prior knowledge of the viruses that are present. However, it does not detect all viruses and consequently, our knowledge of mycoviruses has historically been biased against DNA viruses and ssRNA viruses that do not produce high concentrations of dsRNA. This has been readdressed to some extent in recent years and approximately one-third of the known mycoviruses have ssRNA genomes (Pearson, Beever, Boine, & Arthur, 2009). For example, the discovery of the ssRNA viruses BVF and BVX was highly fortuitous, as they were detected in a resuspended pellet following ultracentrifugation of a Botrytis extract used as a dsRNA negative control (Howitt et al., 1995). If only samples containing dsRNAs had been examined, these two viruses would not have been found. Modern high-throughput sequencing (454, illumina, etc.) should help detect a wider range of mycoviruses and reduce this bias. In addition, studies have tended to focus on isolates showing reduced growth or unusual phenotypes in culture, although it is known that many isolates exhibiting normal growth contain dsRNAs. However, it is worth noting that strains with debilitated or reduced virulence will be harder to spot in the field and may be more difficult to culture in the lab.

2.2. Characterized and partially characterized Botrytis viruses and their phylogenetic relationships to other fungal and plant viruses

The reports of viruses in Botrytis are mostly based on the presence of dsRNAs and in many instances it is not known whether these dsRNAs are associated with virus particles. Where particles have been observed, they are predominantly isometric (~30, 35, 40, 45 nm), although bacilliform particles (25 nm × 63 nm) and flexuous filaments have also been observed (Table 9.1). For example, Howitt et al. (1995) detected a wide range of dsRNA profiles in 143 different Botrytis isolates but only five of these were further investigated. Following chloroform extraction and differential centrifugation, isometric and/or bacilliform particles were found in three of the five samples. If the other 138 dsRNA containing samples had been subjected to the same analysis it seems likely that many of these would also have been found to contain virus particles. Of the viruses detected in Botrytis, only five have been fully sequenced: the two flexuous ssRNA viruses BVF (Howitt et al., 2001) and BVX (Howitt et al., 2006), the mitoviruses Botrytis cinerea mitovirus 1 (BcMV1) and Botrytis cinerea mitovirus 1-S (BcMV1-S) described by Wu et al. (2007), Wu et al. (2010), and the novel dsRNA isometric virus reported by Wu et al. (2012).

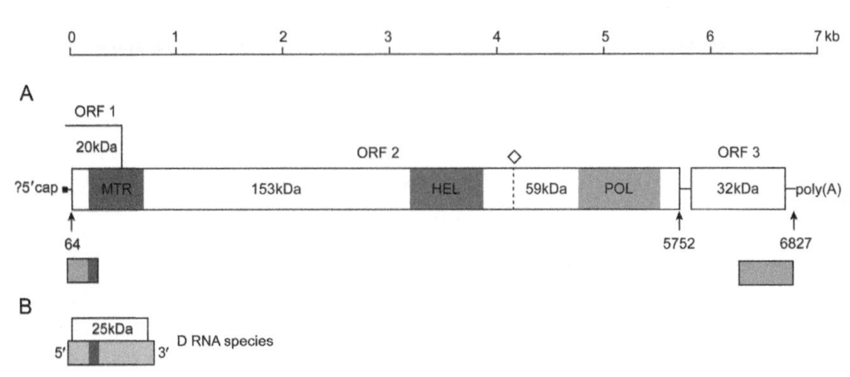

Figure 9.1 The genome organization of BVF (redrawn from Howitt et al., 2001). (A) MTR, methyltransferase; HEL, helicase; POL, RNA-dependent RNA polymerase. The position of the putative read-through codon is indicated with a diamond, the lower shaded boxes at the 5′- and 3′-termini represent regions found in a putative defective RNA (DRNA) species. (B) DRNA species that has an ORF with the potential to encode a replicase-CP fusion protein. (See Page 20 in Color Section at the back of the book.)

BVF (Fig. 9.1) was the first fully sequenced virus from *B. cinerea* and the first flexuous rod-shaped mycovirus to be characterized (Howitt et al., 2001). It has an ssRNA genome of 6827 nucleotides (nts) with a poly(A) tract at the 3′ end. Sequence analysis revealed two potential open reading frames (ORFs) encoding proteins of 212 kDa (ORF1) and 32 kDa (ORF2). ORF1 shows significant sequence identity to the RNA-dependent RNA polymerase (RdRp) of plant "tymo-" and "potex-like" viruses, but with an opal putative readthrough codon between the helicase and RdRp regions. ORF2 shared amino acid similarity with coat proteins of plant "potex-like" viruses. Untranslated regions were found preceding the initiation codon of ORF1 (63 nts), between ORFs 1 and 2 (93 nts), and preceding the 3′ poly(A) tract (70 nts). BVF was subsequently accepted as the type species of a new genus, Mycoflexivirus, in the family *Gammaflexiviridae* (King, Adams, Carstens, & Lefkowitz, 2011).

BVX (Fig. 9.2) was the second fully sequenced virus from *B. cinerea* and also the second flexuous rod-shaped mycovirus to be characterized (Howitt et al., 2006). It has an ssRNA genome of 6966 nts with a poly(A) tract at the 3′ end and sequence analysis revealed five potential ORFs. ORF1 showed 73% identity to the RdRp region of the Allexivirus, Garlic virus A (GarV-A), and the C-terminal region of ORF3 showed amino acid similarity with "potex-like" coat proteins, but ORFs 2, 4, and 5 did not show any significant sequence similarity with known protein sequences. BVX was

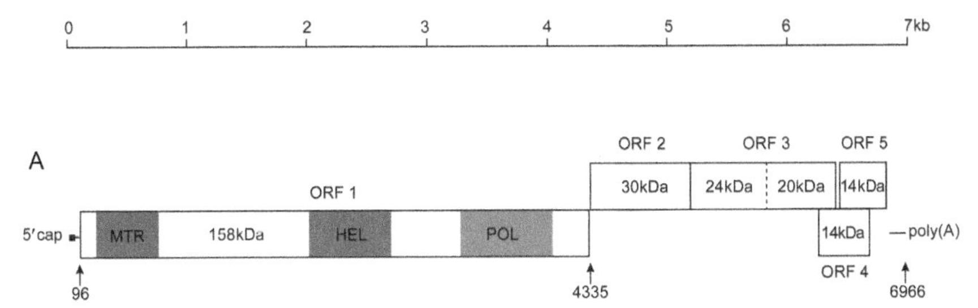

Figure 9.2 The genome organization of BVX (redrawn from Howitt et al., 2006). MTR, methyltransferase; HEL, helicase; POL, RNA-dependent RNA polymerase. The dashed line in ORF3 indicates a putative initiation codon for coat protein synthesis. (For the color version of this figure, the reader is referred to the online version of this chapter.)

subsequently accepted as the type species of a new genus, *Botrexvirus*, in the family *Alphaflexivirdae* (King et al., 2011).

BcMV1 was first described by Wu et al. (2007) as *B. cinerea* debilitation-related virus. It consists of an unencapsidated ~3.0 kb dsRNA, which, based on analysis of a 920 bp cDNA, is closely related to Ophiostoma mitovirus 3b. Following further characterization Wu et al. (2010) provisionally renamed it Botrytis cinerea mitovirus 1 and also identified a variant named Botrytis cinerea mitovirus 1-S. Sequencing revealed that BcMV1 RNA is 2804-nt long and AU-rich (66.8%) and that nts 447–2778 (GenBank acc. no. EF580100) are 95% identical to the full-length genome sequence of OnuMV3b (GenBank acc. no. AM087550), confirming that BcMV1 can be considered a strain of OnuMV3b. BcMV1-S (GenBank acc. no. EF583556) is 633 nt shorter than BcMV1 and relatively AU-rich (66.7%); it showed 100% identity with nts 1–1144 and 1778–2804 of BcMV1. Consequently, the authors concluded that "BcMV1-S is derived from BcMV1 through a single deletion of a 633-nt region (1145–1777)." Sequence analysis of the positive strand of BcMV1 identified one major large ORF, the putative protein sequence from amino acids 1 to 738 being 96% identical to the RdRp of OnuMV3b (GenBank acc. no. CAJ32468), including a conserved motif considered characteristic of the genus *Mitovirus*. A smaller ORF, upstream of the large ORF, which potentially codes for 44 amino acid residues, showed no significant identity to known proteins or polypeptides. However, the authors note that "small ORFs were detected in the 5′-UTR of the genomes of Cryphonectria parasitica mitovirus 1-NB631 and Sclerotinia homoeocarpa mitovirus" and that "small ORFs coding for polypeptides with unknown functions in the upstream region of a major large ORF had been previously been reported in other mycoviruses, including Helminthosporium victoriae 190S virus (Huang & Ghabrial, 1996) and Rosellinia necatrix megabirnavirus 1 (Chiba et al., 2009)." The regions comprising the full length of the two ORFs were also found on the positive strand of BcMV1-S.

Botrytis porri RNA virus 1 (Wu et al., 2012) consists of two dsRNAs of 6,215 bp (dsRNA-1: GenBank acc. no. JF716350) and 5879 bp (dsRNA-2: GenBank acc. no. JF716351), which show 62% and 95% nt sequence identities at the 3′-termini (80 bp) and the 5′-termini (500 bp), respectively. The extreme terminal sequences are strictly conserved and stem-loop structures are present at the 3′- and 5′-terminal regions of dsRNA-1 and the 5′-terminal region of dsRNA-2. The positive strand of each dsRNA contains a single ORF, designated as ORF I (dsRNA-1, 404–6112) and ORF II (dsRNA-2, 405–5771). ORF I encodes a 1902 amino acid polypeptide with a proline-rich region at nt position at 1067–1196 and an RdRp_4 superfamily domain at nt positions 4625–5455. The RdRp sequence includes eight conserved motifs also found in the *Totiviridae*, *Chrysoviridae*, and *Megabirnaviridae* families, although the overall similarity of the RdRp sequence to the RdRps of these families is low (<25%). No other proteins homologous to the remaining part of ORF I-encoded polypeptide were identified. ORF II encodes a 1788 amino acid polypeptide with numerous phosphorylation/glycosylation sites. No conserved proteins or proteases homologous to the polypeptide were found. Peptide analysis indicated that ORF II codes for the p70 protein detected in the purified virus particles of BpRV1 while ORF I codes for the p80 and p85 proteins; the authors concluded that the latter is derived from p85 via posttranslational modification. Analysis of BpRV1 RdRp sequences found no significant homology with the known bipartite dsRNA viral families *Birnaviridae*, *Partitiviridae*, and *Picobirnaviridae*. Consequently, it appears to be a novel virus clustering with two unassigned insect viruses, Spissistilus festinus virus 1 and Circulifer tenellus virus 1. Based on conserved motifs it appears to be most closely related to members of the genera *Totivirus* and *Victorivirus* (family *Totiviridae*).

Viruses that have been partially characterized, but for which no sequence is available, include 40 nm isometric virus particles that copurified with a 8.3-kb dsRNA (Vilches & Castillo, 1997). Electron microscopy of thin sections observed particles in the cytoplasm but not in mitochondria or nuclei. Based on the genome size and particle morphology it was possibly a totivirus. They also detected 2.0 and 1.4 kb dsRNAs, but these were not associated with the particles. Castro et al. (1999) reported a 1.8-kb dsRNA associated with 28 nm particles and while the particles are similar in size to partitiviruses, other partitiviruses have two linear segments of dsRNA. Castro et al. (2003) reported a 6.8-kb dsRNA associated with 33 nm particles, which, based on the genome size and particle morphology, was possibly a totivirus.

3. EFFECTS OF VIRUSES ON BOTRYTIS

3.1. Methods used to study the effects of mycoviruses and their merits and limitations

Experimentally determining the effects of viruses on fungal growth and virulence can be problematic. The growth rates and pathogenicity of different *B. cinerea* isolates can be very variable (Choquer et al., 2007) and different isolates have been shown to react quite differently to virus infection (Boine, 2012). The difficulty in comparing isolates of different genetic backgrounds is illustrated by the results of Howitt et al. (1995) who compared the *in vitro* growth rates, conidial and sclerotial production, and virulence on *Proteus vulgaris* leaves of 12 naturally dsRNA-infected isolates of *B. cinerea* (dsRNA+) with 12 dsRNA-free isolates (dsRNA−). Although, as a group, the dsRNA+ isolates showed statistically significant differences from the dsRNA− group, the differences were small and the ranges overlapped. Consequently, to get meaningful data, it is necessary to compare virus-infected cultures with genetically identical virus-free isolates.

Several methods have been used to produce genetically identical virus-infected and virus-free fungal cultures. These include the selection of naturally infected and uninfected single asexual spores (Wu et al., 2007) and "curing" isolates of viral infection using chemical agents such as cycloheximide (Urayama et al., 2010). The first has the merit that there is no treatment of the fungus that might inadvertently affect its' phenotype, but both methods are limited to the study of naturally infected fungal isolates. The use of cured isolates is also dependent on the availability of highly sensitive and reliable tests for the presence of the eliminated virus. To study the effects of viruses on a wide range of Botrytis isolates, it is necessary to virus-infect previously virus-free isolates. The natural means of horizontal transmission is by hyphal fusion, but since *B. cinerea* has at least 66 compatibility groups (Beever & Weeds, 2004), this is potentially a major obstacle to experimental transmission. For experimental purposes, alternatives to natural transmission include protoplast fusion between donor and recipient lines (Lee, Yu, Son, Lee, & Kim, 2011), *in vitro* transfection with purified virus particles (Boine, 2012; Castro et al., 2003; Kanematsu, Sasaki, Onoue, Oikawa, & Ito, 2010;

Sasaki, Kanematsu, Onoue, Oyama, & Yoshida, 2006), and the use of infectious transcripts derived from full-length viral cDNA clones (Choi & Nuss, 1992; Moleleki, van Heerden, Wingfield, Wingfield, & Preisig, 2003).

3.2. Effects of viruses on Botrytis phenotypes

The reported effects of mycoviruses on Botrytis range from negligible to severe. Castro et al. (1999) did not comment on changes in gross morphology or growth rates but reported that virus-containing cells showed cellular degeneration and loss of organelles. Castro et al. (2003) found that sporulation and laccase activity of dsRNA+ strains was approximately 50% less than that of dsRNA− isolates and that dsRNA− isolates were more aggressive than dsRNA+ isolates in detached bean leaf assays, although no quantitative data are given for the latter. Transfection of sphaeroplasts from dsRNA-free *B. cinerea* isolates with virus particles produced similar results.

Wu et al. (2007, 2010) described "hypovirulence" and "debilitation" of a mitovirus-infected *B. cinerea* isolate when compared to 20 noninfected isolates. In a pathogenicity test on detached leaves of oil seed rape (*Brasica napus*) virus-free isolates produced large necrotic lesions within 72 h whereas the hypovirulent strain produced no lesions. The hypovirulent strain also showed abnormalities in <79% of mitochondria, consisting of swelling, aggregates of fibrous matrix materials, and remnants of degenerated cristae. *In vitro* growth of the hypovirulent isolate was approximately one-third that of virulent isolates, conidial production was sparse, and no sclerotia were produced within 15 days.

As previously mentioned, Howitt et al. (1995) observed small but statistically significant differences in the mean *in vitro* growth rates, conidial and sclerotial production, and virulence on *P. vulgaris* leaves of 12 naturally dsRNA-infected isolates of *B. cinerea* compared with 12 dsRNA-free isolates. Subsequent evaluation of the effects of BVX (Boine, 2012; Pearson, Beever, Boine, & Tan, 2009) found that some naturally BVX-infected *B. cinerea* isolates showed a small, but statistically significant, increase in growth rates on malt extract agar plates compared to BVX-free isolates. In contrast, in an apple assay, BVX-infected isolates grew slightly slower (i.e., were less virulent) than virus-free isolates. However, when Boine (2012) transfected *B. cinerea* isolate B05-10 with BVX and/or BVF, some of the subsequent single spore isolates exhibited substantial effects on the gross morphology, virulence in a bean leaf assay (Fig. 9.3) and metabolism (see Section 3.3).

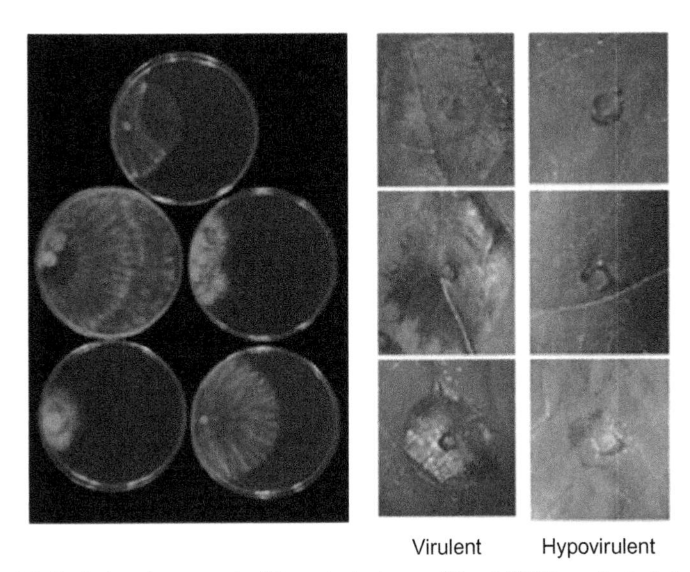

Figure 9.3 Variation in growth (A) and virulence (B) of BVF-transfected *B. cinerea* BO5-10. *From Boine (2012).* (For the color version of this figure, the reader is referred to the online version of this chapter.)

3.3. Effects of viruses on fungal metabolism

Castro et al. (2003) found that the laccase activity of dsRNA+ strains was approximately 50% less than that of dsRNA− isolates, but other than this, there has been little published on the viral effects on metabolism. Boine (2012) carried out a preliminary metabolomics study on BVX- and BVF-transfected and wild-type *B. cinerea* B05-10, grown on Vogel's media agar with cellophane overlay for 10 days at 20 °C. Analysis by "Pathway Activity Profiling" (Aggio, Ruggiero, & Villas-Boas, 2009) revealed significant changes in amino acid, carbohydrate, and fatty acid metabolism. These included decreased levels of the amino acids threonine and glycine and elevated levels of homocysteine, norvaline, and methionine. A similar trend was observed in CHV1-infected *Cryphonectria parasitica*, where amino acids including homocysteine and methionine were increased (Dawe, Van Voorhies, Lau, Ulanov, & Li, 2009). Several lipids, including docosanoic and eichosanoic acid, were also elevated in some BVX/BVF-infected isolates which was also comparable with results from *C. parasitica* infected with CHV1 (Dawe et al., 2009) and which was considered consistent with a disruption of lipid catabolism and biosynthesis and the altered metabolic pathways operating in virus-infected mycelium. Another notable result was that the apparent absence, in some transfected lines, of oxalic acid which is known to play a key role in the pathogenicity of many fungi.

3.4. Insights from Botrytis genome data

Two European isolates of *B. cinerea* have been sequenced and made publically available, strain T4 from a tomato was sequenced to 10× coverage (Amselem et al., 2011) by the Botrytis and Sclerotinia genome project (http://urgi.versailles.inra.fr/Species/Botrytis) and strain B05-10 from a grape vine to 4× coverage by Syngenta (http://www.broadinstitute.org/annotation/genome/botrytis_cinerea). The fungus has a genome of approximately 39.5 Mb, with a predicted gene content of 16,360 and 16,448 for T4 and B05-10, respectively, and with 99.5% sequence identity between the two genomes. The ready availability of this genome data has been of paramount importance in the identification and subsequent analysis of numerous pathways from *B. cinerea*.

Fungi, like most eukaryotes have been shown to have a defensive pathway that can identify and degrade RNA molecules in a sequence-specific manner, including viral RNA molecules. This is known by several names having been independently identified in different organisms, but the most commonly used terms are posttranscriptional gene silencing, RNA silencing, or RNA interference (Cogoni, 2001; Nakayashiki, 2005). The degradation pathway can be triggered in a number of ways, but a common theme is the presence (even transiently) of a dsRNA molecule, something not usually found in the cell. The dsRNA molecule is recognized and cleaved into short fragments (siRNA) by a type III RNase called dicer. Some of the short RNA fragments are then recruited to an enzyme complex called RISC, the RNA-induced silencing complex, where they act as a template and are used to identify other complementary RNA molecules, targeting these for degradation and thus moving from dsRNA degradation to degradation of similar single-stranded molecules. The catalytic portion of RISC is provided by Argonaute proteins that have endonuclease activity and digest the complementary RNA into short pieces. The dsRNA that triggers the silencing pathway can be formed in a number of ways; in normal cell function, aberrant overaccumulation of a particular native mRNA can cause a complementary strand to be synthesized by RdRp. In some organisms, expression of a native natural antisense transcript may hybridize to its cognate mRNA, thereby targeting it for degradation. Particular genes may be targeted in research by deliberate expression of an antisense

RNA, or alternatively by introduction of a self-complementary RNA molecule, which will spontaneously fold to give a dsRNA stem–loop structure.

While some fungi, such as *Ustilago maydis*, lack key genes of the silencing pathway and are unable to perform posttranscriptional gene silencing (Kamper et al., 2006), the majority of fungi have been found to be able to perform gene silencing. *B. cinerea* has been shown to be competent for RNA silencing (Patel, van Kan, Bailey, & Foster, 2008; Patel, van Kan, Bailey, & Foster, 2010), and analysis of the genome has identified all the genes required for effective RNA silencing, including two dicers, two RdRps, and four argonaute-like proteins. Among the fungi for which genome sequence data are available, the majority analyzed to date have two dicer genes. In species such as *Magnaporthe grisea*, *Aspergillus nidulans*, and *C. parasitica*, disruption of the dicer 2 gene leads to impairments in gene silencing (Hammond, Andrewski, Roossinck, & Keller, 2008; Kadotani, Nakayashiki, Tosa, & Mayama, 2004; Segers, Zhang, Deng, Sun, & Nuss, 2007). In *Neurospora crassa*, however, disruption of dicer 2 did not lead to loss of gene silencing; however, when both dicer 1 and dicer 2 were mutated, the fungus became impaired in gene silencing (Catalanotto et al., 2004). We recently knocked out the dicer 2 from *B. cinerea* using targeted gene replacement and found that such a strain was still fully competent to perform silencing (Tauati, 2011; Tauati et al., in preparation). It remains to be seen whether it requires the elimination of both dicers to stop the RNA silencing pathway from functioning.

RNA viruses will either have a dsRNA genome, or go through a dsRNA stage during replication, and it is therefore possible that these may be recognized and targeted for destruction by the hosts RNA silencing machinery. In the case of *C. parasitica*, impaired silencing can lead to higher titres of mycovirus (Segers, van Wezel, Zhang, Hong, & Nuss, 2006), showing that the gene silencing machinery is a controlling factor in regulating viral titre. We would speculate that *B. cinerea* lacking dicer function following disruption of both dicer genes would similarly be impaired in its ability to regulate viral titre. Given that fungi have mechanisms to identify and degrade dsRNA, the paradox remains as to how RNA-genome viruses can persist. It has commonly been observed that plant viruses are able to block gene silencing systems, with several different viral proteins performing the role of silencing-suppressors depending on the type of virus concerned, and similar findings have been reported for other mycoviruses (Hammond et al., 2008; Segers et al., 2006). It is unclear whether mycoviruses of Botrytis are similarly able to suppress RNA silencing of their host, but it is likely that some mechanism may be present to allow dsRNA of the virus to persist in a host that has a posttranscriptional gene silencing system.

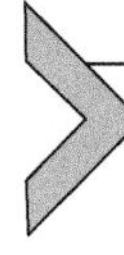

4. VIRUS TRANSMISSION

4.1. Botrytis life cycles and possible routes of virus transmission

B. cinerea has both a sexual and asexual life cycle and it is possible that the viral infection could be propagated by either of these routes. In the asexual cycle, a vegetative mycelium gives rise to numerous multinucleate conidia. It would be expected that these spores would be infected with virus, whether the virus was predominantly within the nucleus, mitochondria or indeed free within the cytoplasm. Dispersal of these spores will give rise to more clonal colonies, thereby propagating the virus; this is known as vertical transmission, the spread of the virus to a new clonal generation of the fungus. While this keeps the virus within the same genetic background of the fungus and aids in viral dispersal, if the viral infection is detrimental to the conidial yield of the fungus, this would have serious selection consequences on the long-term viability of the infected line of the fungus, particularly if in competition with virus-free isolates.

Reciprocal sexual crosses between BVX-free and BVX-infected *B. cinerea* isolates (Colin Tan, unpublished) resulted in BVX transmission to 34% of ascospores when the BVX-infected isolate was the "male" parent and to 53% when it was the "female" parent (Pearson, Beever, Boine, Tan, 2009). No information is presented for transmission through ascospores for the other viruses listed in Table 9.1, but it seems reasonable to assume that at least some of these are also transmitted through ascospores. Sexual reproductive structures of *B. cinerea* are rarely found in nature and consequently virus spread through this route is probably minimal. However, an important aspect of sexual reproduction is that it generates progeny belonging to different compatibility groups, thus increasing the opportunity for the virus to spread more widely within the Botrytis population via hyphal anastomosis.

The wide host range for *B. cinerea* and its ability to grow in necrotic tissues makes it very likely that *B. cinerea* will occur in mixed infections with other fungi, either other individuals of the same species or different species. While this raises the possibility of transmission of viruses during such mixed infections, only a few mycoviruses have been shown to infect more than one fungal species and then using only *in vitro* transfection with infectious cDNA clones (Sasaki et al., 2002), purified virus particles (Kanematsu et al., 2010) or protoplast fusion (Lee et al., 2011). While anastomosis has been observed between different species, at least for *Colletotrichum* sp. (Roca et al., 2004), cross species transmission in nature seems highly unlikely if it requires hyphal fusion.

4.2. Distribution and incidence of Botrytis viruses

The publications listed in Table 9.1 do not provide any extensive data on the incidence or distribution of the Botrytis mycoviruses described. However, testing of international *B. cinerea* isolates in culture collections in New Zealand has provided some data on the incidence and distribution on BVX and BVF (Arthur, 2007; Tan, unpublished). BVF and BVX were first identified from a single isolate of *B. cinerea* collected from field grown strawberry in Auckland, New Zealand in 1994 (Howitt et al., 2001; Howitt et al., 2006). Subsequent testing of 87 *B. cinerea* isolates from New Zealand and overseas, from a range of different host plants, detected an overall incidence of 18.4% for BVF and 28.7% for BVX (Table 9.2), the latter being significantly influenced by the overseas samples with 55.6% (15/27) compared to 16.7% (11/60) for NZ isolates. The incidence of BVF was more consistent being 20% (12/60) for NZ and 14.8% (4/27) for overseas isolates. The viruses were detected in *B. cinerea* isolates from a range of hosts, including lettuce, cucumber, grapevine, tomato, and strawberry. Both viruses were detected in isolates from England, France, and Israel, sometimes as a coinfection, and BVX was also present in isolates from Belgium, Greece, Italy, Portugal, Switzerland, and USA. The sequences for all BVX isolates were

Table 9.2 Overall incidence of BVX and BVF in *B. cinerea* isolates

Survey	Number of isolates tested	BVX positive	BVF positive
Misc NZ ICMP	27	8	1
Overseas	29	15	4
Grape (NZ)	17	4	7
Strawberry (Akld)	20	0	3
		% incidence	% incidence
NZ incidence	64	18.8	17.2
Overseas incidence	29	51.7	13.8
Overall incidence	93	29.0	16.1

Data from Arthur (2007) and Tan (unpublished).

essentially identical, based on a ~570-bp region of the putative coat protein (nt positions 5831–6400) (Arthur, 2007) and a ~700-nt of partial 5′ replicase and the 3′ end of the genome (partial ORF4 and ORF5 and 3′UTR) (Tan, unpublished). In contrast, sequences for BVF (untranslated region between ORFs 1 and 2; nts 5681–5990) fell into three distinct clades with 15–24% nt difference between clades, but less than 6% variance within clades (Arthur, 2007). New Zealand isolates were found in all three clades. In addition, Al Rwahnih, Daubert, Urbez-Torres, Cordero, and Rowhani (2011) detected BVX-related sequences in a grapevine sample in California, the USA, by 454 sequencing of dsRNA.

The detection of both BVF and BVX in samples of wide geographic origin is circumstantial evidence that they are able to spread in natural Botrytis populations, the distinct clades of BVF sequences suggesting that BVF has had a long association with *B. cinerea* with divergence occurring in different geographic locations. The apparent contradiction of members of all clades being present in NZ may be the result of multiple introductions of the fungus into NZ along with plant hosts during the relatively short period since European settlement, since *B. cinerea* has successfully colonized wherever its hosts are grown (Elad et al., 2004). For BVX, the sequences obtained from both the coat protein and methyltransferase and the helicase regions show high levels of conservation, but PCR using primers designed to other regions (based on BVX sequence AY055762) failed to amplify the genome, which suggests that there is greater variation than is apparent from the sequences obtained so far (Tan, unpublished). It is unlikely that these viruses are recently derived from plant viruses given that they have no equivalent of the movement protein found in plant flexiviruses (Howitt et al., 2001, 2006).

4.3. Vegetative compatibility groups and virus transmission

In a growing fungal colony, it is common for separate hyphae to be able to fuse together, a process known as anastomosis (Read, Fleissner, Roca, & Glass, 2010). This is thought to allow the organism to be able to improve the routing of nutrient around the colony, to be able to alter its hyphal network in response to damage or physical obstructions, and to enable some hyphae to be removed by autodigestion. The process of hyphal fusion is tightly regulated so that fusion between hyphae of the same colony or of the same clonal type will occur, while preventing fusion with other unrelated individuals. This is under genetic control in a process known as vegetative or heterokaryon incompatibility. The process has been best characterized in fungi such as *N. crassa*, using readily obtainable auxotrophic mutants to assess heterokaryon formation. By this manner, it has been shown that incompatibility is controlled by several unlinked *het* loci. If the two fungi attempting anastomosis have different alleles for any of the *het* loci, then anastomosis is aborted, leading to incompatibility. Several such loci have been characterized in *N. crassa*, and each of them is identifiable by specific amino acid motifs known as *het* boxes within the encoded proteins (reviewed in Aanen, Debets, Glass, & Saupe, 2010).

There have been several studies from different geographical locations that demonstrate a similar vegetative incompatibility in *B. cinerea* using either chlorate resistance to isolate nitrate nonutilizing mutants or selenate resistance to isolate sulfate nonutilizing mutants (e.g., Beever & Parkes, 2003; Declan & Melgarejo, 2002; Fekete et al., 2012; Fournier, Leroux, Giraud, & Brygoo, 2003; Korolev, Elad, & Katan, 2008). In the most detailed of these studies, more than 66 different VCG types were identified by Beever and Weeds (2004) and even this was far from an exhaustive study. Indeed, comparatively few clonal isolates (as indicated by vegetative compatibility) are to be found within field-collected isolates, suggesting enormous genetic diversity, presumably driven by sexual recombination. Genetic studies have indicated the presence of at least seven loci controlling vegetative incompatibility in *B. cinerea* (Weeds, P.L. and Beever, R.E., personal communication) but analysis of the *B. cinerea* genome has identified 52 different genes encoding het-containing proteins (Aanen et al., 2010). While not all of these will participate in controlling anastomosis, it does show that *B. cinerea* can make use of a well-developed incompatibility system to reduce the likelihood of viral transmission.

As the only known means of horizontal mycovirus transmission is via hyphal anastomosis, the large number of compatibility groups within Botrytis would seem to be a major constraint on the movement of mycoviruses. However, given the large number of *B. cinerea* compatibility groups the incidence of BVX and BVF is greater than would be expected if incompatibility were a complete barrier to horizontal virus transmission. For *Sclerotinia sclerotiorum*, Deng, Melzer, and Boland, (2002) have demonstrated that horizontal virus transmission can occur during some types of incompatible reactions and we hypothesize that this is also true for *B. cinerea*.

5. CONCLUSIONS AND FUTURE RESEARCH

Studies of the phenotypic effects of mycoviruses have, at least in the first instance, usually used natural transmission between fungal isolates by hyphal anastomosis or *in vitro* transfection using purified virus. While these are highly appropriate methods to determine the phenotypic and metabolic effects of virus infection, they are not able to provide specific details of the fungal–virus interaction at the molecular level. The use of infectious cDNA clones has been successfully used for a few fungal species, including *C. parasitica* (Chen & Nuss, 1999) and *Diaporthe ambigua* (Moleleki et al., 2003) but not so far for Botrytis viruses. The use of infectious cDNA clones opens up a number of possibilities including genetic modifications that affect fungal compatibility and using mycoviruses as vectors to introduce genes deleterious to the fungal host. BVX and BVF are potentially good candidates for the latter since the closely related potexviruses have been successfully used as gene vectors in plants (Lu, Martin-Hernandez, Peart, Malcuit, & Baulcombe, 2003).

Along with the possibility of infectious clones to explore the properties of the virus, the recent advances in high-throughput RNA sequencing opens up a host of opportunities to investigate the impact of viral infection on the fungus. Not only will this allow more viruses to be identified (Al Rwahnih et al., 2011), but it should also be possible to extract mRNA from fungal tissues across a time course of infection, or from differentially infected isolates and to then identify the host genes up- or downregulated under each condition. When coupled with the availability of the genome sequence data, this makes such studies feasible from the short reads typically obtained from pyrosequencing. Thus, it should soon be possible to investigate whether different classes of mycoviruses have different effects on the host, and, by inclusion of appropriate gene-disruption mutants, to see whether the host can combat, or at least adapt to, infection by such viruses.

While it has been demonstrated that several mycoviruses can produce hypovirulent phenotypes in Botrytis, the ability of those viruses to spread within the Botrytis population is largely unknown. There is a distinct lack of epidemiological studies and the limited data on virus incidence currently available is not consistent with a single scenario. The highly variable dsRNA profiles observed in *B. cinerea* (Howitt et al., 1995) and the low frequency with which particular profiles were found could indicate that incompatibility is an effective barrier to infection and this would need to be borne in mind when contemplating a control system based on hypovirulence. In contrast, both BVX and BVF are widespread in isolates from different geographical locations, raising the possibility that dsRNA and ssRNA elements behave differently. Incompatibility is not an absolute barrier against viral transmission (Deng et al., 2002), so when anastomosis is aborted after hyphal fusion, there remains the possibility that viral transmission may have occurred before the barrier is complete. This is a key area for study if mycoviruses are to be considered as biocontrol agents for Botrytis. A universal donor strain of Botrytis would be ideal, or at least one that has the characteristics of maximizing the chance of mycovirus transmission even in an incompatible interaction. Even if incompatibility does not prove a major obstacle, the situations in which mycoviruses can be successfully used require careful consideration; after all, this will involve the deliberate release of an isolate, albeit an attenuated one, of a pathogenic fungus. Successful use of microorganisms as biological control agents often requires targeted application to the infection court rather than general release into the crop environment. Consequently, mycoviral control of Botrytis in defined and controlled environments, such as glasshouses, may be a more realistic goal than the control of widespread field epidemics.

REFERENCES

Aanen, D. K., Debets, A. J. M., Glass, N. L., & Saupe, S. J. (2010). Biology and genetics of vegetative incompatability in fungi. In K. A. Borkovich & D. J. Ebbole (Eds.), *Cellular and molecular biology of filamentous fungi* (pp. 274–288). New York: APS Press.

Aggio, R. B. M., Ruggiero, K., & Villas-Boas, S. G. (2009). Pathway Active Profiling (PAPi): From metabolic profile to the metabolic pathway activity. *Bioinformatics*, *26* (23), 2969–2976.

Al Rwahnih, M., Daubert, S., Urbez-Torres, J. R., Cordero, F., & Rowhani, A. (2011). Deep sequencing evidence from single grapevine plants reveals a virome dominated by mycoviruses. *Archives of Virology*, *156*(3), 397–403.

Amselem, J., Cuomo, C. A., van Kan, J. A. L., Viaud, M., Benito, E. P., et al. (2011). Genomic analysis of the necrotrophic fungal pathogens *Sclerotinia sclerotiorum* and *Botrytis cinerea*. *PLoS Genetics*, 7(8), e1002230. http://dx.doi.org/10.1371/journal.pgen.1002230.

Arthur, K. (2007). Detection and incidence of the mycoviruses BVX and BCVF in the plant pathogenic fungus *Botrytis cinerea*. MSc thesis, The University of Auckland, 2007. 158pp. Call number: Thesis 07-244.

Beever, R. E., & Parkes, S. L. (2003). Use of nitrate non-utilising (Nit) mutants to determine vegetative compatibility in *Botryotinia fuckeliana* (*Botrytis cinerea*). *European Journal of Plant Pathology*, *109*(6), 607–613.

Beever, R. E., & Weeds, P. L. (2004). Taxonomy and genetic variation of Botrytis and Botryotinia. In Y. Eladet al. (Ed.)*Botrytis: biology, pathology and control* (pp. 29–52). The Netherlands: Springer.

Boine, B. (2012). A study of the interaction between the plant pathogenic fungus *Botrytis cinerea* and the filamentous ssRNA mycoviruses *Botrytis virus X* and *Botrytis virus F*. PhD Thesis. The University of Auckland, New Zealand, 2012.

Castro, M., Kramer, K., Valdivia, L., Ortiz, S., & Castillo, A. (1999). A new double-stranded RNA mycovirus from *Botrytis cinerea*. *FEMS Microbiology Letters*, *175*(1), 95–99.

Castro, M., Kramer, K., Valdivia, L., Ortiz, S., & Castillo, A. (2003). A double-stranded RNA mycovirus confers hypovirulence-associated traits to *Botrytis cinerea*. *FEMS Microbiology Letters*, *228*(1), 87–91.

Catalanotto, C., Pallotta, M., ReFalo, P., Sachs, M. S., Vayssie, L., Macino, G., et al. (2004). Redundancy of the two dicer genes in transgene-induced posttranscriptional gene silencing in *Neurospora crassa*. *Molecular and Cellular Biology*, *24*, 2536–2545.

Chen, B., & Nuss, D. L. (1999). Infectious cDNA clone of hypovirus CHV1-Euro7: A comparative virology approach to investigate virus-mediated hypovirulence of the chestnut blight fungus *Cryphonectria parasitica*. *Journal of Virology*, *73*, 985–992.

Chiba, S., Salaipeth, L., Lin, Y.-H., Sasaki, A., Kanematsu, S., & Suzukil, N. (2009). A novel bipartite double-stranded RNA mycovirus from the white root rot fungus *Rosellinia necatrix*: Molecular and biological characterization, taxonomic considerations, and potential for biological control. *Journal of Virology*, *83*, 12801–12812.

Choi, G. H., & Nuss, D. L. (1992). Hypovirulence of chestnut blight fungus conferred by an infectious viral cDNA. *Science*, *257*, 800–803.

Choquer, M., Fournier, E., Kunz, C., Levis, C., Pradier, J. M., Simon, A., et al. (2007). Botrytis cinerea virulence factors: New insights into a necrotrophic and polyphageous pathogen. *FEMS Microbiology Letters*, *277*(1), 1–10.

Cogoni, C. (2001). Homology-dependent gene silencing mechanisms in fungi. *Annual Review of Microbiology*, *55*, 381–406.

Dawe, A. L., Van Voorhies, W. A., Lau, T. A., Ulanov, A. V., & Li, Z. (2009). Major impacts on the primary metabolism of the plant pathogen *Cryphonectria parasitica* by the virulence-attenuating virus CHV1-EP713. *Microbiology*, *155*(12), 3913–3921.

Declan, J., & Melgarejo, P. (2002). Mating behavior and vegetative incompatability in Spanish populations of *Botryotina fuckeliana*. *European Journal of Plant Pathology*, *108*, 391–400.

Deng, F., Melzer, M. S., & Boland, G. J. (2002). Vegetative compatibility and transmission of hypovirulence-associated dsRNA in *Sclerotinia homoeocarpa*. *Canadian Journal of Plant Pathology*, *24*, 481–488.

Elad, Y., Williamson, B., Tudzynski, P., & Delen, N. (2004). Botrytis*: biology, pathology and control*. Netherlands: Kluwer Academic Publishers.

Fekete, E., Fekete, E., Irinyi, L., Karaffa, L., Arnyasi, M., Asadollahi, M., et al. (2012). Genetic diversity of a Botrytis cinerea cryptic species complex in Hungary. *Microbiological Research*, *167*, 283–291.

Fournier, E., Leroux, P., Giraud, T., & Brygoo, Y. (2003). Characterisation of Bc-hch, the Botrytis cinerea homolog of the Neurospora crassa het-c vegetative incompatability locus, and its use as a population marker. *Mycologia*, *95*, 251–261.

Hammond, T. M., Andrewski, M. D., Roossinck, M. J., & Keller, N. P. (2008). Aspergillus mycoviruses are targets and suppressors of RNA silencing. *Eukaryotic Cell*, *7*, 350–357.

Howitt, R., Beever, R., Pearson, M. N., & Forster, R. L. S. (1995). Presence of double-stranded RNA and virus-like particles in *Botrytis cinerea*. *Mycological Research*, *99*, 1472–1478.

Howitt, R., Beever, R., Pearson, M. N., & Forster, R. L. S. (2001). Genome characterization of *Botrytis virus F*, a flexuous rod-shaped mycovirus resembling plant 'potex-like' viruses. *The Journal of General Virology*, *82*, 67–78.

Howitt, R., Beever, R., Pearson, M. N., & Forster, R. L. S. (2006). Genome characterization of a flexuous rod-shaped mycovirus, *Botrytis virus X*, reveals high amino acid identity to genes from plant potex-like viruses. *Archives of Virology*, *151*(3), 563–579.

Huang, S., & Ghabrial, S. A. (1996). Organization and expression of the double-stranded RNA genome of Helminthosporium victoriae 190S virus, a totivirus infecting a plant pathogenic filamentous fungus. *Proceedings of the National Academy of Sciences of the United States of America*, *93*, 12541–12546.

Jaspers, M. V., Seyb, A. M., Trought, M. C., & Balasubramaniam, R. (2012). Overwintering grapevine debris as an important source of Botrytis cinerea inoculums. *Plant Pathology*, *62*, 130–138.

Kadotani, N., Nakayashiki, H., Tosa, Y., & Mayama, S. (2004). One of the two Dicer-like proteins in the filamentous fungi Magnaporthe oryzae genome is responsible for hairpin

RNA-triggered RNA silencing and related small interfering RNA accumulation. *The Journal of Biological Chemistry*, *279*, 44467–44474.
Kamper, J., et al. (2006). Insights from the genome of the biotrophic fungal plant pathogen *Ustilago maydis*. *Nature*, *444*, 97–101.
Kanematsu, S., Sasaki, A., Onoue, M., Oikawa, Y., & Ito, T. (2010). Extending the fungal host range of a partitivirus and a mycoreovirus from *Rosellinia necatrix* by Inoculation of protoplasts with virus particles. *Phytopathology*, *100*, 922–930.
King, A. M. Q., Adams, M. J., Carstens, E. B., & Lefkowitz, E. J. (2011). Virus taxonomy: Classification and nomenclature of viruses. Ninth report of the International Committee on taxonomy of viruses. International Union of Microbiological Societies Virology Division. Elsevier Academic Press.
Korolev, N., Elad, Y., & Katan, T. (2008). Vegetative compatability grouping in *Botrytis cinerea* using sulphate non-utilizing mutants. *European Journal of Plant Pathology*, *122*, 369–383.
Lee, K. M., Yu, J., Son, M., Lee, Y. W., & Kim, K. H. (2011). Transmission of *Fusarium boothii* mycovirus via protoplast fusion causes hypovirulence in other phytopathogenic fungi. *PLoS One*, *6*(6), e21629.
Lu, R., Martin-Hernandez, A. M., Peart, J. R., Malcuit, I., & Baulcombe, D. C. (2003). Virus-induced gene silencing in plants. *Methods*, *30*, 296–303.
Macfarlane, H. H. (1968). Plant host-pathogen index to volumes 1–40 (1922–1961), Rev. Appl. Mycol. Commonwealth Mycological Institute, Kew.
Moleleki, N., van Heerden, S. W., Wingfield, S. J., Wingfield, B. D., & Preisig, O. (2003). Transfection of *Diaporthe perjuncta* with Diaporthe RNA Virus. *Applied and Environmental Microbiology*, *69*, 3952–3956.
Morris, T. J., & Dodds, J. A. (1979). Isolation and analysis of double-stranded RNA from virus-infected plant and fungal tissue. *Phytopathology*, *69*, 854–858.
Nakayashiki, H. (2005). RNA silencing in fungi: Mechanisms and applications. *FEBS Letters*, *579*, 5950–5957.
Patel, R. M., van Kan, J. A. L., Bailey, A. M., & Foster, G. D. (2008). RNA-mediated gene silencing of superoxide dismutase (*bcsod1*) in *Botrytis cinerea*. *Phytopathology*, *98*, 1334–1339.
Patel, R. M., van Kan, J. A. L., Bailey, A. M., & Foster, G. D. (2010). Inadvertent gene silencing of argininosuccinate synthase (Bcass1) in *Botrytis cinerea* by the pLOB1 vector system. *Molecular Plant Pathology*, *11*, 613–624.
Pearson, M. N., Beever, R. E., Boine, B., & Arthur, K. (2009). Mycoviruses of filamentous fungi and their relevance to plant pathology. *Molecular Plant Pathology*, *10*(1), 115–128.
Pearson, M. N., Beever, R. E., Boine, B., & Tan, C. (2009). Can mycoviruses be used for the biocontrol of the plant pathogenic fungus Botrytis cinerea? In Yiga lElad, Monika Maurhofer, Christoph Keel, Cesare Gessler & Brion Duffy (Eds.), *Integrated Control of Plant Pathogens. Proceedings of the meeting "Molecular Tools for Understanding and Improving Biocontrol"IOBC/WPRS Bulletin*, Vol. 43, (pp. 7–10). ISBN 978-92-9067-217-3.
Read, N. D., Fleissner, A., Roca, M. G., & Glass, N. L. (2010). Hyphal fusion. In K. A. Borkovich & D. J. Ebbole (Eds.), *Cellular and molecular biology of filamentous fungi* (pp. 260–273). USA: ASM Press.
Roca, M. G., Davide, L. C., Davide, L. M. C., Mendes-Costa, M. C., Schwan, R. F., & Wheals, A. E. (2004). Conidial anastomosis fusion between different Colletotrichum species. *Mycological Research*, *108*, 1320–1326.
Sasaki, A., Kanematsu, S., Onoue, M., Oyama, Y., & Yoshida, K. (2006). Infection of *Rosellinia necatrix* with purified viral particles of a member of *Partitiviridae* (RnPV1-W8). *Archives of Virology*, *151*, 697–707.
Sasaki, A., Onoue, M., Kanematsu, S., Suzaki, K., Miyanishi, M., Suzuki, N., et al. (2002). Extending chestnut blight hypovirus host range within Diaporthales by biolistic delivery of viral cDNA. *Molecular Plant-Microbe Interactions*, *15*, 780–789.

Segers, G. C., van Wezel, R., Zhang, X., Hong, Y., & Nuss, D. L. (2006). Hypovirus papain-like protease p29 suppresses RNA silencing in the natural fungal host and in a heterologous plant system. *Eukaryotic Cell*, *5*, 896–904.

Segers, G. C., Zhang, X., Deng, F., Sun, Q., & Nuss, D. L. (2007). Evidence that RNA silencing functions as an antiviral defense mechanism in fungi. *Proceedings of the National Academy of Sciences of the United States of America*, *104*, 12902–12906.

Staats, M., van Baarlen, P., & van Kan, J. A. L. (2004). Molecular phylogeny of the plant pathogenic genus Botrytis and the evolution of host specificity. *Molecular Biology and Evolution*, *22*, 333–346.

Tauati, S. J. (2011). Investigating RNA silencing and mycoviruses in *Botrytis cinerea*. PhD Thesis, University of Bristol UK.

Tauati, S. J., Pearson, M. N., Choquer, M., Foster, G. D., & Bailey, A. M. (in preparation). Investigating the role of Dicer 2 (*dcr2*) in gene silencing and the regulation of mycoviruses in *Botrytis cinerea*.

Urayama, S., Kato, S., Suzuki, Y., Aoki, N., Le, M. T., Arie, T., et al. (2010). Mycoviruses related to chrysovirus affect vegetative growth in the rice blast fungus *Magnaporthe oryzae*. *The Journal of General Virology*, *91*, 3085–3094.

Urbanek, H., & Zalewska-Sobczak, J. (2003). Multiplicity of cell wall degrading glycosidic hydrolases produced by apple infecting *Botrytis cinerea*. *Journal of Phytopathology*, *110*, 261–271.

Vilches, S., & Castillo, A. (1997). A double-stranded RNA mycovirus in *Botrytis cinerea*. *FEMS Microbiology Letters*, *155*, 125–130.

Williamson, B., Tudzynski, B., Tudzynski, P., & van Kan, J. A. L. (2007). *Botrytis cinerea*: The cause of grey mould disease. *Molecular Plant Pathology*, *8*, 561–580.

Wu, M. D., Jin, F., Zhang, J., Yang, L., Jiang, D. H., & Li, J. (2012). Charcaterization of a novel bipartite double-stranded RNA mycovirus conferring hypovirulence in the pathogenic fungus *Botrytis porri*. *Journal of Virology*, *86*, 6605–6619.

Wu, M. D., Zhang, L., Li, G., Jiang, D., & Ghabrial, S. A. (2010). Genome characterization of a debilitation-associated mitovirus infecting the phytopathogenic fungus *Botrytis cinerea*. *Virology*, *406*(1), 117–126.

Wu, M. D., Zhang, L., Li, G. Q., Jiang, D. H., Hou, M. S., & Huang, H.-C. (2007). Hypovirulence and double-stranded RNA in *Botrytis cinerea*. *Phytopathology*, *97*(12), 1590–1599.

CHAPTER TEN

Insight into Mycoviruses Infecting *Fusarium* Species

Won Kyong Cho, Kyung-Mi Lee, Jisuk Yu, Moonil Son and Kook-Hyung Kim[1]

Department of Agricultural Biotechnology, Center for Fungal Pathogenesis and Research Institute for Agriculture and Life Sciences, Seoul National University, Seoul, Republic of Korea
[1]Corresponding author: e-mail address: kookkim@snu.ac.kr

Contents

1. Introduction 274
2. Isolation and Characterization of dsRNA Mycoviruses Infecting *Fusarium* Species 275
3. Diversity of Fusarium Mycoviruses Based on Genome Organization 277
4. Identification of Fungal Host Factors by Proteomics and Transcriptomics 281
5. Transmission of Fusarium Mycovirus by Protoplast 283
6. Future Directions for Research on Mycovirus–Host Interactions 285

Acknowledgments 285
References 285

Abstract

Most of the major fungal families including plant-pathogenic fungi, yeasts, and mushrooms are infected by mycoviruses, and many double-stranded RNA (dsRNA) mycoviruses have been recently identified from diverse plant-pathogenic *Fusarium* species. The frequency of occurrence of dsRNAs is high in *Fusarium poae* but low in other *Fusarium* species. Most Fusarium mycoviruses do not cause any morphological changes in the host but some mycoviruses like Fusarium graminearum virus 1 (FgV1) cause hypovirulence. Available genomic data for seven of the dsRNA mycoviruses infecting *Fusarium* species indicate that these mycoviruses exist as complexes of one to five dsRNAs. According to phylogenetic analysis, the *Fusarium* mycoviruses identified to date belong to four families: *Chrysoviridae*, *Hypoviridae*, *Partitiviridae*, and *Totiviridae*. Proteome and transcriptome analysis have revealed that FgV1 infection of *Fusarium* causes changes in host transcriptional and translational machineries. Successful transmission of FgV1 via protoplast fusion suggests the possibility that, as biological control agents, mycoviruses could be introduced into diverse species of fungal plant pathogens. Research is now needed on the molecular biology of mycovirus life cycles and mycovirus–host interactions. This research will be facilitated by the further development of omics technologies.

Advances in Virus Research, Volume 86
ISSN 0065-3527
http://dx.doi.org/10.1016/B978-0-12-394315-6.00010-6

1. INTRODUCTION

Viruses are small, infectious pathogens whose hosts include animals, plants, bacteria, Archaea, and fungi. Many kinds of viruses have been identified, and although they can be composed of either DNAs or RNAs (Breitbart & Rohwer, 2005), most have RNA genomes, and plant viruses usually have single-stranded (ss) RNA genomes (Wren et al., 2006). The viruses that infect plant-pathogenic fungi, yeasts, and mushrooms are referred to as mycoviruses and are composed of ssRNAs or dsRNAs. dsRNA mycoviruses have been frequently identified and studied (El-Sherbeini & Bostian, 1987; Fujimura, Esteban, & Wickner, 1986; McCabe, Pfeiffer, & Van Alfen, 1999; Tavantzis, Romaine, & Smith, 1980). According to the International Committee for the Taxonomy of Viruses (ICTV), dsRNA can be divided into eight families. The mycoviruses with dsRNA genomes identified thus far have been classified into four families: *Chrysoviridae*, *Totiviridae*, *Partitiviridae*, and *Reoviridae* (King, Adams, Carstens, & Lefkowitz, 2011).

Although our understanding of mycoviruses is poor relative to our understanding of plant viruses (Pearson, Beever, Boine, & Arthur, 2009), two mycoviruses that infect the chestnut blight pathogen *Cryphonectria parasitica* (Cryphonectria hypovirus 1-EP713 and *Cryphonectria parasitica mitovirus* 1-NB631) have been intensively studied (Nuss, 1992; Polashock & Hillman, 1994). Some mycoviruses have the ability to attenuate the disease caused by fungal plant pathogens in a phenomenon referred to as hypovirulence. A recent study suggested that hypovirulence caused by mycoviruses could be used for the biological control of *C. parasitica* and perhaps other plant pathogens (Nuss, 2005). Another recent study provided evidence that a virus-infected fungus confers thermal tolerance to host plants, suggesting that mycoviruses could participate in three-way symbioses (Márquez, Redman, Rodriguez, & Roossinck, 2007).

The hosts of dsRNA mycoviruses include many *Fusarium* species, which are filamentous fungi and which include important plant pathogens (Parry, Jenkinson, & McLeod, 1995) that cause serious damage to economically important crops like maize, wheat, and barely. *Fusarium* species also produce mycotoxins like fumonisins and trichothecenes that cause public health problems (Marasas, Nelson, & Toussoun, 1984). This chapter discusses the current state of research related to dsRNA mycoviruses infecting *Fusarium* species.

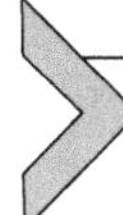

2. ISOLATION AND CHARACTERIZATION OF dsRNA MYCOVIRUSES INFECTING *FUSARIUM* SPECIES

Fusarium-infecting mycoviruses have been detected by the presence of dsRNAs. Many mycoviruses infecting *Fusarium* species that have been detected and identified to date are from *Fusarium graminearum* (Aminian, Azizollah, Abbas, & Naser, 2011; Chu et al., 2002, 2004; Darissa, Willingmann, Schäfer, & Adam, 2011; Yu, Lee, Son, & Kim, 2011). Among the 827 *F. graminearum* isolates obtained from diseased barley and maize in Korea, 19 contained dsRNA fragments (Chu et al., 2004), and 1 dsRNA virus has been identified as Fusarium graminearum virus-DK21 (recently named Fusarium graminearum virus 1 strain DK21, FgV1-DK21) (Chu et al., 2002; Kwon et al., 2007; Yu, Kwon, Lee, Son, & Kim, 2009). The isolated mycoviruses were composed of two to five segments of dsRNAs that ranged from 1.7 to 10 kb (Chu et al., 2004). Two *Fusarium graminearum* isolates (JB33 and JNKY19) seemed to be infected by two different viruses.

Recently, two studies have identified additional mycoviruses infecting *F. graminearum*. One mycovirus, referred to as Fusarium graminearum virus-ch9 (FgV-ch9), consists of five dsRNAs and was identified from 10 *F. graminearum* strains in China (Darissa et al., 2011). In Iran, at least three different dsRNAs, ranging in size from 0.9 to 5 kb, were detected in 12 of 33 *F. graminearum* isolates that were screened for dsRNA (Aminian et al., 2011). The 12 isolates containing dsRNAs in Iran caused less severe disease than dsRNA-free isolates on susceptible wheat in the greenhouse and produced substantially less of the mycotoxin deoxynivalenol (DON).

In an earlier study in Japan, only 1 of 34 strains *Fusarium solani* contained dsRNA (Nogawa, Shimosaka, Kageyama, & Okazaki, 1993). This strain (SUF704) had one dsRNA fragment of 1.9 kb and another of 1.7 kb, and together these fragments produced a single polypeptide of 38 kDa on SDS-PAGE gels (Nogawa et al., 1993). This mycovirus, which was named Fusarium solani virus 1 (FsV1; synonym, FusoV), was subsequently proven to contain two different dsRNA segments with RNA-dependent RNA polymerase (RdRp) domain in one of the dsRNA segments (Nogawa et al., 1996).

In the United States, only 4 of 100 *Fusarium proliferatum* isolates infecting maize and sorghum were found to contain dsRNAs; the dsRNAs ranged from 0.7 to 3.1 kb (Heaton & Leslie, 2004). One isolate contained a single kind of dsRNA, while the other isolates contained multiple kinds of dsRNAs. Interestingly, the multiple dsRNAs in one isolate were associated with mitochondria (Heaton & Leslie, 2004). In another study in the United States, only 6 of 57 isolates of *Fusarium oxysporum* contained dsRNAs, and all six isolates were composed of four dsRNA segments with sizes of 2.2, 2.7, 3.1, and 4.0 kb (Kilic & Griffin, 1998).

A recent survey of 103 isolates of endophytic fungi belonging to 53 species, 12 isolates contained dsRNAs, and one isolate of *Fusarium culmorum* species contained two dsRNAs of 3 and 4.4 kb (Herrero, Sánchez Márquez, & Zabalgogeazcoa, 2009). Whether these two dsRNAs were part of one mycovirus or from two different mycoviruses was not determined.

Among 55 *Fusarium poae* isolates collected from wheat in different geographical regions in the world, all contained dsRNAs and encapsidated virus-like particles (VLP; Fekete, Giczey, Papp, Szabó, & Hornok, 1995). This was surprising because previous surveys had generally indicated that only a low percentage of *Fusarium* isolates were infected by mycoviruses. Furthermore, the patterns of dsRNAs were different in each *F. poae* isolate but were stable after repeated subculturing (Fekete et al., 1995). None of the *F. poae* isolates harboring dsRNAs showed any morphological changes, indicating that the mycoviruses did not greatly harm the host. One of these mycoviruses, which was obtained from *F. poae* isolate A-11, was later named Fusarium poae virus 1 (FpV1; synonym, FuPO-1; Compel, Papp, Bibo, Fekete, & Hornok, 1999).

Only a few of the isolated mycoviruses have been associated with morphological changes in the host. Similarly, only a few studies have identified *Fusarium* mycoviruses that cause hypovirulence. For instance, three mycoviruses infecting *Fusarium oxysporum* were found to confer hypovirulence (Kilic & Griffin, 1998). FgV1 was the first characterized Fusarium mycovirus causing decreased fungal pathogenicity and morphological changes, including reduced mycelial growth and increased pigmentation (Chu et al., 2002). When mango was inoculated with a VLP-infected and a VLP-free isolate of *Fusarium moniliforme*, only the VLP-free isolate caused shoot malformation, suggesting that the mycovirus in *F. moniliforme* might suppress mango shoot malformation.

According to previous studies (Gupta, 1991) dsRNA viruses are clearly present in various *Fusarium* species and the infection rate differs among *Fusarium* species although it is generally low. Previous studies also suggest that only a limited number of mycoviruses cause hypovirulence.

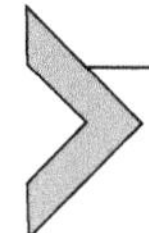

3. DIVERSITY OF FUSARIUM MYCOVIRUSES BASED ON GENOME ORGANIZATION

Although many Fusarium mycoviruses have been identified, the complete genome sequences have been determined for only seven: FgV1, Fusarium graminearum virus 2 (FgV2), Fusarium graminearum virus 3 (FgV3), Fusarium graminearum virus 4 (FgV4), FgV-ch9, FpV1, and FsV1 (Table 10.1). A partial genome sequence is available for FoV1 (Sharzehei, Banihashemi, & Afsharifar, 2007). The genomes of both FpV1 and FsV1 consist of two dsRNA segments, which encode capsid protein and RdRp, respectively (Compel et al., 1999; Nogawa et al., 1996).

The complete genome of FgV1 is 6624-bp long and encodes four proteins (Fig. 10.1; Kwon et al., 2009). FgV1 contains 5′ and 3′ untranslated regions (UTRs) with lengths of 53 and 46 bp, respectively, and has a poly(A) tail at the 3′-terminus (Fig. 10.1). Among the four ORFs in FgV1, only ORF1 is considered as a potential RdRp; ORF1 in FgV1 is similar to that in *Cryphonectria* hypoviruses and Barley yellow mosaic virus (BaYMV), while the other three ORFs in FgV1 do not show sequence identity with other known virus sequences (Kwon et al., 2009).

FgV2 was originally isolated from *F. graminearum* strain 98-8-60 infecting barley (Chu et al., 2004). The genome of FgV2 consists of five dsRNA segments referred to as dsRNA1 to dsRNA5 (Fig. 10.1; Yu et al., 2011). The five FgV2 dsRNA segments range from 2414 to 3580 bp. FgV2 dsRNA1 is the largest segment (3580 bp), whereas dsRNA5 is the smallest (2414 bp). Each dsRNA segment encodes a single ORF and contains 5′ and 3′ UTRs ranging from 78–105 to 84–306 bp, respectively. Interestingly, UTR sequences in FgV2 display a high degree of similarity among dsRNA segments. Only dsRNA1 possesses an RdRp domain, while the other four dsRNA fragments lack a known functional domain. In addition, a recent study reported the complete genome sequence of an *F. graminearum* mycovirus named FgV-ch9, which was obtained from cereals in China (Darissa et al., 2011). FgV-ch9 consists of five dsRNA fragments having similar genome organization, and each dsRNA segment of FgV-ch9 has strong sequence identity (83–98% in amino acids) with FgV2.

Both FgV3 and FgV4 were identified from *F. graminearum* strain DK3 (Fig. 10.1; Chu et al., 2004). The genome of FgV3 is 9098-bp long and encodes two ORFs, which are divided into ORF1 (145 kDa) and ORF2

Table 10.1 The known *Fusarium*-infecting mycoviruses

Mycovirus	Family or propsed family	Host isolated from	RNA segment	Size of nucleotides (bp)	No. of proteins	Accession no.
FgV1	Unassigned	*F. graminearum* isolate DK21	RNA1	6621	4	NC_006937
FgV2	*Chrysoviridae*	*F. graminearum* isolate 98-8-60	RNA1	3580	1	HQ343295
			RNA2	3000	1	HQ343296
			RNA3	2982	1	HQ343297
			RNA4	2748	1	HQ343298
			RNA5	2414	1	HQ343299
FgV3	*Totiviriridae*	*F. graminearum* isolate DK3	RNA1	9098	2	NC_013469
FgV4	*Partitiviridae*	*F. graminearum* isolate DK3	RNA1	2383	1	NC_013470
			RNA2	1739	2	NC_013471
FgV-ch9	*Chrysoviridae*	*F. graminearum* isolate China 9	RNA1	3581	1	HQ228213
			RNA2	2850	1	HQ228214
			RNA3	2830	1	HQ228215
			RNA4	2746	1	HQ228216
			RNA5	2423	1	HQ228217
FpV1	*Partitiviridae*	*F. poae*	RNA1	2185	1	NC_003883
			RNA2	2203	1	NC_003884
FoV1	*Chrysoviridae*	*F. oxysporum* f. sp. *melonis*	RNA1	2574	1	EF152346
			RNA2	648	1	EF152347
			RNA3	994	1	EF152348
FsV1	*Partitiviridae*	*F. solani* f. sp. *robiniae*	RNA1	1645	1	NC_003885
			RNA2	1445	1	NC_003886

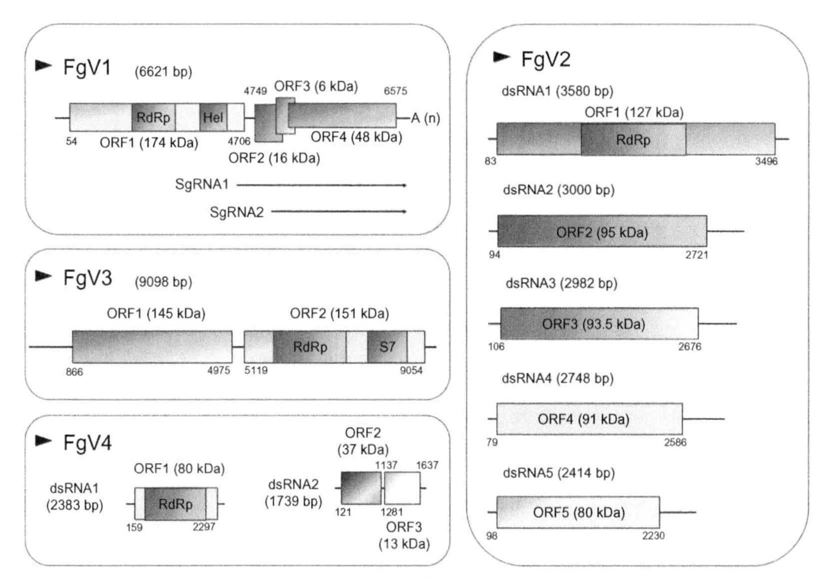

Figure 10.1 Genome organization of four *Fusarium graminearum* mycoviruses. Each open reading frame (ORF) is indicated by an open bar. RdRp, Hel, and S7 indicate the predicted conserved domains of RNA-dependent RNA polymerase, helicase, and S7 protein domain, respectively. The molecular weight of each ORF is indicated. A black, single line indicates non-translated regions. The nucleotide positions for the initiation and termination codons are indicated above the border of each ORF. The genome organization of FgV-ch9 is similar to that of FgV2. Genome organization of four *F. graminearum* mycoviruses were drawn based on previous studies with minor modification (Kwon et al., 2009; Yu et al., 2009, 2011).

(151 kDa). ORF2 is predicted to contain two conserved domains that are similar to RdRp and phytoreovirus S7 protein domain (S7; Yu et al., 2009).

To reveal the genetic relationships among seven mycoviruses infecting *Fusarium* species and 26 other mycoviruses, we used the amino acid sequences of RdRp for phylogenetic tree construction. The phylogenetic tree identified two groups of mycoviruses (Fig. 10.2; Kwon et al., 2009). Group A contains FpV1, FsV1, FgV1, and FgV4, whereas group B contains FgV2, FgV3, and FgV-ch9 (Fig. 10.2). FpV1 and FsV1 are clustered together with members of the *Partitiviridae* family. Group A also contains four "hypoviruses" including Cryphonectria hypovirus 3, Cryphonectria hypovirus 4, Valsa ceratosperma hypovirus 1, and Sclerotinia sclerotiorum hypovirus 1. FgV4 forms a clade in group A with other mycovirus in the family *Partitiviridae*. Although FgV1 belongs to group A, which contains partiviruses and hypoviruses, FgV1 is more closely associated with members of the family *Hypoviridae* like Cryphonectria hypovirus 1–4 (Kwon et al., 2009).

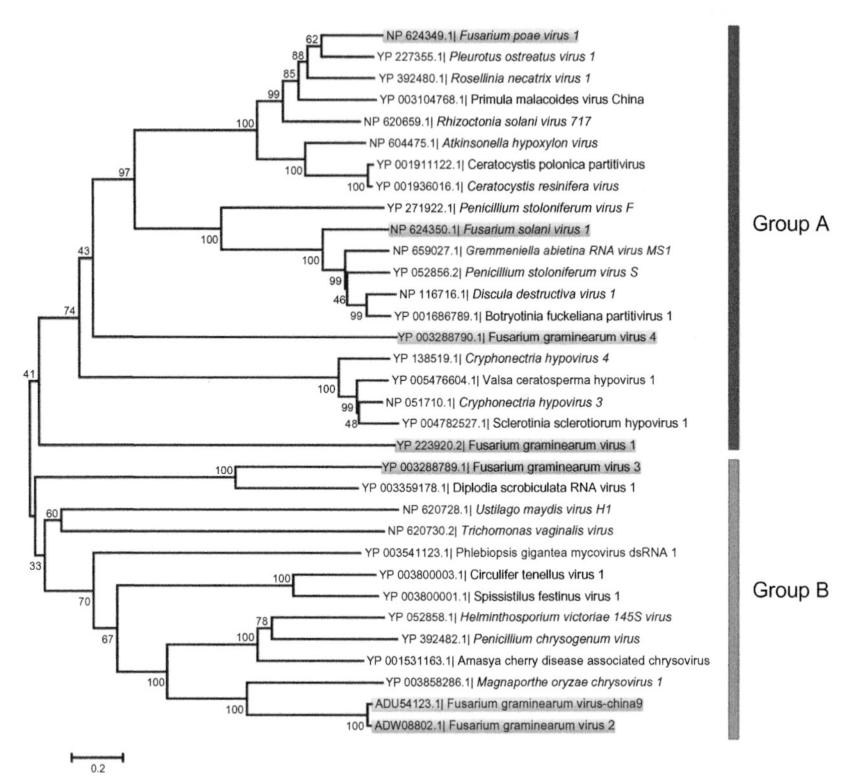

Figure 10.2 The phylogenetic relationship of seven mycoviruses infecting *Fusarium* species. RdRp amino sequences for seven mycoviruses infecting *Fusarium* species as well for 24 other homologous mycoviruses were retrieved. The phylogenetic tree was constructed using the neighbor-joining method with 1000 bootstraps. Numbers at each node indicate bootstrap values. The scale bar represents 0.1 substitutions per amino acid site. (For color version of this figure, the reader is referred to the online version of this chapter.)

FgV2 and FgV-ch9 were isolated from nearby geographical areas (Korea and China) and are clustered together with strong bootstrap support (Fig. 10.2), suggesting that they may have diverged from a common ancestor. Because they are grouped together with Magnaporthe oryzae chrysovirus 1 and contain four dsRNAs, FgV2 and FgV-ch9 might be members of the *Chrysoviridae* family containing five dsRNAs (Urayama et al., 2010). FgV3 is very closely related to Diplodia scrobiculata RNA virus 1, which contains two dsRNA segments and which infects pathogens of conifers like *Diplodia scrobiculata* (De Wet, Preisig, Wingfield, & Wingfield, 2008). Both FgV3 and Diplodia scrobiculata RNA virus 1 are likely to be members of the family

Totiviridae. Thus, genomic data and phylogenetic tree analysis indicate that Fusarium-infecting viruses are members or probable members of the families *Chrysoviridae*, *Hypoviridae*, *Partitiviridae*, and *Totiviridae*.

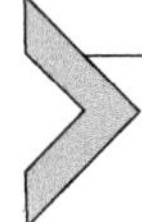

4. IDENTIFICATION OF FUNGAL HOST FACTORS BY PROTEOMICS AND TRANSCRIPTOMICS

An understanding of mycovirus–host interactions and mycovirus life cycles is now being facilitated by various omics technologies, which have recently been developed and used to reveal genes or biological processes at genome-wide levels (Joyce & Palsson, 2006). Based on omics, fungal genes required for pathogenicity can now be identified (Van de Wouw & Howlett, 2011). In the case of mycoviruses, omics-based approaches could be used to identify host genes or proteins involved in mycovirus replication and movement and in hypovirulence. In particular, the complete sequencing of the *F. graminearum* genome provides a useful resource for proteomics and transcriptomics (Cuomo et al., 2007). To elucidate *F. graminearum* proteins or genes that respond to or are involved in mycovirus infection, two studies have used FgV1 as a well-characterized model mycovirus with strong hypovirulent ability (Cho et al., 2012; Kwon et al., 2009).

For proteomics, proteins were extracted from virus-free and FgV1-infected *F. graminearum* and analyzed by two-dimensional gel electrophoresis (2DE) with mass spectrometry (Kwon et al., 2009). Comparison of 2DE gels identified a total of 148 spots indicating differential expression between the infected and noninfected fungus. Among the 148 spots, 33 were selected for ESI-MS/MS analyses, which resulted in the identification of 23 proteins. Expression of seven proteins was highly induced in FgV1-infected *F. graminearum*, and these included the sporulation-specific protein SPS2, triosephosphate isomerase, nucleoside diphosphate kinase, and woronin body major protein. Expression of 16 proteins was downregulated in FgV1-infected *F. graminearum*, and these included enolase, saccharopine dehydrogenase, flavohemoglobin, mannitol dehydrogenase, and malate dehydrogenase (Kwon et al., 2009). To our knowledge, this represents the first proteomics to identify fungal host factors associated with mycovirus infection, but the number of proteins identified was insufficient to obtain a comprehensive understanding of the host response. Therefore, it would be very desirable to examine the *Fusarium* proteome with a high-throughput MS/MS technology or a non-gel-based quantitative method (Scott-Craig, Adhikari, Cuomo, & Walton, 2007; Taylor et al., 2008). Because attenuation of fungal host virulence by mycovirus infection might result from inhibition of the proteins involved in the synthesis and secretion of trichothecenes and other mycotoxins, proteomics should now be used to identify proteins secreted by the *Fusarium* host in response to mycovirus infection (Phalip et al., 2005).

The recent publication of the complete genome sequence for *F. graminearum* provides the information required to construct a 3′-tiling microarray covering whole genes (13,382 genes; Cuomo et al., 2007). In a study determining global transcriptional changes in FgV1-infected versus virus-free *F. graminearum*, the numbers of differentially expressed genes at 36 h after infection were similar to those at 120 h but the gene lists were different, indicating that mycovirus infection causes time-dependent changes in the *Fusarium* transcriptome (Cho et al., 2012). At the early time, expression of genes required for protein synthesis, such as ribosome assembly, as well as genes required for nucleolus RNA- and ribosomal RNA-processing genes, was induced in FgV1-infected *F. graminearum*. The induction of these genes soon after infection might be related to viral replication because the nucleolus is the site for ribosomal RNA synthesis and ribosome assembly, and ribosomes are required for protein production (Nemeth & Langst, 2011). The nucleolus is also actively involved in a variety of biological processes, such as cell cycle regulation, cell growth, stress sensing, and viral infection (Kim, Ryabov, Brown, & Taliansky, 2004).

In the same study (Cho et al., 2012), expression of fungal-specific transcription factors (TFs), such as members of the Zn2Cys6 family, was strongly changed by virus infection at both time points. The Zn2Cys6 TF family contains 309 TFs, making it the largest TF family in *F. graminearum*, and its members have various functions, such as sugar and amino acid metabolism, gluconeogenesis, respiration, vitamin synthesis, cell cycle, chromatin remodeling, nitrogen utilization, peroxisome proliferation, drug resistance, and stress response (Shelest, 2008). At the early time point, transcripts encoding enzymes producing various metabolites like carboxylic acids, aromatic amino acids, nitrogen compounds, and polyamines were highly downregulated in FgV1-infected *F. graminearum*. This result was inconsistent with those of a previous report of *C. parasitica*, which indicated that virus infection resulted in the upregulation of most genes that produce metabolites (Dawe, Van Voorhies, Lau, Ulanov, & Li, 2009). Perhaps the differences in the results reflect differences in sampling time relative to the time of infection. The upregulated metabolic-related genes in *C. parasitica* are also components for host defense. In the study by Cho et al. (2012) with FgV1-infected *F. graminearum*, those genes associated with defense and virulence were upregulated at the late time point, and genes associated with polyamine production showed dramatic decreases in gene expression. Polyamines function in cell growth, development, and responses to various stresses in plants (Gill & Tuteja, 2010). Likewise, it seems that reduced levels of DON might be correlated with the downregulation of genes involved in polyamine biosynthesis during the early stage (36 h) (Cho et al., 2012). Moreover, plasma membrane-associated transcripts were downregulated in FgV1-infected *F. graminearum*, indicating that the host transport system was severely damaged so as to hinder the recruitment of materials for viral replication. This result supports a previous study that indicated that virus infection leads to depolarization of the host cell membrane resulting in reduced solute transport via plasma membranes (Agarkova et al., 2008). Taken together, the genome-wide transcriptomic data indicate that infection by FgV1 causes reprogramming of *F. graminearum* transcriptional and translational machinery.

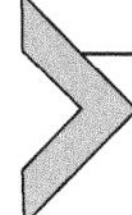

5. TRANSMISSION OF FUSARIUM MYCOVIRUS BY PROTOPLAST

Although fungicides are currently used to control plant fungal diseases, increasing concerns about environmental pollution and food safety have increased the demand for alternative fungal control systems. One alternative is biological control via mycovirus-infected hypovirulent fungal strains. A previous study demonstrated the utility of mycoviruses for managing the plant-pathogenic fungus *C. parasitica* (Nuss, 1992). Successful application of mycoviruses requires several kinds of information and technologies. First, mycoviruses causing strong hypovirulence should be identified. Second, a system to transmit a hypovirulent fungal strain to a virulent fungal strain should be developed.

With respect to transmission, a virus can generally be transmitted between vegetatively compatible fungal strains (but not between vegetatively incompatible strains) by hyphal fusion (anastomosis). Protoplast fusion has been investigated as a method to transmit mycoviruses among different strains of plant-pathogenic fungi such as *Aspergillus* (Van Diepeningen, Debets, & Hoekstra, 1998), *F. poae* (Diepeningen et al., 2000), and *Rosellinia necatrix* (Kanematsu, Sasaki, Onoue, Oikawa, & Ito, 2010). Protoplast fusion also was investigated for the purpose of transmitting FgV1 from a previously isolated hypovirulent fungal strain of *F. graminearum* to virus-free fungal strains of *F. graminearum*, *F. asiaticum*, *F. oxysporum* f. sp. *lycopersici*, and

Cryphonectria parasitica (Lee, Yu, Son, Lee, & Kim, 2011). In that research, plasmid DNAs carrying genes resistant to antibiotics were transformed into donor and recipient fungal strains and were later used as indicators of successful fusion (Fig. 10.3). Young mycelia were prepared from the transformants, and protoplasts were harvested by centrifugation. Equal amounts of the two protoplast suspensions were mixed in a solution containing polyethylene glycol (PEG) and calcium ions (Ca^{2+}). After regeneration from the protoplast-fused transformants, fungal strains harboring mycovirus were selected in a medium containing the appropriate antibiotics. Most FgV1-recipient fungal strains showed changed phenotypes including reduced virulence, altered pigmentation, and reduced growth rates while *F. oxysporum* carrying FgV1 displayed only decreased pathogenicity. These data indicate that protoplast fusion could be used to introduce FgV1 into other fungal species such as *Fusarium* and *Cryphonectria* species and that such transmission can result in morphological changes and hypovirulence.

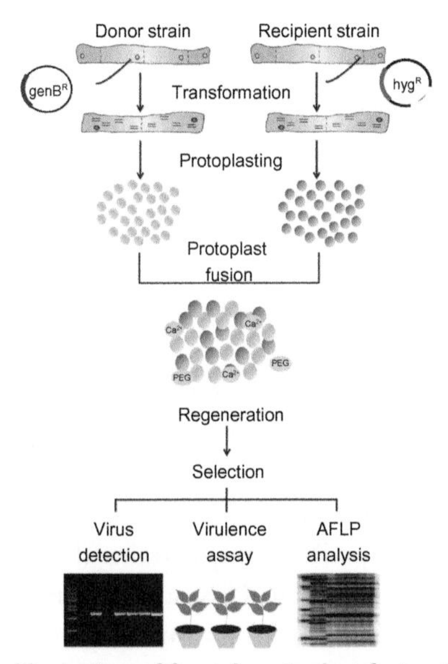

Figure 10.3 Schematic illustration of fungal protoplast fusion for mycovirus transmission. Plasmid DNAs expressing an antibiotic resistance gene are transformed into donor and recipient fungal strains. Protoplasts are isolated and fused in a medium containing PEG and calcium ions. After regeneration, protoplast-fused transformants are selected on the medium carrying the appropriate antibiotic. Transmission of mycovirus can be checked by RT-PCR, a virulence assay, and AFLP (amplified fragment length polymorphism) analysis. (See Page 21 in Color Section at the back of the book.)

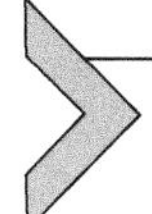

6. FUTURE DIRECTIONS FOR RESEARCH ON MYCOVIRUS–HOST INTERACTIONS

Many mycoviruses infecting *Fusarium* species have been identified and characterized, and some mycovirus genomes have been fully sequenced. Functional characterization of *Fusarium* mycoviruses requires the development of infectious clones. Moreover, the determination of differences between mycoviruses that cause or fail to cause hypovirulence could provide useful information for development of these viruses for the biological control of plant-pathogenic fungi. In addition, the publication of the complete genome sequence for *F. graminearum* and next-generation sequencing now make it possible to examine in detail the transcriptional genome-wide changes in *F. graminearum* upon mycovirus infection. Currently, we are using RNA-Seq to perform a comprehensive transcriptome analysis of *F. graminearum* when infected with different mycoviruses. Infection by a single mycovirus and by combinations of multiple mycoviruses will be used to determine how mycoviruses alter transcriptional regulation in *Fusarium*. *Fusarium* secretome analysis upon mycovirus infection will reveal which fungal proteins are required for hypovirulence. Finally, interactions between viral RNA and host protein or between viral protein and host protein should elucidate the functional roles of host proteins in mycovirus life cycles.

ACKNOWLEDGMENTS

This research was supported by grants from the Center for Fungal Pathogenesis (No. 20110000959) funded by the Ministry of Education, Science, and Technology (MEST) and the Next-Generation BioGreen 21 Program (Nos. PJ00819801 & PJ00798402), Rural Development Administration, Republic of Korea. W. K. C., K. M. L., J. Y., and M. S. were supported by research fellowships from the MEST through Brain Korea 21 Project.

REFERENCES

Agarkova, I., Dunigan, D., Gurnon, J., Greiner, T., Barres, J., Thiel, G., et al. (2008). Chlorovirus-mediated membrane depolarization of chlorella alters secondary active transport of solutes. *Journal of Virology*, *82*, 12181–12190.

Aminian, P., Azizollah, A., Abbas, S., & Naser, S. (2011). Effect of double-stranded RNAs on virulence and deoxynivalenol production of *Fusarium graminearum* isolates. *Journal of Plant Protection Research*, *51*, 29–37.

Breitbart, M., & Rohwer, F. (2005). Here a virus, there a virus, everywhere the same virus? *Trends in Microbiology*, *13*, 278–284.

Cho, W. K., Yu, J., Lee, K. M., Son, M., Min, K., Lee, Y. W., et al. (2012). Genome-wide expression profiling shows transcriptional reprogramming in *Fusarium graminearum* by Fusarium graminearum virus 1-DK21 infection. *BMC Genomics*, *13*, 173.

Chu, Y. M., Jeon, J. J., Yea, S. J., Kim, Y. H., Yun, S. H., Lee, Y. W., et al. (2002). Double-stranded RNA mycovirus from *Fusarium graminearum*. *Applied and Environmental Microbiology*, *68*, 2529–2534.

Chu, Y. M., Lim, W. S., Yea, S. J., Cho, J. D., Lee, Y. W., & Kim, K. H. (2004). Complexity of dsRNA mycovirus isolated from *Fusarium graminearum*. *Virus Genes*, *28*, 135–143.

Compel, P., Papp, I., Bibo, M., Fekete, C., & Hornok, L. (1999). Genetic interrelationships and genome organization of double-stranded RNA elements of *Fusarium poae*. *Virus Genes*, *18*, 49–56.

Cuomo, C. A., Güldener, U., Xu, J. R., Trail, F., Turgeon, B. G., Di Pietro, A., et al. (2007). The *Fusarium graminearum* genome reveals a link between localized polymorphism and pathogen specialization. *Science*, *317*, 1400–1402.

Darissa, O., Willingmann, P., Schäfer, W., & Adam, G. (2011). A novel double-stranded RNA mycovirus from *Fusarium graminearum*: Nucleic acid sequence and genomic structure. *Archives of Virology*, *156*, 647–658.

Dawe, A. L., Van Voorhies, W. A., Lau, T. A., Ulanov, A. V., & Li, Z. (2009). Major impacts on the primary metabolism of the plant pathogen *Cryphonectria parasitica* by the virulence-attenuating virus CHV1-EP713. *Microbiology*, *155*, 3913–3921.

De Wet, J., Preisig, O., Wingfield, B. D., & Wingfield, M. J. (2008). Patterns of multiple virus infections in the conifer pathogenic fungi, *Diplodia pinea* and *Diplodia scrobiculata*. *Journal of Phytopathology*, *156*, 725–731.

Diepeningen, A. D., Debets, A., Slakhorst, S., Fekete, C., Hornok, L., & Hoekstra, R. (2000). Interspecies virus transfer via protoplast fusions between *Fusarium poae* and black *Aspergillus* strains. *Fungal Genetics Newsletter*, *47*, 99–100.

El-Sherbeini, M., & Bostian, K. A. (1987). Viruses in fungi: Infection of yeast with the K1 and K2 killer viruses. *Proceedings of the National Academy of Sciences of the United States of America*, *84*, 4293–4297.

Fekete, C., Giczey, G., Papp, I., Szabó, L., & Hornok, L. (1995). High-frequency occurrence of virus-like particles with double-stranded RNA genome in *Fusarium poae*. *FEMS Microbiology Letters*, *131*, 295–299.

Fujimura, T., Esteban, R., & Wickner, R. B. (1986). In vitro LA double-stranded RNA synthesis in virus-like particles from *Saccharomyces cerevisiae*. *Proceedings of the National Academy of Sciences of the United States of America*, *83*, 4433.

Gill, S. S., & Tuteja, N. (2010). Polyamines and abiotic stress tolerance in plants. *Plant Signaling & Behavior*, *5*, 26–33.

Gupta, S. (1991). Newer evidence to demonstrate mycovirus of *Fusarium moniliforme* var. *glutinans* as causal agent of mango shoot malformation. *Journal of the Entomological Research Society*, *15*, 222–228.

Heaton, L. A., & Leslie, J. F. (2004). Double-stranded RNAs associated with *Fusarium proliferatum* mitochondria. *Mycological Progress*, *3*, 193–198.

Herrero, N., Sánchez Márquez, S., & Zabalgogeazcoa, I. (2009). Mycoviruses are common among different species of endophytic fungi of grasses. *Archives of Virology*, *154*, 327–330.

Joyce, A. R., & Palsson, B. Ø. (2006). The model organism as a system: Integrating 'omics' data sets. *Nature Reviews. Molecular Cell Biology*, *7*, 198–210.

Kanematsu, S., Sasaki, A., Onoue, M., Oikawa, Y., & Ito, T. (2010). Extending the fungal host range of a partitivirus and a mycoreovirus from *Rosellinia necatrix* by inoculation of protoplasts with virus particles. *Phytopathology*, *100*, 922–930.

Kilic, O., & Griffin, G. (1998). Effect of dsRNA-containing and dsRNA-free hypovirulent isolates of *Fusarium oxysporum* on severity of *Fusarium* seedling disease of soybean in naturally infested soil. *Plant and Soil*, *201*, 125–135.

Kim, S., Ryabov, E., Brown, J., & Taliansky, M. (2004). Involvement of the nucleolus in plant virus systemic infection. *Biochemical Society Transactions*, *32*, 557–560.

King, A. M. Q., Adams, M. J., Carstens, E. B., & Lefkowitz, E. J. (2011). *Virus taxonomy: Ninth report of the international committee on taxonomy of viruses*. San Diego: Elsevier.
Kwon, S. J., Cho, S. Y., Lee, K. M., Yu, J., Son, M., & Kim, K. H. (2009). Proteomic analysis of fungal host factors differentially expressed by *Fusarium graminearum* infected with Fusarium graminearum virus-DK21. *Virus Research*, *144*, 96–106.
Kwon, S. J., Lim, W. S., Park, S. H., Park, M. R., & Kim, K. H. (2007). Molecular characterization of a dsRNA mycovirus, Fusarium graminearum virus-DK21, which is phylogenetically related to hypoviruses but has a genome organization and gene expression strategy resembling those of plant potex-like viruses. *Molecules and Cells*, *30*, 304–315.
Lee, K. M., Yu, J., Son, M., Lee, Y. W., & Kim, K. H. (2011). Transmission of *Fusarium boothii* mycovirus via protoplast fusion causes hypovirulence in other phytopathogenic fungi. *PloS One*, *6*, e21629.
Marasas, W. F. O., Nelson, P. E., & Toussoun, T. A. (1984). *Toxigenic* Fusarium *species: Identity and mycotoxicology*. University Park, Pennsylvania: The Pennsylvania State University Press.
Márquez, L. M., Redman, R. S., Rodriguez, R. J., & Roossinck, M. J. (2007). A virus in a fungus in a plant: Three-way symbiosis required for thermal tolerance. *Science*, *315*, 513–515.
McCabe, P. M., Pfeiffer, P., & Van Alfen, N. K. (1999). The influence of dsRNA viruses on the biology of plant pathogenic fungi. *Trends in Microbiology*, *7*, 377–381.
Nemeth, A., & Langst, G. (2011). Genome organization in and around the nucleolus. *Trends in Genetics*, *27*, 149–156.
Nogawa, M., Kageyama, T., Nakatani, A., Taguchi, G., Shimosaka, M., & Okazaki, M. (1996). Cloning and characterization of mycovirus double-stranded RNA from the plant pathogenic fungus, *Fusarium solani* f. sp. *robiniae*. *Bioscience, Biotechnology, and Biochemistry*, *60*, 784–788.
Nogawa, M., Shimosaka, M., Kageyama, T., & Okazaki, M. (1993). A double-stranded RNA mycovirus from the plant pathogenic fungus Fusarium solani f. sp. robiniae. *FEMS Microbiology Letters*, *110*, 153–157.
Nuss, D. L. (1992). Biological control of chestnut blight: An example of virus-mediated attenuation of fungal pathogenesis. *Microbiological Reviews*, *56*, 561–576.
Nuss, D. L. (2005). Hypovirulence: Mycoviruses at the fungal–plant interface. *Nature Reviews. Microbiology*, *3*, 632–642.
Parry, D., Jenkinson, P., & McLeod, L. (1995). Fusarium ear blight (scab) in small grain cereals—a review. *Plant Pathology*, *44*, 207–238.
Pearson, M. N., Beever, R. E., Boine, B., & Arthur, K. (2009). Mycoviruses of filamentous fungi and their relevance to plant pathology. *Molecular Plant Pathology*, *10*, 115–128.
Phalip, V., Delalande, F., Carapito, C., Goubet, F., Hatsch, D., Leize-Wagner, E., et al. (2005). Diversity of the exoproteome of *Fusarium graminearum* grown on plant cell wall. *Current Genetics*, *48*, 366–379.
Polashock, J. J., & Hillman, B. I. (1994). A small mitochondrial double-stranded (ds) RNA element associated with a hypovirulent strain of the chestnut blight fungus and ancestrally related to yeast cytoplasmic T and W dsRNAs. *Proceedings of the National Academy of Sciences of the United States of America*, *91*, 8680–8684.
Scott-Craig, J. S., Adhikari, N. D., Cuomo, C. A., & Walton, J. D. (2007). Comparative proteomics of extracellular proteins in vitro and in planta from the pathogenic fungus *Fusarium graminearum*. *Proteomics*, *7*, 3171–3183.
Sharzehei, A., Banihashemi, Z., & Afsharifar, A. (2007). Detection and characterization of a double-stranded RNA mycovirus in *Fusarium oxysporum* f. sp. *melonis*. *Iranian J. Plant Pathology*, *43*, 9–26.
Shelest, E. (2008). Transcription factors in fungi. *FEMS Microbiology Letters*, *286*, 145–151.

Tavantzis, S. M., Romaine, C. P., & Smith, S. H. (1980). Purification and partial characterization of a bacilliform virus from *Agaricus bisporus*: A single-stranded RNA mycovirus. *Virology*, *105*, 94–102.

Taylor, R. D., Saparno, A., Blackwell, B., Anoop, V., Gleddie, S., Tinker, N. A., et al. (2008). Proteomic analyses of *Fusarium graminearum* grown under mycotoxin-inducing conditions. *Proteomics*, *8*, 2256–2265.

Urayama, S., Kato, S., Suzuki, Y., Aoki, N., Le, M. T., Arie, T., et al. (2010). Mycoviruses related to chrysovirus affect vegetative growth in the rice blast fungus *Magnaporthe oryzae*. *The Journal of General Virology*, *91*, 3085–3094.

Van de Wouw, A. P., & Howlett, B. J. (2011). Fungal pathogenicity genes in the age of 'omics'. *Molecular Plant Pathology*, *12*, 507–514.

Van Diepeningen, A. D., Debets, A. J. M., & Hoekstra, R. F. (1998). Intra-and interspecies virus transfer in *Aspergilli* via protoplast fusion. *Fungal Genetics and Biology*, *25*, 171–180.

Wren, J. D., Roossinck, M. J., Nelson, R. S., Scheets, K., Palmer, M. W., & Melcher, U. (2006). Plant virus biodiversity and ecology. *PLoS Biology*, *4*, e80.

Yu, J., Kwon, S. J., Lee, K. M., Son, M., & Kim, K. H. (2009). Complete nucleotide sequence of double-stranded RNA viruses from *Fusarium graminearum* strain DK3. *Archives of Virology*, *154*, 1855–1858.

Yu, J., Lee, K.-M., Son, M., & Kim, K.-H. (2011). Molecular characterization of Fusarium graminearum virus 2 isolated from *Fusarium graminearum* strain 98-8-60. *Journal of Plant Pathology*, *27*, 285–290.

CHAPTER ELEVEN

Viruses of *Helminthosporium* (*Cochlioblus*) *victoriae*

Said A. Ghabrial*[,1], Sarah E. Dunn†, Hua Li*, Jiatao Xie* and Timothy S. Baker‡

*Department of Plant Pathology, University of Kentucky, Lexington, Kentucky, USA
†Department of Chemistry and Biochemistry, University of California, San Diego, La Jolla, California, USA
‡Department of Chemistry and Biochemistry, and Division of Biological Sciences, University of California, San Diego, La Jolla, California, USA
[1]Corresponding author: e-mail address: saghab00@email.uky.edu

Contents

1. Introduction 290
2. Historical Perspectives 291
3. Viruses of *H. victoriae* 293
   3.1 Helminthosporium victoriae virus 190S (HvV190S)—genus *Victorivirus*, family *Totiviridae* 293
   3.2 Helminthosporium victoriae virus 145S (HvV145S)—family *Chrysoviridae* 300
   3.3 Viral etiology of the *H. victoriae* disease 305
4. Host Genes Upregulated by Virus Infection 306
   4.1 Hv-p68 306
   4.2 *Victoriocin gene (vin)* 308
   4.3 *P30* gene 313
5. HvV190S Capsid Structure 314
   5.1 Comparison of the HvV190S capsid with that of other totiviruses 316
6. Concluding Remarks 320
Acknowledgments 321
References 321

Abstract

The enigma of the transmissible disease of *Helminthosporium victoriae* has almost been resolved. Diseased isolates are doubly infected with two distinct viruses, the victorivirus Helminthosporium victoriae virus 190S and the chrysovirus HvV145S. Mixed infection, however, is not required for disease development. DNA transformation experiments and transfection assays using purified HvV190S virions strongly indicate that HvV190S alone is necessary for inducing disease symptoms. HvV145, like other chrysoviruses, appears to have no effect on colony morphology. This chapter will discuss the molecular biology of the two viruses and summarize recent results of characterization of host gene products upregulated by virus infection. Furthermore, the novel structural features of HvV190S capsid will be highlighted.

Advances in Virus Research, Volume 86
ISSN 0065-3527
http://dx.doi.org/10.1016/B978-0-12-394315-6.00011-8

1. INTRODUCTION

The senior author of this chapter has written several reviews in the past on the viruses that infect the plant pathogenic fungus *Helminthosporium* (*Cochlioblus*) *victoriae*. However, this review is special because it is the last that he will write. In this review, we would like to reflect on the times and people who were involved and most importantly the accomplishments: how much we do know about the *H. victoriae* system since 1959 when the first report on the transmissible disease of *H. victoriae* was published (Lindberg, 1959).

As a virology graduate student (1961–1965) at the Plant Pathology Department, Louisiana State University (LSU), the senior author was intrigued by the etiology of the transmissible disease of *H. victoriae* and by the pathotoxin "victorin," a cyclized pentapeptide of 814 Da (Wolpert et al., 1985) which is produced by virulent isolates of *H. victoriae*. The findings that victorin reproduces the disease symptoms in the absence of the fungus and that cell-free, fungal culture filtrate diluted one million-fold was not only toxic but also exhibited the same specificity toward oat genotypes as the fungus (Meehan & Murphy, 1947) were of considerable interest. During his tenure at LSU, Professors G. D. Lindberg and H. E. Wheeler were directing nationally recognized research programs on the transmissible disease of *H. victoriae* and mode of action of victorin, respectively. Nevertheless, the senior author did his research on another fascinating project dealing with the physiological basis of the virus-induced Tabasco pepper wilt. With some insight from Professor Wheeler's research on the effect of victorin treatment on membrane permeability and $K+$ efflux resulting in severe wilt of oat leaves (Wheeler & Black, 1962), my studies determined that changes in membrane permeability and $K+$ efflux were key events in the wilt syndrome of virus-infected Tabasco pepper (Ghabrial & Pirone, 1967).

The interest in the *H. victoriae* system was rekindled when he joined the Plant Pathology Department at the University of Kentucky (UK) in 1972 as an Assistant Professor. Luckily, Professor Wheeler had moved earlier to UK from LSU and supplied him with various normal and abnormal isolates of *H. victoriae*. Because of lack of funding for fungal virus projects in the 1970s and 1980s, the *H. victoriae* project was secondary to his plant virus research responsibilities. The objectives at the initiation of studies on viruses of *H. victoriae* in 1976 were to address the following three questions:

i. Are viruses associated with the disease phenotype? If so, then isolate the viruses involved and characterize them at the biochemical and molecular levels.

ii. What is the evidence for virus etiology of the transmissible disease of *H. victoriae*?

iii. Is virus infection required for vectorin production? Or does virus infection alter victorin production and thus virulence of *H. victoriae* isolates?

The answers to these questions and more are listed among the highlights of our journey with *H. victoria* viruses under Section 2.

2. HISTORICAL PERSPECTIVES

H. victoriae was first described in 1946 as the causal agent of a new disease in oats called Victoria blight, named after the parent cultivar, Victoria (Meehan & Murphy, 1946). The Victoria blight disease arose after the introduction of the *Pc-2* resistance gene to a completely different disease, crown rust of oats (causal agent: *Puccinia cronata*), which was genetically linked to susceptibility to a previously unknown soil fungus, *H. victoriae* (Litzenberger, 1949). The disease caused by *H. victoriae* rose to epidemic proportions in 1947 and 1948 and resulted in serious yield losses in most oat-growing regions of the United States (Litzenberger, 1949).

Although Victoria-derived oat cultivars were subsequently abandoned in the major oat-cropping areas, they continued to be grown in relatively large acreage in some of the Southern States during the 1950s (Lindberg, 1960; Scheffer & Nelson, 1967). The discovery in 1959 of a transmissible disease of *H. victoriae* was based on observed cultural abnormalities in isolates obtained from blighted "Victorgrain" oat plants in Louisiana (Lindberg, 1959). Lindberg observed that some of the *H. victoriae* colonies isolated from diseased oats were stunted and highly sectored. Following an initial period of normal growth, sectors appeared at the margins of the colonies with concomitant collapse of existing aerial mycelium and almost complete inhibition of colony expansion (Psarros and Lindberg, 1962). Lindberg referred to these abnormalities as a "disease" of the fungus and showed that the disease could be transmitted to normal colonies via hyphal anastomosis with diseased colonies. Since the fields of "Victorgrain" oats in Louisiana from which diseased and normal *H. victoriae* were isolated did not suffer significant yield losses, which was unusual for this serious disease of oats, it was suggested that a reduction in the pathogenicity level of *H. victoriae* had occurred and that this might be attributed to the transmissible disease of the fungus.

Over 50 years ago, Lindberg suggested a viral role in the disease of *H. victoriae*, but it was the work in the senior author's laboratory in the late 1970s that presented evidence for virus involvement. Diseased isolates of *H. victoriae* have been found to contain two serologically and electrophoretically distinct viruses designated according to their sedimentation values as the 190S and 145S viruses (Ghabrial, Sanderlin, & Calvert, 1979; Sanderlin & Ghabrial, 1978). Further studies on these two viruses led to their molecular characterization and the creation of a new family of mycoviruses (*Chrysoviridae*) and a new genus (*Victorivirus*) in the family *Totiviridae*. In addition, some highlights of the *H. victoriae*-virus quest include

- First report of association of two distinct viruses (HvV190S and HvV145S) with diseased isolates of *H. victoriae* (Sanderlin & Ghabrial, 1978).
- Virulent fungal isolates known to be potent producers of victorin were found to be virus free. Thus, virus infection is not required for victorin production. Interestingly, virus-infected isolates produce little or no victorin and are hypovirulent (Ghabrial, 1986).
- Discovery of HvV190S capsid protein (CP) heterogeneity as a consequence of posttranslational modification (phosphorylation and proteolytic processing) and the occurrence of two types of particles that differ in capsid composition and phosphorylation state (Ghabrial, Bibb, Price, Havens, & Lesnaw, 1987; Ghabrial & Havens, 1989, 1992).
- HvV190S was the first totivirus infecting a filamentous fungus to be characterized at the molecular level. HvV190S was the only virus from filamentous fungi that is listed as a member of *Totiviridae* (Huang & Ghabrial, 1996).
- Molecular characterization of HvV190S (Huang & Ghabrial, 1996) and elucidation of the expression strategy (stop-restart) of its RNA-dependent RNA polymerase (RdRp) led to the realization that many HvV190S-like viruses that infect filamentous fungi should be grouped together in a separate genus (*Victorivirus*; derived from the species name of the host, *H. victoriae*; Ghabrial & Nibert, 2009; Li, Havens, Nibert, & Ghabrial, 2011).
- Established that Penicillium chrysogenum virus (PcV), like HvV145S, has a quadripartite genome. This ended the confusion about the taxonomic status of PcV and HvV145S, and helped to create the new family *Chrysoviridae* for mycoviruses with quadripartite genomes (Jiang &

Ghabrial, 2004). Identification and molecular characterization of host gene products that are upregulated by virus infection (e.g., the multifunctional protein Hv-p68 with alcohol oxidase/protein kinase/ RNA-binding activities and the antifungal preproprotein victoriocin; de Sá, Havens, & Ghabrial, 2010; Soldevila & Ghabrial, 2001).

- Presented evidence in support of a viral etiology of the disease of *H. victoriae* (Ghabrial, 2008b; Li, Havens, & Ghabrial, 2013).
- Collaboration with structure biology laboratories revealed novel capsid structural features of mycoviruses (this chapter, Chapters 3 and 4, this volume). This was made possible because of the ability to purify well-characterized victoriviruses and chrysoviruses (e.g., HvV190S and PcV as well as selected partitiviruses) in good quantity and high quality.

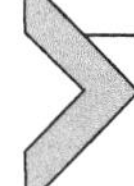

3. VIRUSES OF *H. VICTORIAE*

3.1. Helminthosporium victoriae virus 190S (HvV190S)—genus *Victorivirus*, family *Totiviridae*

Like the majority of members of *Totiviridae*, the nonsegmented HvV190S dsRNA genome comprises two large overlapping open reading frames (ORFs). The 5′ proximal ORF (ORF 1) codes for the CP, whereas the 3′ ORF (ORF 2), which is in the −1 frame relative to ORF 1, encodes the RdRp (Fig. 11.1) with its characteristic eight conserved motifs (Huang & Ghabrial, 1996). Several lines of evidence (Huang & Ghabrial, 1996) indicate that translation of ORF1 starts at the AUG at position 290. This initiation codon, however, resides in an unfavorable sequence context (UCCAUGU). The 5′ end of the positive strand of the dsRNA genome is uncapped and highly structured and contains a relatively long (289 nt) 5′ untranslated region (5′ UTR). These structural features of the 5′ UTR of HvV190S positive sense RNA predict that the CP ORF (with its AUG present in suboptimal context) is possibly translated via a cap-independent mechanism. Site-directed mutagenesis verified that the stop codon (UGA) at position 2606–2608 is the authentic termination codon for the CP ORF (Huang & Ghabrial, 1996; Huang, Soldevila, Webb, & Ghabrial, 1997). The downstream ORF of the HvV190 dsRNA genome, which encodes the RdRp, is in a −1 frame with respect to the CP ORF and its translational start codon (nt positions 2605–2607) overlaps the stop codon for the upstream CP ORF (nt positions 2606–2608) in the tetranucleotide sequence 2605-AUGA-2608 (Fig. 11.1A).

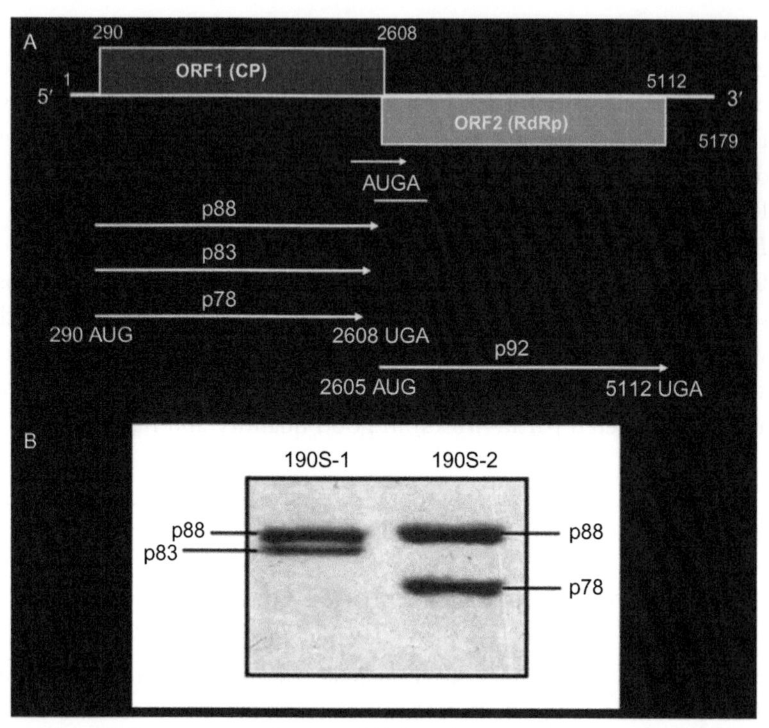

Figure 11.1 (A) Genome organization of Helminthosporium victoriae 190S virus, the type species of the newly recognized genus *Victorivirus*. The dsRNA genome encompasses two large overlapping ORFs with the 5′ ORF encoding a capsid protein (CP) and the 3′ ORF encoding an RNA-dependent RNA polymerase (RdRp). Note that the termination codon of the CP ORF overlaps the initiation codon of the RdRp ORF in the tetranucleotide sequence AUGA. Although the capsid is encoded by a single gene, three related polypeptides (p88, p83, and p78) are resolved by SDS-PAGE of purified virions. Whereas p88 is the primary translation product, p83 and p78 are generated via proteolytic processing of p88 at its C-terminus. (B) Two types of particles (190S-1 and 190S-2) that differ slightly in sedimentation and capsid composition are resolved by sucrose density gradient centrifugation of virion preparations. The 190S-1 capsids contain p88 and p83, occurring in approximately equimolar amounts, and the 190S-2 capsids comprise similar amounts of p88 and p78. The capsids p88 and p83 are phosphoproteins, whereas p78 is nonphosphorylated (Ghabrial & Havens, 1992). (For color version of this figure, the reader is referred to the online version of this chapter.)

Although a single gene encodes the capsid of Hv190SV like other totiviruses, the HvV190S capsid comprises two closely related major CPs, either p88 and p83 or p88 and p78 (Fig. 11.1). Whereas p88 is the primary translation product of ORF1, p83 and p78 are C-terminally cleaved forms of p88, which are virion associated but are not required for capsid assembly (Soldevila, Huang, & Ghabrial, 1998). Capsid heterogeneity and posttranslational modification of the primary translation product may be a common feature of victoriviruses (Nomura, Osahki, Iwanami, Matsumoto, & Ohtsu, 2003). Purified HvV190S virion preparations contain two types of particles, 190S-1 and 190S-2, which differ slightly in sedimentation rates (190S-1 is resolved as a shoulder on the slightly faster sedimenting component 190S-2) and capsid composition (Fig. 11.1B). The 190S-1 capsids contain p88 and p83, occurring in approximately equimolar amounts, and the 190S-2 capsids comprise similar amounts of p88 and p78. Though p78 is nonphosphorylated, p88 and p83 are phosphoproteins (Ghabrial & Havens, 1992). Phosphorylation and proteolytic processing are proposed to play a role in the virus life cycle; phosphorylation of CP may be necessary for its interaction with viral nucleic acid and/or phosphorylation may regulate dsRNA transcription/replication. Phosphorylation/dephosphorylation may regulate viral dsRNA transcription; the finding that the more highly phosphorylated virions of the 190S-1 component were more efficient in transcriptase activity than those of the 190S-2 component is of interest in this regard (Ghabrial & Havens, 1992). Proteolytic processing and cleavage of a C-terminal peptide, which leads to dephosphorylation and the conversion of p88 to p78, may play a role in the release of the plus-strand RNA transcripts from virions (Ghabrial & Havens, 1992).

The overlap regions in the dsRNA genomes of the majority of victoriviruses (Fig. 11.2B) are of the AUGA type, where the initiation codon of the RdRp ORF overlaps the termination codon of the CP ORF. This suggests that the expression of RdRp occurs by a mechanism different from the translational frameshifting utilized by most viruses in the family *Totiviridae* to express their RdRps. It is of interest that the overlap region in one of the newly reported victoriviruses, BbRV1, is of the TAAUG type (with similarity to the overlap region in hypoviruses; Guo, Sun, Chiba, Araki, & Suzuki, 2009). There is no overlap region for another recently reported victorivirus, TcV1, as the CP stop codon and the RdRp initiation codon (TAAAUG) are positioned side by side with no spacer sequence (Herrero & Zabalgogeazcoa, 2011). The complete nucleotide sequences of Hv190SV and 13 other members and probable members of the genus *Victorivirus* have been reported and deposited in GenBank. The genomic, biochemical, and structural properties of HvV190S are the best characterized among all victoriviruses. The RdRp-encoding ORF 2 of HvV190S is expressed via a stop–restart (coupled termination–reinitiation) mechanism (Li et al., 2011; Soldevila & Ghabrial, 2000). The HvV190S RdRp is detectable as a separate, virion-associated component, consistent

A

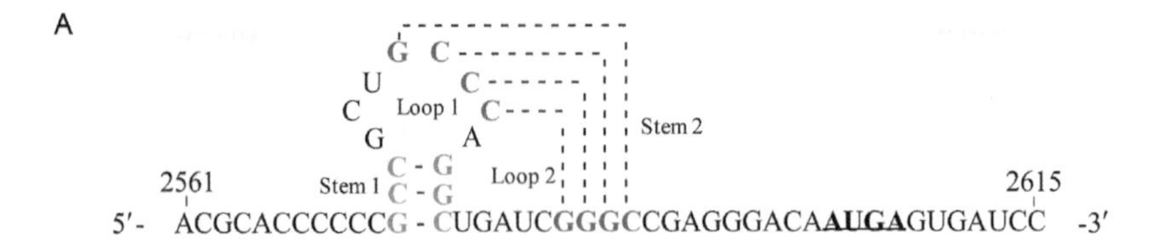

B

Virus	RNA Sequence	MFE
BbRV1	CCCTGTCGCAATTGCCGGTGCTGCCCCACCCGGAGGGCCGAACCCCGAGTAATGG	-14.0
BfTV1	UACCCCUCCGCGUUCCCCUGCUGCCCCAGGUAACGGGCCGGACACGAUAUGAGUG	-13.6
CeRV1	GUAGUAUCUUCUUCUGUGGCUGCCCCCACUAAUGAGGGCCGAAACGAUGUCUAGA	-14.3
CmRV	AGGUGGUGACGUUCACGGUGCCGCCGGAGCCUUACCGGCUCCACAAGUAUGAUCG	-16.0
EfV1	UCCUCCCAUGCCUGAGGCUGCCACCGGUGCCGAACCGGUGCCCCAAGCAUGAUUG	-21.4
GaRV-L1	CGCCCACCUGCCAGCCGAAGCUACUGCUGCUGAAGCAGUGCCCGCUCAAUGAUUG	-16.5
HmTV-17	GGCUGAAGCCAUGCAACAGGACGCAGCCCUGCAAGCUGCGGGGGCUCAAUGAAGG	-20.2
HvV190S	CAUCCACGCACCCCCCGCCGCUGCCCAGGCUGAUCGGGCCGAGGGACAAUGAGUG	-17.0
MoV1	AGGGCCGACCCCCGAACCCCACGGCGCGGCGCCUAGCCUGCACGAAUAGAUAUGG	-13.8
MoV2	AGGCAAUAACAACGAGCAGGCCGCCGGCGCCCAGCCGGCCCCCGCUCAAUGAUUG	-23.4
SsRV1	GCCCCCCCUGUAGCACCCGGCCCCGCCCAGCCUGACGGGCCCGCCAAUGAAUAAG	-16.2
TcV1	CACTGCTCCCCCTGTTGCCGACCCCCAGGCTATGGTGGGCGACCAGATCTAAATG	-16.6

Figure 11.2 (A) H-type pseudoknot predicted for stop–restart region of HvV190S. A predicted RNA pseudoknot structure is encoded near the 3′ end of ORF1 preceding the stop–restart site. The nucleotide positions are indicated, and the stop–restart site is in boldface and underlined. Dashed lines indicate base pairs predicted to form stem 1 (green) and stem 2 (red). (B) Sequences of HvV190S-like fungal viruses (victoriviruses) predicted to form a pseudoknot structure upstream of the stop and restart sites in each. Stem 1 (green) and stem 2 (red) are color coded to match panel A. Virus names and abbreviations: BbRV1, Beauveria bassiana RNA virus; BfTV1, Botryotinia fuckeliana totivirus 1; CeRV1, Chalara elegans RNA virus 1; CmRV, Coniothyrium minitans RNA virus; EfV1, Epichloe festucae virus 1; GaRV-L1, Gremmeniella abietina RNA virus L1; HmTV1–17, Helicobasidium mompa totivirus 1–17; HvV190S, Helminthosporium victoriae virus 190S; MoV1 and MoV2, Magnaporthe oryzae viruses 1 and 2; SsRV1, Sphaeropsis sapinea RNA virus 1 and TcV1, Tolypocladium cylindrosporum virus 1. MFE, minimum free energy value for structure prediction at 37°C. (For interpretation of the references to color in this figure legend, the reader is referred to the online version of this chapter.)

with its independent translation from ORF 2. The tetranucleotide AUGA overlap region, or a very similar structure, is characteristic of the overlap region of all victoriviruses (Fig. 11.2B; Ghabrial & Nibert, 2009; Li et al., 2011). Termination–reinitiation primarily depends on a 32-nt stretch of RNA immediately upstream of the AUGA motif, including a predicted pseudoknot structure (Fig. 11.2A). The close proximity by which this predicted structure is followed by the ORF 1 stop codon appears to be particularly important for promoting translation of the downstream ORF. Similar sequence motifs and predicted RNA structures in other victoriviruses suggest that they all share a related stop–restart strategy for RdRp translation (Fig. 11.2B). Members of genus *Victorivirus* thus provide unique opportunities for exploring molecular mechanisms of translational coupling. It is noteworthy that translation of the downstream RdRp of HvV190S via coupled termination–reinitiation represents the first example of this expression strategy to be identified in a dsRNA virus.

The finding that translation of the ORF 1 and termination at its stop codon are absolute requirements for reinitiation at the victorivirus RdRp start codon distinguishes this process from IRES-mediated initiation (Hellen & Sarnow, 2001), as well as from leaky scanning (Kozak, 2002) and ribosomal shunting (Ryabova & Hohn, 2000). Leaky scanning can be further ruled out as there are >20 AUG codons in reasonably favorable context upstream of the RdRp start codon. The fact that RdRp is expressed from its downstream start codon as a separate, nonfused protein also distinguishes this process from ribosomal frameshifting (Dinman, Icho, & Wickner, 1991) and in-frame read-through of termination codons (Dreher & Miller, 2006), which generate fusion proteins.

The stop–restart strategy for expressing the downstream RdRp ORF allows HvV190S to produce two separate proteins from a single mRNA. This strategy furthermore allows HvV190S to regulate the level of RdRp expression (lower) relative to CP expression (much higher), consistent with the fact that only one or two RdRp molecules are present in each HvV190S virion versus 120 CP molecules (Castón et al., 2006). It is reasonable to predict that the abundance of RdRp molecules must be regulated so that the appropriate copy number is available for proper virion assembly and competence for replication (Dinman & Wickner, 1992). Although it is not known how the plus-strand RNA of HvV190S is packaged into assembling progeny virions, it is thought by analogy to Saccharomyces cerevisiae virus L-A (ScV-L-A), the prototype of both genus *Totivirus* and family *Totiviridae*, that the RdRp, through its binding affinity for single-stranded RNA, would recruit the progeny plus-strand RNA to ensure its packaging into particles (Dinman & Wickner, 1992). The identification of a predicted pseudoknot as a key determinant of the stop–restart mechanism in HvV190S and other victoriviruses is exciting. The predicted pseudoknot must be located closely upstream of the CP stop codon (Li et al., 2011), presumably so that it can tether the terminating ribosome or components thereof, which can then reinitiate with limited efficiency at the next downstream start codon (see Powell, Brown, & Brierley, 2008 for a review).

Among viral mRNAs, the mechanism of translational coupling has perhaps been best studied with feline calicivirus and rabbit hemorrhagic disease virus (Luttermann & Meyers, 2007; Meyers, 2007). In these viruses, 84–87 nt of RNA sequence immediately upstream of the overlapping stop–restart codons, termed the "termination upstream ribosome binding site" (TURBS), are required for efficient reinitiation (Luttermann & Meyers, 2007, 2009; Meyers, 2003, 2007). Two distinct regions (motifs 1 and 2) within the TURBS have been shown to play significant roles. Of these, motif 1 is more important because it is conserved among different caliciviruses and is complementary to a small, single-stranded region at the tip of helix 26 of 18S rRNA, which is juxtaposed to mRNA in the translating ribosome (Matassova, Venjaminova, & Karpova, 1998). This complementary region is thought to tether the mRNA to the 40S subunit, allowing time for the ribosome to acquire the initiation factors necessary for translation of the downstream ORF (Meyers, 2003, 2007). There is also evidence that the TURBS is involved in recruiting eukaryotic initiation factor 3 (eIF3) and eIF3/40S complexes (Pöyry et al., 2007). The multisubunit eIF3 complex plays multiple roles in translation initiation including dissociating the 60S and 40S ribosomal subunits after termination, and it is therefore possible that the TURBS is involved in both ribosome dissociation/recycling and 40S tethering (Powell et al., 2008).

The stop–restart strategy of HvV190S and other victoriviruses is in line with those for the other RNA viruses known to utilize this strategy. Of particular note among all of these viruses is the consistent importance of a region of sequences closely upstream of the stop and restart codons. Interestingly, in the case of victoriviruses, this region (32 nt in length) contains a predicted pseudoknot. Li et al. (2011) presented experimental (via mutational analysis) and genetic evidence for the importance of the predicted pseudoknot sequence in HvV190S. Present results, however, do not rule out a role for sequences upstream of the predicted pseudoknot in the translation of ORF 2. In other viral mRNAs, including ones from other members of the family *Totiviridae*, pseudoknots are known to be structural features that promote translation of an alternative ORF by −1 ribosomal frameshifting (Dinman et al., 1991; Wang, Yang, Shen, & Wang, 1993). Ribosomal pausing induced by the pseudoknot is thought to be important in at least some of those cases (Plant et al., 2003).

The possibility that the predicted pseudoknot sequence upstream of the stop and restart codons of HvV190S and other victoriviruses may interact with the host 18S rRNA was considered, given previous findings with caliciviruses and influenza B viruses (Luttermann & Meyers, 2007, 2009; Powell et al., 2008). Interestingly, a portion of loop 2 in the predicted pseudoknot of HvV190S (Fig. 11.2A) has the sequence CUGAUCG, which is complementary to nt positions 909–915 of the *Cochliobolus* (anamorph: *Helminthosporium*) *sativus* 18S rRNA (Schoch et al., 2006). However, (a) the HvV190S sequence CUGAUCG is not conserved among other victoriviruses (see Fig. 11.2B); (b) the identified region of *Cochliobolus* 18S rRNA does not align with the region of mammalian 18S rRNA implicated in the stop–restart mechanisms of caliciviruses and influenza B viruses; and (c) based on RNA folding predictions, the identified region of *Cochliobolus* 18S rRNA forms part of a stem, unlike the single-stranded region of mammalian 18S rRNA implicated in the stop–restart mechanisms of caliciviruses and influenza B viruses. Therefore, the role of the predicted pseudoknot structure in victoriviruses is likely not to promote base pairing with 18S rRNA. This possibility, however, cannot be ruled out.

The mechanism by which the predicted pseudoknot can bring about coupled translation of victoriviruses is not known. The nonpolyadenylated victorivirus mRNA with a stop codon in the middle of the molecule is reminiscent of a nonsense-containing mRNA without polyA binding protein (PABP) (Amrani et al., 2004). These authors showed that in the absence of PABP the ribosomes tend to remain bound to mRNA in the vicinity of the termination codon. We predict that the secondary structure of the victorivirus mRNA that is upstream of the CP stop codon is present to prevent terminating ribosomes from sliding backward on the mRNA, thus helping to reposition them on the RdRp initiation codon. Although this is an attractive model, the HvV190S mRNA derived from transformants in our experimental system is predicted to be polyadenylated. Furthermore, the required close proximity of the predicted pseudoknot to the CP stop codon would interfere with the reformation of the predicted pseudoknot, particularly stem 2, following termination of CP translation. It is possible that previously proposed unidentified host factors (Soldevila & Ghabrial, 2000) might play a role in the reformation and stabilization of the secondary structure following CP translation.

Phylogenetic analyses of HvV190S and other members of the genus *Victorivirus* reveal that they are more closely related to each other than to any members of other genera in the family *Totiviridae*. Interestingly,

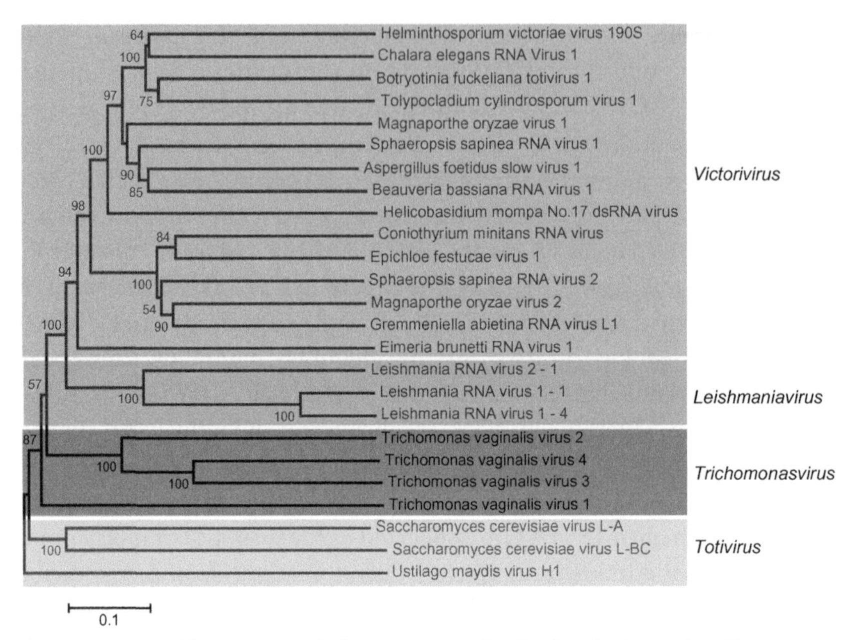

Figure 11.3 Neighbor-joining phylogenetic tree for the family *Totiviridae*. The tree was constructed from complete aa sequences of RdRps of representative members and probable members of the family. The aa sequences were aligned using the program CLUSTAL X2, and the tree was generated for codon positions using the MEGA5 phylogenetic package. Bootstrap percentages out of 2000 replicates are indicated at the nodes. Blue, red, gray, and brown shading, respectively, highlight the assignments of individual viruses to the genera *Victorivirus*, *Leishmaniavirus*, *Trichomonasvirus*, and *Totivirus*. (See Page 21 in Color Section at the back of the book.)

victoriviruses are more closely related to protozoan totiviruses than to the totiviruses that infect yeast and smut fungi (Fig. 11.3). Although HvV190S was originally classified in the genus *Totivirus*, it is obvious that it is phylogenetically distantly related to members of this genus. Thus, removal of HvV190S from *Totivirus* and establishment of the new genus *Victorivirus* to accommodate HvV190S and HvV190S-like viruses that infect filamentous fungi are highly justified (Ghabrial & Nibert, 2009).

3.2. Helminthosporium victoriae virus 145S (HvV145S)—family *Chrysoviridae*

Prior to availability of sequence information, HvV145S and Penicillium chrysogenum virus (PcV) were placed under the genus *Chrysovirus*, family *Partitiviridae* (with PcV designated as the prototype strain of the genus).

The belief was that these viruses (with three or four detectable segments) have bisegmented genomes coding for RdRp and CP and any additional segments were thought to represent satellite or defective RNA molecules. This confusion was clarified when the genomes of HvV145S (Ghabrial, Havens, & Annamalai, 2013; Ghabrial, Soldevila, & Havens, 2002) and PcV (Jiang & Ghabrial, 2004) were completely sequenced and shown to be quadripartite; this eventually led to the creation of the family *Chrysoviridae* (Jiang & Ghabrial, 2004). The genome of HvV145 comprises four dsRNA segments (2.7, 2.9, 3.1, and 3.6 kbp in length); each is monocistronic and has a unique sequence (Ghabrial et al., 2002, 2013). Like the prototype PcV, the HvV145S dsRNAs share common, highly conserved domains at their 5′ (Fig. 11.4A) and 3′ termini (Ghabrial, 2008a; Ghabrial et al., 2002, 2013). Northern hybridization analysis using cloned cDNA probes representing the four dsRNA segments showed that each segment has a unique sequence; no sequence homology to corresponding PcV segments or to the genomic dsRNA of the coinfecting HvV190S was detected (Fig. 11.4B). The 5′ UTRs of HvV145S segments are relatively long, between 200 and 400 nts in length, and their sequences have the potential to form extensive secondary structures. Direct comparison of the nucleotide sequences have revealed that, in addition to the strictly conserved 5′ and 3′ termini, there are regions of high sequence similarity within the 5′ UTRs and 3′ UTRs among all four dsRNAs. A highly conserved 75-nt region with almost 90% identity is present at the 5′ UTR of all four dsRNAs (designated "box 1"). A second region of strong sequence similarity is positioned immediately downstream of the highly conserved box 1. This region consists of a stretch of 30–35 nt (with sequence similarity above 80%) and is composed of a reiteration of the sequence "CAA," which depicts a strong resemblance to the $(CAA)_n$ repeats present in the 5′-UTRs of tobamoviruses (Gallie & Walbot, 1992). The poly (CAA) region of tobamoviruses has been reported to function as translational enhancers (Gallie & Walbot, 1992). Whether the HvV145S "CAA" repeats function like the translational regulatory elements of tobamoviruses has yet to be determined.

HvV145S dsRNA1 encodes the RdRp, dsRNA2 codes for the CP, and dsRNAs 3 and 4 code for proteins of unknown function. Assignment of numbers 1–4 to PcV dsRNAs was made according to their decreasing size. Following the same criterion used for PcV, the dsRNAs associated with HvV145S and other chrysoviruses were accordingly assigned the numbers 1–4. Sequence comparisons, however, indicated that dsRNA3s of Hv145SV, Amasya cherry disease-associated chrysovirus (ACDACV), and

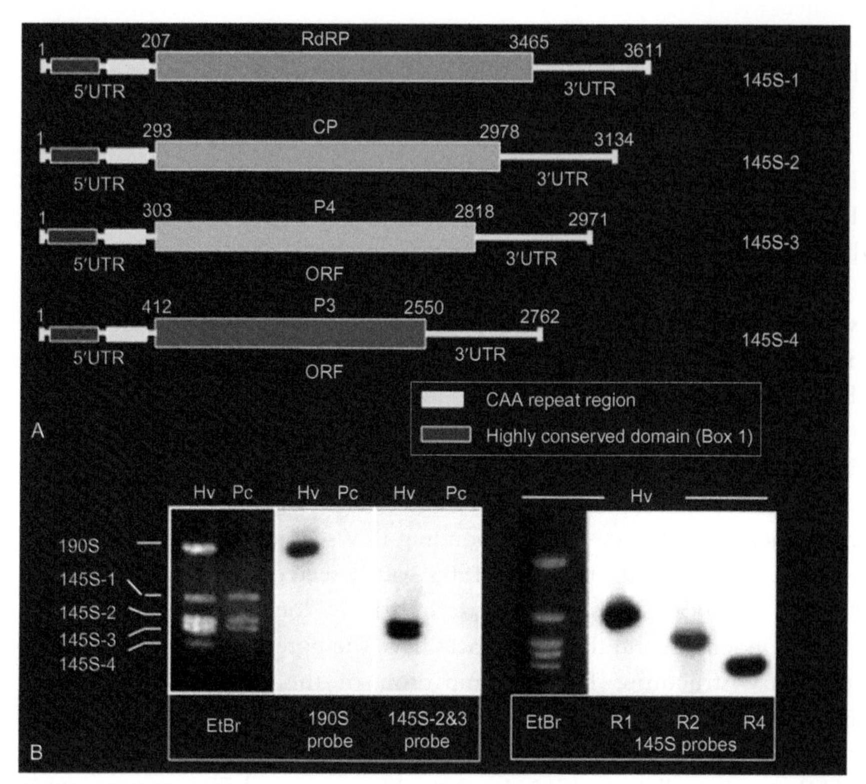

Figure 11.4 (A) Genome organization of Helminthosporium victoriae virus145S (HvV145S). The genome consists of four dsRNA segments, each of which is monocistronic. The RdRp ORF (nt positions 207–3465 on dsRNA1), the CP ORF (nt positions 293–2978 on dsRNA2), the p3 ORF (nt positions 412–2250 on dsRNA4), and the p4 ORF (nt positions 303–2818 on dsRNA3) are represented by rectangular boxes. Note that, unlike PcV, the prototype member of *Chrysovirus*, HvV145S dsRNA 3 codes for P4 and dsRNA 4 codes for P3 (see text). The purple-shaded box near the 5′ end represents a highly conserved domain (box 1), the yellow-shaded box downstream of the purple box is composed of a reiteration of the sequence "CAA." (B) Northern hybridization analysis using cloned cDNA probes representing the four HvV145S dsRNA segments and the HvV190S dsRNA. Results showed that each of the HvV145S dsRNAs has a unique sequence and that they lack sequence homology to both the chrysovirus PcV (lane Pc) and the coinfecting HvV190S. Lanes labeled "Hv" contains all four HvV145S dsRNA segments plus HvV190S dsRNA. (See Page 22 in Color Section at the back of the book.)

possibly Fusarium oxysporum chrysovirus 1 (FoCV1; based on partial sequence) are in fact the counterparts of PcV dsRNA4 rather than dsRNA3. Likewise, dsRNA4 of these chrysoviruses are the counterparts of PcV dsRNA3. Since PcV was the first chrysovirus to be characterized at the molecular level and to avoid confusion, the protein designations P3 and P4 as used for PcV were adopted and referred to as chryso-P3 and chryso-P4 (Ghabrial, 2008a). Thus, whereas the chryso-P3 protein represents the gene product of PcV dsRNA3, it comprises the corresponding gene product of Hv145SVdsRNA4, and so on.

Whereas the prototype PcV dsRNA3 codes for its chryso-P3 protein, Hv145SV, ACDACV, and FoCV1 dsRNA4s encode the corresponding chryso-P3s. Although the functions of chryso-P3 and P4 are not known, sequence analysis and database searches offer some clues. ProDom database searches reveal that chryso-P3 sequences share a "phytoreo S7 domain" with a family (pfam07236:Phytoreo_S7; NBCI) consisting of several phytoreovirus P7 proteins known to be viral core proteins with nucleic acid binding activities. The consensus protein sequence for the three chrysoviruses is [X(V/I)V(M/L) P(A/M)G(C/H) GK(T/S)T-(L/I)]. Phytoreovirus P7 proteins bind to their corresponding P1 (transcriptase/replicase) proteins, which bind to the genomic dsRNAs. It is of interest, in this regard, that the N-terminal regions of all chryso-P3s (encompassing the amino acids within positions 1–500) share significant sequence similarity with comparable N-terminal regions of the putative RdRps encoded by chrysovirus dsRNA1s. The regions in the dsRNA1-encoded proteins with high similarity to chryso-P3 occur upstream of the eight highly conserved motifs characteristic of RdRps of dsRNA viruses of simple eukaryotes. The significance of this sequence similarity to the function of chryso-P3 is not known for certain, but one may speculate that the N-terminal region of these proteins may play a role in viral RNA binding and packaging. The chryso-P4 encoded by chrysoviruses contains the motif PGDGXCXXHX. This motif (I), along with motifs II (with a conserved K), III, and IV (with a conserved H), form the conserved core of the ovarian tumor, gene-like superfamily of predicted cysteine proteases (Covelli et al., 2004). Multiple alignments showed that motifs I–IV are also present in other viruses including Agaricus bisporus Virus 1, a tentative member of the family *Chrysoviridae*. Whether the dsRNAs of these viruses indeed code for the predicted proteases remains to be investigated.

The four HvV145S genome segments are encapsidated separately in identical capsids that are 350–400 Å in diameter (Ghabrial et al., 2002, 2013). BLAST searches of the HvV145S RdRp amino acid sequence showed that it shared significant sequence identity with ACDACV RdRp (42%) followed by the RdRps of Fusarium oxysporum chrysovirus 1 (FoCV1), Cryphonectria nitscheki chrysovirus 1 (CnCV1), PcV, Verticillium dahliae chrysovirus 1 (VdCV1), and Aspergillus fumigatus chrysovirus (AfCV) (41%, 40%, 39%, 39%, and 39%, respectively). High similarities were also found with the RdRps of several members of the family *Totiviridae* (Jiang & Ghabrial, 2004). Interestingly, no significant hits were detected with any of the viruses in the family *Partitiviridae*, another validation for the removal of chrysoviruses from the family *Partitiviridae* and their placement in the recently created family *Chrysoviridae* (Jiang & Ghabrial, 2004). A neighbor-joining phylogenetic tree constructed based on full-length amino acid sequences of RdRps of members and probable members of the family *Chrysoviridae*, as well as selected members of the family *Totiviridae* (Fig. 11.5) is consistent with the sequence identity inferences made earlier. Interestingly, the phylogenetic tree also defines two clusters of chrysoviruses, those in which segment 3 codes for chrysoP3 (PcV, AfCV, CnCV1 and VdCV1, upper cluster) and those in which segment 3 codes for chrysoP4 (HvV145S, ACDACV and FoCV1, lower cluster).

PcV dsRNAs 3 and 4 differ in length by only 74 bp (Jiang & Ghabrial, 2004); therefore, they comigrate when separated by agarose gel electrophoresis.

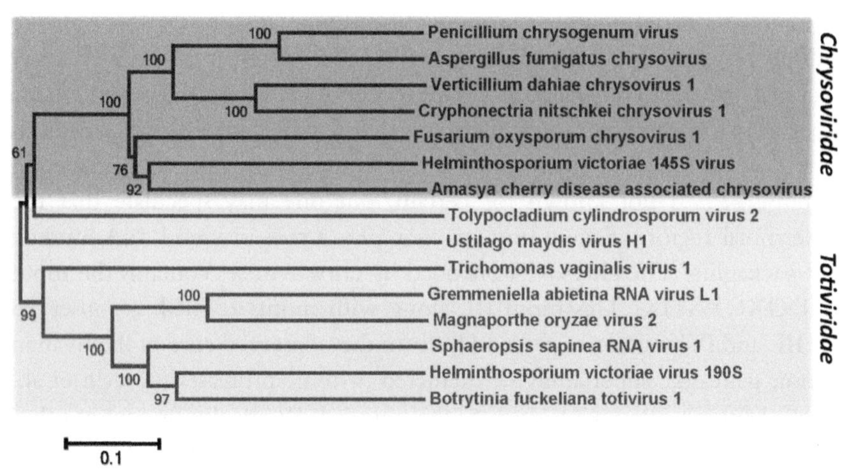

Figure 11.5 Neighbor-joining phylogenetic tree for the family *Chrysoviridae* and selected members of *Totiviridae*. The tree was constructed from complete aa sequences of RdRps of representative members and probable members of the family. The aa sequences were aligned using the program CLUSTAL X2, and the tree was generated for codon positions using the MEGA5 phylogenetic package. Bootstrap percentages out of 2000 replicates are indicated at the nodes. Note that high similarities (BLAST hits of 5e−13 or lower) were found between the HvV145S RdRp and the RdRps of several members of *Totiviridae*, mostly members of the genera *Victorivirus* and *Trichomonasvirus*. Interestingly, no significant hits were evident with any of the viruses in the family *Partitiviridae*, another validation for the removal of chrysoviruses from the family *Partitiviridae* and their placement in the family *Chrysoviridae*. (For color version of this figure, the reader is referred to the online version of this chapter.)

Sequencing analysis and *in vitro* coupled transcription–translation assays showed that each of the four PcV dsRNAs is monocistronic, as each dsRNA contains a single open reading frame and each is translated into a single major product of the size predicted from its deduced amino acid sequence. Thus, the fact that PcV virions contain four distinct dsRNA segments has clearly been established (Jiang & Ghabrial, 2004). Unlike PcV, dsRNAs 3 and 4 from other chrysoviruses including HvV145S are clearly resolved from each other when purified dsRNA preparations are subjected to agarose gel electrophoresis (e.g., see Fig. 11.4B, lanes Hv).

3.3. Viral etiology of the *H. victoriae* disease

In addition to transmitting the disease phenotype and associated viruses by hyphal anastomosis, the disease is also transmitted by incubating fusing protoplasts from virus-free fungal isolates with purified virions containing both HvV190S and HvV145S (Ghabrial & Mernaugh, 1983). The frequency of infection and stability of the newly diseased colonies, however, were very low. We have recently been successful in transfecting *H. victoriae* protoplasts with purified HvV190S particles using the transfection conditions of Sasaki, Kanematsu, Onoue, Oyama, and Yoshida (2006). The transfected colonies exhibited disease symptoms, and virus infection was verified by dsRNA analysis and RT-PCR (unpublished data). Furthermore, purified HvV190S virions were demonstrated to transfect mutant *Cryphonectria parasitica* (Δ*dcl-2*) and confer hypovirulence on infected colonies (Xie, Lin, Suzuki & Ghabrial, unpublished data). Curiously, disruptants Δ*dcl-2* become much more susceptible to hypoviruses and mycoreoviruses compared with the wild-type strain (Ghabrial & Suzuki, 2009). As an alternative approach to provide evidence for viral etiology, a hygromycin B resistance-based transformation system for *H. victoriae* (Li et al., 2011, 2013) was earlier developed and used to transform *H. victoriae* virus-free isolates with full-length cDNA clones of HvV190S or HvV145S dsRNAs. The HvV190S cDNA was inserted downstream of the *Cochliobolus heterostrophus GPD1* promoter and upstream of the *Aspergillus nidulans trpC* terminator signal. The hygromycin-resistant transformants expressed HvV190S CP, as determined by Western blot analysis. Transformation of a normal virus-free fungal isolate with a full-length cDNA of HvV190S dsRNA, but not with cDNAs corresponding to the four HvV145S dsRNAs conferred a disease phenotype in some transformants (Ghabrial, 2008b). Symptom severity varied among the transformants from symptomless to severely stunted and highly sectored (Ghabrial, 2008b, Li et al., 2013). Symptom severity correlated well with the level of viral capsid accumulation in the transformants (Li et al., 2013). Empty capsids accumulated to significantly higher levels in transformants exhibiting the disease phenotype. Although the RdRp was expressed and packaged, no dsRNA was detected inside the virus-like particles or in the total RNA isolated from mycelium. Despite the inability to launch dsRNA replication from the integrated viral cDNA (this is also true with the ScV-L-A virus), the demonstration that ectopic HvV190S cDNA is transcribed and translated and that the resultant transformants exhibit a disease phenotype (Ghabrial, 2008b, Li et al., 2013) provides further convincing evidence for a viral etiology for the disease of *H. victoriae*. Taken together, the results of the transformation/transfection experiments suggest that HvV190S is the primary cause of the disease of *H. victoriae* and that HvV145S has apparently no role in disease development. Like other chrysoviruses, HvV145S has no effect on colony morphology of infected fungal isolates.

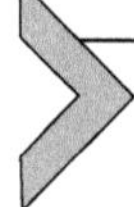

4. HOST GENES UPREGULATED BY VIRUS INFECTION

4.1. Hv-p68

H. victoriae p68 protein (Hv-p68) is a multifunctional protein with alcohol oxidase/protein kinase/and RNA-binding activities. It copurifies with viral dsRNAs, mostly those of HvV145S, in a top component that is resolved as a discrete band near the meniscus when purified virion preparations from diseased *H. victoriae* isolates are subjected to sucrose density gradient centrifugation (Soldevila et al.,2000). SDS-PAGE analysis of the top component reveals a single major protein with an estimated molecular mass of 68 kDa (Hv-p68). The discovery of the Hv-p68 enriched top component was a pleasant surprise as it allowed its purification and characterization (Soldevila et al., 2000).

The gene encoding Hv-p68 has been isolated from a cDNA library generated in lambda phage using mRNA from a virus-infected fungal isolate. The complete nucleotide and deduced amino acid sequences of Hv-p68 have been determined (Soldevila & Ghabrial, 2001). Sequence analysis revealed that Hv-p68 belongs to the large family of flavin-adenine dinucleotide (FAD)-dependent glucose-methanol-choline (GMC) oxidoreductases with 67–70% sequence identity to the alcohol oxidases of methylotrophic yeasts (Soldevila & Ghabrial, 2001). A molecular mass estimate of 550 kDa has been obtained for the native size of Hv-p68 using gradient-purified Hv-p68 and high-performance size-exclusion chromatography (Soldevila et al., 2000). Based on a molecular-mass of 68 kDa, as determined by SDS-PAGE, the native-size estimate of 550 kDa suggested that Hv-p68 is an octomer. The oligomeric arrangement of Hv-p68 has been confirmed by electron microscopic examination (Soldevila et al., 2000), which revealed the presence of octad aggregates of approximately 150–200 Å in diameter comprised of two tetragons aligned face to face. The combined results of SDS-PAGE, size exclusion chromatography, and electron microscopy strongly suggest that Hv-p68, like some other oxidases, is present *in vivo* as an oligomeric protein consisting of eight identical subunits (Soldevila et al., 2000).

Expression of the Hv-p68 gene at the transcriptional level has been examined (Soldevila & Ghabrial, 2001). Elevated levels (10- to 20-fold) of the Hv-p68 transcript were found in the virus-infected fungal isolates compared to the virus-free ones. This finding is in agreement with the higher levels of protein detected by Western blot analysis in a virus-infected isolate (Soldevila et al., 2000; Soldevila & Ghabrial, 2001). Unlike the alcohol oxidases from methylotrophic yeasts, the Hv-p68 purified from fungal extract shows low levels of methanol oxidizing activity. Furthermore, the levels of Hv-p68 transcripts are not significantly different in methanol-supplemented versus glucose-supplemented cultures (Soldevila & Ghabrial, 2001), indicating that Hv-p68 transcription is neither inducible by methanol nor repressed by glucose. Altogether, these data suggest that the Hv-p68 promoter may not be regulated in the same fashion as the promoters for the alcohol oxidases of yeast and that the Hv-p68 promoter is upregulated during the course of viral infection of *H. victoriae*.

The natural substrate for Hv-p68 is not known, but the structurally similar alcohol oxidases are known to oxidize primary alcohols irreversibly to toxic aldehydes. Overexpression of Hv-p68 and subsequent putative accumulation of toxic intermediates have been hypothesized as a possible mechanism underlying the disease phenotype of virus-infected *H. victoriae* isolates. Overexpression of Hv-p68 in virus-free fungal isolates, however, resulted in a significant increase in colony growth and did not induce a disease phenotype (Zhao, Havens, & Ghabrial, 2006). The finding that colonies overproducing the Hv-p68 protein did not exhibit the disease phenotype and grew more rapidly than the nontransformed wild type suggests that accumulation of toxic aldehydes did not occur and that such aldehydes were probably assimilated into carbohydrates via the xylulose monophosphate pathway (Reid & Fewson, 1994).

Hv-p68, which copurifies with viral dsRNAs as a top component in sucrose gradient centrifugation of virus preparations, has been examined for RNA-binding activity by gel retardation and Northwestern analysis

(Soldevila et al., 2000; Soldevila and Ghabrial, 2001). When ^{32}P end-labeled 190S and 145S dsRNAs were incubated with increasing amounts of Hv-p68 followed by agarose gel electrophoresis, a band of free dsRNA and a band of retarded dsRNA (RNA-protein complex) were detected at the lower protein concentrations. However, at higher protein concentrations, only the band of retarded dsRNA was observed. HvV190S and HvV145S dsRNA and yeast tRNA in molar excess amounts were tested as competitors in binding reactions. A 5- to 10-fold molar excess of the unlabeled 190S/145S dsRNAs, when added to the standard binding reaction, completely abolished binding to the probe. Yeast tRNA at 250-fold molar excess only partially competed with binding to the dsRNA probe (Soldevila et al., 2000). On the basis of Northwestern analysis and deletion mutants of bacterially expressed Hv-p68, the RNA-binding domain of Hv-p68 was mapped to the N-terminal region that contains a canonical ADP-binding $\beta-\alpha-\beta$ fold motif. Southern analysis of genomic DNA from several species of the genus *Cochliobolus* (anamorph, *Helminthosporium*) indicated that a single copy of the *Hv-p68* gene is present in all species of the genus *Cochliobolus* examined. The *Hv-p68* gene, however, was detected neither in *Penicillium chrysogenum* nor in two nonmethylotrophic yeasts, *S. cerevisiae* and *S. pombe* (Soldevila & Ghabrial, 2001).

Recent evidence (Ghabrial & Havens, 2013) strongly suggests that Hv-p68 is the cellular protein kinase responsible for phosphorylating the p88 HvV190S CP. From phosphoamino acid analysis, we determined that Hv-p68 is a serine/threonine protein kinase that catalyzes autophosphorylation and phosphorylation of casein and HvV190S CP. The cofactor flavin adenine dinucleotide (FAD), which is required for Hv-p68 activities, was demonstrated to be a strong competitive inhibitor of Hv-p68 kinase activity (Ghabrial & Havens, 2013).

4.2. *Victoriocin gene (vin)*

Victoriocin is a novel broad-spectrum, secreted, antifungal protein that is overproduced in virus-infected *H. victoriae* isolates (de Sá, Havens & Ghabrial, 2010). Culture filtrates of virus-infected *H. victoriae* isolates exhibit antifungal activity against a broad range of fungal and oomycete plant pathogens, including *Fusarium solani*, *Phomopsis longicola*, *C. parasitica*, *Phytophthora parasitica* var. *nicotianae*, *Peronospora tabacina*, and the opportunistic human pathogens *Aspergillus fumigatus* and *Candida albicans*. de Sá, Havens, et al. (2010) purified victoriocin from culture filtrates using a multistep procedure that includes ultrafiltration and reverse-phase high-performance liquid chromatography (RP-HPLC). A well assay similar to that shown in Fig. 11.6A is used to test culture filtrates or RP-HPLC fractions for antifungal activity. SDS/PAGE and Western blot analysis of active RP-HPLC fractions, as determined by the well assay (Fig. 11.6B), reveals the presence of two major proteins, P10 (the mature form of victoriocin) and P30 (a cell wall protein; see Section 4.3).

The fact that the yeast and smut strains of fungi that secrete killer proteins (killer toxins) are infected with totiviruses similar to HvV190S stimulated the search for viral origin for the antifungal activity associated with culture

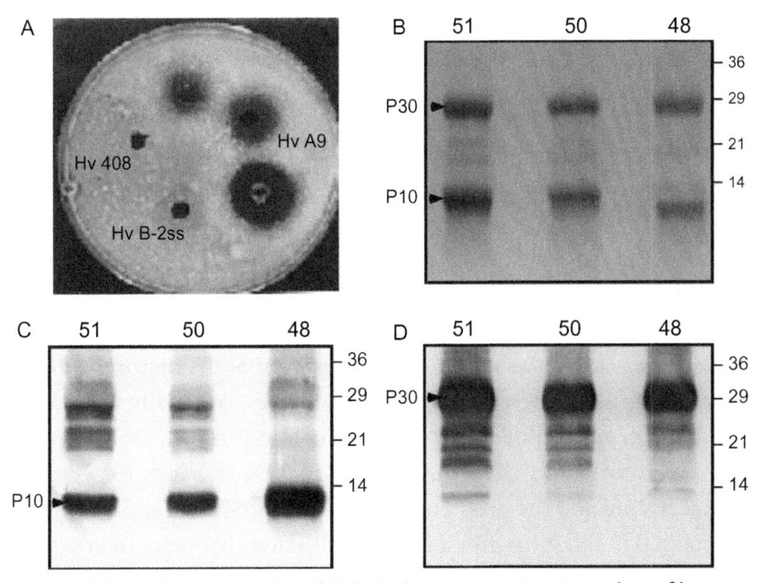

Figure 11.6 (A) Antifungal activity of *Helminthosporium victoriae* culture filtrates using the "well assay." Plates of potato dextrose agar, prepared in 0.05 M citrate buffer, pH 4.5, were seeded with spore suspensions of *Penicillium chrysogenum*, and wells were cut with a cork borer and filled with the assay solutions. (A) Antifungal activity of twofold serial dilutions of culture filtrates of the virus-infected strain A-9 is demonstrated in wells 1–3. No antifungal activity was detected with culture filtrates of virus-free strains 408 and B-2ss (wells 4 and 5, respectively). (B) Sodium dodecyl sulfate polyacrylamide gel electrophoresis of reverse-phase high-performance liquid chromatography retention fractions collected at 48, 50, and 51 min. The gel was stained with Coomassie blue and the positions of P10 and P30 proteins are indicated to the left. (C, D), Western blot analysis of two major proteins, P10 (the mature victoriocin protein) and P30 (a cell wall protein) purified from culture filtrates of the virus-infected *Helminthosporium victoriae* strain A-9. Proteins were transferred from replicate gels to polyvinylidene difluoride membranes, and the membranes were probed with the antiserum to P10 (panel C) or P30 (panel D). Note that the antiserum prepared against P10 reacted strongly with its homologous P10 protein and with larger, possibly related precursor proteins (see text), but not with P30. *Adapted from P. B. de Sá, W. M. Havens and S. A. Ghabrial, 2010, Characterization of a novel broad-spectrum antifungal protein from virus-infected* Helminthosporium (Cochliobolus) victoriae. Phytopathology, *100, 880–889.*

filtrates of *H. victoriae*. The killer proteins from virus-infected yeast and smut strains are encoded by satellite dsRNAs that are dependent on helper totivirus for replication and encapsidation (Schmitt & Breinig, 2006; Wickner, 1996). No satellite dsRNAs, however, were detected in association with HvV190S infection (Ghabrial & Nibert, 2009). Nevertheless, victoriocin, which is encoded by a host chromosomal gene designated *victoriocin* (or *vin*), is structurally similar to killer proteins. The mature victoriocin (P10) is expressed *in vivo* as a 183-amino-acid preprotoxin precursor. This ~20-kDa preprovictoriocin has a predicted, secretory pathway signal peptide, and cleavage between Ala^{23} and Val^{24} releases a predicted proprotein of 17 kDa into the secretory pathway. Further cleavage of the proprotein by a subtilisin/kexin-like proprotein convertase in the latter part of the Golgi complex is predicted to release the mature victoriocin of 86 amino acids with a molecular mass of ~10 kDa (Fig. 11.7). The protease Kex2 (kexin) of *S. cerevisiae* is the prototype of a family of serine proteases that process precursor proteins to active proteins in eukaryotes. It is a Ca^{2+}-dependent transmembrane protease present in a trans-compartment of the yeast Golgi that contains late-processing enzymes (Bryant & Boyd, 1993). Kex2 is necessary for the production and secretion of the mature alpha factor and killer toxins in *S. cerevisiae* at paired basic sites, optimally at Lys-Arg or Arg-Arg (Rockwell & Fuller, 1998). Kexin-like proteins are ubiquitously found in eukaryotes and their occurrence in *H. victoriae* is expected. The mature victoriocin and the predicted precursors (pro- and preprotoxin) were clearly detected in the virus-free *H. victoriae* isolates that were overexpressing the *vin* gene (de Sá, Li, Havens, Farman, & Ghabrial, 2010).

The similarities between victoriocin and the killer protein zygocin, a monomeric protein toxin secreted by a totivirus-infected killer strain of the osmotolerant spoilage yeast *Zygosaccharomyces bailii* (Weiler & Schmitt, 2005), are of considerable interest. Whereas the killer proteins of *S. cerevisiae* (K1, K2, and K28) and *Ustilago maydis* (KP1, KP4, and KP6) are active only against yeasts and smuts, respectively (Bruenn, 2004; Wickner, 1996), zygocin, like victoriocin, has a broad-spectrum antifungal activity against human and phytopathogenic yeasts and filamentous fungi (de Sá, Havens, et al., 2010). On the other hand, zygocin, like other killer proteins, is encoded by a satellite dsRNA that is dependent on a totivirus, Zygosaccharomyces bailii virus, for replication and encapsidation (Schmitt & Breinig, 2006). Although there is a limited primary structure similarity (11% identity) between the two toxins, it is intriguing that the virally encoded zygocin and the chromosome-encoded victoriocin are so similar

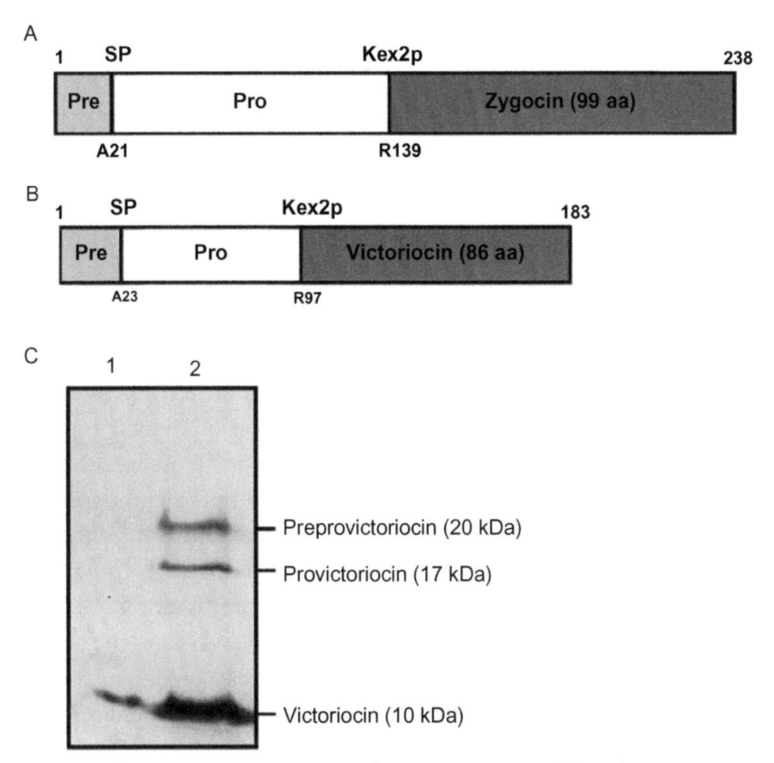

Figure 11.7 Schematic representations of preprovictoriocin (A) and preprozygocin (B). Processing sites are indicated by vertical lines with the name of the pertinent protease indicated above the diagram and the amino acid residue where cleavage occurs indicated below the diagram; signal peptidase (SP) and kexin-like protease (Kex2p). (C) Western blot analysis of proteins extracted from a virus-free *Helminthosporium victoriae* isolate transformed with the victoriocin (*vin*) construct pCB-*vin* (lane 2; pCB-*vin* contains full-length *vin* coding region and flanking 5′ and 3′ untranslated regions) or empty vector (lane 1; pCB1004) (de Sá, Havens,et al., 2010; de Sá, Li, et al., 2010). ppV, preprovictoriocin; pV, provictoriocin; and V, mature victoriocin. Note that *vin* is an endogenous host gene and its translation product can be detected at low levels (lane 1).

in their overall structure (Fig. 11.7) and hydrophobicity profiles (de Sá, Havens, et al., 2010). Each preprotoxin consists of a hydrophobic N-terminal secretion signal, followed by a potentially N-glycosylated pro-region and terminating in a classical Kex2p endopeptidase cleavage site that generates the N-terminus of the mature and biologically active protein toxin in a late Golgi compartment (Fig. 11.7; de Sá, Havens, et al., 2010; Schmitt & Breinig, 2006; Weiler & Schmitt, 2005). The conspicuous similarity of these two proteins (both secreted and both processed by signal peptidase and kexin-like protease) suggests that the zygocin toxin might have been captured by the virus from a cellular transcript because the victoriocin protein is encoded by a nuclear gene.

It is worth noting that victoriocin has a short scorpion toxin family signature motif at its carboxyl terminal region (Bontems, Roumestand, Gilquin, Ménez, & Toma, 1991). Furthermore, it exhibits the conserved CSαβ- and γ-core motifs common to members of the scorpion toxin-like family (de Sá, Havens, et al., 2010). Scorpion toxins are part of one of the functionally most diverse structural superfamilies in single-domain proteins, most of whose functions are associated with host defense mechanisms. Membrane channels are a major target for the various toxins produced by scorpions and other venomous animals (Dauplais et al., 1997). Furthermore, victoriocin shares structural and functional motifs with defensins, which represent a diverse set of antimicrobial peptides known to be part of the innate immune systems of eukaryotes. Defensins are active against fungi, bacteria, and viruses. These small proteins are characterized by having an N-terminal signal sequence, by being highly divergent, mature proteins with the exception of conserved cysteine residues, and by including defensin motifs. Victoriocin is also characterized by an N-terminal signal peptide and shares the conserved CSαβ- and γ-core consensus motifs, characteristics of defensins (de Sá, Havens, et al., 2010; Yount and steadman, 2004). Because defensin expression can be induced by pathogen inoculation and environmental stress, victoriocin may be viewed as a fungal defensin that is induced by virus infection. Insect and mammalian defensins, but not plant defensins, are known to form voltage-regulated multimeric channels in the membranes of susceptible cells (de Sá, Havens, et al., 2010). The mode of action of victoriocin, like many antifungal proteins, is not known. However, considering its similarities to some killer proteins (e.g., zygocin), defensins, and neurotoxins, it is expected to target cytoplasmic membrane functions by forming cation-selective ion channels.

In BLAST searches, the only known protein with significant amino acid sequence identity to the 20-kDa protein (preprovictoriocin) was the IDI-2 precursor of *Podospora anserina* (accession no. AF500213) showing 31% identity along a stretch of 161 amino acids. The IDI-2 protein, like victoriocin, is a small, cysteine-rich protein with a signal peptide. Interestingly, expression of three *idi* genes including *idi-2* in *P. anserina* was shown to correlate with the cell death reaction associated with nonallelic vegetative incompatibility (Bourges, Groppi, Barreau, Clave, & Begueret, 1998).

We have recently constructed an *H. victoriae* genomic DNA library in a cosmid vector and a cosmid clone carrying the *vin* gene and flanking sequences was isolated and used to generate constructs for transformation of virus-free and virus-infected *H. victoriae* isolates with the *vin* gene. Culture filtrates of the virus-free *vin* transformants exhibited high levels of antifungal activity compared with that detected for the nontransformed virus-free wild-type strain, which exhibited little or no antifungal activity. Moreover, transformation of a wild-type virus-infected *H. victoriae* strain with the *vin* gene resulted in still higher production of victoriocin and higher antifungal activity in the culture filtrates of the *vin* transformants compared with the virus-infected wild-type strain. Taken together, these results indicate that victoriocin is the primary protein responsible for the antifungal activity in culture filtrates of virus-infected *H. victoriae* isolates and that virus infection upregulates the expression of victoriocin. Overproduction of victoriocin may give the slower-growing virus infected fungal strains some competitive advantage by inhibiting the growth of other fungi (de Sá, Li, et al., 2010).

4.3. *P30* gene

In addition to the broad-spectrum, antifungal protein, victoriocin, a cell wall protein, designated P30, was also overproduced in the culture filtrates of virus-infected *H. victoriae* isolates (Fig. 11.6B). Full-length cDNA and genomic clones of *P30* coding region have been constructed and sequenced (de Sá, Havens, et al., 2010; Xie & Ghabrial, 2013). Sequence analysis of *P30* indicated that it lacks introns and that the P30 protein is a secreted protein with a predicted signal peptide that is cleaved between amino acid positions 18 and 19 (Ala/Lys). BLASTP search of the deduced amino acid sequence revealed similarity between P30 and PhiA cell wall proteins (Melin, Schnuürer, & Wagner, 2003) from several *Aspergillus* spp. including *Emericella* (*Aspergillus*) *nidulans* (37% identity), *A. fumigatus* (37% identity), *A. clavatus* (35% identity), and also to the alkaline foam protein A precursor protein (AfpA) of *Fusarium culmorum* (37% identity). Furthermore, P30 contains the sequence motifs SGMGQG and ACP present in the highly conserved core region of "fungispumins," a recently described class of proteins (Zapf et al., 2007). PhiA is involved in phialide development in *A. nidulans* and is overexpressed in response to treatment with bafilomycin with antifungal and antibacterial activities (Melin et al., 2003). The natural function of fungispumins remains to be elucidated. The role of P30, if any, in the antifungal activity exhibited by culture filtrates of virus-infected *H. victoriae* isolates is not clear.

The overexpression of P30 may represent an indirect response to the presence of the secreted antifungal protein victoriocin in a manner comparable to the overexpression of the homologous PhiA proteins by *Aspergillus* spp. following treatment with the antifungal agent bafilomycin and the resultant inhibition of vacuolar ATPases (Melin et al., 2003). It was of interest to determine whether overexpression of P30, a homologue of PhiA, plays a role in the development of the disease phenotype (short, swollen hyphae) of virus-free *H. victoriae* isolates, considering the similarity in hyphal morphology between the disease phenotype in *H. victoriae* (Ghabrial et al., 1979) and *A. nidulans* that overexpresses PhiA following treatment with the antifungal agent concanamycin (Melin, Schnuürer, & Wagner, 2002). Colonies generated from virus-free protoplasts transformed with a *P30* overexpression vector were stunted with significantly smaller diameter than control colonies (Xie & Ghabrial, 2013). It is not known, however, whether the disease phenotype in *H. victoriae* is associated with inhibition of vacuolar ATPases.

5. HvV190S CAPSID STRUCTURE

HvV190S virions have isometric capsids that are relatively smooth and featureless with no obvious protrusions, as seen in transmission electron micrographs of unstained, vitrified specimens (Fig. 11.8A) (Dunn et al., 2013). A three-dimensional cryo-reconstruction of the HvV190S virion, computed from 20,904 images and estimated at a resolution of 7.1 Å, showed that the icosahedral capsid reaches a maximum diameter of 462 Å at the fivefold vertices and diameters of 356 and 368 Å at the threefold and twofold axes, respectively (Fig. 11.8B). The thickness of the capsid is ~35 Å on average but drops to a minimum of just 6 Å at the threefold axes. The inner surface of the capsid is also quite smooth, except for small, mushroom-shaped cavities that extend radially outward along the fivefold axes (Fig. 11.8C and D). The dsRNA genome, protected by the capsid, is arranged in five, roughly spherical, concentric shells whose average spacing (~30 Å) compares favorably with the genome spacings observed in ScV-L-A and other dsRNA fungal viruses (Naitow, Tang, Canady, Wickner, & Johnson, 2002; Ochoa et al., 2008, Pan et al., 2009; Tang et al., 2008). The outermost shell of dsRNA lies ~17 Å away from the inner wall of the capsid, except near the twofold axes, where some diffuse density may represent potential points of contact between genome and capsid. Cross-section (equatorial) views of the HvV190S virion demonstrate that the capsid contains numerous punctate and linear density features (Fig. 11.8D). These are ascribed to α-helical secondary-structure elements that lie perpendicular to or in the plane of the section, respectively. Spherical density sections of the

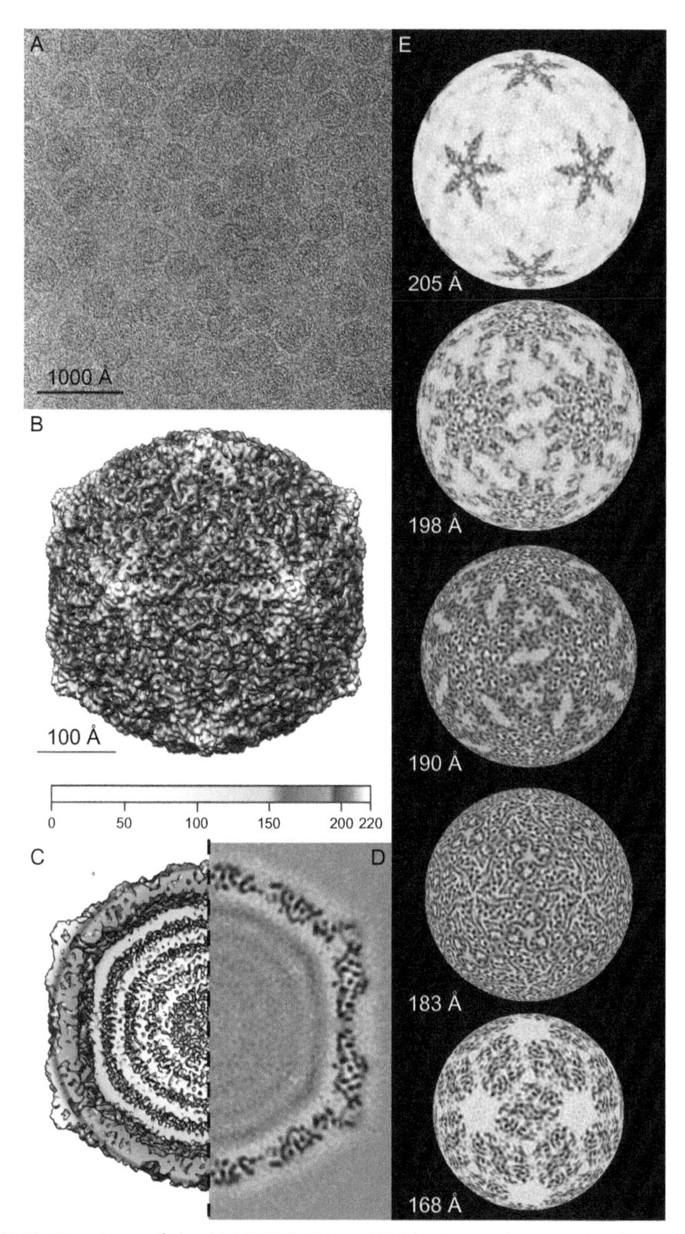

Figure 11.8 Structure of the HvV190S virion. (A) Electron micrograph of an unstained, vitrified sample of HvV190S virions. (B) Radial, color-coded, surface view along a twofold axis of the HvV190S virion cryoreconstruction. (C) Same as (B), with the front half of the density map removed to show the particle interior and only showing the left quadrant. (D) Planar, central density section (~1 Å thick) from the right half of the HvV190S virion cryo-reconstruction (high- and low-density features are depicted in black and white, respectively). (E) Spherical density projections (each ~1 Å thick) from the HvV190S virion cryo-reconstruction at select radii, as indicated in each panel, with highest and lowest density features depicted in black and white, respectively. Intersubunit interactions are most prevalent as well as dense at a radius of 183 Å. (See Page 23 in Color Section at the back of the book.)

HvV190S capsid structure confirm that these same features are distributed throughout the capsid at all radii (Fig. 11.8E).

Close inspection of the HvV190S capsid surface (Fig. 11.7B) reveals the presence of 120 morphological features that are consistent with a "$T=2$" icosahedral arrangement of subunits as found in the capsids or inner capsids of most other dsRNA viruses (Castón et al., 1997; Ochoa et al., 2008; Pan et al., 2009; Tang, Ochoa, et al., 2010). The high radii density features in the capsid adopt a staggered arrangement, with 60 copies of one "unit" clustered in groups of 5 about the 12 fivefold vertices and 60 copies of a similarly shaped second unit clustered in groups of 3 about the 20 threefold vertices. Both surface-exposed units have similar, globular head domains that project toward the fivefolds and similar, anchor-like tails that lie near the twofolds. Segmentation of the entire HvV190S capsid cryoreconstruction (Dunn et al., 2012) made it possible to assign all density features to either an "A" or a "B" subunit, labeled according to the distinct locations they occupy in the icosahedron (A nearest the fivefolds and B nearest the threefolds; Fig. 11.9A). Hence, the asymmetric unit of the icosahedron is an AB dimer (Fig. 11.9B), which consists of two chemically identical monomers (772 aa; calculated molecular mass of 81 kDa).

The segmentation process showed that both monomers have an overall morphology that, when viewed from outside the capsid, it has the shape of a complex quadrilateral (Fig. 11.9B). With the exception of the proximal and distal tips of the subunits, the segmented density maps for the A and B subunits superimpose quite closely, including nearly all of the recognizable secondary structural elements (Dunn et al., 2013). This strongly suggests that the tertiary structures of the subunits are nearly the same and they differ most in regions where intersubunit interactions are nonequivalent.

5.1. Comparison of the HvV190S capsid with that of other totiviruses

The capsid subunits of HvV190S and ScV-L-A share a striking resemblance in their overall morphologies. Given that ScV-L-A is the only totivirus whose high-resolution crystal structure has been determined to date (Naitow et al., 2002), an attempt was made to quantitatively fit as rigid bodies the A and B subunit structures into the segmented density maps of the corresponding HvV190S subunits (Dunn et al., 2013). As expected, because the ScV-L-A subunit is smaller than the HvV190S subunit (680 vs. 772 aa) and 29 residues at the C-termini were disordered and hence not resolved in the ScV-L-A X-ray crystal structure (Naitow et al., 2002), this fitting process

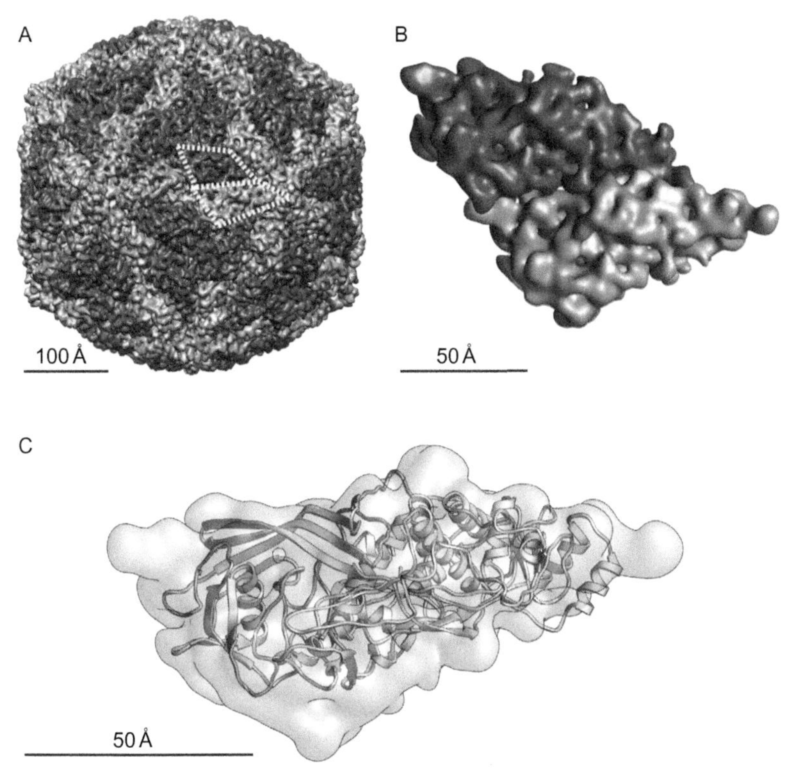

Figure 11.9 Subunit structure of HvV190S and comparison with ScV-L-A. (A) View along a twofold axis of a segmented version of the HvV190S cryoreconstruction with the A- and B-subunits colored in blue and orange, respectively. The A and B subunits in one asymmetric dimer are outlined with quadrilateral polygons. (B) Enlarged view of one asymmetric dimer in an orientation similar to that shown in panel (A). (C) View of the ScV-L-A A-subunit crystal structure (Naitow et al., 2002), represented as a purple ribbon model, fit as a rigid body into the segmented density map of the HvV190S A-subunit (blue transparent envelope), which is oriented like the A-subunit of panel (B). (See Page 24 in Color Section at the back of the book.)

left several regions of the HvV190S subunits unassigned. This included the proximal tip of the A-subunit and a portion of the proximal tip of B, the entire distal tips of both subunits, and a large portion of the long edge of each subunit. In addition, one side of the B-subunit density also remained unassigned. Of even more significance, with few exceptions, the tertiary folds of the HvV190S subunits differ substantially from those of the corresponding ScV-L-A subunits, as evidenced, for example, by misalignment of most of the prominent secondary structural features (Fig. 11.9C). Nonetheless, three separate segments of the ScV-L-A subunit structure did correlate quite well with densities in the HvV190S map. These included in ScV-L-A two α-helices (helix 5: aa 120–139 and helix 13: aa 358–383) and an antiparallel β-sheet comprising three main (aa 26–38, 41–54, 587–600) and three small (aa 302–304, 487–491, 604–606) strands (Fig. 11.10). The helices form a central core region in each ScV-L-A subunit and correspond quite closely in size and location to tubular density features in each of the two HvV190S subunits. The β-sheet in each ScV-L-A subunit model nicely correlates with a large, plate-like density feature that occurs near the distal end and left side of each HvV190S subunit.

In the absence of a crystal structure for the HvV190S capsid or isolated capsid subunit, the secondary structure of the HvV190S subunit was assessed by means of prediction methods and, for validation, these same methods were used to assess the ScV-L-A subunit (Dunn et al., 2013). This validation led to a prediction for ScV-L-A that completely matched all the secondary structure

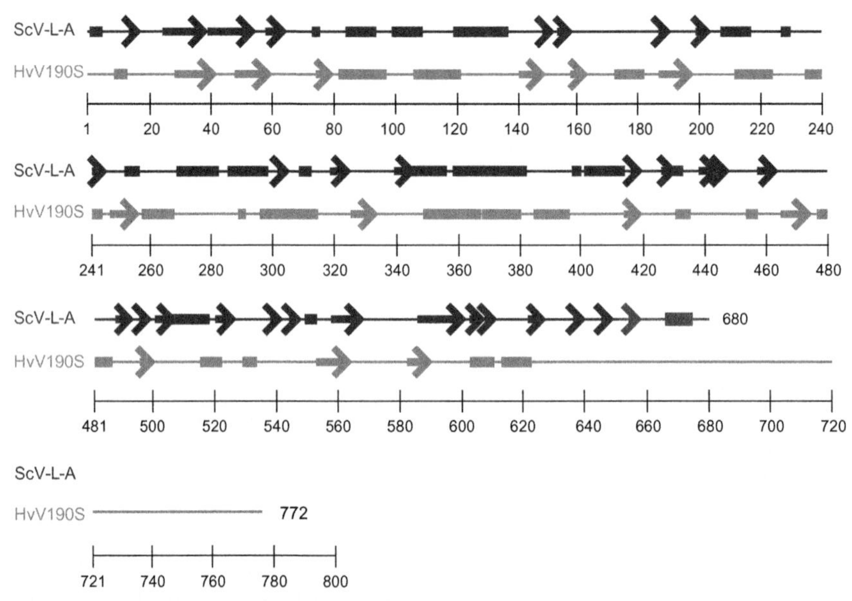

Figure 11.10 Comparison of the secondary structure of the ScV-L-A capsid protein to the predicted secondary structure of the HvV190S capsid protein. For the ScV-L-A subunit, the secondary structure for the first 651 residues (purple) was derived from the crystal structure (PDB ID 1M1C), and the structure of the remaining 29 residues (red) was predicted using the PSIPRED server (Dunn et al., 2013). PSIPRED was used to predict the entire secondary structure of the HvV190S subunit (blue). Rectangles, arrows, and lines are used, respectively, to represent schematically α-helix, β-strand, and random coil segments in both virus capsid proteins. (For interpretation of the references to color in this figure legend, the reader is referred to the online version of this chapter.)

elements observed in the crystal structure (24.4% helix, 20.4% sheet, and 55.2% coil). For HvV190S, the predicted elements included 21% helix, 12% sheet, and 67% random coil. The result of these prediction and three-dimensional fitting analyses confirmed the earlier suggestion that the folds of the subunits of ScV-L-A and HvV190S differ significantly (Castón et al., 2006).

The apparent conservation of a pair of helices in the cores of the ScV-L-A and HvV190S subunits raises the question of whether this core structure might be shared by other or perhaps all totiviruses. To address this, secondary structure predictions were made for the CPs of several other members of the family *Totiviridae* (Dunn et al., 2013). These included six additional members of the genus *Victorivirus* (BfTV1, CmRV, GaRV-L1, SsRV2, CeRV-1, and SsRV1), single representative members of *Leishmaniavirus, Trichomonasvirus,* and *Giardiavirus*, (LRV-1, TVV-1 and GLV), and two unclassified members (EbRV1 and IMNV). The results of this analysis demonstrated high conservation among all victoriviruses, of not just the central core helices but also nearly every secondary element including the β sheets. This is completely consistent with the high levels of sequence identity that each victorivirus shares with HvV190S, which ranges from a maximum of 62% for BfTV1 to a minimum of 37% for CmRV (Dunn et al., 2013). Given this high level of conservation, all victoriviruses are expected to have nearly identical, smooth-shaped, "$T=2$" capsids like HvV190S. Though a similar helix core appears to be present in the other totiviruses, like ScV-L-A, there is no compelling evidence from the secondary structure analysis to suggest that the subunits of totiviruses outside the genus *Victorivirus* adopt the HvV190S tertiary structure.

Results of phylogenetic analysis based on complete amino acid sequences of the CP (Ghabrial & Nibert, 2009) or the RdRp (Fig. 11.3; this chapter) further support the conclusion that all victoriviruses are phylogenetically closely related to each other as they form one major cluster in the phylogenetic tree. On the other hand, victoriviruses are distantly related to members of the genus *Totivirus* (ScV-L-A and UmV-H1) that infect yeast and smut fungi. Interestingly, victoriviruses are more closely related to protozoan totiviruses than to the yeast and smut totiviruses. In this regard, the report that a protozoan totivirus, Eimeria brunetti RNA virus 1 (EbRV1), shares the victorivirus strategy for RdRp expression as a separate protein, with the RdRp start codon overlapping the CP stop codon in the −1 frame at an AUGA tetranucleotide motif, is of considerable interest (Fraga et al., 2006). EbRV is identified in the phylogenetic trees as the virus next most closely related to victoriviruses. Therefore, results of phylogenetic analyses are in line with those of secondary and tertiary structure assessments of victoriviruses and other totiviruses.

6. CONCLUDING REMARKS

- Diseased isolates of *H. victoriae* are doubly infected with the victorivirus HvV190S and the chrysovirus HvV145S. Mixed infection, however, is not required for development of the disease phenotype. DNA transformation and transfection assays with purified virions strongly suggest that HvV190S is the major cause of the disease of *H. victoriae* and that HvV145S, like other chrysoviruses, does not appear to affect colony morphology.
- HvV190S, the prototype of the genus *Victorivirus*, family *Totiviridae*, is the best-characterized victorivirus. It utilizes a novel coupled termination–reinitiation strategy for expression of its RdRp from the downstream ORF. With elucidation of the RNA sequence determinants of the stop–restart strategy of HvV190S, it became apparent that the other 13 victoriviruses, even though less well characterized than HvV190S, utilize similar strategy for expressing their RdRps.
- With the availability of transformation/transfection systems for *H. victoriae*, the questions of whether virus infection affects victorin production and thus virulence could now be directly addressed.
- The multifunctional activities (alcohol oxidase, RNA binding, and protein kinase) of the cellular Hv-p68 have been demonstrated. Whereas the alcohol oxidase activity of Hv-p68 does not appear to play a role in disease development, the RNA binding and kinase activities are thought to be important for virus replication and RNA packaging.
- Victoriocin, the preprotoxin that is overexpressed in virus-infected *H. victoriae*, should be exploited for expression in economically important plants to provide broad protection against fungal and oomycete plant pathogens. Genomic as well as full-length cDNA clones of victoriocin are available for constructing plant expression vectors. It might be necessary, however, to replace the fungal signal peptide in victoriocin with a plant signal peptide for efficient secretion.
- The HvV190S virion structure was determined by electron cryomicroscopy and three-dimensional image reconstruction methods at 7.1-Å resolution. The capsid was found to be relatively thin and featureless, and the 120 CPs form a "$T=2$" icosahedral lattice. Segmentation of the capsid portion of the HvV190S cryo-reconstruction made it possible to assign all density features to either an "A" or a "B" subunit, labeled

according to the distinct locations they occupy in the icosahedron. Hence, the asymmetric unit of the icosahedron is an AB dimer, which consists of two chemically identical monomers.

ACKNOWLEDGMENTS

This work was supported in part by Kentucky Science and Engineering Foundation grant KSEF-2178-RDE-013 (S. A. G.) and by NIH grants R37-GM033050 and 1S10-RR020016 as well as UCSD and the Agouron Foundation (T. S. B.).

REFERENCES

Amrani, N., Ganesan, R., Kervestin, S., Mangus, D. A., Ghosh, S., & Jacobson, A. (2004). A *faux* 3' -UTR promotes mRNA decay. *Nature*, *432*, 112–118.

Bontems, F., Roumestand, C., Gilquin, B., Ménez, A., & Toma, F. (1991). Refined structure of charybdotoxin: Common motifs in scorpion toxins and insect defensins. *Science*, *25*, 1521–1523.

Bourges, N., Groppi, A., Barreau, B., Clave, C., & Begueret, J. (1998). Regulation of gene expression during the vegetative incompatibility reaction in *Podospora anserina*: Characterization of three induced genes. *Genetics*, *150*, 633–641.

Bruenn, J. (2004). The *Ustilago maydis* killer toxins. *Topics in Current Genetics*, *11*, 157–174.

Bryant, N. J., & Boyd, A. (1993). Immunoisolation of Kex2-containing organelles from yeast demonstrates co-localisation of three processing proteinases to a single Golgi compartment. *Journal of Cell Science*, *106*, 815–822.

Castón, R. J., Luque, D., Trus, B. L., Rivas, G., Alfonso, C., González, J. M., et al. (2006). Three-dimensional structure and stoichiometry of Helminthosporium victoriae 190S totivirus. *Virology*, *347*, 323–332.

Castón, J. R., Trus, B. L., Booy, F. P., Wickner, R. B., Wall, J. S., & Steven, A. C. (1997). Structure of L-A virus: A specialized compartment for the transcription and replication of double-stranded RNA. *The Journal of Cell Biology*, *138*, 975–985.

Covelli, L., Coutts, R. H. A., Di Serio, F., Citir, A., Acikgoz, S., Hernandez, C., et al. (2004). Cherry chlorotic rusty spot and Amasya cherry diseases are associated with a complex pattern of mycoviral-like double-stranded RNAs. I. Characterization of a new species in the genus Chrysovirus. *Journal of General Virology*, *85*, 3389–3397.

Dauplais, M., Lecoq, A., Song, J., Cotton, J., Jamin, N., Gilquin, B., et al. (1997). On the convergent evolution of animal toxins. *The Journal of Biological Chemistry*, *272*, 4302–4309.

de Sá, P. B., Havens, W. M., & Ghabrial, S. A. (2010). Characterization of a novel broad-spectrum antifungal protein from virus-infected *Helminthosporium* (*Cochliobolus*) *victoriae*. *Phytopathology*, *100*, 880–889.

de Sá, P. B., Li, H., Havens, W. M., Farman, M. L., & Ghabrial, S. A. (2010). Overexpression of the victoriocin gene in *Helminthosporium victoriae* enhances the antifungal activity of culture filtrates. *Phytopathology*, *100*, 890–896.

Dinman, J. D., Icho, T., & Wickner, R. B. (1991). A -1 ribosomal frameshift in a double-stranded RNA virus of yeast forms a gag-pol fusion protein. *Proceedings of the National Academy of Sciences of the United States of America*, *88*, 174–178.

Dinman, J. D., & Wickner, R. B. (1992). Ribosomal frameshifting efficiency and gag/gag-pol ratio are critical for yeast M1 double-stranded RNA virus propagation. *Journal of Virology*, *66*, 3669–3676.

Dreher, T. W., & Miller, W. A. (2006). Translational control in positive strand RNA plant viruses. *Virology*, *344*, 185–197.

Dunn, S. E., Li, H., Cardone, G., Nibert, M. L., Ghabrial, S. A., & Baker, T. S. (2013). Three-dimensional structure of victorivirus HvV190S suggests that the coat proteins in all totiviruses may share a conserved core. PLoS Pathogens (in press).

Fraga, J. S., Katsuyama, A. M., Fernandez, S., Madeira, A. M. B. N., Briones, M. R. S., & Gruber, A. (2006). The genome of the Eimeria brunetti RNA virus 1 is more closely related to fungal than to protozoan viruses. GenBank Accession No. AF356189. Unpublished

Gallie, D. R., & Walbot, V. (1992). Identification of the motifs within the tobacco mosaic virus 5'-leader responsible for enhancing translation. *Nucleic Acids Research*, *20*, 4631–4638.

Ghabrial, S. A. (1986). A transmissible disease of *Helminthosporium victoriae*: Evidence for a viral etiology. In K. W. Buck (Ed.), *Fungal virology* (pp. 353–369). Boca Raton: CRC Press.

Ghabrial, S. A. (2008a). Chrysoviruses. In B. W. J. Mahy & M. H. V. Van Regenmortel (Eds.), (3rd ed.). *Encyclopedia of virology*, vol. 1, (pp. 503–513). Oxford: Elsevier.

Ghabrial, S. A. (2008b). Totiviruses. In B. W. J. Mahy & M. H. V. Van Regenmortel (Eds.), (3rd ed.). *Encyclopedia of virology*, vol. 5, (pp. 163–174). Oxford: Elsevier.

Ghabrial, S. A., Bibb, J. A., Price, K. H., Havens, W. M., & Lesnaw, J. A. (1987). The capsid polypeptides of the 190S virus of *Helminthosporium victoriae*. *Journal of General Virology*, *68*, 1791–1800.

Ghabrial, S. A., & Havens, W. M. (1989). Conservative transcription of *Helminthosporium victoriae* 190*S* virus dsRNA *in vitro*. *Journal of General Virology*, *70*, 1025–1035.

Ghabrial, S. A., & Havens, W. M. (1992). The *Helminthosporium victoriae* 190*S* mycovirus has two forms distinguishable by capsid protein composition and phosphorylation state. *Virology*, *188*, 657–665.

Ghabrial, S. A., & Havens, W. M. (2013). Characterization of the protein kinase activity of the multifunctional protein Hv-p68 from Helminthosporium victoriae. Unpublished manuscript.

Ghabrial, S. A., Havens, W. M., Xie, J., & Annamalai P. (2013). Molecular characterization of the chrysovirus Helminthosporium victoriae virus 145S. Submitted manuscript.

Ghabrial, S. A., & Mernaugh, R. L. (1983). Biology and transmission of *Helminthosporium victoriae* mycoviruses. In R. W. Compans & D. H. L. Bishop (Eds.), *Double-stranded RNA viruses* (p. 441). New York: Elsevier.

Ghabrial, S. A., & Nibert, M. L. (2009). *Victorivirus*, a new genus of fungal viruses in the family *Totiviridae*. *Archives of Virology*, *154*, 373–379.

Ghabrial, S. A., & Pirone, T. P. (1967). Physiology of tobacco etch virus-induced wilt of Tabasco peppers. *Virology*, *31*, 154–162.

Ghabrial, S. A., Sanderlin, R. S., & Calvert, L. A. (1979). Morphology and virus content of *Helminthosporium victoriae* colonies regenerated from protoplasts of normal and diseased isolates. *Phytopathology*, *69*, 312–315.

Ghabrial, S. A., Soldevila, A. I., & Havens, W. M. (2002). Molecular genetics of the viruses infecting the plant pathogenic fungus *Helminthosporium victoriae*. In S. Tavantzis (Ed.), *Molecular biology of double-stranded RNA: Concepts and applications in agriculture, forestry and medicine* (pp. 213–236). Boca Raton, FL: CRC Press.

Ghabrial, S., & Suzuki, N. (2009). Viruses of plant pathogenic fungi. *Annual Review of Phytopathology*, *47*, 353–384.

Guo, L. H., Sun, L., Chiba, S., Araki, H., & Suzuki, N. (2009). Coupled termination/reinitiation for translation of the downstream open reading frame B of the prototypic hypovirus CHV1 EP713. *Nucleic Acids Research*, *37*, 3645–3659.

Hellen, C. U., & Sarnow, P. (2001). Internal ribosome entry sites in eukaryotic mRNA molecules. *Genes & Development*, *15*, 1593–1612.

Herrero, N., & Zabalgogeazcoa, I. (2011). Mycoviruses infecting the endophytic and entomopathogenic fungus Tolypocladium cylindrosporum. *Virus Research*, *160*, 409–413.

Huang, S., & Ghabrial, S. A. (1996). Organization and expression of the double-stranded RNA genome of Helminthosporium victoriae 190S virus, a totivirus infecting a plant pathogenic filamentous fungus. *Proceedings of the National Academy of Sciences of the United States of America*, *93*, 12541–12546.

Huang, S., Soldevila, A. I., Webb, B. A., & Ghabrial, S. A. (1997). Expression, assembly and proteolytic processing of *Helminthosporium victoriae* 190S totivirus capsid protein in insect cells. *Virology*, *234*, 130–137.

Jiang, D., & Ghabrial, S. A. (2004). Molecular characterization of Penicillium chrysogenum virus: Reconsideration of the taxonomy of the genus *Chrysovirus*. *Journal of General Virology*, *85*, 2111–2121.

Kozak, M. (2002). Pushing the limits of the scanning mechanism for initiation of translation. *Gene*, *299*, 1–34.

Li, H., Havens, W. M., Xie, J., & Ghabrial, S. A. (2013). Transformation of virus-free Helminthosporium victoriae with a full-length cDNA clone of HvV190S-CP confers a disease phenotype. Unsubmitted manuscript.

Li, H., Havens, W. M., Nibert, M. L., & Ghabrial, S. A. (2011). RNA sequence determinants of a coupled termination-reinitiation strategy for downstream open reading frame translation in *Helminthosporium victoriae* virus 190S and other victoriviruses (Family *Totiviridae*). *Journal of Virology*, *85*, 7343–7352.

Lindberg, G. D. (1959). A transmissible disease of *Helminthosporium victoriae*. *Phytopathology*, *49*, 29–32.

Lindberg, G. D. (1960). Reduction in pathogenicity and toxin production in diseased Helminthosporium victoriae. *Phytopathology*, *50*, 457.

Litzenberger, S. C. (1949). Nature of susceptibility to *Helminthosporium victoriae* and resistance to *Puccinia coronata* in Victoria oats. *Phytopathology*, *39*, 300–318.

Luttermann, C., & Meyers, G. (2007). A bipartite sequence motif induces translation reinitiation in feline calicivirus RNA. *The Journal of Biological Chemistry*, *282*, 7056–7065.

Luttermann, C., & Meyers, G. (2009). The importance of inter- and intramolecular base pairing for translation reinitiation on a eukaryotic bicistronic mRNA. *Genes & Development*, *23*, 331–344.

Matassova, N. B., Venjaminova, A. G., & Karpova, G. G. (1998). Nucleotides of 18S rRNA surrounding mRNA at the decoding site of translating human ribosome as revealed from the cross-linking data. *Biochimica et Biophysica Acta*, *1397*, 231–239.

Meehan, F., & Murphy, H. C. (1946). A new Helminthosporium blight of oats. *Science*, *104*, 413–414.

Meehan, F., & Murphy, H. C. (1947). Differential phytotoxicity of metabolic by-products of *Helminthosporium victoriae*. *Science*, *106*, 270–271.

Melin, P., Schnuürer, J., & Wagner, E. G. H. (2002). Proteome analysis of *Aspergillus nidulans* reveals proteins associated with the response of the antibiotic concanamycin A, produced by Streptomyces species. *Molecular Genetics and Genomics*, *267*, 695–702.

Melin, P., Schnuürer, J., & Wagner, E. G. H. (2003). Characterization of *phiA*, a gene essential for phialide development in *Aspergillus nidulans*. *Fungal Genetics and Biology*, *40*, 234–241.

Meyers, G. (2003). Translation of the minor capsid protein of a calicivirus isinitiated by a novel termination-dependent reinitiation mechanism. *The Journal of Biological Chemistry*, *278*, 34051–34060.

Meyers, G. (2007). Characterization of the sequence element directing translation reinitiation in RNA of the calicivirus rabbit hemorrhagic disease virus. *Journal of Virology*, *81*, 9623–9632.

Naitow, H., Tang, J., Canady, M., Wickner, R. B., & Johnson, J. E. (2002). L-A virus at 3.4 Å resolution reveals particle architecture and mRNA decapping mechanism. *Nature Structural Biology*, *9*, 725–728.

Nomura, K., Osahki, H., Iwanami, T., Matsumoto, N., & Ohtsu, Y. (2003). Cloning and characterization of a totivirus double-stranded RNA from the plant pathogenic fungus, *Helicobasidium mompa* Tanaka. *Virus Genes, 26*, 219–226.

Ochoa, W. F., Havens, W. M., Sinkovits, R. S., Nibert, M. L., Ghabrial, S. A., & Baker, T. S. (2008). Partitivirus structure reveals a 120-subunit, helix-rich capsid with distinctive surface arches formed by quasisymmetric coat-protein dimers. *Structure, 16*, 776–786.

Pan, J., Dong, L., Lin, L., Ochoa, W. F., Sinkovits, R. S., Havens, W. M., et al. (2009). Atomic structure reveals the unique capsid organization of a dsRNA virus. *Proceedings of the National Academy of Sciences of the United States of America, 106*, 4225–4230.

Plant, E. W., Jacobs, K. M., Harger, J. W., Meskauskas, A., Jacobs, J. L., Baxter, J. L., et al. (2003). The 9-Å solution: How mRNA pseudoknots promote efficient programmed -1 ribosomal frameshifting. *RNA, 9*, 168–174.

Powell, M. L., Brown, T. D., & Brierley, I. (2008). Translational terminationre-initiation in viral systems. *Biochemical Society Transactions, 36*, 717–722.

Pöyry, T. A., Kaminski, A., Connell, E. J., Fraser, C. S., & Jackson, R. J. (2007). The mechanism of an exceptional case of reinitiation after translationof a long ORF reveals why such events do not generally occur in mammalian mRNA translation. *Genes & Development, 21*, 3149–3162.

Psarros, E. E., & Lindberg, G. D. (1962). Morphology and respiration of diseased and normal Helminthosporium victoriae. *Phytopathology, 52*, 693.

Reid, M. F., & Fewson, C. A. (1994). Molecular characterization of microbial alcohol dehydrogenases. *Critical Reviews in Microbiology, 20*, 13–56.

Rockwell, N. C., & Fuller, R. S. (1998). Interplay between S1 and S4 subsites in the Kex2 protease: Kex2 exhibits dual specificity for the P4 side chain. *Biochemistry, 37*, 3386–3391.

Ryabova, L. A., & Hohn, T. (2000). Ribosome shunting in the cauliflower mosaic virus 35S RNA leader is a special case of reinitiation of translation functioning in plant and animal systems. *Genes & Development, 14*, 817–829.

Sanderlin, R. S., & Ghabrial, S. A. (1978). Physicochemical properties of two distinct types of virus-like particles from *Helminthosporium victoriae. Virology, 87*, 142–151.

Sasaki, A., Kanematsu, S., Onoue, M., Oyama, Y., & Yoshida, K. (2006). Infection of *Rosellinia necatrix* with purified viral particles of a member of *Partitiviridae* (RnPV1-W8). *Archives of Virology, 151*, 697–707.

Scheffer, R. P., & Nelson, R. R. (1967). Geographical distribution and prevalence of Helminthosporium victoriae. *The Plant Disease Reporter, 51*, 110.

Schmitt, M. J., & Breinig, F. (2006). Yeast viral killer toxins: Lethality and self-protection. *Nature Reviews Microbiology, 4*, 212–221.

Schoch, C. L., Shoemaker, R. A., Seifert, K. A., Hamleton, S., Spatafora, J. W., & Crous, P. W. (2006). A multigene phylogeny of the Dothideomycetes using four nuclear loci. *Mycologia, 98*, 1041–1052.

Soldevila, A. I., & Ghabrial, S. A. (2000). Expression of the totivirus *Helminthosporium victoriae 190S Virus* RNA-dependent RNA polymerase from its downstream open reading frame in dicistronic constructs. *Journal of Virology, 74*, 997.

Soldevila, A. I., & Ghabrial, S. A. (2001). A novel alcohol oxidase/RNAbinding protein with affinity for mycovirus double-stranded RNA from the filamentous fungus *Helminthosporium* (*Cochliobolus*) *victoriae. The Journal of Biological Chemistry, 276*, 4652–4661.

Soldevila, A. I., Huang, S., & Ghabrial, S. A. (1998). Assembly of the Hv190S totivirus capsid is independent of posttranslational modification of the capsid protein. *Virology, 251*, 327–333.

Soldevila, A., Havens, W. M., & Ghabrial, S. A. (2000). A cellular protein with RNA-binding activity co-purifies with viral dsRNA from mycovirus infected *Helminthosporium victoriae. Virology, 272*, 183–190.

Tang, J., Ochoa, W. F., Sinkovits, R. S., Poulos, B. T., Ghabrial, S. A., Lightner, D. V., et al. (2008). Infectious myonecrosis virus has a totivirus-like, 120-subunit capsid, but with fiber complexes at the fivefold axes. *Proceedings of the National Academy of Sciences of the United States of America, 105*, 17526–17531.

Tang, J., Ochoa, W. F., et al. (2010). Structure of Fusarium poae virus 1 shows conserved and variable elements of partitivirus capsids and evolutionary relationships to picobirnavirus. *Journal of Structural Biology, 172*, 363–371.

Wang, A. L., Yang, H. M., Shen, K. A., & Wang, C. C. (1993). Giardiavirus double-stranded RNA genome encodes a capsid polypeptide and a gag-pol like fusion protein by a translation frameshift. *Proceedings of the National Academy of Sciences of the United States of America, 90*, 8595–8599.

Weiler, F., & Schmitt, M. J. (2005). Zygocin—a monomeric protein toxin secreted by virus infected Zygosaccharomyces bailli. *Topics in Current Genetics, 11*, 175–187.

Wheeler, H., & Black, H. S. (1962). Change in permeability induced by victorin. *Science, 137*, 983–984.

Wickner, R. B. (1996). Double-stranded RNA viruses of Saccharomyces cerevisiae. *Microbiological Reviews, 60*, 250–265.

Wolpert, T. J., Macko, V., Acklin, W., Jaun, B., Seibl, J., Meili, J., et al. (1985). Structure of victorin C, the major hostselective toxin from *Cochliobolus victoriae. Experientia, 41*, 1524–1529.

Xie, J., & Ghabrial, S. A. (2013). Molecular characterization of the P30 gene encoding a cell wall protein overexpressed in virus-infected Helminthosporium victoriae. Unpublished manuscript.

Yount, N. Y., & Yeaman, M. R. (2004). Multidimensional signatures in antimicrobial peptides. *Proceedings of the National Academy of Sciences of the United States of America, 101*, 7363–7368.

Zapf, M. W., Theisen, S., Rohde, S., Rabenstein, F., Vogel, R. F., & Niessen, L. (2007). Characterization of AfpA, an alkaline foam protein from cultures of *Fusarium culmorum* and its identification in infected malt. *Journal of Applied Microbiology, 103*, 36–52.

Zhao, T., Havens, W. M., & Ghabrial, S. A. (2006). Disease phenotype of virus-infected *Helminthosporium victoriae* is independent of overexpression of the cellular alcohol oxidase/RNA binding protein Hv-p68. *Phytopathology, 96*, 326–332.

CHAPTER TWELVE

Phytophthora Viruses

Guohong Cai[1] and Bradley I. Hillman
Department of Plant Biology and Pathology, Rutgers University, New Brunswick, New Jersey, USA
[1]Corresponding author: e-mail address: cai@aesop.rutgers.edu

Contents

1. Introduction to Oomycetes—Phylogeny, Habitats, and Properties 328
2. History of Oomycete Virus Research 330
   2.1 Identification of viruses from various oomycetes 330
   2.2 An endornavirus from *Phytophthora* taxon Douglas fir and other likely viral elements in *Phytophthora* spp 331
   2.3 Identification of viral elements in *P. infestans* 332
3. Properties of Four Viruses of *P. infestans* 333
   3.1 PiRV-1, a virus with a unique genome structure 333
   3.2 PiRV-2 and its potential role in late blight epidemiology 338
   3.3 PiRV-3, a member of a proposed new virus genus 339
   3.4 PiRV-4 and its comparison to narnaviruses 342
4. Transmission of Phytophthora Viruses 344
5. Molecular Manipulations of Phytophthora Viruses 345
6. Conclusions, Future Directions, and Closing Remarks 345
References 346

Abstract

Phytophthora sp. is a genus in the oomycetes, which are similar to filamentous fungi in morphology and habitat, but phylogenetically more closely related to brown algae and diatoms and fall in the kingdom Stramenopila. In the past few years, several viruses have been characterized in *Phytophthora* species, including four viruses from *Phytophthora infestans*, the late blight pathogen, and an endornavirus from an unnamed *Phytophthora* species from Douglas fir. Studies on *Phytophthora* viruses have revealed several interesting systems. Phytophthora infestans RNA virus 1 (PiRV-1) and PiRV-2 are likely the first members of two new virus families; studies on PiRV-3 support the establishment of a new virus genus that is not affiliated with established virus families; PiRV-4 is a member of *Narnaviridae*, most likely in the genus *Narnavirus*; and *Phytophthora endornavirus 1* (PEV1) was the first nonplant endornavirus at the time of reporting. Viral capsids have not been found in any of the above-mentioned viruses. PiRV-1 demonstrated a unique genome organization that requires further examination, and PiRV-2 may have played a role in late blight resurgence in 1980s–1990s.

Advances in Virus Research, Volume 86
ISSN 0065-3527
http://dx.doi.org/10.1016/B978-0-12-394315-6.00012-X

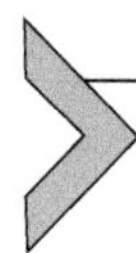

1. INTRODUCTION TO OOMYCETES—PHYLOGENY, HABITATS, AND PROPERTIES

Phytophthora infestans caused the Irish potato famine in 1840s and continues to devastate potatoes and tomatoes worldwide, resulting in billions of dollars annually in losses and control costs (Fry & Goodwin, 1997). During the past three decades, the difficulty in suppressing late blight has increased because of the recent worldwide distribution of diverse populations of *P. infestans* from Mexico, the center of genetic diversity of the organism. These populations contain both the A1 and A2 mating types (previously, only the A1 mating type had a worldwide distribution; Fry et al., 1993). Metalaxyl, a pesticide that was once very effective against late blight, has been overcome by resistant strains in most locations (Goodwin, Sujkowski, & Fry, 1996).

P. infestans belongs to the oomycetes, a group that includes many serious plant pathogens, as well as pathogens of animals and insects, and some saprophytes. The approximately 60 species of *Phytophthora* arguably constitute the most devastating single genus of pathogens of dicotyledonous plants (Erwin & Ribeiro, 1996). A newly emerged *Phytophthora* species, *P. ramorum*, is responsible for a disease called sudden oak death in California and Oregon (Rizzo, Garbelotto, & Hansen, 2005). It affects not only the live oaks that are the keystone species of the ecosystem but also a large variety of woody shrubs that inhabit the oak ecosystems. *P. capsici* is now devastating many vegetable crops across the nation.

Other than *Phytophthora*, the oomycetes include important plant pathogens. *Albugo*, *Bremia*, and *Peronospora* species cause white rust and downy mildew on several crops (Agrios, 1997). Members of the genus *Pythium* include more than a hundred species that are abundantly present in water and soil habitats and cause a diversity of plant diseases, mainly in root tissue (Vanderplaatsniterink, 1981). Animal-pathogenic oomycetes, such as species in the genus *Saprolegnia*, can cause severe losses in aquaculture and fisheries (Bruno & Wood, 1999). *Pythium insidiosum* is known to infect various mammals, including humans, horses, and dogs (Mendoza, Hernandez, & Ajello, 1993).

Oomycetes are similar to filamentous fungi in morphology and habitat and were traditionally classified as fungi. However, these two groups have little taxonomic affinity and have many fundamental differences. For example, the cell walls of oomycetes are composed of cellulose rather than chitin (Bartnicki-Garcia, 1970, 1987), as in fungi; in the vegetative state, oomycetes have diploid nuclei, whereas fungi have haploid nuclei; oomycetes use diaminopimelate (or DAP) pathway for lysine synthesis (Barr, 1983), while fungi use the aminoadipate (or AAA) pathway (Vogel, 1965); promoters from plants, fungi, and animals have no detectable activity in oomycetes (Judelson, Tyler, & Michelmore, 1992); most cloned genes in oomycetes lack the TATA element in their promoter regions; and instead, a 19-nt core sequence (McLeod, Smart, & Fry, 2004; Pieterse et al., 1994) similar to the initiator element (Quon, Delgadillo, Khachi, Smale, & Johnson, 1994) was overrepresented.

Based on modern molecular and biochemical analyses, oomycetes are more closely related to brown algae and diatoms and fall in the kingdom Stramenopila (Fig. 12.1) (Baldauf et al., 2000; Cavalier-Smith, 2000;

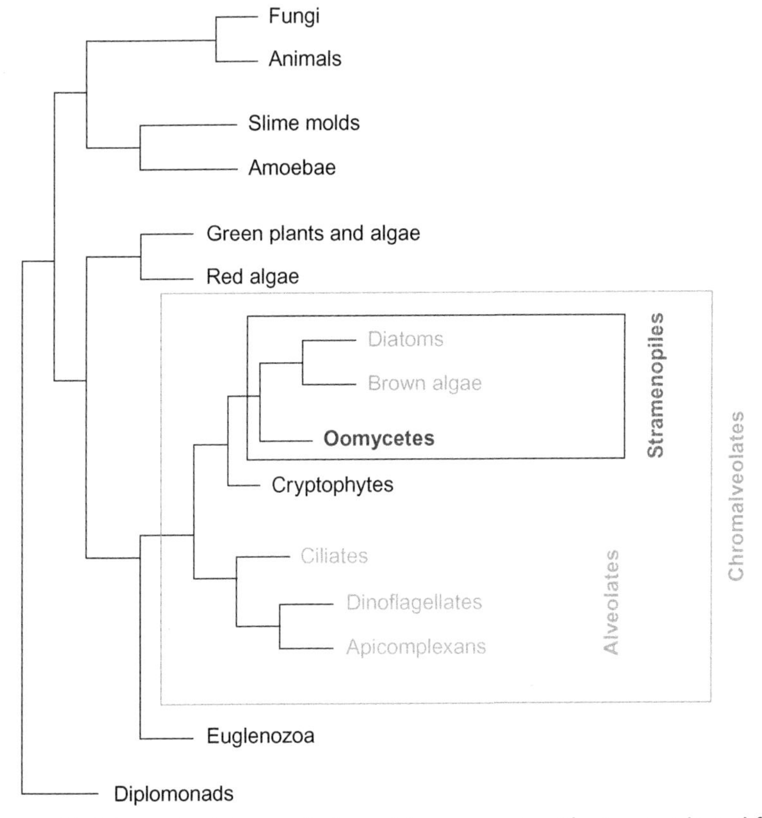

Figure 12.1 Schematic phylogenetic tree of the eukaryotes. The tree is adapted from that of Baldauf, Roger, Wenk-Siefert, and Doolittle (2000) that is based on a concatenation of six highly conserved proteins. (For color version of this figure, the reader is referred to the online version of this chapter.)

Sogin & Silberman, 1998). The algal stramenopiles are secondarily photosynthetic, having engulfed a red alga and adopted its plastid approximately 1300 million years ago (Yoon, Hackett, Pinto, & Bhattacharya, 2002). Nonphotosynthetic stramenopiles, such as the oomycetes, do not even have the vestigial plastids found in apicomplexan and euglenoid parasites that originated from phototrophs. However, analysis of the genomes of *P. sojae* and *P. ramorum* found numerous genes with likely phototroph origins, suggesting that, including oomycetes, the stramenopiles evolved from a photosynthetic ancestor (Tyler et al., 2006).

2. HISTORY OF OOMYCETE VIRUS RESEARCH

2.1. Identification of viruses from various oomycetes

Honkura, Shirako, Ehara, and Yamanaka (1983) reported two morphologically and serologically distinct virus-like particles (VLPs) in rice samples infected by the downy mildew pathogen *Sclerophthora macrospora*. Both were found to be viruses with positive-sense single-strand (ss) RNA genomes (Shirako & Ehara, 1985; Yokoi, Takemoto, Suzuki, Yamashita, & Hibi, 1999; Yokoi, Yamashita, & Hibi, 2003). Sclerophthora macrospora virus A (SmV A) is a 30-nm icosahedral virion with spikes of 4 nm on the periphery, harboring three genomic segments. RNA1 is 2928 nucleotides (nt) in length and has two open reading frames (ORFs), with ORF1a encoding an RNA-dependent RNA polymerase (RdRp) and ORF1b encoding a protein with unknown function. RNA2 consists of 1981 nt and a single ORF (ORF2) that encodes the capsid protein. RNA3 consists of 977 nt with no apparent ORF (Yokoi et al., 2003) and is considered a satellite RNA. In various extractions, relative quantities of RNA1 and RNA2 were stable, while RNA3 varied greatly (Shirako & Ehara, 1985). SmV A has two capsid proteins (p43 and p39, respectively), both encoded by ORF2. p39 is believed to derive from p43 through proteolytic digestion at N-terminus (Yokoi et al., 2003).

Sclerophthora macrospora virus B (SmV B) is a 32-nm icosahedral, monopartite virus with a single ssRNA genome of 5533 nt. It has two ORFs, with ORF1 encoding a putative polyprotein containing the motifs of chymotrypsin-related serine protease and RdRp and ORF2 encoding a capsid protein (Yokoi et al., 1999).

The taxonomy positions of SmV A and SmV B remain to be determined. SmV A RdRp and overall genome structure show similarity to members of *Nodaviridae*, while its capsid protein is more closely related to *Tombusviridae*.

SmV B showed limited similarity to members of *sobemovirus*, *barnavirus*, and *luteovirus* in the order of its functional domains and the sequences of serine protease and RdRp.

A virus related to SmV A was found in the sunflower downy mildew pathogen *Plasmopara halstedii* (Heller-Dohmen, Gopfert, Pfannstiel, & Spring, 2011). Isometric virus particles of Plasmopara halstedii virus (PhV) was first found in a North American pathotype (Gulya, 1990; Gulya, Freeman, & Mayhew, 1992), and the genome contained two ssRNA segments (Mayhew, Cook, & Gulya, 1992). Later, PhV was found in all isolates belonging to eight pathotypes in a worldwide collection of *Plasmopara halstedii* (Heller-Dohmen, Göpfert, Hammerschmidt, & Spring, 2008). Both the RdRp encoded by RNA1 and the capsid protein encoded by RNA2 showed strong similarity to the corresponding proteins of SmV A. However, there were no counterparts in PhV for ORF1b and RNA3 of SmV A (Heller-Dohmen et al., 2011).

VLPs were observed in some isolates of *Albugo candida* (Buck, 1986). In the genus *Pythium*, dsRNAs but not VLPs were reported in *Pythium butleri* (Buck, 1986), while VLPs harboring dsRNAs were found in *Pythium irregulare* (Gillings, Tesoriero, & Gunn, 1993; Klassen, Kim, Barr, & Desaulniers, 1991).

2.2. An endornavirus from *Phytophthora* taxon Douglas fir and other likely viral elements in *Phytophthora* spp

Hacker, Brasier, and Buck (2005) reported the identification of Phytophthora endornavirus 1 (PEV1) from an unnamed *Phytophthora* isolate from Douglas fir. PEV1 contains an unencapsitated dsRNA genome of 13,883 bp with a single ORF encoding a polyprotein with motifs characteristic of an RdRp, an RNA helicase, and a UDP glycosyltransferase. Its genome organization, and composition, order, and sequences of the functional motifs are highly similar to plant endornaviruses such as Oryza sativa endornavirus (OsEV) (Moriyama, Nitta, & Fukuhara, 1995), Oryza rufipogon endornavirus (OrEV) (Moriyama, Horiuchi, Koga, & Fukuhara, 1999), and Vicia faba endornavirus (VfEV) (Pfeiffer, 1998).

PEV1 is probably widespread in the unnamed *Phytophthora* taxon Douglas fir as it was found in all four isolates obtained from Douglas fir in Oregon, USA, and from carrot and white cockle in New York, USA (Kozlakidis, Brown, Jamal, Phoon, & Coutts, 2010). This virus was also found in 7 (5 from UK and 2 from the Netherlands) out of 19 European isolates of *P. ramorum*, but not from any of the 15 isolates from the United States. No phenotype was reported for PEV1 in any host. Smaller dsRNAs were found in three out of four isolates of *Phytophthora* taxon Douglas fir, but were not further characterized (Kozlakidis et al., 2010).

Mathews, Gu, Johnston, and Coffey (2011) screened 200 accessions in World *Phytophthora* Collection (http://phytophthora.ucr.edu/) and found dsRNAs in over a dozen isolates. Five isolates harbored a large dsRNA that they hypothesized to be PEV1. Using electron microscopy, Styer and Corbett (1978) observed intranuclear VLPs in *P. parasitica* var. *parasitica*, Roos and Shaw (1985) observed VLPs in the nuclei of *P. drechsleri*, and VLPs were also found in *P. nicotianae* (Buck, 1986).

2.3. Identification of viral elements in *P. infestans*

Styer and Corbett (1978) reported the observation of two morphologically distinct intranuclear VLPs in *P. infestans*. In a screening of dsRNA in 40 isolates from Mexico and 20 isolates from the United States and Europe, Tooley, Hewings, and Falkenstein (1989) found dsRNA segments of four different sizes, estimated to be 2.75, 2.5, 1.6, and 1.5 kb, respectively. The two larger segments always appeared in the same isolates (top doublets), as did the two smaller segments (bottom doublets). The top doublets and bottom doublets appeared in the same isolates or separately. DsRNAs were found in 36% of Mexican isolates but none in the isolates from the United States or Europe. The authors used a variety of protocols in an attempt to purify VLPs from isolates containing dsRNAs, but no VLPs were found.

In a follow-up study, Newhouse, Tooley, Smith, and Fishel (1992) screened 79 isolates from Mexico, the Netherland, Peru, Israel, and the United States. In addition to the top doublets and bottom doublets, they identified three additional dsRNAs estimated to be 11.10, 3.15, and 1.35 kb, respectively. The estimated sizes of the top doublets and the bottom doublets were revised to 3.25, 2.80, 1.67, and 1.54 kb, respectively. These dsRNAs formed nine patterns. Only dsRNAs of sizes 11.10, 3.15, and 1.35 kb were found in Dutch and Peruvian isolates, and they were also found in Mexican isolates. DsRNAs of the same sizes from different locations cross-hybridized in northern blot analysis.

In a cDNA subtraction study, Judelson and Fabritius (2000) found a cDNA from a *P. infestans* strain that hybridized to total RNA but not to total DNA of the host strain. It did not hybridize to either total RNA or total DNA from other *P. infestans* strains. Further characterization showed that this cDNA represented a linear ssRNA replicon of 625 nt in length in the host strain, with its complementary strand found at approximately

1/130 concentration. This ssRNA, later named Phytophthora infestans extrachromosomal RNA element 1 (PiERE1), has no apparent ORF with polyU and polyA tracts at its 5′ and 3′ termini, respectively. Two-thirds of the replicons were found in the nucleus, one-third in cytoplasm, and none in the mitochondrion. It exhibited predominant maternal inheritance in crosses and had little impact on the growth and pathogenicity of its host. PiERE1 only causes subtle changes in its host, resulting in greater thermotolerance during growth and an increase in secondary homothallism (Judelson, Ah-Fong, & Fabritius, 2010).

Through a project initiated in the lab of Dr. Bill Fry, our group screened 22 isolates of *P. infestans* from Mexico (13), the United States (5), Estonia (3), and South Africa (1) for dsRNA (Cai, Fry, Hillman, & Myers, 2009) and found dsRNAs of five different sizes in four patterns in 9 isolates from Mexico and the United States (Fig. 12.2). The 3.3- and 2.9-kb segments in pattern I always appeared together and were found in three Mexican isolates, the 11.2-kb segment was found in two isolates from the United States, the 8.3- and 3.0-kb segments in pattern III were found in a single isolate from the United States, and a 3.0-kb segment that is homologous (>98% nucleotide sequence identity) to the 3.0-kb segment in pattern III was also found without the 8.3-kb segment in three Mexican isolates (pattern IV). The five dsRNAs belong to four different viruses, tentatively named Phytophthora infestans RNA viruses 1–4 (PiRVs 1–4), respectively (Fig. 12.2). These viruses will be discussed in detail in the next section.

The isolates from the studies by Tooley et al. (1989) and Newhouse et al. (1992) were not available for comparison to our findings. However, the 3.3- and 2.9-kb segments in pattern I matched their top doublets in pattern and sizes, and the 11.2- and 3.0-kb segments in our findings are similar in sizes to the 11.10 and 3.15 segments in their reports.

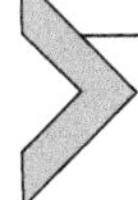

3. PROPERTIES OF FOUR VIRUSES OF *P. INFESTANS*

3.1. PiRV-1, a virus with a unique genome structure

PiRV-1 genome consists of two RNA segments (Fig. 12.2, pattern I). RNA1 is 3160 nt and RNA2 is 2776 nt, each with an ORF spanning most of the sequences (Fig. 12.3A). ORF1 on RNA1 encodes an RdRp, and ORF2 on RNA2 encodes a protease at its C-terminus. There is a potential ORF3 on RNA2 predicted to encode a protein of unknown function. A unique feature of this virus is that we found two forms of RNA2. In the minor form (RNA2B), a 19-nt stretch found in the major form

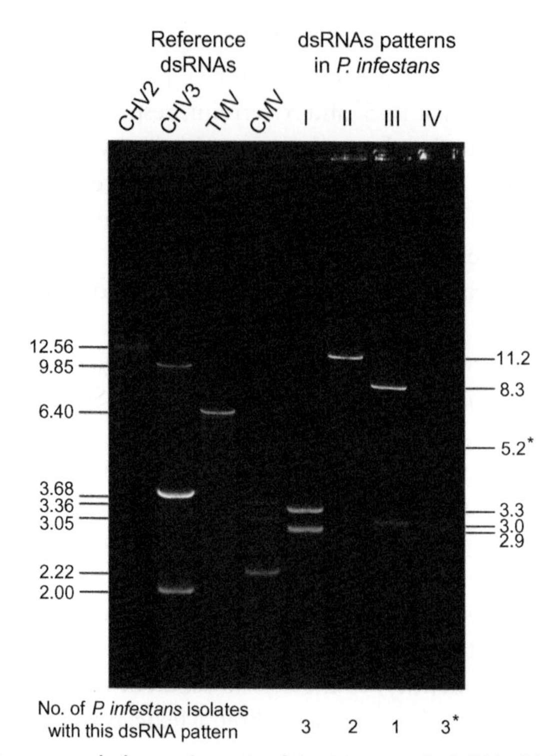

Figure 12.2 Agarose gel electrophoresis of double-stranded RNAs (dsRNAs) in *Phytophthora infestans* and dsRNAs of reference viruses: CHV2, Cryphonectria hypovirus 2 from *C. parasitica* strain NB58 (Hillman, Halpern, & Brown, 1994); CHV3, Cryphonectria hypovirus 3 from *C. parasitica* strain GH2 (Smart et al., 1999); TMV, tobacco mosaic virus (Goelet et al., 1982); and CMV, cucumber mosaic virus (Owen, Shintaku, Aeschleman, Bentahar, & Paulukaitis, 1990; Rizzo & Palukaitis, 1988, 1989). Sizes of reference dsRNAs (in kb) are labeled on the left. CHV2 and CHV3 dsRNAs have poly(A) tails of varying lengths and they were assumed to be 50 nt. Sizes of *P. infestans* dsRNAs (in kb) were estimated using appropriate reference dsRNAs and they are labeled on the right. *All three pattern IV isolates contained the 3.0 kb dsRNA; the weak 5.2-kb segment was visible only in some but not all extractions of one isolate. *Reprinted from Cai, Myers, Hillman, and Fry (2009), courtesy of Elsevier.*

(RNA2A) is replaced by a 9-nt sequence, and this replacement breaks ORF2 into ORF2B-1 and ORF2B-2, the latter encoding the protease.

We believe that RNA2A and RNA2B represent true sequence heterogeneity of PiRV-1 genome rather than cloning artifacts for several reasons: (1) the two forms represent the replacement of a sequence block rather than single nucleotide polymorphism, (2) the minor form was observed in 4 out of 17 clones and the 4 clones were from two independent RT-PCRs, and (3) the specific sequence replacement in RNA2B has profound predicted

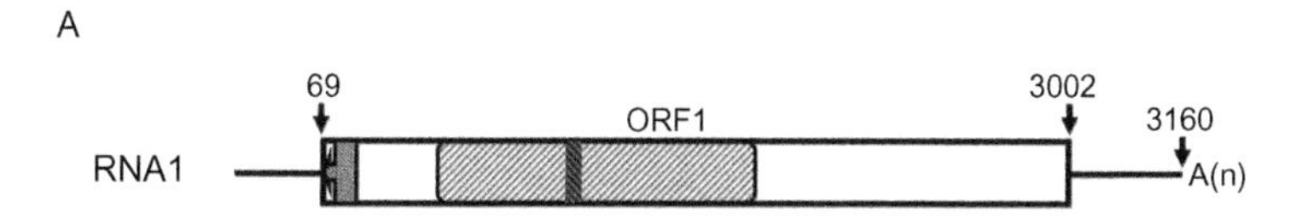

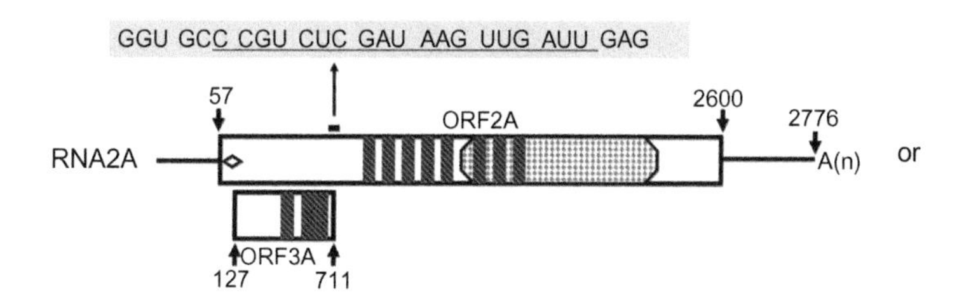

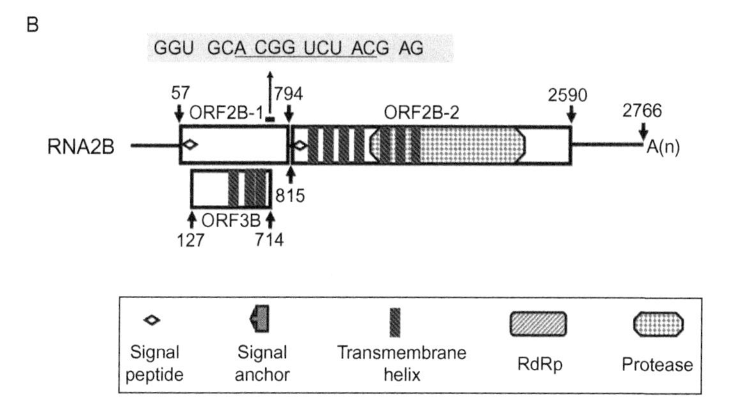

Figure 12.3 (A) Genome structure of PiRV-1 based on sequence analysis. The numbers indicate nucleotide positions. Two forms of RNA2 were found. A sequence heterogeneity (underlined) in RNA2 breaks ORF2A into ORF2B-1 and ORF2B-2. (B) Neighbor-joining tree based on the RdRps of PiRV-1 and related viruses. Bootstrap values (1000 replicates) in percentage are labeled for branches with more than 50% support. Full virus names and EMBL/GenBank/DDBJ accession numbers for the protein sequences (in parentheses) are APMoV, Andean potato mottle virus (Q02941); BatAstV, Bat astrovirus (ACF75855); BaYMV, barley yellow mosaic virus (Q04574); BEV, bovine enterovirus (P12915); BRSV, beet ringspot virus (P18522); CHV-1, Cryphonectria hypovirus 1 (AAA67458); CHV-2, Cryphonectria hypovirus 2 (AAA20137); CHV-3, Cryphonectria hypovirus 3 (AAF13604); CHV-4, Cryphonectria hypovirus 4 (AAQ76546); CPMV, cowpea mosaic virus (P03600); CPSMV, cowpea severe mosaic virus (P36312); CV-A21, Coxsackievirus A21 (P22055); CV-A9, Coxsackievirus A9 (P21404); EMCV, encephalomyocarditis virus (P03304); EV-70, human enterovirus 70 (P32537); FCV, feline calicivirus (P27407); FgV-1, Fusarium graminearum virus 1 (AAT07067);

continued

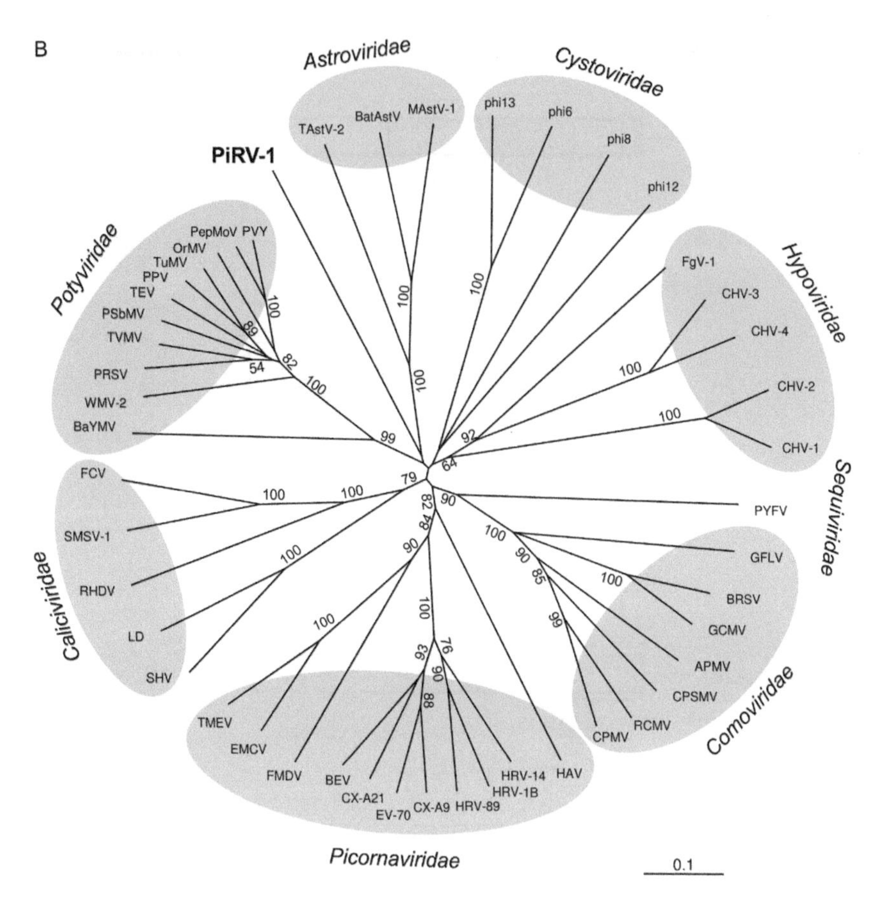

Figure 12.3—cont'd FMDV, foot-and-mouth disease virus (P03306); GCMV, grapevine chrome mosaic virus (P13025); GFLV, grapevine fanleaf virus (P29149); HAV, hepatitis A virus (P26580); HRV-14, human rhinovirus 14 (P03303); HRV-1B, human rhinovirus 1B (P12916); HRV-89, human rhinovirus 89 (P07210); LD, Lordsdale virus (P54634); MAstV-1, mink astrovirus 1 (AAO32082); OrMV, Ornithogalum mosaic virus (P20234); PepMoV, pepper mottle virus (Q01500); phi12, Pseudomonas phage phi12 (Q94M06); phi13, Pseudomonas phage phi13 (Q9FZT2); phi6, Pseudomonas phage phi6 (P11124); phi8, Pseudomonas phage phi8 (Q9MC15); PPV, plum pox virus (P13529); PRSV, papaya ringspot virus (Q01901); PSbMV, pea seed-borne mosaic virus (P29152); PVY, Potato virus Y (Q02963); PYFV, parsnip yellow fleck virus (Q05057); RCMV, red clover mottle virus (P35930); RHDV, rabbit hemorrhagic disease virus (P27410); SHV, Southampton virus (Q04544); SMSV-1, San Miguel sea lion virus serotype 1 (P36286); TAstV-2, turkey astrovirus 2 (AAF60952); TEV, tobacco etch virus (P04517); TMEV, Theiler's murine encephalomyelitis virus (P08544); TuMV, turnip mosaic virus (Q02597); TVMV, tobacco vein mottling virus (P09814); and WMV-2, *Watermelon mosaic virus 2* (P18478). *Reprinted from Cai, Myers, et al. (2009), courtesy of Elsevier.*

implications of PiRV-1 gene expression. The 9-nt sequence replacement is part of a 20-nt sequence stretch (nt 689–708 on RNA2B) upstream of ORF2B-2 (the protease) that is the reverse complement of nt 323–342 of hepatitis C virus (HCV) subtype 1a, which forms an essential part of domain IV of the HCV internal ribosomal entry site (IRES) and can interact with 40S ribosomal subunit directly (Kieft, Zhou, Jubin, & Doudna, 2001; Kolupaeva, Pestova, & Hellen, 2000; Lytle, Wu, & Robertson, 2002). Thus, a logical hypothesis is that RNA2B expresses the functional form of the protease that is necessary for PiRV-1, such as processing of polyproteins encoded by ORF1 and ORF2. Furthermore, the sequence replacement in RNA2B locates at the end of ORF3 thus has minimal impact on the coding sequence of this ORF.

There are several hypotheses for the coexistence of RNA2A and RNA2B. First, both RNA2A and RNA2B are indispensable components of the PiRV-1 genome that are required for replication. In this scenario, for example, RNA2B could be responsible for expression of functional protease that is necessary for the processing of other viral proteins. Transfection experiments using *in vitro* transcripts of RNA1 and RNA2A established only transient, not permanent, virus infection (data not shown), which supports this hypothesis. However, this experiment need to be repeated and other factors may account for the outcome.

Sequence heterogeneity could also reflect two forms of the same segment that have selective advantage in particular situations. Such heterogeneity is often seen at the single nucleotide level due to a quasispecies effect, in which a "cloud" of potentially beneficial mutations affords the virus a greater probability to evolve and adapt to new environments (Vignuzzi, Stone, Arnold, Cameron, & Andino, 2006). If certain variants have complementary selective advantage, they may accumulate to high frequency. For example, in tomato aspermy virus strain V, a single point mutation (C to A) at position 100 of RNA3 led to the abolishment of virus cell-to-cell movement but increased virus replication. When used individually, the "C" variant but not the "A" variant established systemic infection in the host, but the "A" variant accumulated to high frequency (76%) by the complementation of the "C" variant (24%) (Moreno, Malpica, Rodriguez-Cerezo, & Garcia-Arenal, 1997). Similar observation was reported in Rift Valley fever virus (Morrill et al., 2010). While in these cases the heterogeneity was single nucleotide polymorphisms, it is much more likely that the replacement of the large sequence block in RNA2 of PiRV-1 resulted from recombination event rather than point mutation.

Finally, related to the above scenario, it is possible that both RNA2A and RNA2B are independently and competitively functional deriving from two variants of PiRV-1 coinfecting the same isolate of *P. infestans*. In this scenario, we would expect one PiRV-1 variant to be replaced by the other, either by chance or because it is less competitive than the other strain. The *P. infestans* isolate MX980400, from which the PiRV-1 was extracted, was obtained from a potato sample in Mexico in 1998 and transferred many times during the 10 years before PiRV-1 characterization, arguing against the third hypothesis.

Based on Blast searches using the RdRp sequences, PiRV-1 is most similar to members of the positive-sense ssRNA, mammal-infecting virus family *Astroviridae*, followed by that of *Potyviridae* and *Picornaviridae*. In phylogenetic analysis, PiRV-1 clustered with *Astroviridae*, but with only 44% bootstrap support. Our efforts to extract PiRV-1 virus particles turned out negative, and there is no detectable capsid protein identifiable from its sequence. PiRV-1 differs from viruses in the *Astroviridae* family in its genome organization. Astroviruses have icosahedral and nonenveloped particles enclosing monopartite positive-sense ssRNA genomes ranging in size from 6.4 to 7.3 kb that are polyadenylated at the 3′ end (Mendez & Arias, 2007). It is worth noting that the combined size of RNA1 and RNA2 of PiRV-1 is similar to the genome sizes of astroviruses. There is an interesting parallel between PiRV-1 and the astroviruses, and the relationship among the genera of the *Potyviridae*: PiRV-1 has a bipartite genome and is basal to the *Astroviridae*, just as the bipartite *Bymovirus* genus is basal to the monopartite potyviruses (Koonin, Wolf, Nagasaki, & Dolja, 2008) (Fig. 12.3B). Given the absence in other genomic regions of strong similarities to previously characterized viruses, PiRV-1 likely is the first member of a new virus family.

3.2. PiRV-2 and its potential role in late blight epidemiology

PiRV-2 has an RNA genome of 11,170 bp (Fig. 12.2, pattern II) predicted to encode a single polyprotein of 3710 amino acids. Both 5′ and 3′ UTRs are AU-rich, and the 3′ UTR contains a sequence similar to the canonical mRNA polyadenylation signal, but it has no polyA tail (Cai, Fry, et al., 2009). A motif characteristic for cysteine protease is identified at the N-terminus and motifs for RdRp are found close to the C-terminus. However, the RdRp of PiRV-2 only matches RdRp profiles in databases such as Prosite (Hofmann, Bucher, Falquet, & Bairoch, 1999; Hulo et al., 2006) and Conserved Domain (Marchler-Bauer & Bryant, 2004), but no detectable similarity with known sequences in public databases, suggesting that PiRV-2 is only distantly related to known viruses and likely belongs to a new virus family. No virus particle has been found for PiRV-2 in our preliminary experiments.

PiRV-2 has been found in two isolates of *P. infestans* that were isolated from the United States 10 years apart. These two isolates sporulate profusely in repeated transfers on rye agar, while *P. infestans* isolates often lose much of sporulation capability in *in vitro* cultures after repeated transfers. We have cured PiRV-2 from one isolate. The virus-bearing isolate produces several times more sporangia compared to the isogenic, cured counterparts (not published), suggesting that PiRV-2 is responsible for the abundant sporulation.

Both *P. infestans* isolates harboring PiRV-2 belong to the US-8 lineage, the only isolates in this lineage that we screened. Beginning in Europe in the 1980s, resurgence of late blight epidemics swept potato and tomato crops worldwide arriving in North America by late 1980s and early 1990s (Faber, 1994). This wave of late blight epidemics was caused by the introduction of exotic pathogen strains that are more virulent and resistant to Metalaxyl. The US-8 lineage was particularly aggressive, especially on potato. After being detected only in New York in 1992 and in Maine in 1993 (Goodwin et al., 1995), US-8 swept 23 states during 1994 and 1995 and has been dominant nationally since 1996 (Fry & Goodwin, 1997). US-8 not only replaced the old resident lineage US-1 but also displaced other newly introduced lineages, such as US-6 and US-7. We are currently screening more *P. infestans* isolates, including more US-8 isolates, for PiRV-2 and testing the hypothesis that PiRV-2 was an important factor for the high aggressiveness of US-8 lineage and contributed to the resurgence of late blight epidemics in North America.

Interestingly, a dsRNA of 11.10 kb was also found in five Mexican isolates and all four Dutch isolates in the study by Newhouse et al. (1992). The isolates are not available for us to test whether this dsRNA represented genomes of PiRV-2, but if so, then by extension of the above hypothesis, this virus may also have played an important role in late blight epidemics in Europe.

3.3. PiRV-3, a member of a proposed new virus genus

PiRV-3 was found in a single isolate (FLa2005) of *P. infestans* that also harbored PiRV-4. This isolate was obtained from an infected tomato plant from Florida in 2005. The genome of PiRV-3 is 8112 bp with two overlapping ORFs linked by a potential −1 frame shift sequence. ORF1 encodes a protein of unknown function and ORF2 encodes an RdRp. PiRV-3 shares extensive similarities with five unclassified dsRNA viruses in filamentous fungi: Phlebiopsis gigantea virus 2 (PgV2) (Kozlakidis et al., 2009), Fusarium virguliforme RNA virus 1 (FvV1) and FvV2 (Acc.# JN671444 and JN67443), Fusarium graminearum virus 3 (FgV3) (Yu, Kwon, Lee, Son, & Kim, 2009), and grapevine-associated totivirus-2 (GaTV-2) (Al Rwahnih, Daubert, Úrbez-Torres, Cordero, & Rowhani, 2011). The group forms a new proposed virus genus (New genus 2, Fig. 12.4) (Cai et al., 2013). Only partial sequence of GaTV-2 was obtained from a high-throughput sequencing of grapevine-associated dsRNAs. It was assumed to be from a grapevine-associated fungus, but that has not been

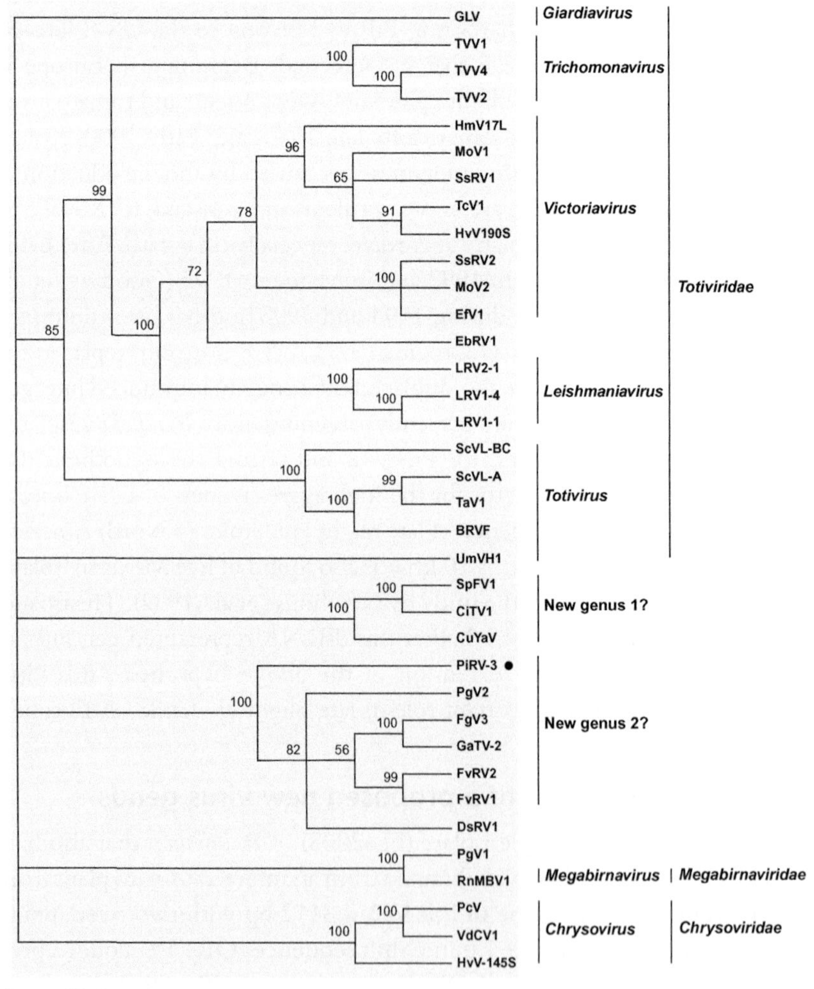

Figure 12.4

continued

confirmed. Thus, GaTV-2 is considered as a tentative member of this newly proposed genus. Diplodia scrobiculata RNA virus 1 (DsRV1), although grouped with new proposed genus 2 based on RdRp, is likely a recombinant virus, obtaining its ORF1 from a fungal host. On the first glance, these groups of viruses look like totiviruses. Indeed, PgV2 and GaTV-2 are listed in GenBank as members of *Totiviridae*, but only some of their properties fit that family. Members of this newly proposed genus have nonsegmented dsRNA genomes of 8–10 kb, longer than totiviruses. Each viral genome has two ORFs, with ORF2 in the -1 reading frame relative to ORF1. They have long 5′ UTRs with multiple AUG codons upstream from the first AUG codon in ORF1. In each instance, some of the AUG codons are in contexts predicted to be favorable for translation initiation (Kozak, 1986; Lütcke et al., 1987), suggesting that these viruses do not use a cap-dependent scanning mechanism for translation initiation of ORF1. There is some evidence suggesting ORF1 expression through IRES. The first in-frame AUG codon in ORF2 is downstream of the stop codon in ORF1, but ORF1 and ORF2

Figure 12.4 Neighbor-joining tree based on alignments of RdRp sequences of PiRV-3 and related viruses. Numbers above the branches indicate bootstrap support in percentile in 1000 replicates. Branches with less than 50% support were condensed. Virus species (acronym, GenBank accession) included in analysis were black raspberry virus F (BRVF, NC_009890), Circulifer tenellus virus 1 (CiTV1, NC_014360), cucurbit yellows-associated virus (CuYaV, X92203), Diplodia scrobiculata RNA virus 1 (DsRV1, NC_013699), *Eimeria brunetti* RNA virus 1 (EbRV1, NC_002701), Epichloe festucae virus 1 (EfV1, AM261427), Fusarium graminearum virus 3 (FgV3, NC_013469), Fusarium virguliforme RNA virus 1 (FvRV1, JN671444), Fusarium virguliforme RNA virus 2 (FvRV2, JN671443), Giardia lamblia virus (GLV, NC_003555), grapevine-associated totivirus-2 (GaTV-2, GU108594), Helicobasidium mompa virus 17L (HmV17L, NC_005074), Helminthosporium victoriae 145S virus (HvV-145S, NC_005978), Helminthosporium victoriae virus 190S (HvV190S, NC_003607), Leishmania RNA virus 1-1 (LRV1-1, M92355), Leishmania RNA virus 1–4 (LRV1–4, NC_003601), Leishmania RNA virus 2–1 (LRV2-1, NC_002064), Magnaporthe oryzae virus 1 (MoV1, NC_006367), Magnaporthe oryzae virus 2 (MoV2, NC_010246), Penicillium chrysogenum virus (PcV, NC_007539), Phlebiopsis gigantea mycovirus dsRNA 1 (PgV1, NC_013999), Phlebiopsis gigantea mycovirus dsRNA2 (PgV2, AM111097), Rosellinia necatrix megabirnavirus 1 (RnMBV1, NC_013462), Saccharomyces cerevisiae virus L-A (ScVL-A, NC_003745), Saccharomyces cerevisiae virus L-BC (ScVL-BC, NC_001641), Sphaeropsis sapinea RNA virus 1 (SsRV1, NC_001963), Sphaeropsis sapinea RNA virus 2 (SsRV2, NC_001964), Spissistilus festinus virus 1 (SpFV1, NC_014359), Tolypocladium cylindrosporum virus 1 (TcV1, NC_014823), Trichomonas vaginalis virus 1 (TVV1, NC_003824), Trichomonas vaginalis virus 2 (TVV2, AF127178), Trichomonas vaginalis virus 4 (TVV4, HQ607522), Tuber aestivum virus 1 (TaV1, HQ158596), Ustilago maydis virus H1 (UmVH1, NC_003823), and Verticillium dahliae chrysovirus virus 1 (VdCV1, HM004067). *Courtesy of Elsevier.*

have a short overlap when in-frame codons upstream of the first AUG codon in ORF2 are considered. A candidate slippery sequence for −1 frameshift was found at the same location in each virus—immediately before the ORF1 stop codon, and followed by a predicted pseudoknot or stem loop.

In totiviruses, ORF1 encodes a capsid protein. In this new virus group, ORF1-encoded proteins share no detectable similarity to viral capsid proteins. In fact, they only share similarity to each other but not to any other proteins in BlastP searches of GenBank nonredundant database. Our efforts to purify virus particles of PiRV-3 using various protocols have not yielded any virus particles thus far. Similarly, Kozlakidis et al. (2009) also failed to purify virus particles for PgV2. In FvV1 and FvV2, ORF1 is speculated to be a structural/gag protein; however, virus particle purifications have not yet confirmed that prediction (Dr. Les Domier, personal communication).

Spear, Sisterson, Yokomi, and Stenger (2010) proposed a new virus genus consisting of Spissistilus festinus virus 1, Circulifer tenellus virus 1, and Cucurbit yellows-associated virus (New genus 1, Fig. 12.4). Their study included some members of new genus 2 and they proposed that two new genera form a new virus family. Our RdRp phylogeny provides less than 50% bootstrap support for this contention (Fig. 12.4). In ORF1-encoded sequences, these two groups did not show similarity to each other in GenBank BlastP searches, but they do have something in common—all rich in tryptophan, proline, serine, and alanine, but underrepresented by isoleucine, glutamic acid, and methionine. The International Committee for the Taxonomy of Viruses plans to examine the taxonomic status of these viruses.

PiRV-3 but not PiRV-4 has been cured from isolate FLa2005. PiRV-3 only causes minor change in colony morphology and does not impact growth rate or sporulation on rye agar.

3.4. PiRV-4 and its comparison to narnaviruses

The small dsRNAs from *P. infestans* isolates MX980317, US970001, and FLa005 all were found to be isolates of the same virus, a member of the virus family *Narnaviridae* (Cai, Myers, Fry, & Hillman, 2012) (Fig. 12.5). A similar sized dsRNA was identified in one other isolate, MX980211, but that isolate was lost in storage and the dsRNA was not sequenced.

PiRV-4 was found alone in isolates MX980211, MX980317, and US970001 and was found together with PiRV-3 in isolate FLa005. Attempts to cure PiRV-4 from infected hosts have not been successful,

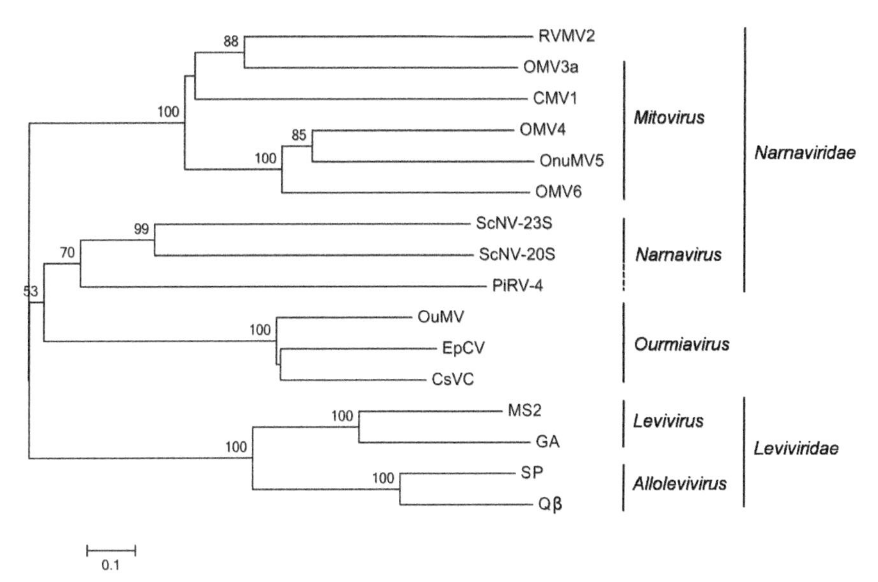

Figure 12.5 Neighbor-joining tree based on RdRp sequences of PiRV-4 and related viruses. Bootstrap support values above 50% from 1000 replicates are labeled above the branches. Virus species (acronym, GenBank accession) included in analysis were Cryphonectria mitovirus 1 (CMV1, NC_004046), Cassava virus C (CsVC, FJ157981), Enterobacteria phage GA (GA, X03869), Enterobacteria phage MS2 (MS2, NC_001417), Enterobacteria phage Qβ (Qβ, NC_001890), Enterobacteria phage SP (SP, X07489), Epirus cherry virus (EpCV, EU770620), Ophiostoma mitovirus 3a (OMV3a, NC_004049), Ophiostoma mitovirus 4 (OMV4, NC_004052), Ophiostoma mitovirus 5 (OMV5, NC_004053), Ophiostoma mitovirus 6 (OMV6, NC_004054), Ourmia melon virus (OuMV, EU770623), Rhizoctonia virus M2 (RVMV2, U51331), Saccharomyces narnavirus 20S RNA (ScNV-20S, NC_004051), and Saccharomyces narnavirus 23S RNA (ScNV-23S, U90136). *Reprinted from Cai et al. (2012), courtesy of SpringerLink.*

preventing direct comparison to an isogenic uninfected *P. infestans* strain and therefore preventing unequivocal determination of effects of the virus on its host. Isolates harboring PiRV-4 produce few sporangia on rye agar; however, this is not uncommon for *P. infestans* after extended maintenance on agar medium.

PiRV-4 has been placed tentatively in the genus *Narnavirus* of the family *Narnaviridae*. Its closest relatives are the two accepted members of that genus, the yeast-infecting viruses Saccharomyces cerevisiae narnavirus 20S (ScNV-20S) and ScNV-23S (Hillman & Esteban, 2011). Like PiRV-4, neither of these viruses has been successfully eliminated from their yeast hosts. However, both ScNV-20S and ScNV-23S have been launched as replication-competent RNA molecules from full-length cDNA constructs integrated into the yeast genome (Esteban & Fujimura, 2003; Esteban, Vega, & Fujimura, 2005), allowing for unambiguous examination of host phenotype in response to these elements. For more detailed treatment of this family, please see the associated chapters in this volume on the virus family *Narnaviridae* (Chapter 6) and on viruses of *S. cerevisiae* (Chapter 1).

There is as yet no cytological or genetic evidence regarding whether the site of replication and accumulation of PiRV-4 is in the cytosol or in the mitochondria. This is an important distinction not only in terms of transmission considerations and the relative ease or difficulty of future molecular manipulation (e.g., Polashock, Bedker, & Hillman, 1997) but also because it is the sole biological feature determining taxonomic placement of viruses in the genus *Narnavirus* or the genus *Mitovirus* of the family *Narnaviridae* (Hillman & Esteban, 2011). PiRV-4 was placed tentatively in the genus *Narnavirus* on the basis of its phylogenetic position, which is close to the two characterized members of the genus, and because it lacked the feature found in all accepted members of the genus *Mitovirus*, that is, UGA codons that are presumably read as tryptophan rather than as translation terminators in mitochondria. Thanatephorus cucumeris mitovirus (TcMV; also called Rhizoctonia solani virus M2; Lakshman, Jian, & Tavantzis, 1998) has been found in both the cytosol and mitochondria of its basidiomycete host. It has been tentatively placed in the genus *Mitovirus* because it is found at least to some extent in mitochondria and because phylogenetic analysis places it within the genus.

Because of its small size and its close relationship to the two yeast elements that have been successfully launched as cDNA constructs, it will be of interest to examine whether PiRV-4 infection can be initiated either in its *Phytophthora* host or perhaps in yeast from cDNA clones.

4. TRANSMISSION OF PHYTOPHTHORA VIRUSES

The four viruses in *P. infestans* are quite stable within the host. Hyphal-tipping in combination with Ribavirin treatment did not eliminate PiRV-1 or PiRV-4 in repeated transfers during 6-month period. PiRV-2 and PiRV-3 were eliminated only after 4 and 5, respectively, consecutive transfers. PiRV-1 exhibited 100% inheritance in over 30 single-zoospore progenies (unpublished), as did PiRV-2 in over two-dozen single-sporangium and single-zoospore progenies (unpublished).

Virus transmission through anastomosis has only been attempted to some extent with PiRV-2 because it has a microscopically observable phenotype.

Initial attempts at transmission by hyphal anastomosis have been unsuccessful, but these are being repeated using different cultural conditions. Virus transmission through sexual recombination has yet to be conducted, mainly due to difficulty in isolating oospores without mycelium contamination. We are not aware of any report on PEV1 transmission.

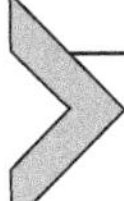

5. MOLECULAR MANIPULATIONS OF PHYTOPHTHORA VIRUSES

In our molecular manipulation of PiRV-1, we did not include RNA2B. Using *in vitro* transcripts of RNA1 and RNA2A, we established transient but not stable PiRV-1 infection in a virus-free isolate. Furthermore, these transcripts drove transient expression of GFP sandwiched between 5′ and 3′ UTRs of RNA1. In *in vitro* translation experiments, RNA2A, alone or in combination with RNA1, yielded a protein smaller than expected, and no protease activity was detected. A possible explanation is that RNA2B produces the active protease, which is needed for processing of polyproteins produced by RNA1 or RNA2A, and/or other activities.

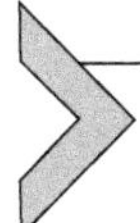

6. CONCLUSIONS, FUTURE DIRECTIONS, AND CLOSING REMARKS

Study of *Phytophthora* viruses is only beginning, but several interesting systems are already apparent. PiRV-1 and PiRV-2 are likely the first members of two new virus families. Our study on PiRV-3 supports the establishment of a new virus genus that is not affiliated with established virus families. PEV1 was the first nonplant endornavirus at the time of reporting.

The unique feature of PiRV-1 requires further study. Current evidence suggests that both RNA2A and RNA2B are required, but direct test of this hypothesis is needed. If indeed both forms are needed, what are the roles of each form in virus life cycle?

The capability of PiRV-2 to increase sporulation of *P. infestans* suggests that it can be a hypervirulence factor. The fact that it was found in the aggressive US-8 lineage in the U.S. and European isolates collected during the late blight resurgence suggested it could be an important factor contributing to late blight epidemiology. If this hypothesis is true, we have one more important angle to look at late blight. Studying the interaction between PiRV-2 and *P. infestans* will enhance and improve our knowledge of *P. infestans* sporulation regulation. These studies will lay the foundation of new control strategies.

Expression of foreign genes and knockout of indigenous genes in *Phytophthora*, and in oomycetes in general, are much more difficult than in fungi and many other organisms, even though much effort and progress have been reported (e.g., Judelson & Michelmore, 1991; Judelson, Tyler, & Michelmore, 1991; Vijn & Govers, 2003; Whisson, Avrova, Van West, & Jones, 2005). Oomycetes are diploid except briefly during sexual recombination. Controlled crosses to study specific traits are very difficult in many species, including *P. infestans*, and purification of a trait through traditional genetics is either difficult or infeasible in these species. Traditional transformation is laborious and inefficient, partly because homologous recombination does not work in these organisms (Judelson et al., 1991). *Agrobacterium tumefaciens*-mediated transformation is more time consuming and also inefficient (Vijn & Govers, 2003). Some success using dsRNA for transient gene silencing has been reported (Whisson et al., 2005). Virus-based vectors have been used as an effective and high-throughput tool for gene expression in mammals (Nakanishi, 1995) and for gene silencing or expression in plants (Pogue, Lindbo, Garger, & Fitzmaurice, 2002). Understanding viruses in *Phytophthora* may lead to a highly functional virus-based gene expression/knockout system in these organisms.

REFERENCES

Agrios, G. N. (1997). *Plant pathology*. San Diego, CA: Academic Press.

Al Rwahnih, M., Daubert, S., Úrbez-Torres, J., Cordero, F., & Rowhani, A. (2011). Deep sequencing evidence from single grapevine plants reveals a virome dominated by mycoviruses. *Archives of Virology, 156*, 397–403.

Baldauf, S. L., Roger, A. J., Wenk-Siefert, I., & Doolittle, W. F. (2000). A kingdom-level phylogeny of eukaryotes based on combined protein data. *Science, 290*, 972–977.

Barr, D. J. S. (1983). The zoosporic grouping of plant pathogens—Entity or non-entity? In S. T. Buczacki (Ed.), *Zoosporic plant pathogens: A modern perspective* (pp. 43–83). London: Academic Press.

Bartnicki-Garcia, S. (1970). Cell wall composition and other biochemical markers in fungal phylogeny. In J. B. Harborne (Ed.), *Phytochemical phylogeny* (pp. 81–103). London: Academic Press.

Bartnicki-Garcia, S. (1987). The cell wall: A crucial structure in fungal evolution. In A. D. M. Rayner, C. M. Brasier & D. Moore (Eds.), *Evolutionary biology of the fungi* (pp. 389–403). Cambridge: Academic Press.

Bruno, D. W., & Wood, B. P. (1999). *Saprolegnia* and other oomycetes. In P. T. K. Woo & D. W. Bruno (Eds.), *Fish diseases and disorders* (pp. 599–659). Wallingford, UK: CAB International.

Buck, K. W. (1986). Fungal virology—An overview. In K. W. Buck (Ed.), *Fungal virology* (pp. 1–84). Boca Raton, FL: CRC Press Inc.

Cai, G., Fry, W. E., Hillman, B. I., & Myers, K. (2009). An RNA virus from *Phytophthora infestans* with no apparent similarity to known viruses. *Phytopathology, 99*, S18.
Cai, G., Krychiw, J. F., Myers, K., Fry, W. E., & Hillman, B. I. (2013). A new virus from the plant pathogenic oomycete *Phytophthora infestans* with an 8 kb dsRNA genome: The sixth member of a proposed new virus genus. *Virology, 435*, 341–349.
Cai, G., Myers, K., Fry, W., & Hillman, B. (2012). A member of the virus family *Narnaviridae*, from the plant pathogenic oomycete *Phytophthora infestans*. *Archives of Virology, 157*, 165–169.
Cai, G., Myers, K., Hillman, B. I., & Fry, W. E. (2009). A novel virus of the late blight pathogen, *Phytophthora infestans*, with two RNA segments and a supergroup 1 RNA-dependent RNA polymerase. *Virology, 392*, 52–61.
Cavalier-Smith, T. (2000). Membrane heredity and early chloroplast evolution. *Trends in Plant Science, 5*, 174–182.
Faber, H. (1994). A virulent potato fungus is killing the northeast crop. *New York Times*. (November 12).
Erwin, C. D., & Ribeiro, O. K. (1996). Phytophthora *diseases worldwide*. St. Paul: APS Press.
Esteban, R., & Fujimura, T. (2003). Launching the yeast 23S RNA Narnavirus shows 5′ and 3′ *cis-acting* signals for replication. *Proceedings of the National Academy of Sciences of the United States of America, 100*, 2568–2573.
Esteban, R., Vega, L., & Fujimura, T. (2005). Launching of the yeast 20 S RNA Narnavirus by expressing the genomic or antigenomic viral RNA *in vivo*. *The Journal of Biological Chemistry, 280*, 33725–33734.
Fry, W. E., & Goodwin, S. B. (1997). Re-emergence of potato and tomato late blight in the United States. *Plant Disease, 81*, 1349–1357.
Fry, W. E., Goodwin, S. B., Dyer, A. T., Matuszak, J. M., Drenth, A., Tooley, P. W., et al. (1993). Historical and recent migrations of *Phytophthora infestans*: Chronology, pathways, and implications. *Plant Disease*, 77, 653–661.
Gillings, M. R., Tesoriero, L. A., & Gunn, L. V. (1993). Detection of double-stranded RNA and virus-like particles in Australian isolates of *Pythium irregulare*. *Plant Pathology, 42*, 6–15.
Goelet, P., Lomonossoff, G. P., Butler, P. J. G., Akam, M. E., Gait, M. J., & Karn, J. (1982). Nucleotide sequence of tobacco mosaic virus RNA. *Proceedings of the National Academy of Sciences of the United States of America, 79*, 5818–5822.
Goodwin, S. B., Sujkowski, L. S., Dyer, A. T., Fry, B. A., & Fry, W. E. (1995). Direct-detection of gene flow and probable sexual reproduction of *Phytophthora infestans* in northern North America. *Phytopathology, 85*, 473–479.
Goodwin, S. B., Sujkowski, L. S., & Fry, W. E. (1996). Widespread distribution and probable origin of resistance to metalaxyl in clonal genotypes of *Phytophthora infestans* in the United States and western Canada. *Phytopathology, 86*, 793–800.
Gulya, T. J. (1990). Virus-like particles in *Plasmopara halstedii*, sunflower downy mildew. *Phytopathology, 80*, 1032.
Gulya, T. J., Freeman, T. P., & Mayhew, D. E. (1992). Ultrastructure of virus-like particles in *Plasmopara halstedii*. *Canadian Journal of Botany, 70*, 334–339.
Hacker, C. V., Brasier, C. M., & Buck, K. W. (2005). A double-stranded RNA from a *Phytophthora* species is related to the plant endornaviruses and contains a putative UDP glycosyltransferase gene. *The Journal of General Virology, 86*, 1561–1570.
Heller-Dohmen, M., Göpfert, J. C., Hammerschmidt, R. A. Y., & Spring, O. (2008). Different pathotypes of the sunflower downy mildew pathogen *Plasmopara halstedii* all contain isometric virions. *Molecular Plant Pathology, 9*, 777–786.
Heller-Dohmen, M., Gopfert, J. C., Pfannstiel, J., & Spring, O. (2011). The nucleotide sequence and genome organization of Plasmopara halstedii virus. *Virology Journal, 8*, 123.

Hillman, B. I., & Esteban, R. (2011). Family *Narnaviridae*. In A. M. Q. King, M. J. Adams, E. B. Castens & E. J. Lefkowitz (Eds.), *Virus taxonomy: Ninth report of the international committee for the taxonomy of viruses*. New York: Elsevier.

Hillman, B. I., Halpern, B. T., & Brown, M. P. (1994). A viral dsRNA element of the chestnut blight fungus with a distinct genetic organization. *Virology*, *201*, 241–250.

Hofmann, K., Bucher, P., Falquet, L., & Bairoch, A. (1999). The PROSITE database, its status in 1999. *Nucleic Acids Research*, *27*, 215–219.

Honkura, R., Shirako, Y., Ehara, Y., & Yamanaka, S. (1983). Two types of virus-like particles isolated from downy mildew diseased rice plants. *Annals of the Phytopathological Society of Japan*, *49*, 653–658.

Hulo, N., Bairoch, A., Bulliard, V., Cerutti, L., De Castro, E., Langendijk-Genevaux, P. S., et al. (2006). The PROSITE database. *Nucleic Acids Research*, *34*, D227–D230.

Judelson, H. S., Ah-Fong, A. M. V., & Fabritius, A. L. (2010). An RNA symbiont enhances heat tolerance and secondary homothallism in the oomycete *Phytophthora infestans*. *Microbiology (UK)*, *156*, 2026–2034.

Judelson, H. S., & Fabritius, A. L. (2000). A linear RNA replicon from the oomycete *Phytophthora infestans*. *Molecular & General Genetics*, *263*, 395–403.

Judelson, H. S., & Michelmore, R. W. (1991). Transient expression of genes in the oomycete *Phytophthora infestans* using *Bremia lactucae* regulatory sequences. *Current Genetics*, *19*, 453–459.

Judelson, H. S., Tyler, B. M., & Michelmore, R. W. (1991). Transformation of the oomycete pathogen, *Phytophthora infestans*. *Molecular Plant–Microbe Interactions*, *4*, 602–607.

Judelson, H. S., Tyler, B. M., & Michelmore, R. W. (1992). Regulatory sequences for expressing genes in oomycete fungi. *Molecular & General Genetics*, *234*, 138–146.

Kieft, J. S., Zhou, K., Jubin, R., & Doudna, J. A. (2001). Mechanism of ribosome recruitment by hepatitis C IRES RNA. *RNA*, 7, 194–206.

Klassen, G. R., Kim, W. K., Barr, D. J. S., & Desaulniers, N. L. (1991). Presence of double-stranded RNA in isolates of *Pythium irregulare*. *Mycologia*, *83*, 657–661.

Kolupaeva, V. G., Pestova, T. V., & Hellen, C. U. T. (2000). An enzymatic footprinting analysis of the interaction of 40S ribosomal subunits with the internal ribosomal entry site of hepatitis C virus. *Journal of Virology*, *74*, 6242–6250.

Koonin, E. V., Wolf, Y. I., Nagasaki, K., & Dolja, V. V. (2008). The Big Bang of picorna-like virus evolution antedates the radiation of eukaryotic supergroups. *Nature Reviews. Microbiology*, *6*, 925–939.

Kozak, M. (1986). Point mutations define a sequence flanking the AUG initiator codon that modulates translation by eukaryotic ribosomes. *Cell*, *44*, 283–292.

Kozlakidis, Z., Brown, N. A., Jamal, A., Phoon, X., & Coutts, R. H. A. (2010). Incidence of endornaviruses in *Phytophthora* taxon douglasfir and *Phytophthora ramorum*. *Virus Genes*, *40*, 130–134.

Kozlakidis, Z., Hacker, C. V., Bradley, D., Jamal, A., Phoon, X., Webber, J., et al. (2009). Molecular characterisation of two novel double-stranded RNA elements from *Phlebiopsis gigantea*. *Virus Genes*, *39*, 132–136.

Lakshman, D. K., Jian, J., & Tavantzis, S. M. (1998). A double-stranded RNA element from a hypovirulent strain of *Rhizoctonia solani* occurs in DNA form and is genetically related to the pentafunctional AROM protein of the shikimate pathway. *Proceedings of the National Academy of Sciences of the United States of America*, *95*, 6425–6429.

Lütcke, H. A., Chow, K. C., Mickel, F. S., Moss, K. A., Kern, H. F., & Scheele, G. A. (1987). Selection of AUG initiation codons differs in plants and animals. *The EMBO Journal*, *6*, 43–48.

Lytle, J. R., Wu, L., & Robertson, H. D. (2002). Domains on the hepatitis C virus internal ribosome entry site for 40s subunit binding. *RNA*, *8*, 1045–1055.

Marchler-Bauer, A., & Bryant, S. H. (2004). CD-Search: Protein domain annotations on the fly. *Nucleic Acids Research*, *32*, W327–W331.
Mathews, D. M., Gu, D., Johnston, B. S., & Coffey, M. D. (2011). Screening of the world *Phytophthora* collection for viruses. *Phytopathology*, *101*, S116.
Mayhew, D. E., Cook, A. L., & Gulya, T. J. (1992). Isolation and characterization of a mycovirus from *Plasmopara halstedii*. *Canadian Journal of Botany*, *70*, 1734–1737.
McLeod, A., Smart, C. D., & Fry, W. E. (2004). Core promoter structure in the oomycete *Phytophthora infestans*. *Eukaryotic Cell*, *3*, 91–99.
Mendez, E., & Arias, C. F. (2007). Astroviruses. In D. M. Knipe, P. M. Howley, D. E. Griffin, R. A. Lamb, M. A. Martin, B. Roizman & S. E. Straus (Eds.), *Fields virology* (pp. 981–1000). (5th ed.). Philadelphia, PA: Lippincott Williams & Wilkins.
Mendoza, L., Hernandez, F., & Ajello, L. (1993). Life cycle of the human and animal oomycete pathogen *Pythium insidiosum*. *Journal of Clinical Microbiology*, *31*, 2967–2973.
Moreno, I., Malpica, J., Rodriguez-Cerezo, E., & Garcia-Arenal, F. (1997). A mutation in tomato aspermy cucumovirus that abolishes cell-to-cell movement is maintained to high levels in the viral RNA population by complementation. *Journal of Virology*, *71*, 9157–9162.
Moriyama, H., Horiuchi, H., Koga, R., & Fukuhara, T. (1999). Molecular characterization of two endogenous double-stranded RNAs in rice and their inheritance by interspecific hybrids. *The Journal of Biological Chemistry*, *274*, 6882–6888.
Moriyama, H., Nitta, T., & Fukuhara, T. (1995). Double-stranded RNA in rice: A novel RNA replicon in plants. *Molecular & General Genetics*, *248*, 364–369.
Morrill, J. C., Ikegami, T., Yoshikawa-Iwata, N., Lokugamage, N., Won, S., Terasaki, K., et al. (2010). Rapid accumulation of virulent Rift Valley fever virus in mice from an attenuated virus carrying a single nucleotide substitution in the M RNA. *PLoS One*, *5*, e9986.
Nakanishi, M. (1995). Gene introduction into animal tissues. *Critical Reviews in Therapeutic Drug Carrier Systems*, *12*, 263–310.
Newhouse, J. R., Tooley, P. W., Smith, O. P., & Fishel, R. A. (1992). Characterization of double-stranded RNA in isolates of *Phytophthora infestans* from Mexico, the Netherlands, and Peru. *Phytopathology*, *82*, 164–169.
Owen, J., Shintaku, M., Aeschleman, P., Bentahar, S., & Paulukaitis, P. (1990). Nucleotide sequence and evolutionary relationships of cucumber mosaic virus (CMV) strains: CMV RNA 3. *The Journal of General Virology*, *71*, 2243–2249.
Pfeiffer, P. (1998). Nucleotide sequence, genetic organization and expression strategy of the double-stranded RNA associated with the '447' cytoplasmic male sterility trait in *Vicia faba*. *The Journal of General Virology*, *79*, 2349–2358.
Pieterse, C. J., van West, P., Verbakel, H. M., Brasse, P. W. H. M., van den Berg-Velthuis, G. C. M., & Govers, F. (1994). Structure and genomic organization of the *ipiB* and *ipiO* gene clusters of *Phytophthora infestans*. *Gene*, *138*, 67–77.
Pogue, G. P., Lindbo, J. A., Garger, S. J., & Fitzmaurice, W. P. (2002). Making an ally from an enemy: Plant virology and the new agriculture. *Annual Review of Phytopathology*, *40*, 45–74.
Polashock, J. J., Bedker, P. J., & Hillman, B. I. (1997). Movement of a small mitochondrial double-stranded RNA element of *Cryphonectria parasitica*: Ascospore inheritance and implications for mitochondrial recombination. *Molecular & General Genetics*, *256*, 566–571.
Quon, D. V. K., Delgadillo, M. G., Khachi, A., Smale, S. T., & Johnson, P. J. (1994). Similarity between a ubiquitous promoter element in an ancient eukaryote and mammalian initiator elements. *Proceedings of the National Academy of Sciences of the United States of America*, *91*, 4579–4583.

Rizzo, D. M., Garbelotto, M., & Hansen, E. A. (2005). *Phytophthora ramorum*: Integrative research and management of an emerging pathogen in California and Oregon forests. *Annual Review of Phytopathology, 43*, 309–335.

Rizzo, T. M., & Palukaitis, P. (1988). Nucleotide sequence and evolutionary relationships of cucumber mosaic virus (CMV) strains: CMV RNA 2. *The Journal of General Virology, 69*, 1777–1787.

Rizzo, T. M., & Palukaitis, P. (1989). Nucleotide sequence and evolutionary relationships of cucumber mosaic virus (CMV) strains: CMV RNA 1. *The Journal of General Virology, 70*, 1–11.

Roos, U. P., & Shaw, D. S. (1985). Intranuclear virus-like particles in a laboratory strain of *Phytophthora drechsleri. Transactions of the British Mycological Society, 84*, 340–344.

Shirako, Y., & Ehara, Y. (1985). Composition of viruses isolated from *Sclerophthora macrospora* infected rice plants. *Annals of the Phytopathological Society of Japan, 51*, 459–464.

Smart, C. D., Yuan, W., Foglia, R., Nuss, D. L., Fulbright, D. W., & Hillman, B. I. (1999). *Cryphonectria hypovirus 3*, a virus species in the family *Hypoviridae* with a single open reading frame. *Virology, 265*, 66–73.

Sogin, M. L., & Silberman, J. D. (1998). Evolution of the protists and protistan parasites from the perspective of molecular systematics. *International Journal for Parasitology, 28*, 11–20.

Spear, A., Sisterson, M. S., Yokomi, R., & Stenger, D. C. (2010). Plant-feeding insects harbor double-stranded RNA viruses encoding a novel proline-alanine rich protein and a polymerase distantly related to that of fungal viruses. *Virology, 404*, 304–311.

Styer, E. L., & Corbett, M. K. (1978). Intranuclear virus-like particles of *Phytophthora infestans* and *P. parasitica* var. *parasitica. Mycovirus Newsletter, 6*, 16–18.

Tooley, P. W., Hewings, A. D., & Falkenstein, K. F. (1989). Detection of double-stranded RNA in *Phytophthora infestans. Phytopathology, 79*, 470–474.

Tyler, B. M., Tripathy, S., Zhang, X., Dehal, P., Jiang, R. H. Y., Aerts, A., et al. (2006). *Phytophthora* genome sequences uncover evolutionary origins and mechanisms of pathogenesis. *Science, 313*, 1261–1266.

Vanderplaatsniterink, A. J. (1981). Monograph of the genus *Pythium. Studies in Mycology, 21*, 1–242.

Vignuzzi, M., Stone, J. K., Arnold, J. J., Cameron, C. E., & Andino, R. (2006). Quasispecies diversity determines pathogenesis through cooperative interactions in a viral population. *Nature, 439*, 344–348.

Vijn, I., & Govers, F. (2003). *Agrobacterium tumefaciens* mediated transformation of the oomycete plant pathogen *Phytophthora infestans. Molecular Plant Pathology, 4*, 459–467.

Vogel, H. J. (1965). Lysine biosynthesis and evolution. In V. Bryson & H. J. Vogel (Eds.), *Evolving genes and proteins* (pp. 25–40). London: Academic Press.

Whisson, S. C., Avrova, A. O., Van West, P., & Jones, J. T. (2005). A method for double-stranded RNA-mediated transient gene silencing in *Phytophthora infestans. Molecular Plant Pathology, 6*, 153–163.

Yokoi, T., Takemoto, Y., Suzuki, M., Yamashita, S., & Hibi, T. (1999). The nucleotide sequence and genome organization of Sclerophthora macrospora virus B. *Virology, 264*, 344–349.

Yokoi, T., Yamashita, S., & Hibi, T. (2003). The nucleotide sequence and genome organization of *Sclerophthora macrospora* virus A. *Virology, 311*, 394–399.

Yoon, H. S., Hackett, J. D., Pinto, G., & Bhattacharya, D. (2002). The single, ancient origin of chromist plastids. *Proceedings of the National Academy of Sciences of the United States of America, 99*, 15507–15512.

Yu, J., Kwon, S.-J., Lee, K.-M., Son, M., & Kim, K.-H. (2009). Complete nucleotide sequence of double-stranded RNA viruses from *Fusarium graminearum* strain DK3. *Archives of Virology, 154*, 1855–1858.

INDEX

Note: Page numbers followed by "*f*" indicate figures, and "*t*" indicate tables.

A

ACDACV. *See* Amasya cherry disease-associated chrysovirus (ACDACV)
Amasya cherry disease-associated chrysovirus (ACDACV)
  chryso-P3s, 303
  and FoCV1, 301–303

B

BcMV1. *See Botrytis cinerea* mitovirus 1 (BcMV1)
Birnaviruses *vs.* partitiviruses
  CP, 74
  structural level, 74
Botrytis cinerea
  detection methods, virus, 254–255
  fungal metabolism
    BVX/BVF-infected isolates, 261–262
    description, 261–262
    "pathway activity profiling", 261–262
  genome data
    antisense transcript, 262–263
    degradation pathway, 262–263
    description, 262
    dsRNA genome, 263–264
    Ustilago maydis, 263
  mycoviruses (*see* Mycoviruses)
  pathogen
    aerial mycelia, 251
    agricultural fungicides, 252
    cultural control methods, 252
    cup-shaped fruiting body/apothecium, 251–252
    gray mold, 250
    plant pathogenic fungi, 250
    population analyses, *B. cinerea*, 251
    sexual reproduction, 251–252
    symptomless infection, 252
  phenotypes
    bean leaf assay, 260, 261*f*
    BVX/BVF, 260
    hypovirulence and debilitation, 260
    in vitro growth rates, 260
    morphology/growth rates, 260
  phylogenetic relationships, fungal and plant viruses, 255–259
  transmission, virus, 264–267
  viruses
    and dsRNAs detected, 253–254, 253*t*
    fungicide resistance and residues, 254
    isometric/bacilliform particles, 253–254
Botrytis cinerea mitovirus 1 (BcMV1)
  and BcMV1-S, 255
  debilitation-related virus, 257–258
  strain, OnuMV3b, 257–258
Botrytis virus F (BVF). *See* Botrytis virus X (BVX)
Botrytis virus X (BVX)
  and BVF-infected isolates, 261–262, 265*t*
  Flexiviridae family, 253–254
  free isolates, 260
  genome organization, 256–257, 257*f*
  geographic origin, 266
BVX. *See* Botrytis virus X (BVX)

C

Capsid proteins (CPs)
  birnaviruses, 74, 78
  chrysoviruses, 78
  CP ORF, 293
  dsRNA viruses (*see* Double-stranded RNA (dsRNA))
  FpV1, 61–63
  inner (ICPs), 63–64
  PcV (*see* Penicillium chrysogenum virus (PcV))
  picobirnaviruses, 74–76
  PsV-S and PsV-F, 61, 64–65
  stop codon, 297–298
  structural comparison, CnCV, PcV, and L-A, 96–99, 98*f*
  totivirus and reovirus, 77–78

Capsid proteins (CPs) (*Continued*)
X-ray and cryo-EM-based structures, 90–91, 91*f*
Capsid structure, HvV190S
dsRNA genome, 314–316
icosahedron, 316
morphological features, 316
segmentation process, 316
subunit structure, 316, 317*f*
vs. totiviruses, 316–320
Chrysovirus
capsid structure, 89
CnCV1 (*see* Cryphonectria nitschkei chrysovirus 1 (CnCV1))
description, 88
dsRNA viruses, 90–92
icosahedral viruses, 89–90
PcV (*see* Penicillium chrysogenum virus (PcV))
quasi-equivalence laws, 89–90
totiviruses (*see* Totiviruses and chrysovirus)
Cis-acting signals
description, 15–17, 16*f*
5¢/3¢ end site, 17–18
23S RNA/p104 complex, 18
CnCV1. *See* Cryphonectria nitschkei chrysovirus 1 (CnCV1)
Coinfection, *S. sclerotiorum*
conidial spores, 241
description, 240
replication, 240
SsRV-L and SsDRV, 240
CpMV-1. *See* Cryphonectria mitovirus 1 (CpMV-1)
CPs. *See* Capsid proteins (CPs)
Cryphonectria mitovirus 1 (CpMV-1)
cDNA cloning and sequencing experiments, 154
description, 152–154
mitochondrial isolation experiments, 154–155
NB631, 152–154
virulence, 155
Cryphonectria nitschkei chrysovirus 1 (CnCV1)
CP amino acid sequences, 95–96
description, 95–96
PcV, and L-A, structural comparison, 96–99, 98*f*
structure, 95–96, 96*f*
Cryphonectria parasitica. See Hypoviruses/*Cryphonectria parasitica*
CThTV. *See* Curvularia thermal tolerance virus (CThTV)
Curvularia thermal tolerance virus (CThTV)
characterisation, 46
C. protuberata stains, 50–51
description, 45–46
functions, 45–46
mutualistic viruses, 50

D

DaRV. *See* Diaporthe ambigua RNA virus (DaRV)
Dendrolimus punctatus tetravirus (DpTV), 225–226
Diaporthe ambigua RNA virus (DaRV), 226–227
Double-stranded RNA (dsRNA)
chrysovirus
architectural and functional principles, 90
birnaviruses, 92
description, 90
genome packaging densities, 101, 102*t*
interactions, 101–104
L-A and P4 totiviruses, 90–91, 91*f*
packaged chrysovirus RNA, 101–104
PcV and CnCV1 capsids, 92
resolution, 92
RNA/protein contacts, 104
electron microscopy, 332
genomes, PiRV-2, 339
mycoviruses
description, 275
endophytic fungi, 276
F. graminearum, 275
Fusarium oxysporum, 275–276
Fusarium poae, 276
hypovirulence, 276
SDS-PAGE gels, 275
Phytophthora infestans, 333
screening, 332
DpTV. *See* Dendrolimus punctatus tetravirus (DpTV)

E

EfV1. *See* Epichloë festucae virus 1 (EfV1)
eIF3. *See* Eukaryotic initiation factor 3 (eIF3)
Endornavirus
  characteristics, 331
  electron microscopy, 332
  PEV1, 331–332
Epichloë festucae virus 1 (EfV1), 40, 45
Eukaryotic initiation factor 3 (eIF3)
  and eIF3/40S complexes, 298
  translation initiation, 298

F

FAD. *See* Flavin-adenine dinucleotide (FAD)
Flavin-adenine dinucleotide (FAD), 306–307
FMDV. *See* Foot-and-mouth disease virus (FMDV)
Foot-and-mouth disease virus (FMDV), 335*f*, 338
FpV1. *See* Fusarium poae virus 1 (FpV1)
Fungal antiviral defense mechanisms, *C. parasitica*
  description, 128–129
  RNA silencing
    components, 129–130, 130*f*, 131*f*
    description, 129
    induction and suppression, 130–131
    viral RNA recombination, 131–132
  vegetative incompatibility *(vic)* genes
    barrage formation, 137
    candidate *vic2* locus, 136
    candidate *vic4* locus, 136
    candidate *vic6* locus, 134–135, 135*f*
    candidate *vic7* locus, 135–136
    EP146 genome, 134–135
    genetic system, 132–133
    interactions, linked genes *pix6* and *vic6*, 137–138, 138*f*
    polymorphic *vic* alleles, 136–137
    strain EP146, 133–134, 134*f*
    virus transmission, 138
Fungal endophytes
  classification
    class 1, 38–39
    class 2, 39
    class 3, 39
    class 4, 39–40
  mutualistic plant-microbe interactions, 38
  role, agricultural system, 53
  symbiosis, 37–38
  viruses (*see* Viruses, fungal endophytes)
Fungus, partitiviruses. *See* Partitiviruses
Fusarium mycoviruses
  complete genome sequences, 277, 278*t*
  FgV2, 277
  FgV3 and FgV4, 277–279
  FgV1 genome, 277
  genome organization, 277, 279*f*
  hypoviruses, 279
  Magnaporthe oryzae chrysovirus 1, 280–281
  phylogenetic tree construction, 279, 280*f*
  transmission, protoplast, 283–284
Fusarium poae virus 1 (FpV1)
  arch domain, 68–69
  CPs, 61–63
  description, 61–63
  diameter, 65
  genomic dsRNA, 72, 73*f*
  sequence-based phylogenetic comparisons, 61–63

G

Glucose-methanol-choline (GMC), 306–307
GMC. *See* Glucose-methanol-choline (GMC)

H

HaSV. *See* Helicoverpa armigera stunt virus (HaSV)
Helicoverpa armigera stunt virus (HaSV), 225–226
Helminthosporium victoriae
  capsid structure, HvV190S, 314–320
  description, 290
  diseased isolates, 292–293
  fungal virus projects, 290–291
  host genes, 306–314
  HvV145S (*see* Helminthosporium victoriae virus 145S (HvV145S))
  HvV190S, 293–300

Helminthosporium victoriae (*Continued*)
molecular characterization and creation, 292–293
pathotoxin victorin, 290
Victoria blight disease, 291
Victoria-derived oat cultivars, 291–292
viral etiology, 305–306
Helminthosporium victoriae virus 145S (HvV145S)
ACDACV, 301–303
agarose gelelectrophoresis, 304–305
5' and 3' UTRs, 300–301
BLAST searches, 303–304
chryso-P3 and P4, 303
dsRNA segments, 300–301
genome organization, 300–301, 302*f*
neighbor-joining phylogenetic tree, 303–304, 304*f*
Partitiviridae, 303–304
PcV dsRNAs, 301–303
Helminthosporium victoriae virus 190S (HvV190S)
capsid structure, 314–320
Cochliobolus, 299
CP ORF, 293
CP stop codon and PABP, 299
description, 293
dsRNA genome, 295–297, 296*f*
genome organization, 293, 294*f*
neighbor-joining phylogenetic tree, 299, 300*f*
nonpolyadenylated victorivirus, 299
ORFs, 293
p88 and p83/p88 and p78, 294–295
pseudoknot, 298
RdRp ORF, 295–297
190S-1 and 190S-2, 294–295
18S rRNA, 299
stop-restart strategy, 297–298
termination-reinitiation, 295–297
Totivirus, 297–298
translational coupling, 298
TURBS and eIF3, 298
Hepatitis E virus (HEV)
HaSV and DpTV, 226
HEV-swine, 225–226
HEV. *See* Hepatitis E virus (HEV)
HGT. *See* Horizontal gene transfer (HGT)
Horizontal gene transfer (HGT), 236
Host genes
Hv-p68 (*see H. victoriae* p68 protein (Hv-p68))
P30 gene, 313–314
vin, 308–313
Host signaling pathways and gene expression, hypovirus infection
calcium signaling, 123–124
cDNA microarray system, 123
Cpst12, transcription factors, 124
G protein-signaling system, 123
identification, Vir2, 122
laccase, 122
MAPKs, 125
microarray data, 123–124
pleiotropic effects, 125–126
Reoviridae, 124–125
transformation protocol, 122–123
vegetative colony morphology, 124–125
H. victoriae p68 (Hv-p68) protein
FAD and GMC, 306–307
gene encoding, 306–307
multifunctional protein, 306
overexpression, 307
^{32}P end-labeled 190S and 145S dsRNAs, 307–308
phosphoamino acid analysis, 308
SDS-PAGE, 306–307
transcriptional level, 307
yeast tRNA, 307–308
Hv-p68 protein. *See H. victoriae* p68 (Hv-p68) protein
HvV145S. *See* Helminthosporium victoriae virus 145S (HvV145S)
HvV190S. *See* Helminthosporium victoriae virus 190S (HvV190S)
Hypoviridae
CHV-3 and CHV-4 species, 114–115
CHV-1/EP713, 111–112, 112*f*
genome organization, comparison, 113–114, 114*f*
hypovirulent *C. parasitica* strains, 113
Hypovirus, 111–112
SsHV1 and VcHV1, 115
taxonomy, 113
Hypoviruses/*Cryphonectria parasitica*
description, 110

fungal antiviral defense mechanisms, 128–138
host physiology, 126–128
Hypoviridae, 113–115
hypovirulent strains, 121–122
molecular characterization, 110–112
mycoviruses, experimental system, 115–117
signaling pathways and gene expression, fungal host, 122–126
translation and gene expression, 117–121
Hypovirus translation and gene expression
CHV1-EP713 replication, 120–121
membrane vesicles, 118
Mfold constrained predicted structure, 118–120, 119*f*
nuclease mapping approach, 118–120
nucleotide sequence, 117–118
p48, 121
roles, viral protein products, 120
untranslated regions (UTRs), 118

I

IAA-Leucine resistant 2 protein (ILR2), 236
ILR2. *See* IAA-Leucine resistant 2 protein (ILR2)

L

L-A virus
cap-snatching
description, 9, 10*f*
eukaryotic mRNA, 9
guanylyltransferase, 10–11
mutagenesis analyses, 11
mutation, His-154, 9–10
N7 methylation, 9–10, 11
description, 3
encapsidation, 11–12
K1 and K28, 3
and M-encoded proteins, 4, 5*f*
M replication cycle, 4–7, 6*f*
replication, 12
strains, 3
transcription, 7
translation
N-acetylation, Gag, 8–9
ribosomal frameshifting, 7–8
60S ribosomal subunits, 8
superkiller (SKI) genes, 8
virion structure, 3, 4*f*

M

MAPKs. *See* Mitogen-activated protein kinases (MAPKs)
Marine_PP. *See* Marine putative protein (Marine_PP)
Marine putative protein (Marine_PP), 219–221
Megabirnaviridae
genome organization, 186–189, 187*f*
phylogenetic tree of viruses, 186–189, 188*f*
RnMBV1 (*see* Rosellinia necatrix megabirnavirus 1 (RnMBV1))
Mitogen-activated protein kinases (MAPKs), 125
Mitoviruses
ascomycetes and basidiomycetes, 162–163
BcMV-1, 163–164
characterization, 163
CpMV-1 (*see* Cryphonectria mitovirus 1 (CpMV-1))
description, 152
filamentous fungi, 162
host effects, 156–158
in vitro delivery, 167
latent infection, 168
Ophiostoma ulmi and *Ophiostoma novo-ulmi*, 152
phylogenetic tree, genus, 152, 153*f*
plant pathogens, 162–163
Rhizoctonia solani M2 virus, 155–156
satellite and defective RNAs, 168–169
Sclerotinia homoeocarpa, 163
Sclerotinia sclerotiorum
BcMV1, 233
description, 231
horizontal transfer, 233
hypovirulent strain, 231, 232*f*
mitochondrial codon usage, 231
phylogenetic analysis, 233
screening and bioinformatics approach, 152–154
sequences, 162
SsMV-1/KL1 and CcMV-2a, 164
terminal sequences, RNA, 168

Molecular characterization, hypovirulence-associated dsRNAs
cDNA cloning, sequence analysis, and assembly efforts, 111
description, 110–111
genome organization, CHV-1/EP713, 111–112, 112*f*
polypeptides, p29 and p40, 111–112
sequence alignment analyses, 111–112
structural properties, analysis, 111
Mycoreovirus (MyRV)
MyRV1
C. parasitica, 194–195
genome segments, 195–196
MyRV2
C. parasitica, 194–195
genome segments, 195–196
MyRV3
functional analyses, 195
genome segments, 195–196
hypovirulent strain, W370, 194–195
S8, functional role, 195–196
Mycoviruses
Botrytis cinerea
cycloheximide, 259–260
donor and recipient lines, 259–260
Proteus vulgaris, 259
Cryphonectria parasitica, 290
DNAs/RNAs, 290
fungal host factors (*see* Proteomics and transcriptomics)
Fusarium mycoviruses (*see* Fusarium mycoviruses)
Fusarium species, 290–291
host interactions, 285
hypovirus/*C. parasitica* experimental system
description, 115–116
Koch's postulates, 116–117
plasmid DNA, 116
technical challenges, 115–116
transfection protocol, 116
transformation efficiency, 117
transgenic hypovirulent strain, 116–117
isolation and characterization, dsRNA, 275–276
R. necatrix
double-stranded (ds) RNAs, 184
etiological analysis, 184–185
inoculation, 185
virocontrol, 183–184
S. sclerotiorum
description, 236–238
hypovirulence and XG36-1strain, 238
mitoviruses, 231–233
SsDRV, 222–225
SsHADV-1, 218–222
SsHV-1 (*see Sclerotinia sclerotiorum* hypovirus 1 (SsHV-1))
SsPV-S, 233–236
SsRV-L, 225–227
strain XG36-1, 236–238
virus-like particles, 238
transmission, Fusarium mycovirus (*see* Protoplast)
MyRV. *See* Mycoreovirus (MyRV)

N

Narnaviridae
description, 150–151
genera, 150–151
genome organizations, 150–151, 150*f*
identification
Phytophthora infestans, 151–152
Saccharomyces cerevisiae, 151
mitochondrial transformation and transfection, 166–167
mitoviruses (*see* Mitoviruses)
population-level studies of members, 164–166
reverse genetics system, 166
RNA silencing, 159–162
RNA viruses
Levivirus and *Allolevivirus*, 158–159
members, 158–159, 160*t*
mitoviruses, 158
ourmiaviruses, 158
ScNV-23S and ScNV-20S RNAs, 166
Narnaviruses, 20S and 23S RNAs
description, 13
genomes, 13, 14*f*
launching systems
cis-acting signals (*see Cis*-acting signals)
description, 15
negative strand, 17
ribozyme, 15

terminal repair system, 15–17
top-half domain, tRNA, 15–17
replication intermediates
de novo synthesis, 19–20
sporulation conditions, 19–20
20S RNA positive strand synthesis, 19–20, 21*f*
SKI1 antiviral activity, 18–19
sucrose gradients, 14–15

O

Oomycetes
endornavirus, 331–332
morphology, 328–329
P. infestans (*see Phytophthora infestans*)
plant pathogens, 328
Pythium, 328
schematic phylogenetic tree, 329–330, 329*f*
stramenopiles, 329–330
virus research
ORFs, 330
PhV, 331
SmV B, 330
taxonomy positions, 330–331
VLPs, 331
Open reading frames (ORFs)
CP ORF, 293
ORF1 and ORF2, 293
p83 and p78, 294–295
RdRp ORF, 295–297
ORFs. *See* Open reading frames (ORFs)

P

Partitiviridae
bornavirus-infected cells, 192–194
RnPV (*see* Rosellinia partitivirus (RnPV))
Partitiviruses
Alphacryptovirus and *Betacryptovirus*, 79, 81
vs. birnaviruses, 74
capsid structures, 63–70
vs. chrysoviruses CP, 78
description, 60
dsRNA genome structures, 72–74, 73*f*
FpV1 (*see* Fusarium poae virus 1 (FpV1))
genera, 61
genome segments, 60
α-helices, 78
vs. Megabirnaviridae, 77
natural transmission, 60
neighbor-joining phylogenetic tree, *Partitiviridae*, 79, 80*f*
vs. picobirnaviruses (*see* Picobirnaviruses *vs.* partitiviruses)
PsV-F (*see* Penicillium stoloniferum virus F (PsV-F))
PsV-S (*see* Penicillium stoloniferum virus S (PsV-S))
vs. Quadriviridae, 77
RdRp (*see* RNA-dependent RNA polymerase (RdRp))
taxonomy and properties, 61, 62*f*
three-dimensional (3D) structures, 59–60
vs. totivirus and reovirus CP, 77–78
Partitiviruses capsid structures
arch domain, 68–69
assembly pathways, 69–70
diameter, 65
domain swapping, 68
dsRNA molecules, 70
genomic dsRNA, 63–64
protein (CP) molecules, 63–64
quasisymmetric dimer, 66–68
reovirus and totivirus, comparison, 66–68, 67*f*
single-layer, 63–64
virions, PsV-S and PsV-F, 64–65
PcV. *See* Penicillium chrysogenum virus (PcV)
PEG. *See* Polyethylene glycol (PEG)
Penicillium chrysogenum virus (PcV)
analytical ultracentrifugation analysis, 94
Chrysovirus, 300–301
CP monomers, 94
3D cryo-EM reconstruction, 92, 93–94, 93*f*
gene duplication, 95
HvV145S dsRNAs, 300–301
quadripartite genome, 292
sequence-based SSE, 94–95, 97*f*
Penicillium stoloniferum virus F (PsV-F)
arch domain, 68–69
assembly pathways, 69–70
description, 61
diameter, 65
domain swapping, 68

Penicillium stoloniferum virus F (PsV-F) (*Continued*)
dsRNA molecules, 70
genomic dsRNA, 72, 73*f*
virions, structural analyses, 64–65, 65*f*, 66*f*
Penicillium stoloniferum virus S (PsV-S)
arch domain, 68–69
assembly pathways, 69–70
description, 61
diameter, 65
domain swapping, 68
dsRNA molecules, 70
genomic dsRNA, 72, 73*f*
virions, structural analyses, 64–65, 65*f*, 66*f*
PEV1. *See* phytophthora endornavirus 1 (PEV1)
Phytophthora endornavirus 1 (PEV1), 331–332
PhV. *See* Plasmopara halstedii virus (PhV)
Phylogenetic relationships, *Botrytis cinerea*
BcMV1, 257–258
BVF and BVX, 255
description, 255
dsRNAs, 258
electron microscopy, 258
Gammaflexiviridae, 256
genome organization, BVX, 256, 257*f*
merits and limitations, 254–255
ORF II, 258
RdRp, OnuMV3b, 257–258
tymo-and potex-like viruses, 256
Phytophthora infestans
agarose gel electrophoresis, 333, 334*f*
cDNA, 332–333
dsRNA, 332
PiRV-1 (*see* Phytophthora infestans RNA virus 1 (PiRV-1))
PiRV-3 (*see* Phytophthora infestans RNA virus 3 (PiRV-3))
PiRV-2 and late blight
agarose gel electrophoresis, 334*f*, 338–339
virus-bearing isolate, 339
PiRV-4 and narnaviruses (*see* PiRV-4 and narnaviruses)
top and bottom doublets, dsRNAs, 332
Phytophthora infestans RNA virus 1 (PiRV-1)
agarose gel electrophoresis, 333–334, 334*f*
characterization, 338
FMDV, 335*f*, 338
genome structure, 333–334, 335*f*
heterogeneity, 334–337
quasispecies effect, 337
transfection, 337
Phytophthora infestans RNA virus 3 (PiRV-3)
colony morphology, 342
description, 339–342
neighbor-joining tree, 339–342, 340*f*
ORF1-encoded proteins, 342
RdRp phylogeny, 342
Phytophthora viruses
late blight, 345–346
molecular manipulations, 345
oomycetes (*see* Oomycetes)
P. infestans (*see Phytophthora infestans*)
PiRV-1, PiRV-2 and PiRV-3, 345
RNA2A and RNA2B, 345
transmission, 344–345
Picobirnaviruses *vs.* partitiviruses
assembly pathway, 74–76
capsid organizations and CP folds, 74–76, 75*f*
extracellular transmission, 76
packaging strategy, 76–77
rabbit picobirnavirus (RaPBV), 74–76
structural level, 74
PiRV-1. *See* Phytophthora infestans RNA virus 1 (PiRV-1)
PiRV-4 and narnaviruses
cytosol/mitochondria, 344
neighbor-joining tree, 342, 343*f*
P. infestans strain, 342–343
Plasmopara halstedii virus (PhV), 331
Polyethylene glycol (PEG)
and calcium ions, 284*f*
description, 283–284
Population structure and life cycle, *R. necatrix*
ascospore progenies, 183
description, 180, 181*f*
mycelial compatibility groups (MCGs), 179–180

mycovirus transmission, 179
reproductive strategy, 180–183
Proteomics and transcriptomics
2DE gels, 281–282
metabolic-related genes, 282–283
mycovirus-host interactions, 281
omics-based approaches, 281
plasma membrane-associated transcripts, 282–283
ribosomal RNA synthesis, 282
TFs, 282–283
3'-tiling microarray, 282
Protoplast
description, 283
fungal fusion, 283–284, 284*f*
PEG, 283–284
plant-pathogenic fungi, 283–284
PsV-F. *See* Penicillium stoloniferum virus F (PsV-F)
PsV-S. *See* Penicillium stoloniferum virus S (PsV-S)

Q

Quadriviridae. *See* Rosellinia necatrix quadrivirus 1 (RnQV1)

R

RdRp. *See* RNA-dependent RNA polymerase (RdRp)
Reoviridae
description, 194–195
MyRV (*see* Mycoreovirus (MyRV))
Rhizoctonia solani M2 virus. *See* Thanatephorus cucumeris mitovirus (TcMV)
RNA-dependent RNA polymerase (RdRp)
ACDACV RdRp, 303–304
amino acid sequences, SsRV-L, 224*f*
chrysovirus dsRNA1s, 303
conserved motifs, 225–226
CP ORF, 295–297
ML phylogenetic trees, 235*f*
ORF1 and ORF2, 293
partitiviruses
cystovirus procapsids, 71
and dsRNA genome structures, 72–74, 73*f*
location, 71
structural elements, 70–71
transcription, 71–72
transcripts, 71
Rubi-and Tobamo-like viruses, 228*f*
SsDRV, 222–223
stop-restart strategy, 295–297
RNA silencing and *Narnaviridae*
description, 159
Dicer and Argonaute genes, 159–162
reconstitution, *S. cerevisiae* strains, 159–162
RnMBV1. *See* Rosellinia necatrix megabirnavirus 1 (RnMBV1)
RnMBV2. *See* Rosellinia necatrix megabirnavirus 2 (RnMBV2)
RnPV. *See* Rosellinia partitivirus (RnPV)
RnQV1. *See* Rosellinia necatrix quadrivirus 1 (RnQV1)
Rosellinia necatrix
controls, 179
Cryphonectria parasitica, 202
description, 178
epidemiological analysis of viruses, 185–186
experimental host ranges, mycoviruses
investigation, 199
protoplast fusion, 201–202
reproducible transfection systems, 199–201, 200*f*
genus, 178
infectivity of viruses, other fungi, 203, 203*t*
mycovirus infection (*see* Mycoviruses)
population structure and life cycle, 179–183
prevalence, 178
symptoms, 178–179
virocontrol, 205–206, 206*f*
viruses
description, 186
Fusarium graminearum, 198–199
Megabirnaviridae, 186–190
Partitiviridae, 190–194
Phlebiopsis gigantea, 198–199
Quadriviridae, 196–198
Reoviridae, 194–196
Totiviridae, 198

Rosellinia necatrix megabirnavirus 1 (RnMBV1)
- approaches, etiology, 190
- colony morphology and virulence, 190, 191*f*
- description, 186–189
- genome organization, 186–189, 187*f*
- mutant strains (RnMBV1-M), 189
- phylogenetic tree of viruses, 186–189, 188*f*
- 5'-untranslated region (UTR), 189–190

Rosellinia necatrix megabirnavirus 2 (RnMBV2), 190–192

Rosellinia necatrix quadrivirus 1 (RnQV1)
- agarose gel profile, 196
- description, 196
- dsRNAs, 196–197
- encoded protein sequences, 197
- W1075 strain, 197
- W1118 strain, 197–198

Rosellinia partitivirus (RnPV)
- description, 190–191
- RnPV1
  - genome organization, 190–191
  - RnMBV2, 190–192
  - transfection, strain W97, 191–192
- RnPV2
  - CP sequence, 192
  - dsRNA1:DI-dsRNA1, 194
  - genome segments,-W57, 190–192
  - partitivirus CP-like sequences (PCLSs), 192, 193*f*

S

Saccharomyces cerevisiae
- dsRNA viruses
  - cDNA clones, 12
  - L-BC, 12
  - *ski* mutants, 12–13
  - and ssRNA viruses, 2
- viruses and prions
  - endornaviruses, 2
  - fungal prion [Het-s], 27
  - L-A virus, 3–12
  - narnaviruses, 13–20
  - yeast prions, 20–26

Sclerophthora macrospora virus B (SmV B), 330

Sclerotinia sclerotiorum
- coinfection, 240–241
- description, 216
- fungicides, 216–218
- gene hypothesis, 216–218
- hypovirulence-associated mycoviruses, 218
- infected plants, 216
- microspores, 216
- mycoviruses, 218–238
- and mycovirus interaction, 238–240
- necrotrophic fungal pathogen, 216–218
- QTL genes, 216–218
- uses, mycoviruses
  - chestnut blight disease, 241–242
  - *C. parasitica*, 241–242
  - hypovirulence-associated mycoviruses, 242
  - transmission, 242

Sclerotinia sclerotiorum debilitation-associated RNA virus (SsDRV)
- *Allexivirus S. sclerotiorum*, 225
- description, 222
- Ep-1PN, 222
- full-length genome, 222–223
- mycoviruses, 222, 223*f*
- phylogenetic analysis, replication protein, 223–225, 224*f*
- RdRp, 222–223

Sclerotinia sclerotiorum hypovirulence-associated DNA virus 1 (SsHADV-1)
- coat protein, 219–221
- description, 218–219
- DT-8 and isometric particles, 218–219, 220*f*
- geminiviruses, 219
- genome structure and length, 218–219
- Marine_PP, 219–221
- MmFV, 221–222
- PEG, 222
- RNA and DNA elements, 218–219

Sclerotinia sclerotiorum hypovirus 1 (SsHV-1)
- CHV3/GH2 and CHV4/SR2, 229, 230*f*
- dsRNA segment, SZ-150, 229
- phylogenetic analysis, 227, 228*f*
- *vs.* strain SZ-150, 230–231
- SZ-150 strain, 227
- totivirus and a hypovirus, 231
- VcHV1, 229–230

Sclerotinia sclerotiorum mitovirus 1 (SsMV1). *See* Mycoviruses
Sclerotinia sclerotiorum mitovirus 2/KL-1 (SsMV-1/KL1), 164
Sclerotinia sclerotiorum partitivirus S (SsPV-S)
 description, 233–234
 dsRNA segments, 234
 ILR2 and HGT, 236, 237*f*
 ML phylogenetic trees, RdRps, 234, 235*f*
 Sunf-M, 233–234, 234*f*
Sclerotinia sclerotiorum RNA virus L (SsRV-L)
 alpha-like viruses, 226
 DaRV, 226–227
 description, 225–226
 HaSV and DpTV, 225–226
 HEV, 226
 RNA replicase, 225–226
 RnQV1, 226–227
Secondary structure element (SSE)
 CnCV, PcV, and L-A CP, comparison, 96–99, 98*f*
 sequence-based analysis, 94–95, 97*f*
SmV B. *See* Sclerophthora macrospora virus B (SmV B)
SsDRV. *See* Sclerotinia sclerotiorum debilitation-associated RNA virus (SsDRV)
SSE. *See* Secondary structure element (SSE)
SsHADV-1. *See* Sclerotinia sclerotiorum hypovirulence-associated DNA virus 1 (SsHADV-1)
SsHV-1. *See* Sclerotinia sclerotiorum hypovirus 1 (SsHV-1)
SsPV-S. *See* Sclerotinia sclerotiorum partitivirus S (SsPV-S)
SsRV-L. *See* Sclerotinia sclerotiorum RNA virus L (SsRV-L)

T

TcMV. *See* Thanatephorus cucumeris mitovirus (TcMV)
Termination upstream ribosome binding site (TURBS)
 eIF3, 298
 reinitiation, 298
TFs. *See* Transcription factors (TFs)
Thanatephorus cucumeris mitovirus (TcMV)
 description, 155
 features, system, 156
 homology, 156
 R. solani system, 155–156
 tractable host system, 156–157
Totiviridae, 198
Totiviruses
 and chrysovirus
  conserved fold, 100–101
  CP structural halves, 99
  icosahedral viruses, 100
  monocistronic segments, 99–100
  spatial distribution, 100
  structural matching, 99–100
 phylogenetic analysis, 319–320
 RdRp expression, 319–320
 ScV-L-A subunit, 318–319
 secondary structure, ScV-L-A, 316–318, 318*f*
 Totiviridae and *Victorivirus*, 319
 x-ray crystal structure, ScV-L-A, 316–318
Transcription factors (TFs)
 description, 282–283
 Zn2Cys6 family, 282–283
TURBS. *See* Termination upstream ribosome binding site (TURBS)

V

Valsa ceratosperma hypovirus 1 (VcHV1), 115, 229–230
VcHV1. *See* Valsa ceratosperma hypovirus 1 (VcHV1)
Victoriocin (*Vin*) gene
 Ala^{23} and Val^{24}, 309–310
 antifungal activity, 308–309, 309*f*
 BLAST searches, 312
 CSab-and g-core motifs, 312
 description, 308–309
 H. victoriae genomic DNA library, 312–313
 killer toxins, 309–310
 preprotoxin, 310–312
 preprovictoriocin, 309–310, 311*f*
 S. cerevisiae, 309–310

Victoriocin (*Vin*) gene (*Continued*)
scorpion toxin family, 312
totivirus-infected killer strain, 310–312
Zygosaccharomyces bailii virus, 310–312
Vin gene. *See* Victoriocin (Vin) gene
Viruses, fungal endophytes
class 2 and 3 endophytes
Alternaria alternata, 47
Cladosporium, 46
CThTV, 45–46
Curvularia and *Alternaria*, 47
Curvularia protuberata, 45–46
Phoma and *Stemphylium*, 46
class 1 endophytes
Atkinsonella hypoxylon, 45
EfV1, 40, 45
Epichloë festucae, 40
class 4 endophytes
dark septate, 47–48
Phialophora radicicola var. *graminicola*, 47–48
VLPs, *P. graminicola*, 47–48
mycoviruses and putative viruses, 40, 41*t*
roles
and adaptation, stress, 48–49, 49*f*
CThTV, 50–51
epigenetic role, 52
growth rate, 51
incidence, dsRNAs, 48–49
mutualistic viruses, 50
transmission rates, mycoviruses, 51–52
Virus-like particles (VLPs), 47–48
Virus transmission, *B. cinerea*
distribution and incidence
BVX and BVF, 265–266, 265*t*
description, 265–266
detection, BVF and BVX, 266
life cycles
description, 264
growth, necrotic tissues, 264–265
reciprocal sexual crosses, 264
and vegetative compatibility groups
anastomosis, 266–267
description, 266–267
genetic studies, 267
geographical locations, 267
horizontal mycovirus transmission, 267
VLPs. *See* Virus-like particles (VLPs)

W

White root rot fungus, viruses. *See Rosellinia necatrix*

Y

Yeast narnaviruses. *See* Narnaviruses, 20S and 23S RNAs
Yeast prions
amyloid-based, 20–21, 22*t*, 23
benefit/detriment, 26
chaperones, 25
domains, 26
parallel in-register beta-sheets, 23, 24–25, 24*f*
transmission barriers, 25–26
[URE3] and [PSI+], 22–23
Ure2p prion domain, 23
variants, 23

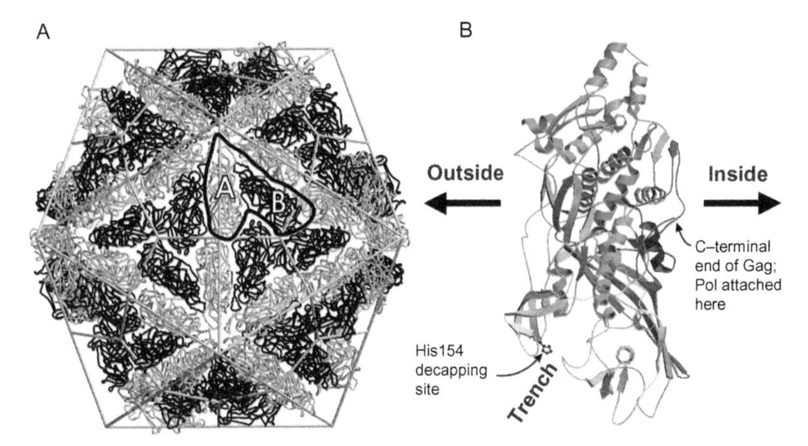
A
B
Outside
Inside
C–terminal
end of Gag;
Pol attached
here
His154
decapping
site
Trench

Figure 1.1, Reed B. Wickner *et al.* (See Page 4 of this Volume)

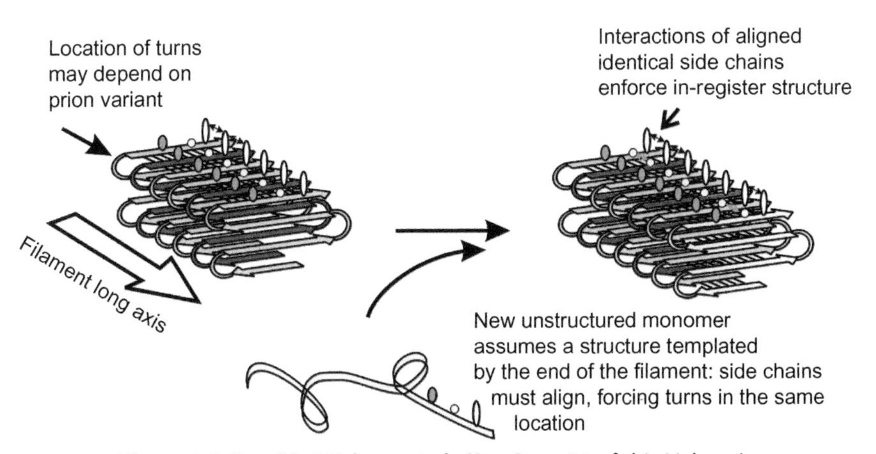
Location of turns
may depend on
prion variant
Interactions of aligned
identical side chains
enforce in-register structure
Filament long axis
New unstructured monomer
assumes a structure templated
by the end of the filament: side chains
must align, forcing turns in the same
location

Figure 1.8, Reed B. Wickner *et al.* (See Page 24 of this Volume)

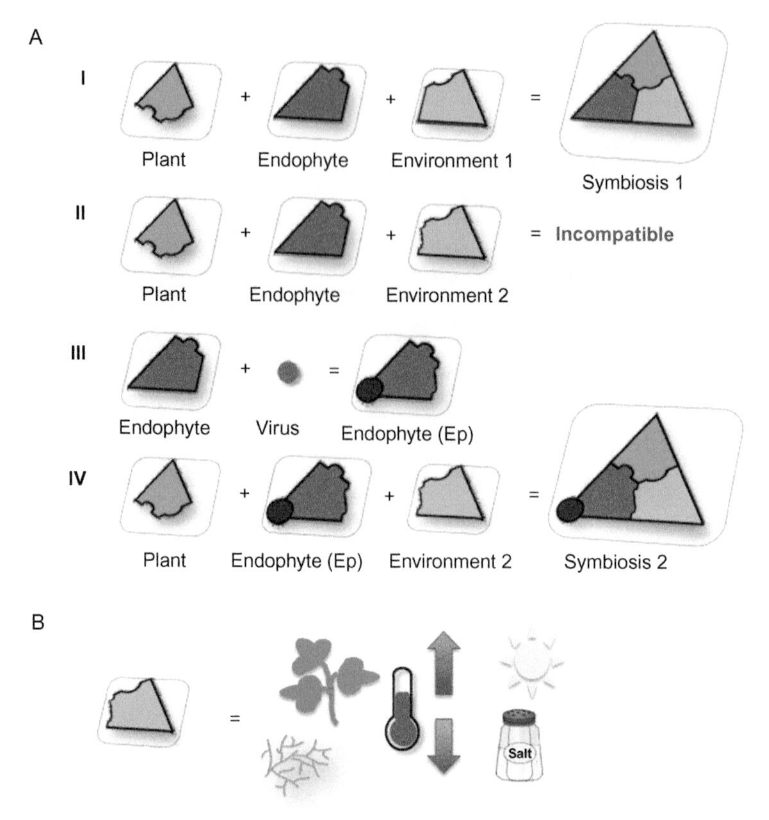

Figure 2.1, Xiaodong Bao and Marilyn J. Roossinck (See Page 49 of this Volume)

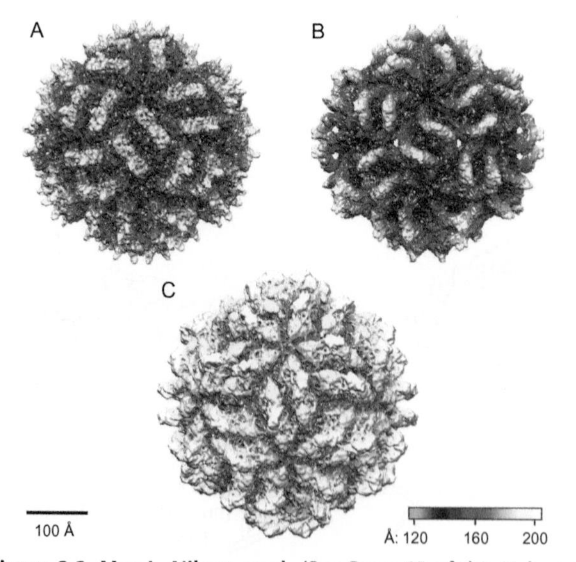

Figure 3.2, Max L. Nibert *et al.* (See Page 65 of this Volume)

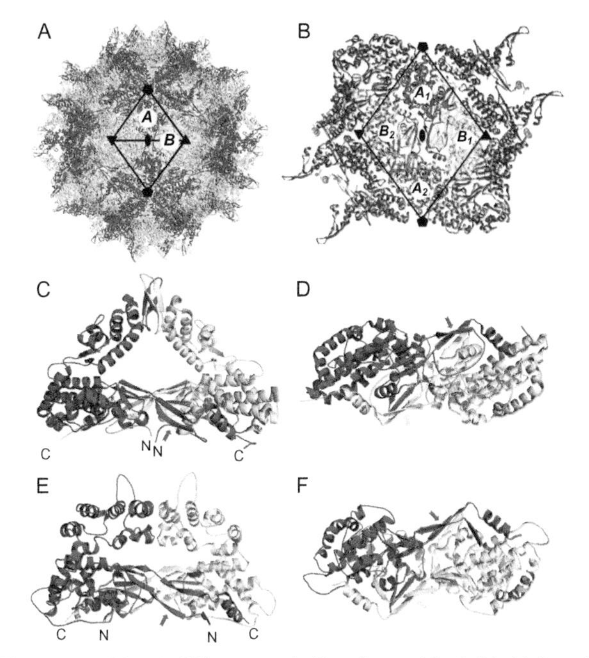

Figure 3.3, Max L. Nibert *et al.* (See Page 66 of this Volume)

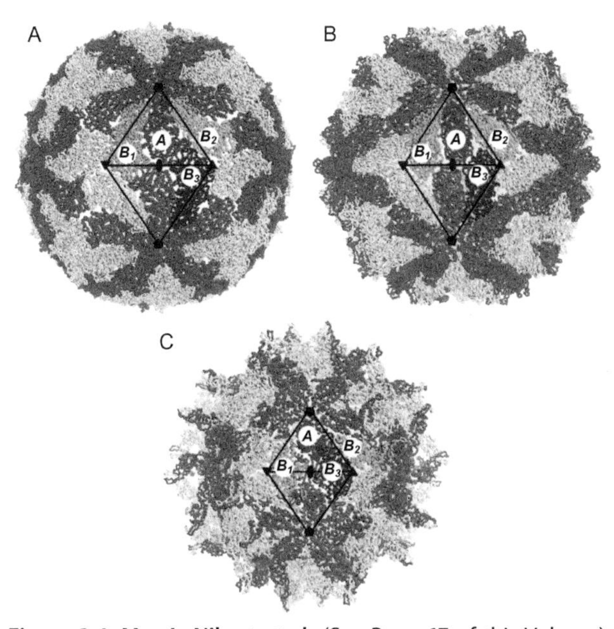

Figure 3.4, Max L. Nibert *et al.* (See Page 67 of this Volume)

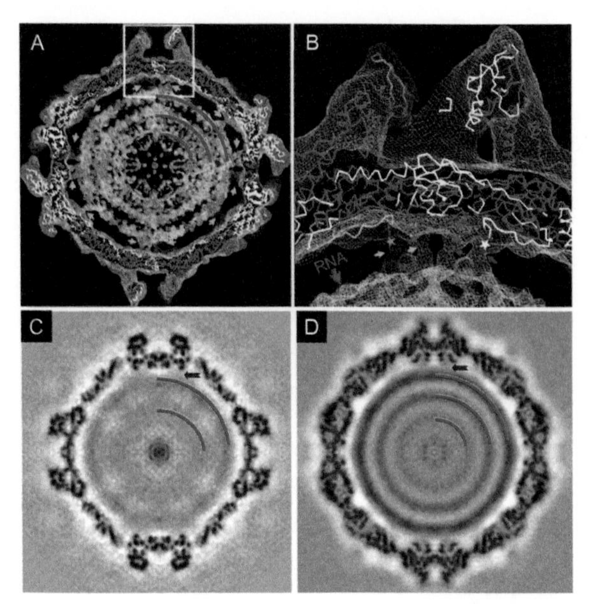

Figure 3.5, Max L. Nibert *et al.* (See Page 73 of this Volume)

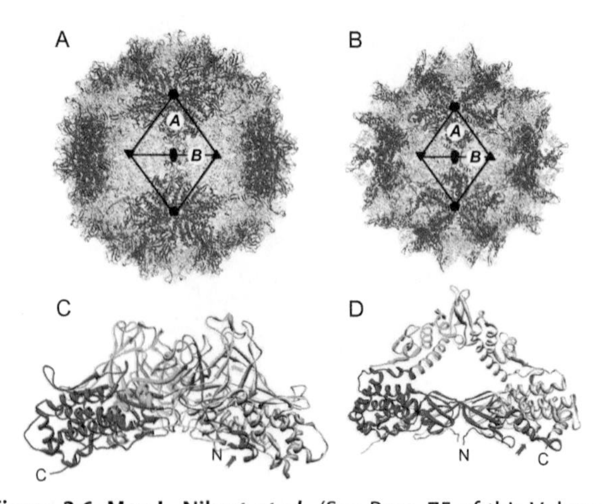

Figure 3.6, Max L. Nibert *et al.* (See Page 75 of this Volume)

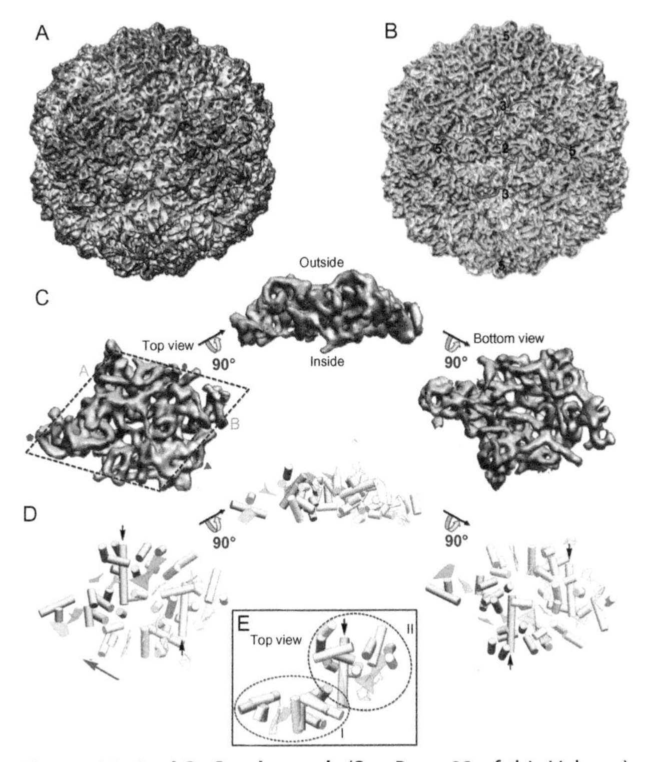

Figure 4.2, José R. Castón *et al.* (See Page 93 of this Volume)

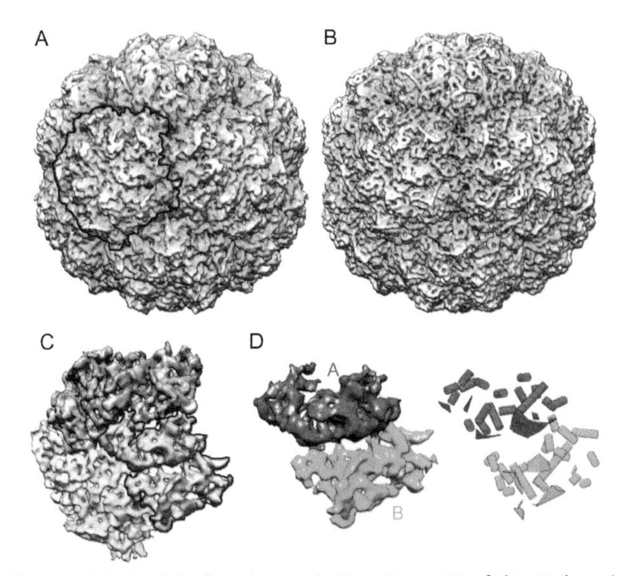

Figure 4.3, José R. Castón *et al.* (See Page 96 of this Volume)

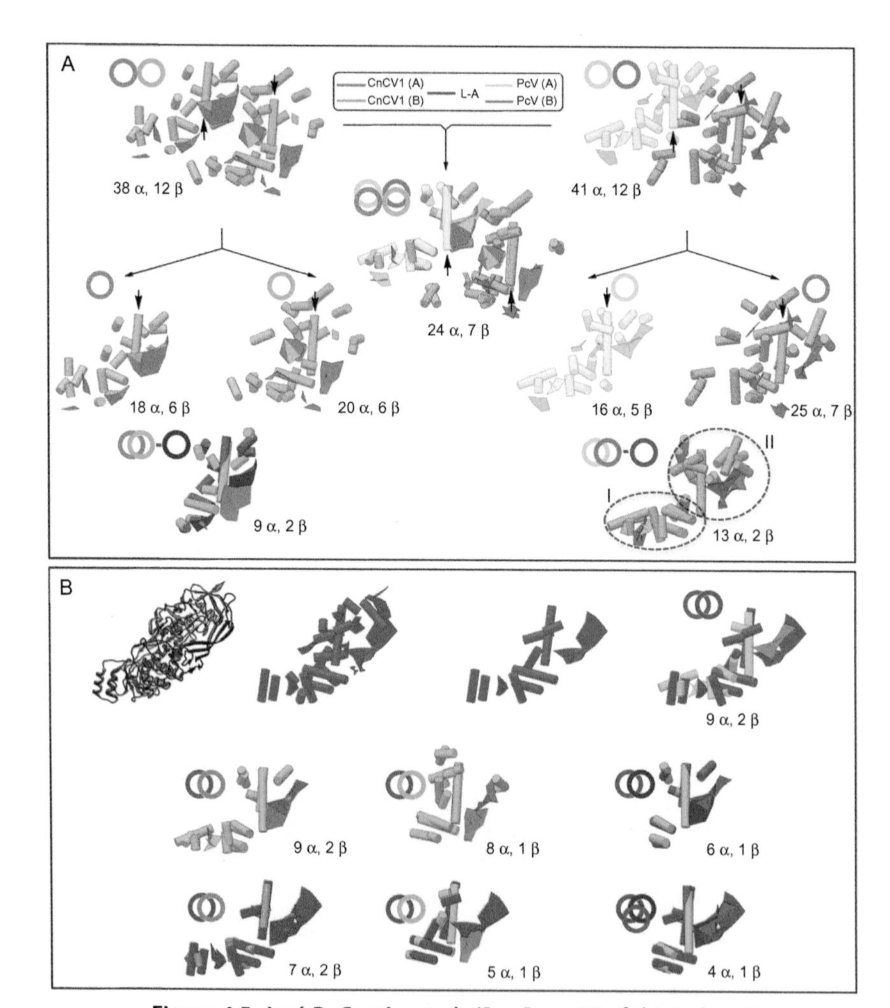

Figure 4.5, José R. Castón *et al.* (See Page 98 of this Volume)

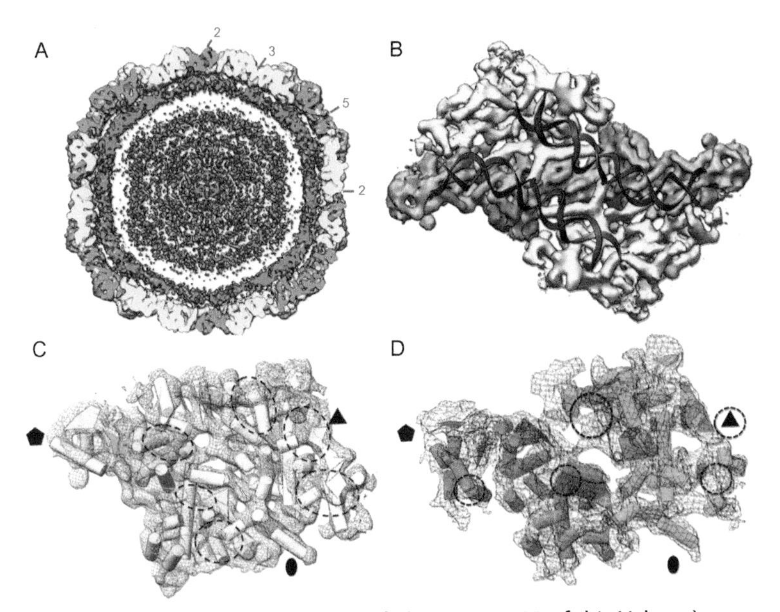

Figure 4.6, José R. Castón *et al.* (See Page 103 of this Volume)

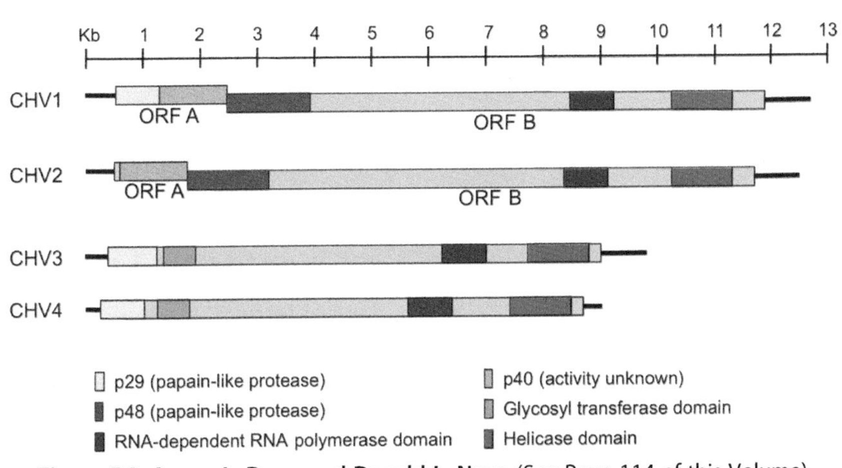

Figure 5.2, Angus L. Dawe and Donald L. Nuss (See Page 114 of this Volume)

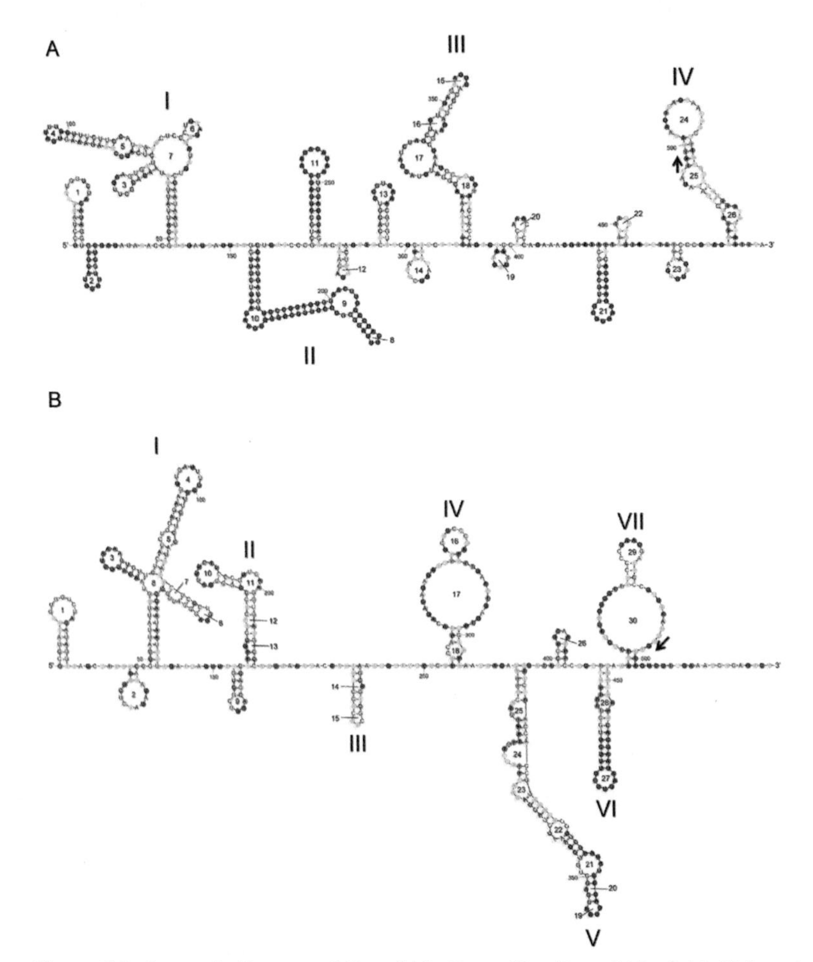

Figure 5.3, Angus L. Dawe and Donald L. Nuss (See Page 119 of this Volume)

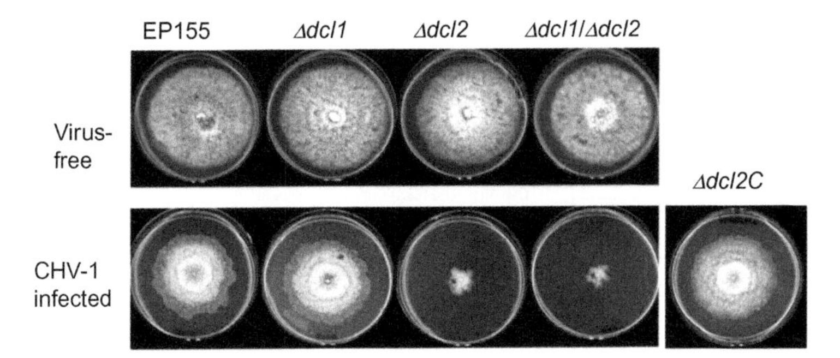

Figure 5.5, Angus L. Dawe and Donald L. Nuss (See Page 130 of this Volume)

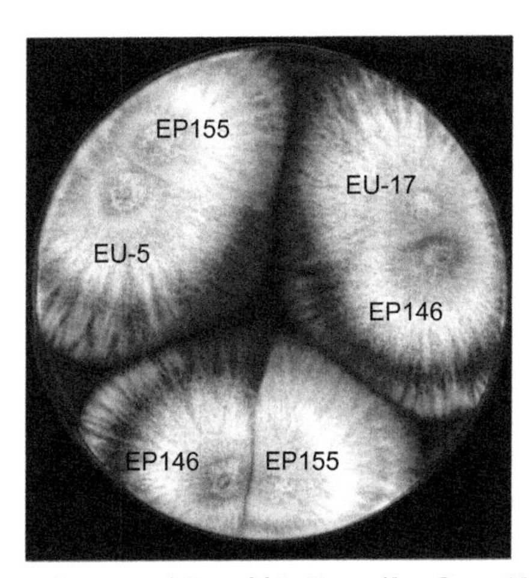

Figure 5.7, Angus L. Dawe and Donald L. Nuss (See Page 134 of this Volume)

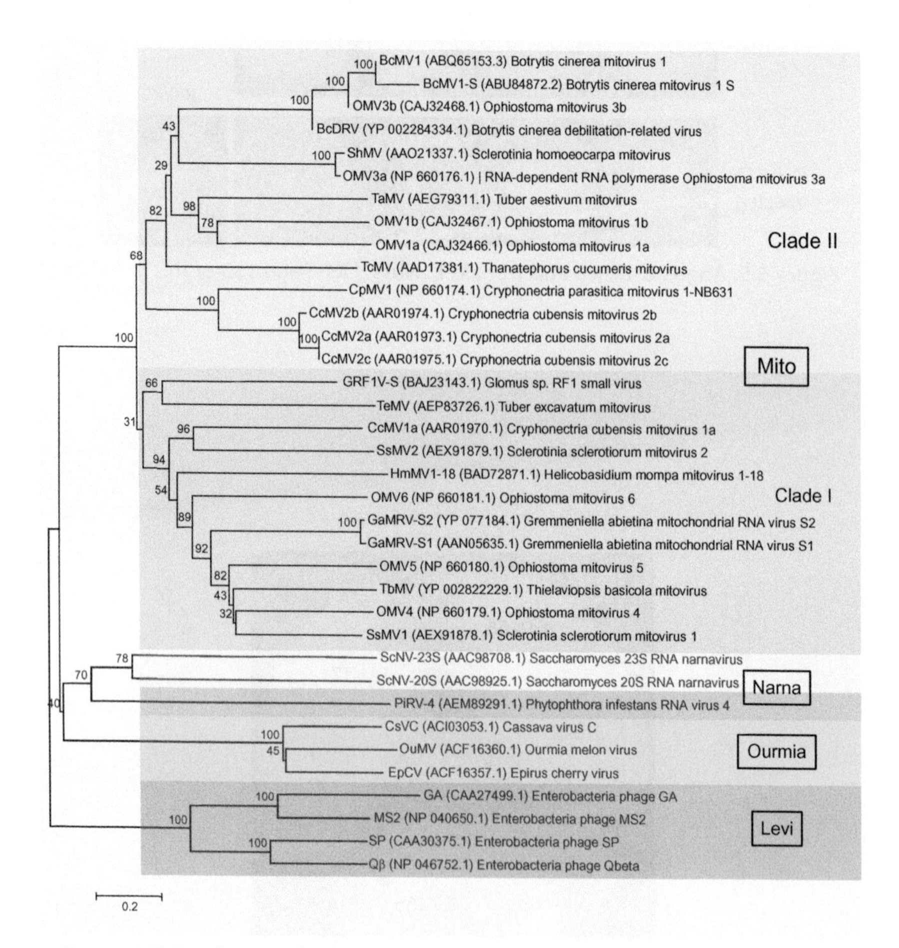

Figure 6.2, Bradley I. Hillman and Guohong Cai (See Page 153 of this Volume)

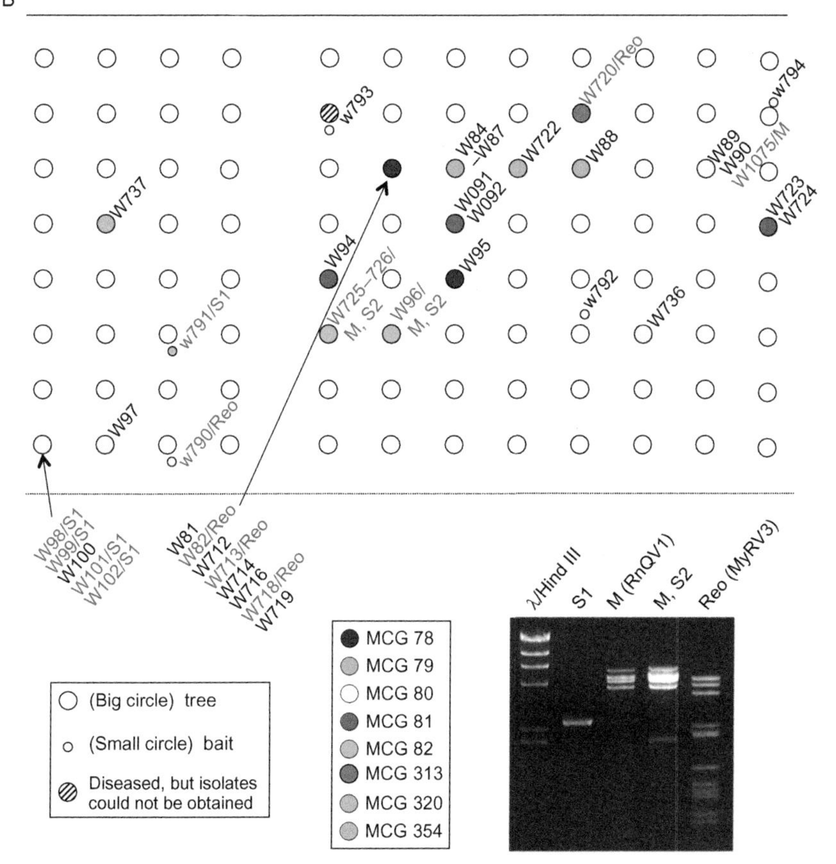

Figure 7.1, Hideki Kondo *et al.* (See Page 182 of this Volume)

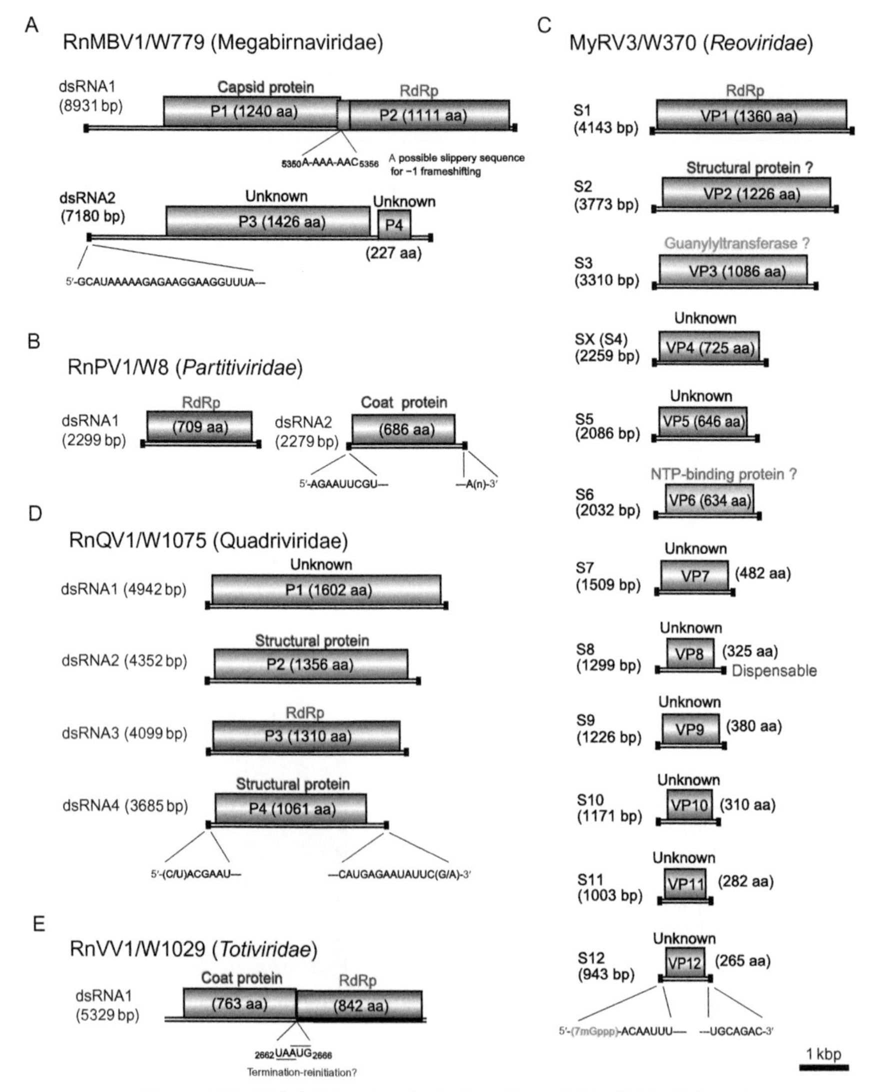

Figure 7.2, Hideki Kondo *et al.* (See Page 187 of this Volume)

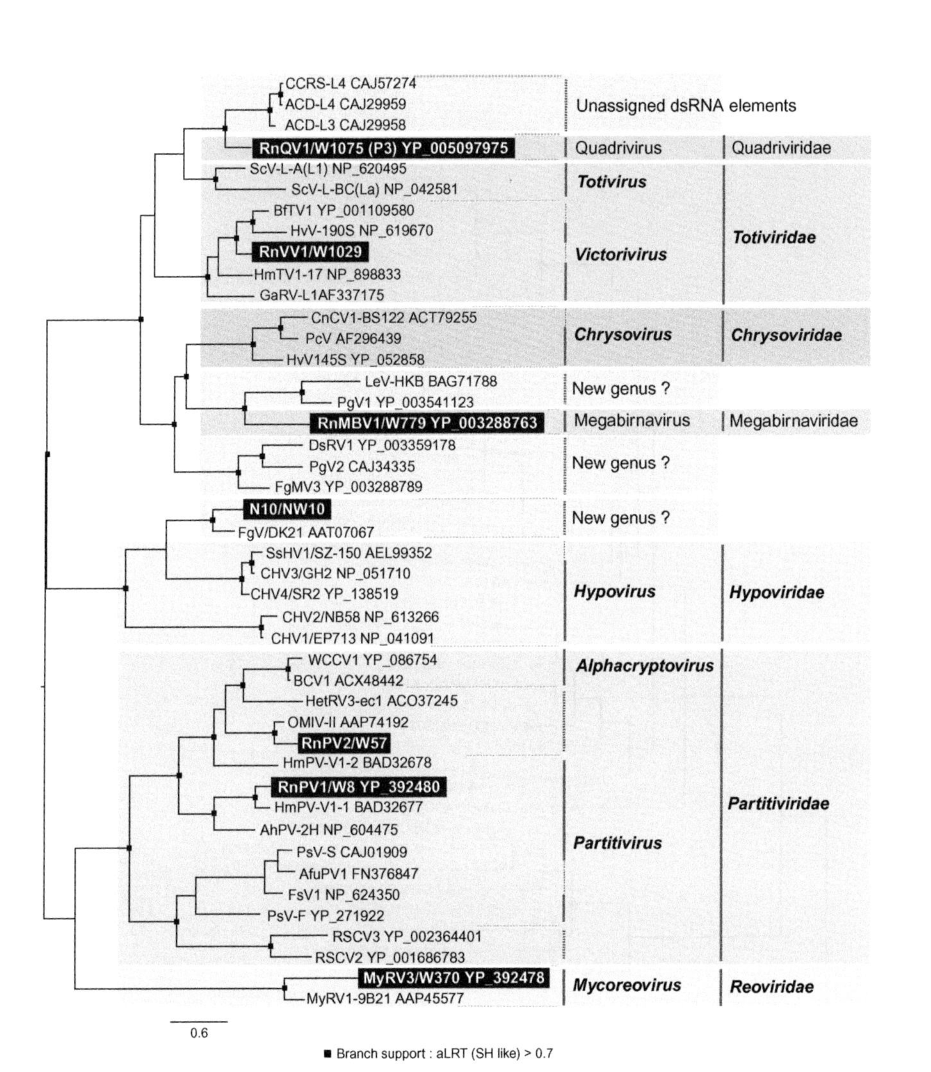

Figure 7.3, Hideki Kondo *et al.* (See Page 188 of this Volume)

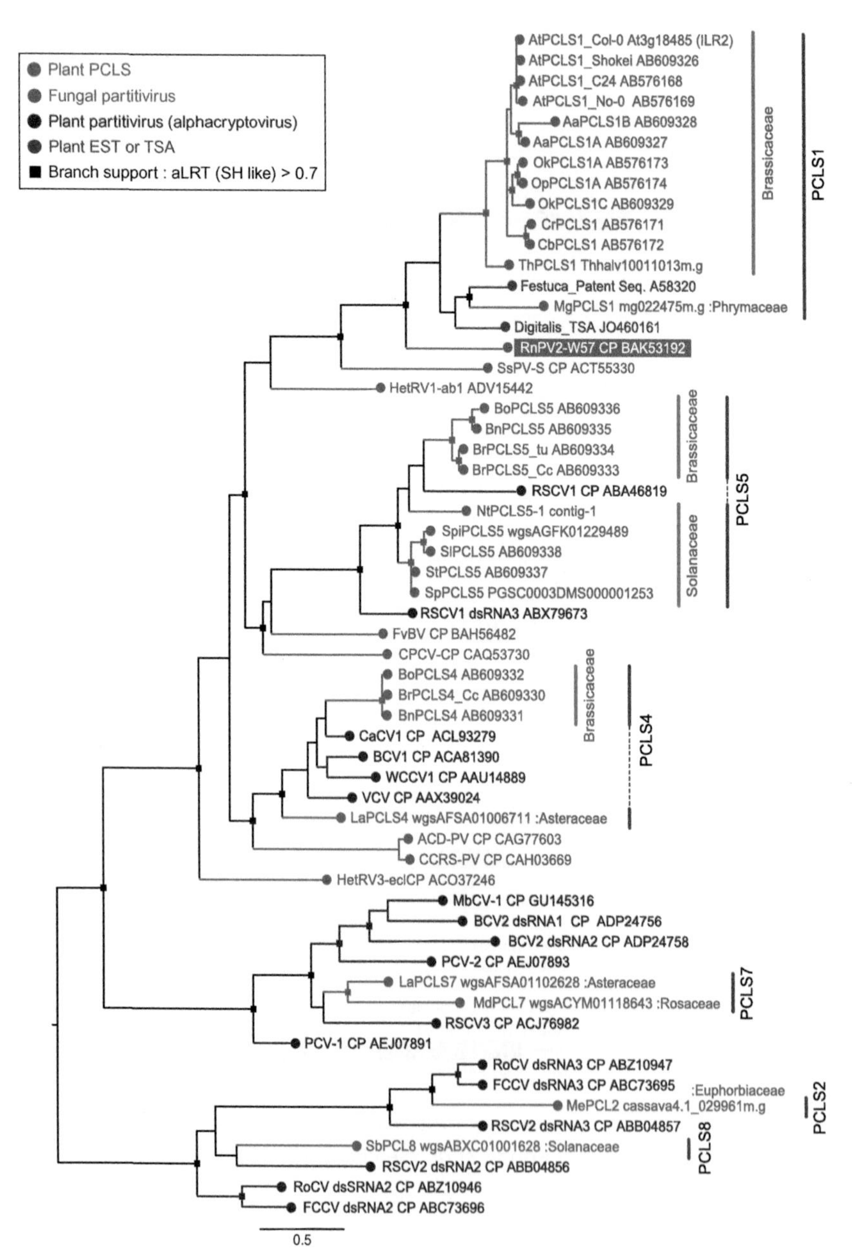

Figure 7.5, Hideki Kondo *et al.* (See Page 193 of this Volume)

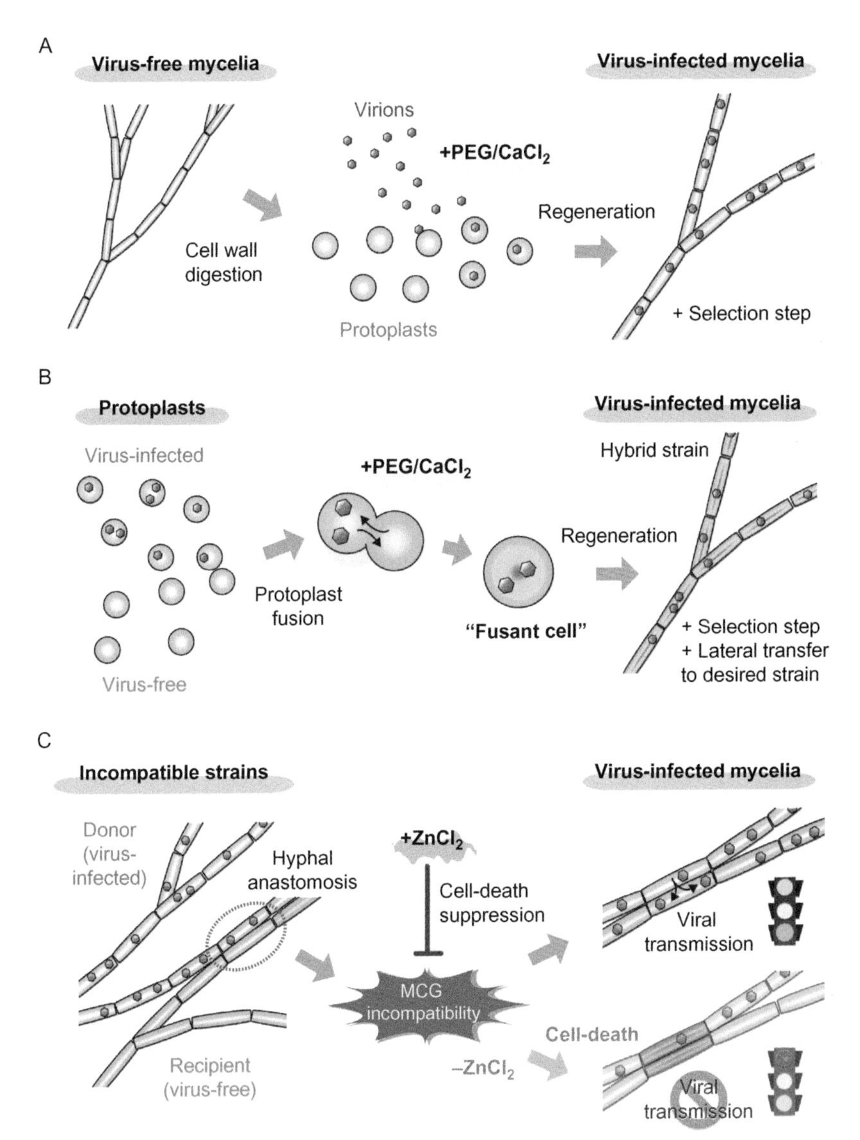

Figure 7.6, Hideki Kondo *et al.* (See Page 200 of this Volume)

Figure 8.1, Daohong Jiang *et al.* (See Page 217 of this Volume)

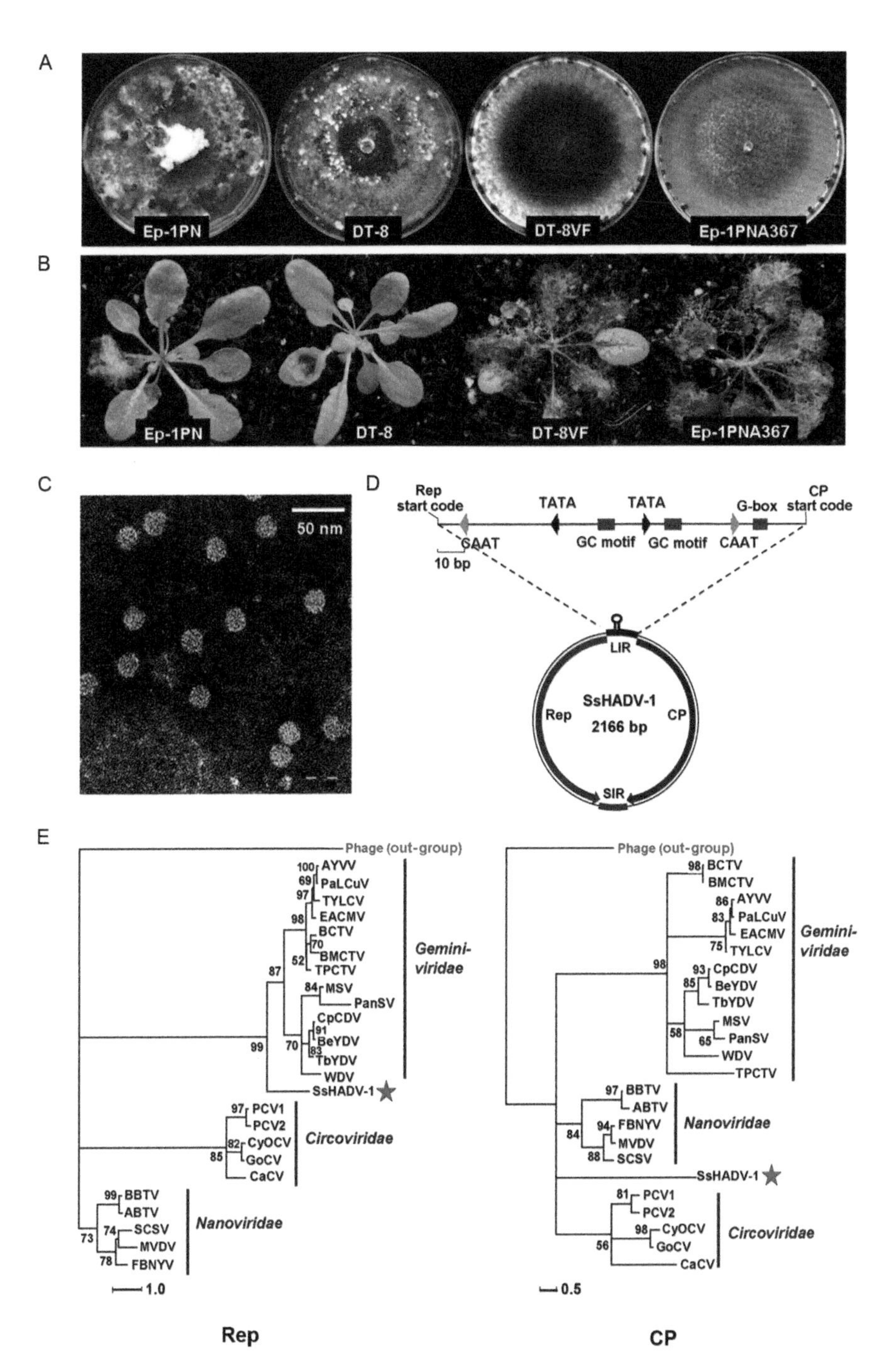

Figure 8.2, Daohong Jiang *et al.* (See Page 220 of this Volume)

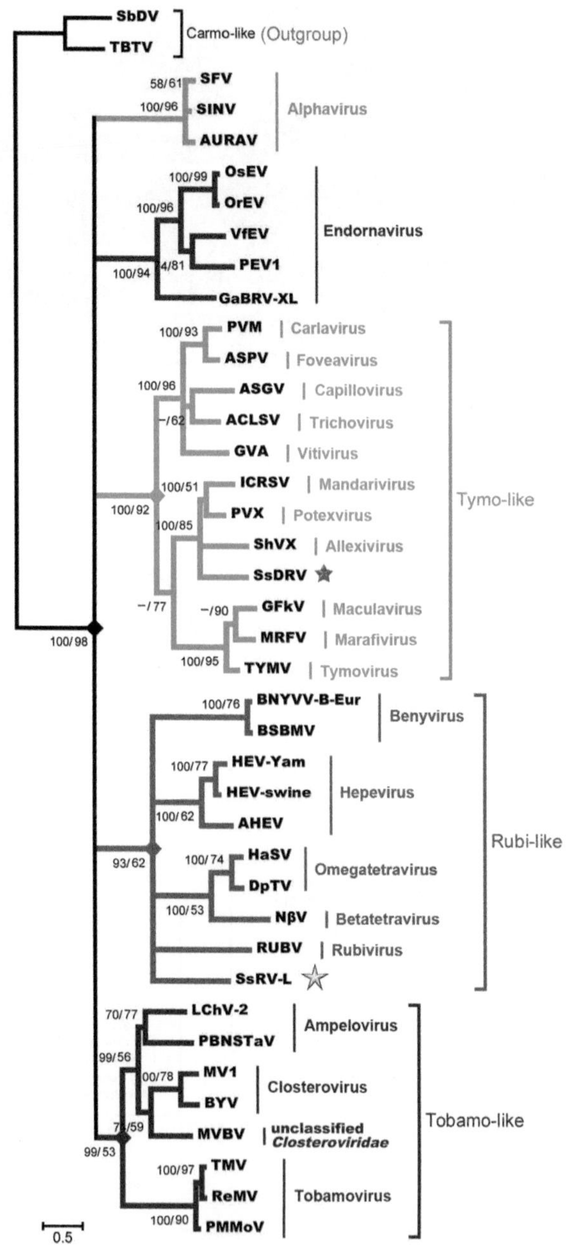

Figure 8.4, Daohong Jiang *et al.* (See Page 224 of this Volume)

A

B

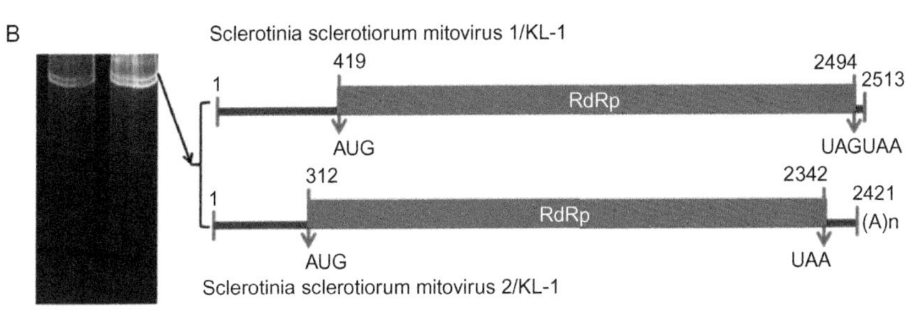

C

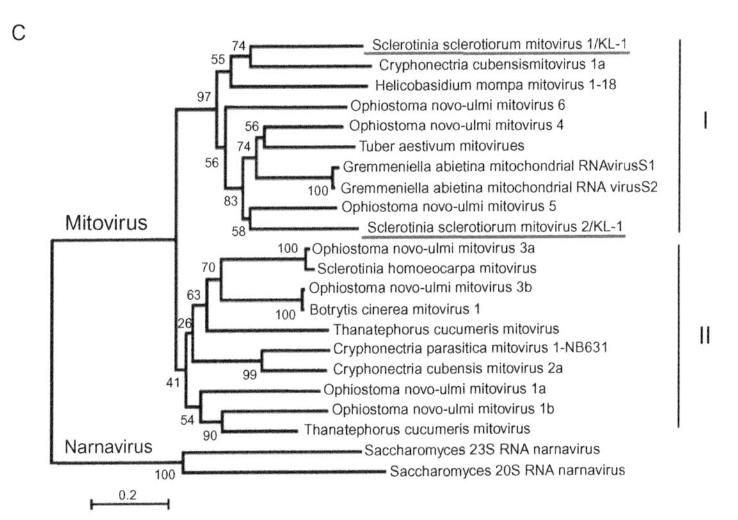

Figure 8.7, Daohong Jiang *et al.* (See Page 232 of this Volume)

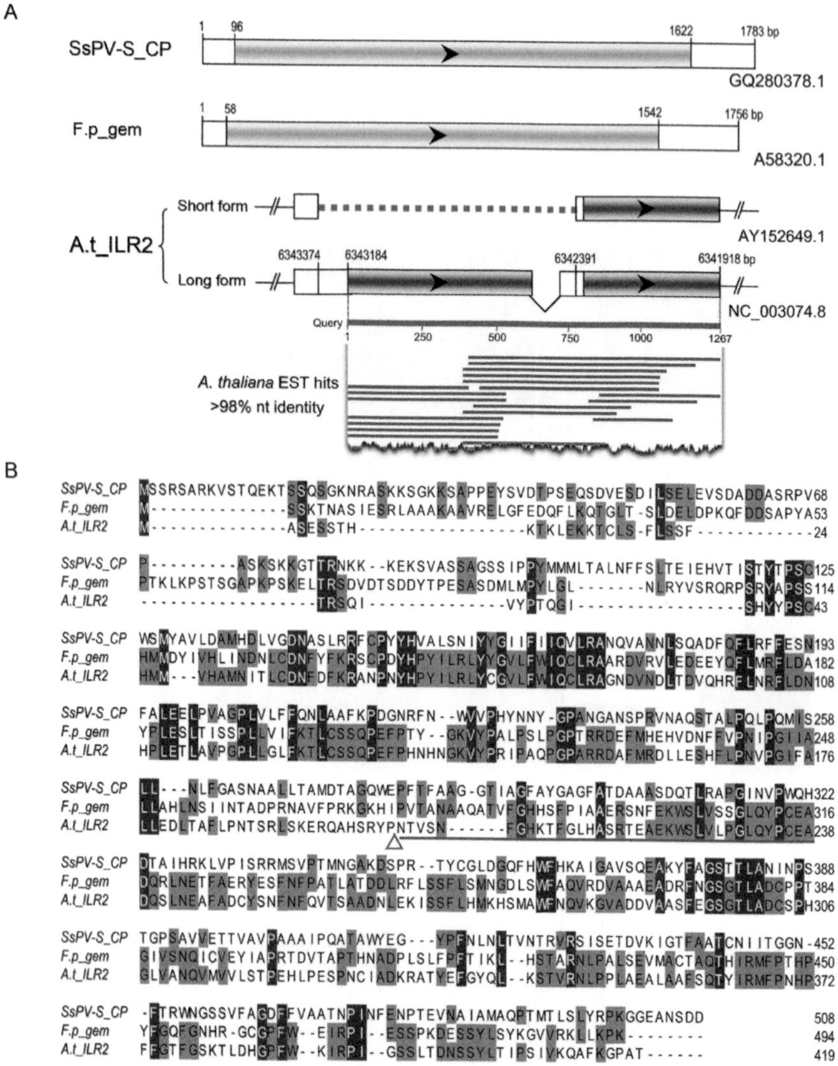

Figure 8.10, Daohong Jiang *et al.* (See Page 237 of this Volume)

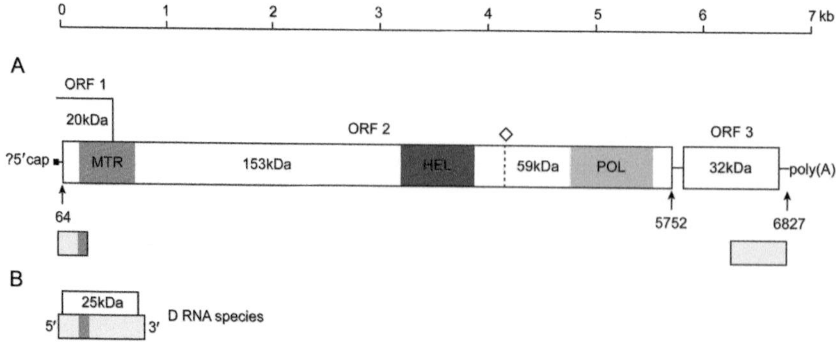

Figure 9.1, Michael N. Pearson and Andrew M. Bailey (See Page 256 of this Volume)

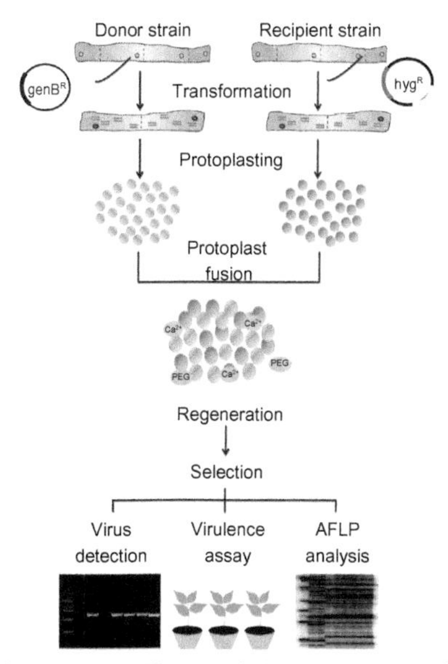

Figure 10.3, Won Kyong Cho *et al.* (See Page 284 of this Volume)

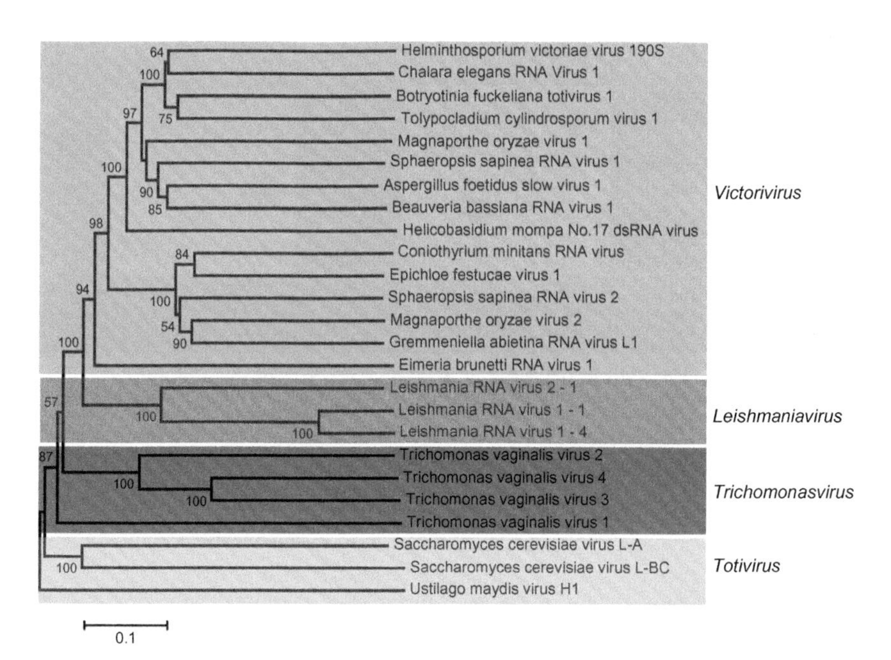

Figure 11.3, Said A. Ghabrial *et al.* (See Page 300 of this Volume)

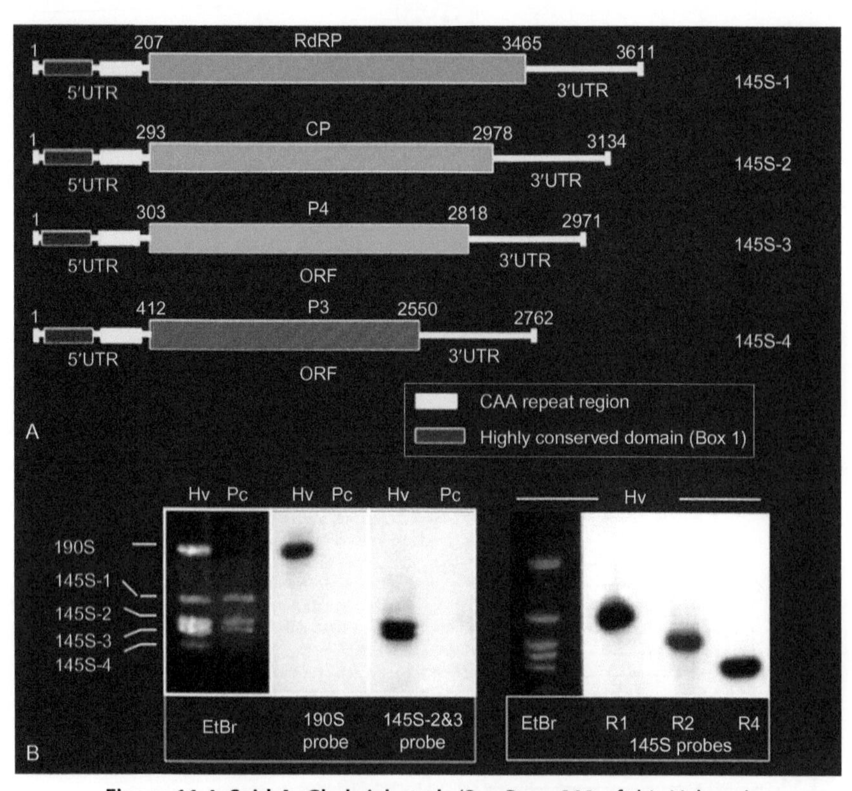

Figure 11.4, Said A. Ghabrial *et al.* (See Page 302 of this Volume)

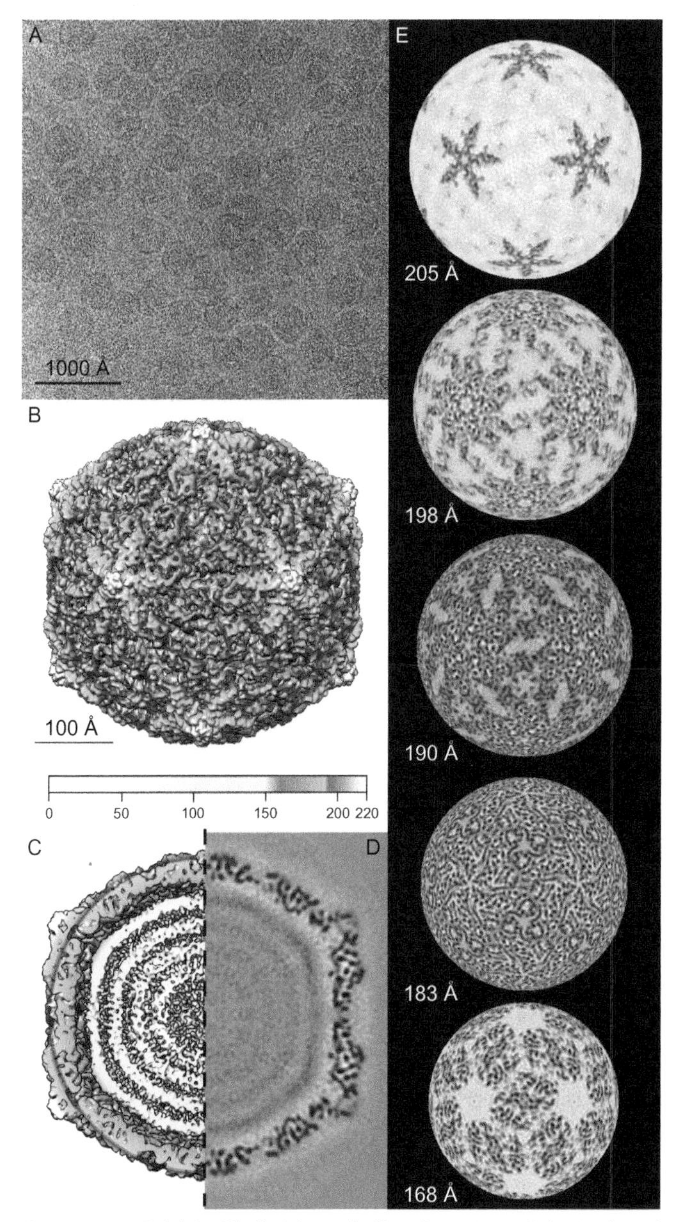

Figure 11.8, Said A. Ghabrial *et al.* (See Page 315 of this Volume)

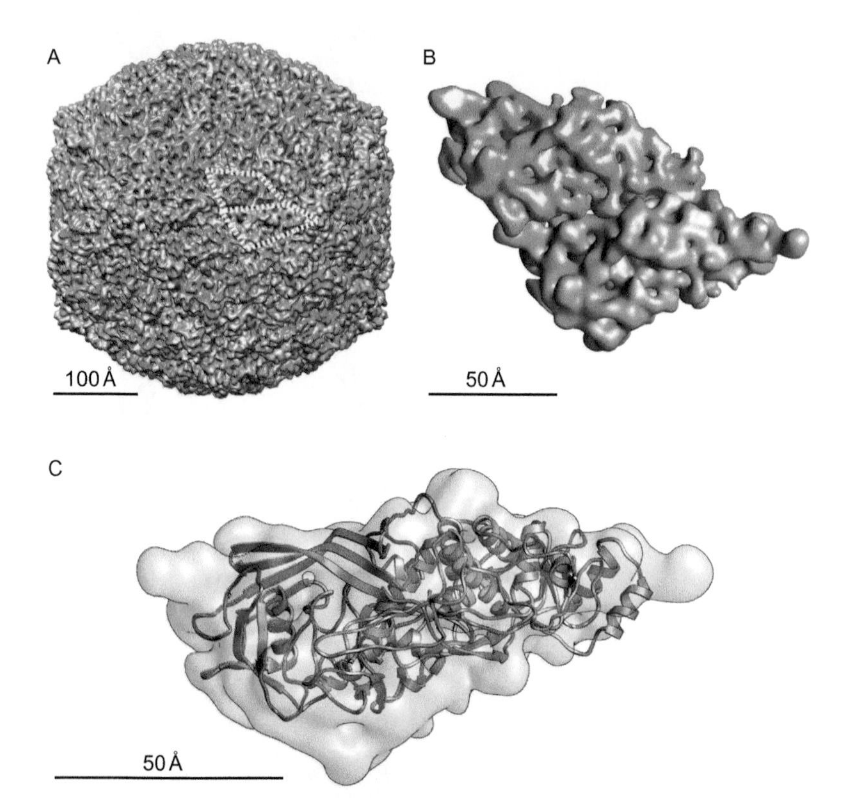

Figure 11.9, Said A. Ghabrial *et al.* (See Page 317 of this Volume)

Printed and bound by CPI Group (UK) Ltd, Croydon, CR0 4YY
08/05/2025
01864955-0003